Interpreting and Visualizing Regression Models Using Stata

Second Edition

Interpreting and Visualizing Regression Models Using Stata

Second Edition

Michael N. Mitchell

A Stata Press Publication
StataCorp LLC
College Station, Texas

Published by Stata Press, 4905 Lakeway Drive, College Station, Texas 77845
Typeset in LaTeX 2_ε
Printed in the United States of America

10 9 8 7 6 5 4 3 2 1

Print ISBN-10: 1-59718-321-0
Print ISBN-13: 978-1-59718-321-5
ePub ISBN-10: 1-59718-322-9
ePub ISBN-13: 978-1-59718-322-2
Mobi ISBN-10: 1-59718-323-7
Mobi ISBN-13: 978-1-59718-323-9

Library of Congress Control Number: 2020950108

Contents

Tables

Figures

Preface to the Second Edition

It was back in March of 2012 that I penned the preface for the first edition of this book. That was over eight years and four Stata versions ago (using Stata 12.1). The techniques illustrated in this book are as relevant today as they were back in 2012. Over this time, Stata has grown considerably. A key change that impacts the interpretation of statistical results (a focus of this book) is that the levels of factor variables are now labeled using value labels (instead of group numbers). For example, a two-level version of marital status might be labeled as `Married` and `Unmarried` instead of using numeric values such as `1` and `2`. All the output in this new edition capitalizes on this feature, emphasizing the interpretation of results based on variables labeled using intuitive value labels. Stata now includes features that allow you to customize output in ways that increase the clarity of the results, aiding interpretation. This new edition includes a new appendix (appendix A) that illustrates how you can customize the output of estimation commands for maximum clarity.

The `margins`, `contrast`, and `pwcompare` commands also reflect this new output style, defaulting to labeling groups according to their value labels. Results of these commands are easier to interpret than ever. For instance, a contrast regarding marital status might be labeled as `widowed vs. married`, making it very clear which groups are being compared. This new edition uses this labeling style and also includes appendices that describe how to customize such output. Appendix B is on the `margins` command, appendix D is on the `contrast` command, and appendix E is on the `pwcompare` command—each illustrate how you can customize the display of output produced by these commands. Additionally, appendix C on the `marginsplot` command illustrates new graphical features that have been recently introduced, including using transparency to more clearly visualize overlapping confidence intervals.

Among the other new features introduced since the last edition of this book, the `mixed` and `contrast` commands now include options for computing estimates for small-sample sizes. Chapter 17 describes these techniques and illustrates how the `mixed` and `contrast` commands can use small-sample size methods to analyze a longitudinal dataset with a small-sample size.

As with the first edition, I hope the examples shown in this book help you understand the results of your regression models so you can interpret and present them with clarity and confidence.

Ventura, California
November 2020

Michael N. Mitchell

Preface to the First Edition

Think back to the first time you learned about simple linear regression. You probably learned about the underlying theory of linear regression, the meaning of the regression coefficients, and how to create a graph of the regression line. The graph of the regression line provided a visual representation of the intercept and slope coefficients. Using such a graph, you could see that as the intercept increased, so did the overall height of the regression line, and as the slope increased, so did the tilt of the regression line. Within Stata, the `graph twoway lfit` command can be used to easily visualize the results of a simple linear regression.

Over time, we learn about and use fancier and more abstract regression models—models that include covariates, polynomial terms, piecewise terms, categorical predictors, interactions, and nonlinear models such as logistic. Compared with a simple linear regression model, it can be challenging to visualize the results of such models. The utility of these fancier models diminishes if we have greater difficulty interpreting and visualizing the results.

With the introduction of the `marginsplot` command in Stata 12, visualizing the results of a regression model, even complex models, is a snap. As implied by the name, the `marginsplot` command works in tandem with the `margins` command by plotting (graphing) the results computed by the `margins` command. For example, after fitting a linear model, the `margins` command can be used to compute adjusted means as a function of one or more predictors. The `marginsplot` command graphs the adjusted means, allowing you to visually interpret the results.

The `margins` and `marginsplot` commands can be used following nearly all Stata estimation commands (including `regress`, `anova`, `logit`, `ologit`, and `mlogit`). Furthermore, these commands work with continuous linear predictors, categorical predictors, polynomial (power) terms, as well as interactions (for example, two-way interactions, three-way interactions). This book uses the `marginsplot` command not only as an interpretive tool but also as an instructive tool to help you understand the results of regression models by visualizing them.

Categorical predictors pose special difficulties with respect to interpreting regression models, especially models that involve interactions of categorical predictors. Categorical predictors are traditionally coded using dummy (indicator) coding. Many research questions cannot be answered directly in terms of dummy variables. Furthermore, interactions involving dummy categorical variables can be confusing and even misleading. Stata 12 introduces the `contrast` command, a general-purpose command that can be used to precisely test the effects of categorical variables by forming contrasts among the levels of the categorical predictors. For example, you can compare adjacent groups, compare each group with the overall mean, or compare each group with the mean of the previous groups. The `contrast` command allows you to easily focus on the comparisons that are of interest to you.

The `contrast` command works with interactions as well. You can test the simple effect of one predictor at specific levels of another predictor or form interactions that involve comparisons of your choosing. In the parlance of analysis of variance, you can test simple effects, simple contrasts, partial interactions, and interaction contrasts. These kinds of tests allow you to precisely understand and dissect interactions with surgical precision. The `contrast` command works not only with the `regress` command but also with commands such as `logit`, `ologit`, `mlogit`, as well as random-effects models like `xtmixed`.

As you can see, the scope of the application of the `margins`, `marginsplot`, and `contrast` commands is broad. Likewise, so is the scope of this book. It covers continuous variables (modeled linearly, using polynomials, and piecewise), interactions of continuous variables, categorical predictors, interactions of categorical predictors, as well as interactions of continuous and categorical predictors. The book also illustrates how the `margins`, `marginsplot`, and `contrast` commands can be used to interpret results from multilevel models, models where time is a continuous predictor, models with time as a categorical predictor, nonlinear models (such as logistic regression or ordinal logistic regression), and analyses that involve complex survey data. However, this book does not contain information about the theory of these statistical models, how to perform diagnostics for the models, the formulas for the models, and so forth. The summary section concluding each chapter includes references to books and articles that provide background for the techniques illustrated in the chapter.

My goal for this book is to provide simple and clear examples that illustrate how to interpret and visualize the results of regression models. To that end, I have selected examples that illustrate large effects generally combined with large sample sizes to create patterns of effects that are easy to visualize. Most of the examples are based on real data, but some are based on hypothetical data. In either case, I hope the examples help you understand the results of your regression models so you can interpret and present them with clarity and confidence.

Simi Valley, California
March 2012

Michael N. Mitchell

Acknowledgments

This book was made possible by the help and input of many people. I want to thank Bill Rising for his detailed and perceptive feedback, which frequently helped me think more deeply about what I was really trying to say. I want to thank Adam Crawley for such excellent editing, smoothing the rough edges and sharp corners in my writing. I also want to thank Kristin MacDonald for her insightful technical editing. I am grateful to Annette Fett for the brilliant cover design of the first edition and to Eric Hubbard for the amazing cover for this edition that is unique yet retains the inspiration of the original cover. I want to also give deep, heartfelt thanks to Lisa Gilmore for all the amazing things she does to transform a manuscript into a fully realized book. Without her and the amazing Stata Press team, this would remain a pile of words aspiring to be a book.

This book contains numerous corrections and clarifications thanks to Professor Bruce Weaver and the students of his Psychology 5151 class (Multivariate Statistics for Behavioural Research), namely, Dani Rose Adduono, Dylan Antoniazzi, Brooke Bigelow, Stephanie Campbell, Kristen Chafe, Lauren Dalicandro, Jane A. Harder, Joshua Ryan Hawkins, Chiao-En Kao, Nayoung Sabrina Kim, Kristy R. Kowatch, Rachel Kushnier, Tiffany See-Yan Leung, Jessie Lund, Angela MacIsaac, Brittany Mascioli, Laura McGeown, Shakira Mohammed, and Flavia Spiroiu. I am very grateful for all of your help in noting errors and explanations that were murky and needed clarification.

I want to thank the National Opinion Research Center (NORC) for granting me permission to use the General Social Survey (GSS) dataset for this book. My thanks to Jibum Kim for facilitating this process and keeping me up to date on the newest GSS developments.

I want to give a tip of my hat to the Stata team who created the `contrast`, `margins`, and `marginsplot` commands. Without this impressive and unique toolkit, this book would not have been possible.

Finally, I want to thank the statistics professors who taught me so much. I am grateful to Professors Donald Butler, Ron Coleman, Linda Fidell, Robert Dear, Jim Sidanius, and Bengt Muthén. I am also deeply grateful to Professor Geoffrey Keppel, whose book built a foundation for so much of my statistical knowledge. This book is a reflection of and dedication to their teaching.

1 Introduction

1.1 Read me first

I encourage you to download the example datasets and run the examples illustrated in this book. All example datasets and programs used in this book can be downloaded from within Stata using the following commands.

```
. net from https://www.stata-press.com/data/ivrm2/
. net install ivrm2
. net get ivrm2
```

The `net install` command downloads the `showcoding` program (used later in the book). The `net get` command downloads the example datasets. I encourage you to download these example datasets so you can reproduce and extend the examples illustrated in this book. These datasets are described in this chapter; see sections 1.2, 1.3, 1.4, 1.5, and 1.6. Those sections provide background about the example datasets, especially the GSS dataset, which is used throughout the book. The other datasets are briefly described in the following sections and are described in more detail in the chapter in which they are used.

After reading this introduction, I encourage you to read chapter 2 on continuous linear predictors. This provides important information about the use of the `margins` and `marginsplot` commands. I would next suggest reading chapter 7. This provides important information about the use of the `contrast` command for interpreting categorical predictors. Many of the other chapters build upon what is covered in those two key chapters.

In fact, the chapters in this book are highly interdependent, and many chapters build upon the ideas of previous chapters. Such chapters include cross-references to previous chapters. For example, chapter 11 illustrates interpreting polynomial by categorical interactions. That chapter cross-references chapter 3 regarding continuous variables modeled using polynomials as well as chapter 7 on categorical variables. It might be tempting to try to read chapter 11 without reading chapters 3 and 7, but I think it will make much more sense having read the cross-referenced chapters first.

I would also like to call your attention to the appendices that are contained in part V. You might get the impression that those topics are unimportant because of their placement at the back of the book in an appendix. Actually, I am trying to underscore the importance of those topics by placing them at the end of the book where they can be quickly referenced. These appendices show how to customize the output from estimation commands and provide details about the `margins`, `marginsplot`, `contrast`, and `pwcompare` commands that are not specific to any particular type of variable or type of model. I think that you will get the most out of the book (and these commands) by reading the appendices sooner rather than later.

Note! Using the set command to control reporting of base levels

I prefer output that displays the base (reference) category for factor variables. All the output that you will see in this book uses that style of output. To make that the default, you can type

```
. set baselevels on
```

and the base (reference) categories will be displayed by default. By adding the `permanently` option (shown below), that setting will be the default each time you start Stata.

```
. set baselevels on, permanently
```

You can revert back to the default settings, turning off the display of the base (reference) category with the `set baselevels` command below.

```
. set baselevels off
```

You can add the `permanently` option to make that setting the default each time you invoke Stata. For more details, see appendix A.

Finally, I would like to note that the approach of the writing of this book differs in some key ways from the way that you would approach your own research. In this book, I take a discovery learning perspective, showing the results of a model and then taking you on a journey exploring how we can use Stata to interpret and understand the results. This contrasts with the kind of approach that would commonly be used in research where a theoretical rationale is used to form a research plan, which is translated into a series of analyses to test previously articulated research questions. Although I think the approach I have used is effective as a teaching tool, it may convey three bad research habits that I would not want you to emulate.

Bad research habit #1: You let the pattern of the data guide further analysis. The examples frequently illustrate a regression analysis, show the pattern of results, and then use the pattern of results to motivate further exploration. When analyzing your own data, I encourage you to develop an analysis plan based on your research questions. For example, if your analysis plan involves testing an interaction, I recommend that you describe the predicted pattern of results and the particular method that will be used to test whether the pattern of the interaction conforms to your predictions.

Bad research habit #2: The results should be dissected in every manner possible. This issue is particularly salient in the chapters involving interactions. Those chapters illustrate the multiple ways that you can dissect an interaction to show you the different options you can choose from. However, this is not to imply that you should dissect your interactions using every method illustrated. Instead, I would encourage you to develop an analysis plan that dissects the interaction in the way that answers your research question.

Bad research habit #3: No attention should be paid to the overall type I error rate. Each chapter illustrates a variety of ways that you can understand and dissect your results. Sometimes, many methods are illustrated, resulting in many statistical tests being performed without any adjustments to the type I error rate. For your research, I suggest that your analytic plan considers the number of statistical tests that will be performed and includes, as needed, methods for properly controlling the type I error rate.

Tip! Schemes used for displaying graphs

Unless otherwise specified, all the graphs shown in the book were produced with the s2mono scheme. You can create graphs with the same look by adding scheme(s2mono) to the end of commands that create graphs or using the following set scheme command below to change your default scheme to s2mono:

```
. set scheme s2mono
```

In some instances, I display graphs using the scheme(s1mono) option, which displays multiple lines using of different line patterns. You can find more details on customizing the look of graphs created by the marginsplot command in appendix C.

1.2 The GSS dataset

The most frequently used dataset in this book is based on the General Social Survey (GSS). The GSS dataset is collected and created by the National Opinion Research Center (NORC). You can learn more about NORC and the GSS by visiting the website https://gss.norc.org. The GSS is a unique survey and dataset. It contains numerous variables measuring demographics and societal trends from 1972 to 2018 (and continues to add data year after year). This is a cross-sectional dataset; thus, for each year the data represents different respondents. (Note that the GSS does have a panel datasets, but this is not used here.) In some years, certain demographic groups were oversampled. For simplicity, I am overlooking this and treating the sample as though simple random sampling was used.

Tip! Complex survey sampling

Datasets from surveys often involve complex survey sampling designs. In such cases, the svyset command and svy prefix are needed to obtain proper estimates and standard errors. The tools illustrated in this book can all be used in combination with such complex surveys, as illustrated in chapter 19.

The version of the dataset we will be using for the book was accessed from the NORC website by downloading the dataset titled *Entire 1972–2010 Cumulative Data Set (Release 1.1, Feb. 2011)*. I created a Stata do-file that subsets and recodes the variables to create the analytic data file we will use, named gss_ivrm.dta. This dataset is used below.

```
. use gss_ivrm
```

The describe command shows that the dataset contains 55,087 observations and 34 variables.

```
. describe, short
Contains data from gss_ivrm.dta
  obs:          55,087
 vars:              34                          14 Aug 2020 14:04
Sorted by:
```

Let's have a look at the main variables that are used from this dataset. The main outcome variable is realrinc (income), and the main predictors are age (age), educ (education), and female (gender).

1.2.1 Income

The variable realrinc measures the annual income of the respondent in real dollars. This permits comparisons of income across years. The incomes are normed to the year 1986 and are adjusted using the Consumer Price Index–All Urban Consumers (CPI-U). For those interested in more details, see *Getting the Most Out of the GSS Income Measures* available from the NORC website. You can find this by searching the Internet for *GSS income adjusted inflation.*

Incomes generally have a right-skewed distribution, and this measure of income is no exception. Using the histogram command, we can see that the variable realrinc shows a considerable degree of right skew (see figure 1.1).

```
. histogram realrinc
(bin=45, start=259, width=10664.122)
```

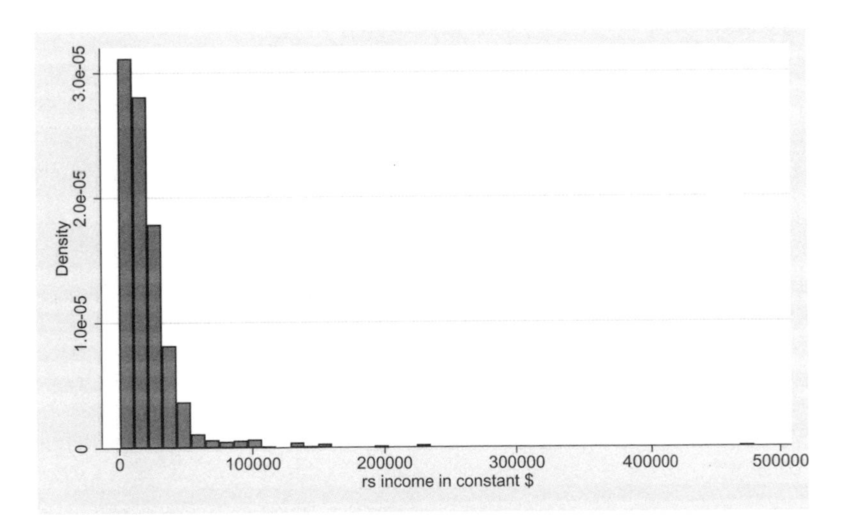

Figure 1.1. Histogram of income

This will be the main outcome measure for many of the examples in this book. There are a variety of methods that might be used for handling the right skewness of this measure. Examples include top-coding the extreme values, using robust regression, or performing a log transformation. For the analyses in this book, I would like to remain true to the incomes as measured (because these values are presumably accurate) and would like to use a simple and common method of analysis. The simplest and most common analysis method is ordinary least-squares regression. Another reasonably simple method is the use of linear regression with robust standard errors (in Stata parlance, adding the `vce(robust)` option). This permits us to analyze the variable `realrinc` as it is (without top-coding or transforming it) and accounting for the right skewness in the dataset. The regression coefficients from such an analysis are the same as the ones that would be obtained from ordinary least-squares analysis, but the standard errors are replaced with robust standard errors. I am sure that a case could be made for the superiority of other analytic methods (such as taking the log of income), but the use of robust standard errors provides familiar point estimates using a familiar metric, while still providing a reasonable analytic strategy.

1.2.2 Age

The variable `age` is used as a predictor of `realrinc`. The values of `age` can range from 18 to 89, where the value of 89 represents being age 89 or older. Rather than showing the entire distribution of `age`, let's look at the distribution of ages for the youngest and oldest respondents. The `tabulate` command below shows the distribution of `age` for those aged 18 to 25. This shows relatively few 18-year-olds (compared with the other ages).

```
. tabulate age if (age<=25)

     age of |
 respondent |      Freq.     Percent        Cum.
------------+-----------------------------------
         18 |        194        2.78        2.78
         19 |        757       10.86       13.64
         20 |        799       11.46       25.11
         21 |        899       12.90       38.01
         22 |        939       13.47       51.48
         23 |      1,100       15.78       67.26
         24 |      1,082       15.52       82.78
         25 |      1,200       17.22      100.00
------------+-----------------------------------
      Total |      6,970      100.00
```

Let's now look at the tabulation of ages for those aged 75 to 89. We can see that the sample sizes are comparatively small for those in their late 80s.

```
. tabulate age if (age>=75)
```

age of respondent	Freq.	Percent	Cum.
75	425	10.98	10.98
76	422	10.90	21.88
77	394	10.18	32.05
78	353	9.12	41.17
79	309	7.98	49.15
80	274	7.08	56.22
81	273	7.05	63.27
82	238	6.15	69.42
83	214	5.53	74.95
84	179	4.62	79.57
85	163	4.21	83.78
86	141	3.64	87.42
87	116	3.00	90.42
88	92	2.38	92.79
89 or older	279	7.21	100.00
Total	3,872	100.00	

Many examples in this book look at the relationship between income and age. As you might expect, incomes rise with increasing age until reaching a peak and then incomes decline. Figure 1.2 illustrates the relationship between income and age by showing the mean of `realrinc` at each level of `age`.

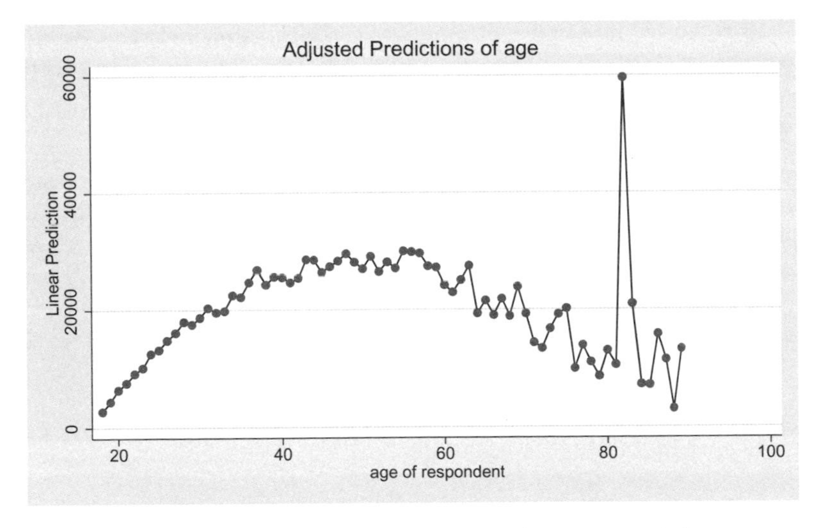

Figure 1.2. Mean income by age (ages 18 to 89)

After around age 70, the mean income as a function of age is more variable and especially so after age 80. This is probably due in large part to the decreasing sample sizes in these age groups. This could also be due to the increasing variability of whether one works or not and the variability of retirement income sources. Let's look at this graph again, but we will include only respondents who are at most 80 years old. This is shown in figure 1.3.

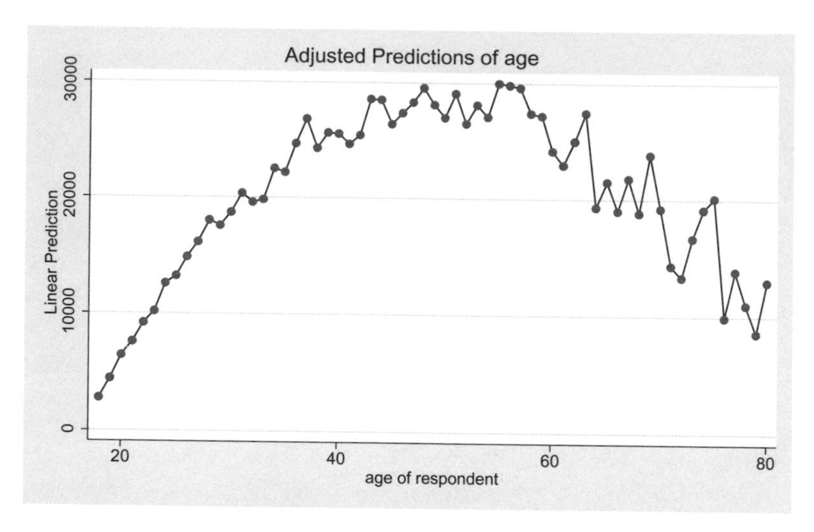

Figure 1.3. Mean income by age (ages 18 to 80)

Figure 1.3 clearly shows that the relationship between age and income is curvilinear for the ages 18 to 80. Chapter 3 will model the relationship between income and age using a quadratic model focusing on those who are 18 to 80 years old. In chapter 11, we will examine the interaction of age with college graduation status (1 = yes, 0 = no). For those examples, we will focus on ages ranging from 22 to 80.

Looking at figure 1.3, we might conclude that it would be inappropriate to fit a linear relationship between age and income. This would be unfortunate because I think that examples based on a linear relationship between age and income can be intuitive and compelling. Suppose we focused on the ages ranging from 22 to 55 (years in which people are commonly employed full time). Figure 1.4 shows a line graph of the average income at each level of age overlaid with a linear fit predicting income from age for this age range.

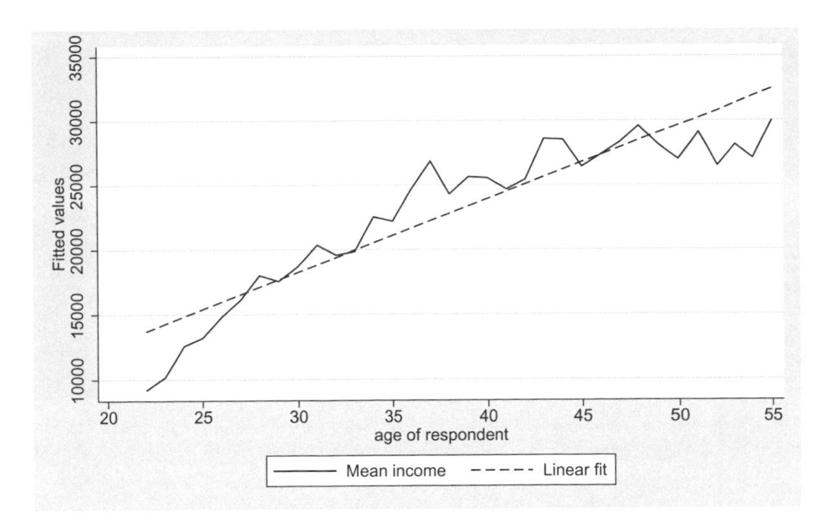

Figure 1.4. Mean income by age (solid line) with linear fit (dashed line) for ages 22 to 55

The linear fit reasonably depicts the association between age and income, with some minor deviations. In fact, linear regression analysis shows that the linear component of age explains 4.1% of the variance in income. Adding a quadratic term leads to only a minor increase in the explained variance, increasing it to 4.7%. For the sake of the examples in this book, I will consider the relationship between age and income to be linear when focusing on the ages of 22 to 55. This will permit me to create examples that look at the linear relationship between age and income in a way that can be both intuitive and justifiably linear.

Note! Analyses involving age

I present many examples depicting the relationship between income and age, like the graph shown in figure 1.4. These examples might connote that the analyses are longitudinal where the relationship between age and income is being depicted for a cohort of people studied over time. The GSS dataset used for these examples is completely cross-sectional. This particular GSS dataset that accompanies this book includes surveys that were conducted in the years 1972 to 2010, with an independent sample drawn each year. So a graph like figure 1.4 is showing the cross-sectional association between age and income from this GSS dataset. Such an association reflects a combination of cohort effects (when one was born) and changes as one gets older. In my presentation, I will be presenting the associations and ways to understand the associations, much as you would find in a results section. I will forego exploring underlying explanation of such associations, the kind of as you would find in a discussion section.

1.2.3 Education

Another variable that will be used as a predictor of **realrinc** is **educ** (education). In the GSS dataset, education is measured as the number of years of education, ranging from 0 to 20. A tabulation of the variable **educ** is shown below. The missing-value code .d indicates *don't know* and .n indicates *no answer*.

```
. tabulate educ, missing
```

highest year of school completed	Freq.	Percent	Cum.
0	148	0.27	0.27
1	39	0.07	0.34
2	139	0.25	0.59
3	232	0.42	1.01
4	299	0.54	1.56
5	382	0.69	2.25
6	725	1.32	3.57
7	837	1.52	5.08
8	2,550	4.63	9.71
9	1,873	3.40	13.11
10	2,576	4.68	17.79
11	3,295	5.98	23.77
12	16,953	30.77	54.55
13	4,579	8.31	62.86
14	5,909	10.73	73.59
15	2,414	4.38	77.97
16	6,681	12.13	90.10
17	1,604	2.91	93.01
18	1,885	3.42	96.43
19	719	1.31	97.73
20	1,086	1.97	99.71
.d	68	0.12	99.83
.n	94	0.17	100.00
Total	55,087	100.00	

The relationship between income and education is one that has been studied at great depth and one that people commonly understand. The average of income at each level of education is graphed in figure 1.5.

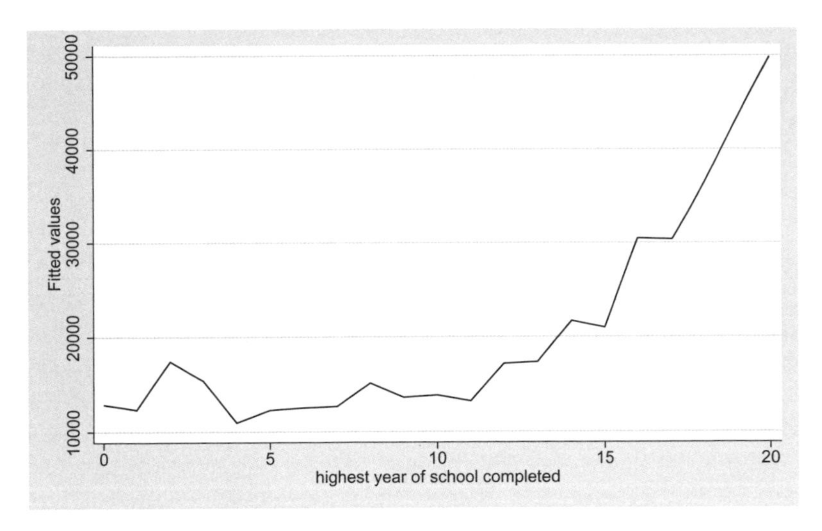

Figure 1.5. Mean income by education

Higher education is associated with higher income, but this relationship is not linear. However, the relationship appears to have linear components. Between 0 and 11 years of education, the relationship appears linear, as does the relationship for the span of 12 to 20 years of education. This figure is repeated in figure 1.6, showing a separate fitted line for these two spans of education.

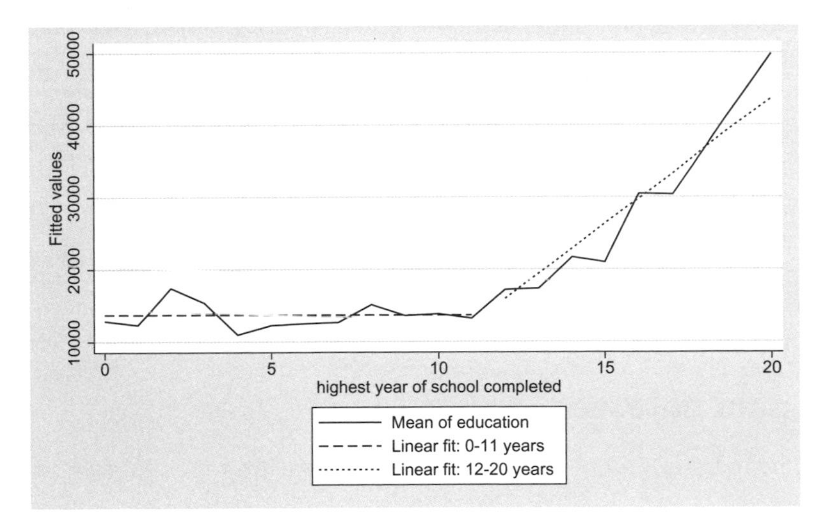

Figure 1.6. Mean income by education with linear fit for educations of 0–11 and 12–20 years

Figure 1.6 illustrates that although the relationship between education and income may not be linear, a piecewise linear approach can provide an effective fit. In fact, chapter 4 uses piecewise regression to model the relationship between income and education.

The graph in figure 1.6 seems to preclude the possibility of including education as a linear predictor of income. If we focus on those with 12 to 20 years of education, the relationship between education and income is reasonably linear. For some examples, education will be considered a linear predictor by focusing on educations ranging from 12 to 20 years.

There may be other times where it would be useful, for the sake of illustration, to treat `educ` as a categorical variable. Some examples will use a two-level categorical version of the variable `educ` called `cograd`, that indicates whether the respondent is a college graduate. This variable is coded 1 if the person has 16 or more years of education and 0 if the person has fewer than 16 years of education. Another two-level variable, `hsgrad`, will sometimes be used to indicate whether the person has graduated high school. Some examples will use a three-level version of `educ` called `educ3`. This variable is coded 1 if the respondent is not a high school graduate, 2 if the respondent is a high school graduate, and 3 if the respondent is a college graduate.

1.2.4 Gender

Some analyses will include the gender of the respondent as a predictor of `realrinc`. The variable `female` is coded 1 if the respondent is a female and 0 if the respondent is a male. When the purpose of this variable is to act as a control variable, I will specify `female`, and it will function as a dummy variable in the model. By contrast, some analyses will actively explore the role of gender, and interactions with gender, on outcome variables. In such cases, I will use the variable `gender`[1] (coded: 1 = Male and 2 = Female). In those cases, I chose to use `gender` (instead of `female`) because it leads to output that clearly distinguishes the variable name (that is, `gender`) and its values (that is, `Male`, and `Female`). By contrast, the dummy variable `female` has values labeled `Male` and `Female`, thus creating confusion between the variable `female` and the value `Female`. In chapters where I use `gender` instead of `female`, I will specifically note and explain why that variable is being used.

1.3 The pain datasets

Chapter 7 includes examples that assess the relationship between medication dosage and the amount of pain a person experiences. Two hypothetical datasets are used: `pain.dta` and `pain2.dta`. In both of these examples, the variable `pain` represents the patient's rating of pain on a scale of 0 (no pain) to 100 (worst pain).

1. In the GSS dataset, this variable is named `sex`.

1.4 The optimism datasets

The examples illustrated in chapters 8 and 9 are based on hypothetical studies comparing the effectiveness of different kinds of psychotherapy for increasing a person's optimism. The examples in chapter 8 illustrate the interaction of two categorical variables using the datasets named `opt-2by2.dta`, `opt-2by3-ex1.dta`, `opt-2by3-ex2.dta`, and `opt-3by3.dta`. Chapter 9 illustrates models involving the interactions of three categorical variables and uses the datasets named `opt-2by2by2.dta`, `opt-3by2by2.dta`, and `opt-3by3by4.dta`. These datasets are described in more detail as they are used in chapters 8 and 9.

1.5 The school datasets

Chapter 15 illustrates the interpretation of multilevel models. The examples are based on hypothetical studies looking at performance on different standardized tests. The datasets allow us to explore how to interpret cross-level interactions of school and student characteristics. These datasets are named `school_math.dta`, `school_read.dta`, `school_science.dta`, and `school_write.dta` and are described in more detail in the sections in which they are used.

1.6 The sleep datasets

The examples used in chapters 16 and 17 are based on hypothetical longitudinal studies of how many minutes people sleep at night. Chapter 16 presents four examples that treat time as a continuous predictor. The datasets are named `sleep_conlin.dta`, `sleep_conpw.dta`, `sleep_cat3conlin.dta`, and `sleep_cat3pw.dta`. The examples from chapter 17 treat time as a categorical predictor. Three example datasets are used in this chapter: `sleep_cat3.dta`, `sleep_catcat23.dta`, and `sleep_catcat33.dta`. In each of these examples, the outcome variable is named `sleep`, which contains the number of minutes the person slept at night.

1.7 Overview of the book

This book illustrates how to interpret and visualize the results of regression models using an example-based approach. The way we interpret the effect of a predictor depends on the nature of the predictor. For example, the strategy we use to interpret and visualize the contribution of a linear continuous predictor is different from focusing on, say, the interaction of two categorical variables.

The first three parts of this book illustrate how to interpret the results of linear regression models, classifying examples based on whether the predictors are continuous, categorical, or continuous by categorical interactions. Part I of the book focuses on the interpretation of continuous predictors, including two-way and three-way interactions of continuous predictors. Part II focuses on the interpretation of categorical predictors, including two-way interactions and three-way interactions. Part III focuses on the interpretation of interactions that combine continuous and categorical predictors. The examples from parts I to III focus on linear models (for example, models fit using the `regress` or `anova` commands). The first three parts of the book are described in more detail below.

Part I focuses on continuous predictors. This part begins with chapter 2, which focuses on a linear continuous predictor. Even though you are probably familiar with such models, I encourage you to read this chapter because it introduces the `margins` and `marginsplot` commands. Furthermore, this chapter addresses models that include covariates and how you can compute margins and marginal effects while holding covariates constant at different values. It also describes how to check for nonlinearity in the relationship between the predictor and outcome using graphical and analytic techniques. Chapter 3 covers polynomial terms, including not only quadratic and cubic terms but also fractional polynomial models. Part I concludes with chapter 4 on piecewise models. Such models permit you to account for nonlinearities in the relationship between the predictor and outcome by fitting two or more line segments that can have separate slopes or intercepts. All examples from part I are illustrated using the GSS dataset (described in section 1.2).

Part II focuses on categorical predictors. This includes models with one categorical predictor (see chapter 7), the interaction of two categorical predictors (see chapter 8), and interactions of three categorical predictors (see chapter 9). The examples in chapter 7 use the GSS dataset (see section 1.2) as well the pain datasets (described in section 1.3). The examples from chapters 8 and 9 are based on the optimism datasets (described in section 1.4).

Part III focuses on interactions of continuous and categorical variables. Such models both blend and build upon the examples from parts I and II. Chapters 10 to 12 illustrate interactions of a continuous predictor with a categorical variable. Chapter 10 illustrates the interaction of a linear continuous variable with a categorical variable. Chapter 11 covers continuous variables fit using polynomial terms interacted with a categorical variable. Interactions of a categorical variable with a continuous variable fit via a piecewise model are covered in chapter 12. Chapters 13 and 14 cover three-way interactions of continuous and categorical variables. Chapter 13 illustrates the interaction of two continuous predictors with a categorical variable. This includes linear by linear by categorical interactions and linear by quadratic by categorical interactions. Chapter 14 illustrates the interaction of a linear continuous predictor and two categorical variables. All examples from part III are illustrated using the GSS dataset (described in section 1.2).

Part IV covers topics that go beyond linear regression. Chapter 15 covers multilevel models (also known as hierarchical linear models), such as models where students are nested within classrooms. The examples from this chapter are based on the school datasets (described in section 1.5). Chapters 16 and 17 cover longitudinal models. Chapter 16 focuses on models in which time is treated as a continuous predictor, and chapter 17 covers models where time is treated as a categorical predictor. These examples are illustrated using the sleep datasets, described in section 1.6. Chapter 18 covers nonlinear models. This includes logistic regression, multinomial logistic regression, ordinal logistic regression, and Poisson regression. These examples are illustrated using the GSS dataset (see section 1.2). Finally, chapter 19 illustrates the interpretation of the results of models that include complex survey data (that is, models fit using the `svy` prefix).

The book concludes with part V, which contains five appendices. Appendix A describes options that you can use to customize the output from estimation commands (such as the `regress` or `logistic` command). My aim is to give you options for customizing the display of such results to aid in the interpretation of results. The following four appendices (appendices B to E) provide more details about the `margins`, `marginsplot`, `contrast`, and `pwcompare` commands (respectively). These appendices describe features of these commands that were not covered in the previous parts of the book.

Part I

Continuous predictors

This part of the book focuses on the interpretation of continuous predictors. This includes one continuous predictor (see chapters 2 to 4), interactions of two continuous predictors (see chapter 5), and interactions of three continuous predictors (see chapter 6).

Chapters 2 to 4 focus on one continuous predictor. Chapter 2 focuses on one linear continuous predictor. Polynomial models (including quadratic, cubic, and fractional polynomial models) are covered in chapter 3. A wide variety of piecewise regression models are covered in chapter 4.

Chapter 5 covers models including an interaction of two continuous predictors. This includes models with an interaction of two linear continuous predictors as well as an interaction of a linear and quadratic predictor.

Chapter 6 focuses on models involving interaction of three continuous predictors (linear by linear by linear interactions).

2 Continuous predictors: Linear

2.1 Chapter overview

This chapter focuses on how to interpret the coefficient of a continuous predictor in a linear regression model. This chapter begins with a simple linear regression model with one continuous variable predicting a continuous outcome (see section 2.2). This is followed by a multiple regression model with multiple predictors, but still focusing on one of the continuous predictors (see section 2.3). The next two sections illustrate methods for checking the linearity of the relationship between the predictor and outcome, illustrating graphical methods for assessing linearity (see section 2.4) as well as analytic methods for assessing linearity (see section 2.5).

2.2 Simple linear regression

This section illustrates the use of a continuous predictor for predicting a continuous outcome using ordinary least-squares regression. This section illustrates how to interpret and graph the results of such models. The examples in this section are based on the gss_ivrm.dta dataset. In this section, let's focus only on males who were interviewed in 2008. Below, gss_ivrm.dta is used and the relevant observations are kept.

```
. use gss_ivrm
. keep if female==0 & yrint==2008
(54,157 observations deleted)
```

Tip! Combining the use command and the if qualifier

Rather than issuing the use command followed by the keep if command, you can instead include the if specification as part of the use command, as shown below.

```
. use gss_ivrm if female==0 & yrint==2008
```

This saves you a little bit of typing and can execute faster because only the observations that meet the if condition are read into memory.

Let's use the educational level of the respondent as the outcome variable. Below, we see a frequency distribution of this variable. We can see that the variable ranges from 0 years of education to 20 years of education. The missing value code .n indicates *no answer*.

```
. tabulate educ, missing
```

highest year of school completed	Freq.	Percent	Cum.
0	3	0.32	0.32
3	2	0.22	0.54
4	5	0.54	1.08
5	3	0.32	1.40
6	15	1.61	3.01
7	7	0.75	3.76
8	21	2.26	6.02
9	22	2.37	8.39
10	29	3.12	11.51
11	51	5.48	16.99
12	264	28.39	45.38
13	66	7.10	52.47
14	117	12.58	65.05
15	45	4.84	69.89
16	155	16.67	86.56
17	29	3.12	89.68
18	48	5.16	94.84
19	22	2.37	97.20
20	25	2.69	99.89
.n	1	0.11	100.00
Total	930	100.00	

The dataset includes several variables that can be used as predictors of the respondent's education, including the education of the respondent's father, the education of the respondent's mother, and the age of the respondent. Summary statistics for these variables are shown below. The education of the father and mother range from 0 to 20, and the age of the respondent ranges from 18 to 89.

```
. summarize paeduc maeduc age
    Variable |        Obs        Mean    Std. Dev.       Min        Max
-------------+---------------------------------------------------------
      paeduc |        697    11.20373    4.304347          0         20
      maeduc |        815    11.50429    3.696358          0         20
         age |        927    47.37864    16.52631         18         89
```

As we interpret the meaning of these education variables, let's assume that having 12 years of education corresponds to graduating high school and having 16 years of education corresponds to completing a four-year college degree.

Terminology: Continuous and categorical variables

In statistics books and classes, four levels of measurement are often described: nominal, ordinal, interval, and ratio. When I use the term continuous variable, I am referring to a variable that is measured on an interval or ratio scale. By contrast, when I speak of a categorical variable, I am referring to either a nominal variable or an ordinal/interval/ratio variable that we wish to treat as though it were a nominal variable. Later in this chapter, we will analyze a variable named `agedec`, which is one's age converted into a decade (for example, a value of 2 indicates someone is in their 20s). Such a variable could be analyzed either as a continuous variable or as a categorical variable. When using the `regress` command, if we specify `agedec` as a predictor, this variable will be treated as a continuous variable. However, specifying `i.agedec` means this variable is to be analyzed as a *factor* variable, and it will be analyzed as a categorical variable. This will be discussed in more detail in section 2.5.2.

Let's run a simple regression model in which we predict the education of the respondent from the education of the respondent's father.

```
. regress educ paeduc
      Source |       SS           df       MS      Number of obs   =       696
-------------+----------------------------------   F(1, 694)       =    228.14
       Model |  1649.70181         1  1649.70181   Prob > F        =    0.0000
    Residual |  5018.43038       694  7.23116769   R-squared       =    0.2474
-------------+----------------------------------   Adj R-squared   =    0.2463
       Total |  6668.13218       695   9.5944348   Root MSE        =    2.6891

------------------------------------------------------------------------------
        educ |      Coef.   Std. Err.      t    P>|t|     [95% Conf. Interval]
-------------+----------------------------------------------------------------
      paeduc |   .3594331   .0237969    15.10   0.000     .3127106    .4061555
       _cons |   9.740211   .2857915    34.08   0.000     9.179092    10.30133
------------------------------------------------------------------------------
```

The regression equation can be written as shown in (2.1).

$$\widehat{\texttt{educ}} = 9.74 + 0.36\texttt{paeduc} \tag{2.1}$$

The intercept is 9.74 and the coefficient for paeduc is 0.36. The intercept is the predicted mean of the respondent's education when the father's education is 0. For every one-year increase in the education of the father, we would predict that the education of the respondent increases by 0.36 years. The output from the regress command shows that the coefficient for paeduc is statistically significant.

> **Note! A note about significance**
>
> In this book, I will use the term *significant* as a shorthand for statistically significant using an alpha of 0.05. Such references to an effect being significant does not speak to matters such as clinical significance or practical significance.

2.2.1 Computing predicted means using the margins command

We could use the regression equation (2.1) to compute the predicted mean of the respondent's education for any given level of the father's education. For example, if the father had 8 years of education, we would substitute 8 for paeduc, yielding $9.74 + 0.36 \times 8$, which results in the predicted mean of 12.62. Rather than doing this computation by hand, we can use the margins command. The margins command below computes the predicted mean of the outcome when paeduc equals 8 by specifying the at(paeduc=8) option.

```
. margins, at(paeduc=8)
Adjusted predictions                              Number of obs     =        696
Model VCE    : OLS
Expression   : Linear prediction, predict()
at           : paeduc          =           8

------------------------------------------------------------------------------
             |            Delta-method
             |     Margin   Std. Err.      t    P>|t|     [95% Conf. Interval]
-------------+----------------------------------------------------------------
       _cons |   12.61568   .1275167    98.93   0.000     12.36531    12.86604
------------------------------------------------------------------------------
```

The **margins** command produces the same predicted mean we computed by hand, but with greater precision. The **margins** command also displays the standard error of the predicted mean (0.13) as well as a 95% confidence interval [12.37, 12.87]. The **margins** command also includes a statistical test of whether the predicted mean significantly differs from 0 ($z = 98.93$, $p = 0.000$). In this case, this statistical test is not useful, but in other cases this test can be very useful.

Suppose we wanted to compute the predicted mean of education for the respondent, assuming separately that the father had 8, 12, or 16 years of education. Instead of running the **margins** command three separate times, we can run it once specifying `at(paeduc=(8 12 16))`, as shown below.

```
. margins, at(paeduc=(8 12 16)) vsquish
Adjusted predictions                              Number of obs     =        696
Model VCE    : OLS
Expression   : Linear prediction, predict()
1._at        : paeduc          =           8
2._at        : paeduc          =          12
3._at        : paeduc          =          16

------------------------------------------------------------------------------
             |            Delta-method
             |     Margin   Std. Err.      t    P>|t|     [95% Conf. Interval]
-------------+----------------------------------------------------------------
         _at |
          1  |   12.61568   .1275167    98.93   0.000     12.36531    12.86604
          2  |   14.05341   .1036064   135.64   0.000     13.84999    14.25683
          3  |   15.49114   .1527395   101.42   0.000     15.19125    15.79103
------------------------------------------------------------------------------
```

Note! The vsquish option

The `vsquish` option vertically squishes the output by omitting extra blank lines. I frequently use this option in this book to save space and produce more compact output. You do not need to use this option when you run the examples for yourself.

The margins output includes a legend indicating that the margins are computed for three _at values, when father's education is 8, 12, and 16 (respectively). Thus, the first line of output shows the predicted mean is 12.62 when father's education is 8, the second line of output shows the predicted mean is 14.05 when father's education is 12, and the third line shows the predicted mean is 15.49 when father's education is 16.

Sometimes, we might want to compute the predicted means given a range of values for a predictor. For example, we might want to compute the predicted means when father's education is 0, 4, 8, 12, 16, and 20. Rather than typing all of these values, we can specify 0(4)20, which tells Stata that we mean 0 to 20 in 4-unit increments. (For more information, see [U] **11.1.8 numlist**.)

```
. margins, at(paeduc=(0(4)20)) vsquish
Adjusted predictions                            Number of obs     =        696
Model VCE    : OLS

Expression   : Linear prediction, predict()
1._at        : paeduc          =           0
2._at        : paeduc          =           4
3._at        : paeduc          =           8
4._at        : paeduc          =          12
5._at        : paeduc          =          16
6._at        : paeduc          =          20

------------------------------------------------------------------------------
             |            Delta-method
             |     Margin   Std. Err.      t    P>|t|     [95% Conf. Interval]
-------------+----------------------------------------------------------------
         _at |
          1  |   9.740211   .2857915    34.08   0.000     9.179092    10.30133
          2  |   11.17794   .1997699    55.95   0.000     10.78572    11.57017
          3  |   12.61568   .1275167    98.93   0.000     12.36531    12.86604
          4  |   14.05341   .1036064   135.64   0.000     13.84999    14.25683
          5  |   15.49114   .1527395   101.42   0.000     15.19125    15.79103
          6  |   16.92887   .2324774    72.82   0.000     16.47243    17.38532
------------------------------------------------------------------------------
```

2.2.2 Graphing predicted means using the marginsplot command

We can use the marginsplot command to create a graph showing the predicted means and confidence intervals based on the most recent margins command. The last margins command used the at(paeduc=(0(4)20)) option to compute the predicted mean given that the father had 0, 4, 8, 12, 16, or 20 years of education. The marginsplot command below graphs the predicted means as a function of the father's education, as shown in figure 2.1. The graph shows the predicted means of the respondent's education as a function of the father's education (spanning from 0 to 20). Confidence intervals are shown for each value specified by the at() option (0, 4, 8, 16, and 20).

```
. marginsplot
  Variables that uniquely identify margins: paeduc
```

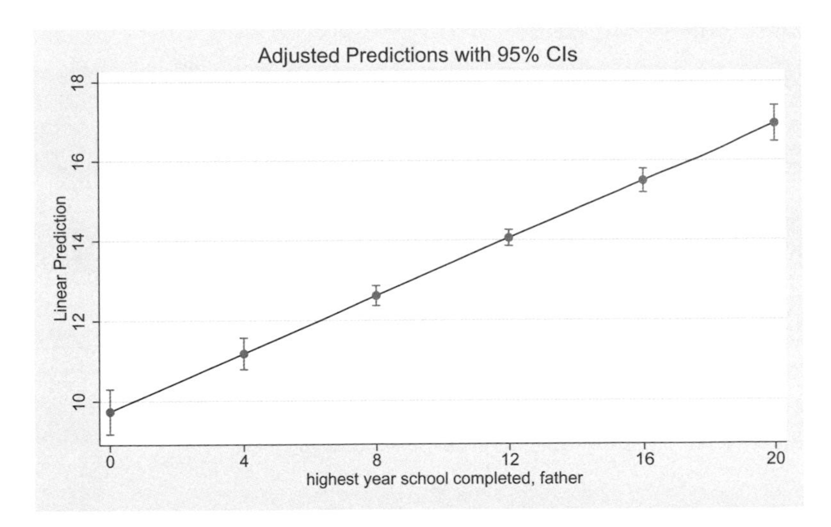

Figure 2.1. Linear regression predicting education from father's education

> **Note! The margins and marginsplot commands are a team**
>
> The `margins` and `marginsplot` commands work together as a team. Following the `margins` command, the `marginsplot` command graphs the values produced by the `margins` command. But if you execute any other command after `margins`, but before `marginsplot`, you will get the following error:
>
> ```
> . marginsplot
> previous command was not margins
> r(301);
> ```
>
> This error is saying that the last command you issued was not the `margins` command. The solution is to run the `margins` command followed by the `marginsplot` command (with no other commands in between).

Let's consider another example of using the `margins` command followed by the `marginsplot` command. In this example, the `margins` command is used with the `at(paeduc=(0 20))` option. This computes the predicted mean of the respondent's education given the father's education equals 0, and then again given that the father's education equals 20. Then, the `marginsplot` command graphs the predicted means computed by the `margins` command. This creates the graph shown in figure 2.2, showing the predicted mean of the respondent's education as the father's education ranges from 0 to 20 years. This graph does not include a confidence interval because the `noci` option was used on the `marginsplot` command.

```
. margins, at(paeduc=(0 20))
Adjusted predictions                         Number of obs    =        696
Model VCE      : OLS

Expression     : Linear prediction, predict()
1._at          : paeduc          =           0
2._at          : paeduc          =          20

------------------------------------------------------------------------------
             |            Delta-method
             |     Margin   Std. Err.      t    P>|t|     [95% Conf. Interval]
-------------+----------------------------------------------------------------
         _at |
          1  |   9.740211   .2857915    34.08   0.000     9.179092    10.30133
          2  |   16.92887   .2324774    72.82   0.000     16.47243    17.38532
------------------------------------------------------------------------------

. marginsplot, noci
  Variables that uniquely identify margins: paeduc
```

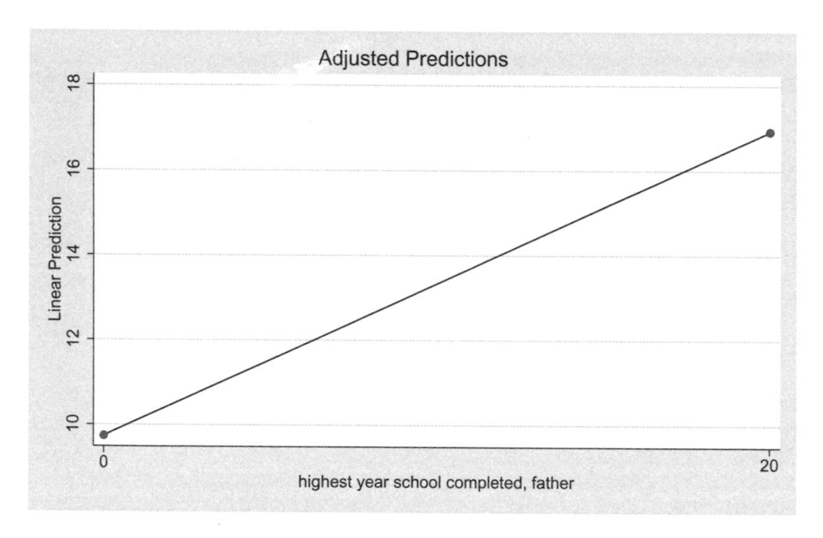

Figure 2.2. Predicted means from simple regression model predicting education from father's education

Let's now create a graph that shows the fitted line with a shaded confidence interval. We can create such a graph by first rerunning the `margins` command and specifying `at(paeduc=(0(1)20))` to compute the predicted means at each level of father's education. The output is omitted to save space. We then create the graph using the `marginsplot` command. The `recast()` option specifies that the fitted line should be displayed as a `line` graph (suppressing the markers). The `recastci()` option specifies that the confidence interval should be displayed as an `rarea` graph, displaying a shaded area for the confidence region. The resulting graph is shown in figure 2.3.

```
. margins, at(paeduc=(0(1)20))
  (output omitted)
. marginsplot, recast(line) recastci(rarea)
  Variables that uniquely identify margins: paeduc
```

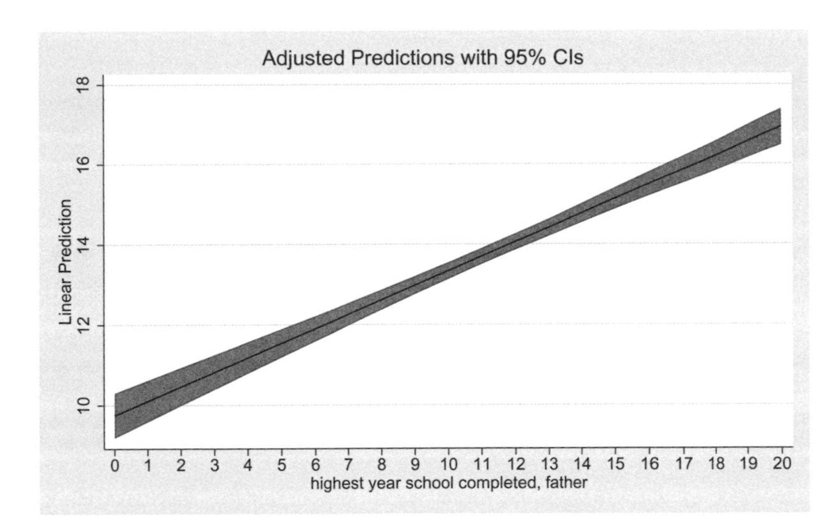

Figure 2.3. Predicted means and confidence interval shown as shaded region

2.3 Multiple regression

So far, all the examples have illustrated simple linear regression. Let's now turn to a multiple regression model that predicts the respondent's education from the father's education (**paeduc**), the mother's education (**maeduc**), and the age of the respondent (**age**). This multiple regression model is fit below using the **regress** command.

```
. regress educ paeduc maeduc age

      Source |       SS           df       MS      Number of obs   =       650
-------------+----------------------------------   F(3, 646)       =     92.93
       Model |  1822.26082         3  607.420272   Prob > F        =    0.0000
    Residual |  4222.37918       646  6.53619069   R-squared       =    0.3015
-------------+----------------------------------   Adj R-squared   =    0.2982
       Total |     6044.64       649  9.31377504   Root MSE        =    2.5566

------------------------------------------------------------------------------
        educ |      Coef.   Std. Err.      t    P>|t|     [95% Conf. Interval]
-------------+----------------------------------------------------------------
      paeduc |   .2580701   .0330141     7.82   0.000     .1932421    .3228981
      maeduc |   .2075304   .0378868     5.48   0.000     .1331342    .2819266
         age |   .0344424   .0065219     5.28   0.000     .0216356    .0472492
       _cons |    6.96181   .5106454    13.63   0.000     5.959085    7.964535
------------------------------------------------------------------------------
```

The multiple regression equation predicting the respondent's education from the father's education, mother's education, and age can be written as shown in (2.2).

$$\widehat{\texttt{educ}} = 6.96 + 0.26\texttt{paeduc} + 0.21\texttt{maeduc} + 0.03\texttt{age} \tag{2.2}$$

The coefficients from this multiple regression model reflect the association between each predictor and the outcome after adjusting for all the other predictors. For example, the coefficient for paeduc is 0.26, meaning that for every one-year increase in the education of the father, we would expect the education of the respondent to be 0.26 years higher, holding the mother's education and the age of the respondent constant.

2.3.1 Computing adjusted means using the margins command

As an aid to interpreting the coefficients from the multiple regression model, we can use the margins command to compute adjusted means of the outcome as a function of one or more predictors from the model. For example, to help interpret the coefficient for paeduc, we can use the margins command to compute the adjusted mean of education given different values of paeduc, adjusting for the other predictors in the model. The margins command below computes the adjusted mean of the respondent's education when the father's education equals 8, 12, and 16, adjusting for the other predictors (mother's education and age of the respondent).

```
. margins, at(paeduc=(8 12 16)) vsquish
Predictive margins                              Number of obs     =        650
Model VCE    : OLS

Expression   : Linear prediction, predict()
1._at        : paeduc          =           8
2._at        : paeduc          =          12
3._at        : paeduc          =          16

------------------------------------------------------------------------------
             |            Delta-method
             |     Margin   Std. Err.      t    P>|t|     [95% Conf. Interval]
-------------+----------------------------------------------------------------
         _at |
          1  |   13.03035   .1478844    88.11   0.000     12.73996    13.32075
          2  |   14.06263   .1029638   136.58   0.000     13.86045    14.26482
          3  |   15.09491   .1849627    81.61   0.000     14.73171    15.45812
------------------------------------------------------------------------------
```

Terminology: Adjusted means

Looking at the output from the previous `margins` command, we see that when a respondent's father has 8 years of education, the respondent is predicted to have 13.03 years of education, after adjusting for education of the mother and age of the respondent. What do we call the quantity 13.03? We can call this a predicted mean after adjusting for all other predictors. For example, we can say that the predicted mean, given the father has 8 years of education, is 13.03 after adjusting for all other predictors. We could also call this an adjusted mean. When the father has 8 years of education, the adjusted mean is 13.03. (The term *adjusted mean* implies after adjusting for all other predictors in the model.) When using nonlinear models (such as a logistic regression model), we will use a more general term, such as predictive margin.

The `margins` command shows the adjusted means at three different levels of father's education. When father's education is 8, the adjusted mean is 13.03. When father's education increases to 12, the adjusted mean increases to 14.06, and when father's education is 16, the adjusted mean is 15.09.

The `margins` command allows us to hold more than one variable constant at a time. In the example below, we compute the adjusted means when the father's education equals 8, 12, and 16, while holding the mother's education constant at 14.

```
. margins, at(paeduc=(8 12 16) maeduc=14) vsquish
Predictive margins                                Number of obs     =        650
Model VCE    : OLS
Expression   : Linear prediction, predict()
1._at        : paeduc          =           8
               maeduc          =          14
2._at        : paeduc          =          12
               maeduc          =          14
3._at        : paeduc          =          16
               maeduc          =          14

------------------------------------------------------------------------------
             |            Delta-method
             |     Margin   Std. Err.      t    P>|t|     [95% Conf. Interval]
-------------+----------------------------------------------------------------
         _at |
          1  |   13.54407   .2110229    64.18   0.000      13.1297    13.95844
          2  |   14.57635   .1281276   113.76   0.000     14.32476    14.82795
          3  |   15.60863   .1522517   102.52   0.000     15.30966     15.9076
------------------------------------------------------------------------------
```

Note how the legend informs us that the first `_at` value corresponds to father's education equal to 8 and mother's education equal to 14. The second and third `_at` values correspond to father's education equal to 12 and 16 (respectively), with mother's education equal to 14. (The variable **age** is not mentioned, which indicates that the predicted means are adjusted for age by averaging across all levels of age. As we will see shortly, for a linear model this is equivalent to holding age constant at its mean.)

Compared with the results of the previous **margins** command, we can see that the adjusted means are higher when the mother's education is held constant at 14. However, the effect of the father's education remains the same. For example, in the previous **margins** command, the change in the adjusted means due to increasing the father's education from 8 to 16 years was 2.06 (15.09 − 13.03). When the mother's education is held constant at 14, the change due to increasing the father's education from 8 to 16 years is the same (aside from rounding), 2.07 (15.61 − 13.54). Although the adjusted means are higher when the mother's education is held constant at 14 years, the difference in the adjusted means due to increasing the father's education remains the same.

2.3.2 Some technical details about adjusted means

A traditional ordinary least-squares approach computes adjusted means by holding other predictors constant at their mean value. The adjusted mean computed using this approach is called the marginal value at the mean. The **margins** command, by default, uses a different approach; it adjusts for the other predictors by averaging across the values of the other predictors. An adjusted mean computed in this fashion is called the average marginal value.

In a linear model, these two methods yield the same results. We can see this for ourselves by computing adjusted means using both methods. The **margins** command below computes the adjusted mean given that father's education is 8, averaging across the other predictors (average marginal value).

```
. margins, at(paeduc=8) vsquish
Predictive margins                          Number of obs     =        650
Model VCE    : OLS

Expression   : Linear prediction, predict()
at           : paeduc          =           8

------------------------------------------------------------------------------
             |            Delta-method
             |     Margin   Std. Err.      t    P>|t|     [95% Conf. Interval]
-------------+----------------------------------------------------------------
       _cons |   13.03035   .1478844    88.11   0.000     12.73996    13.32075
------------------------------------------------------------------------------
```

Now, let's repeat this command and add the **atmeans** option (shown below). This command computes the adjusted means at the marginal value at the mean. That is, the adjusted means are computed by holding the mother's education at the mean and the respondents age at the mean (as indicated in the legend of the **margins** output).

```
. margins, at(paeduc=8) atmeans vsquish
Adjusted predictions                          Number of obs     =        650
Model VCE    : OLS
Expression   : Linear prediction, predict()
at           : paeduc          =           8
               maeduc          =    11.52462 (mean)
               age             =    46.81077 (mean)

------------------------------------------------------------------------------
             |            Delta-method
             |     Margin   Std. Err.      t    P>|t|     [95% Conf. Interval]
-------------+----------------------------------------------------------------
       _cons |   13.03035   .1478844    88.11   0.000     12.73996    13.32075
------------------------------------------------------------------------------
```

We can see that adding the **atmeans** option yields the same adjusted means and standard errors as the previous **margins** command where this option was omitted.[1]

2.3.3 Graphing adjusted means using the marginsplot command

Let's graph the adjusted means of **educ** as a function of **paeduc**, adjusting for all other predictors. We can create such a graph using the **margins** and **marginsplot** commands below. The **marginsplot** command creates the graph shown in figure 2.4.

1. This is not the case for nonlinear models such as logistic regression, ordinal logistic regression, Poisson regression, and so forth. See chapter 18 for more details about computing predictive margins for such models.

```
. margins, at(paeduc=(0(4)20))
  (output omitted)
. marginsplot
  Variables that uniquely identify margins: paeduc
```

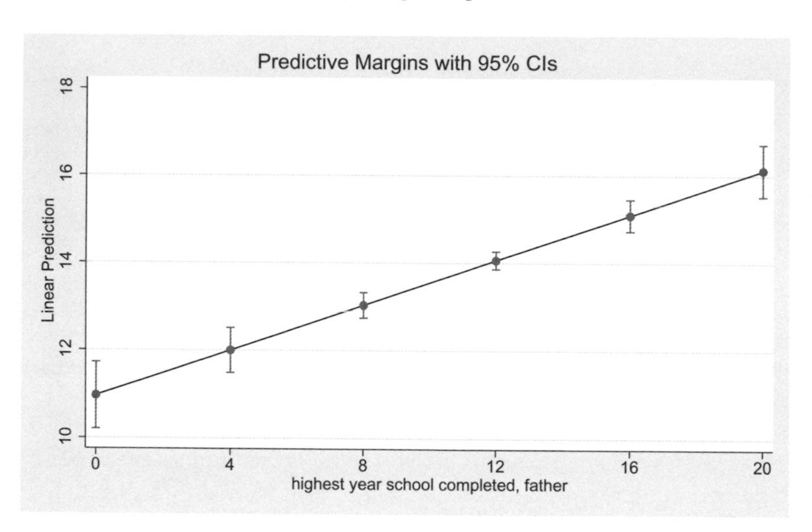

Figure 2.4. Predicted means by father's education, adjusting for mother's education and age

> **Tip! More on the margins and marginsplot commands**
>
> This chapter has only scratched the surface about the use of the `margins` and `marginsplot` commands. Later chapters will provide more examples about the use of these commands in the context of more complex regression models. In addition, appendix B provides more details about the `margins` command, and appendix C provides more details about the `marginsplot` command. Although we have not seen these commands yet, you can find more details about the `contrast` command in appendix D and about the `pwcompare` command in appendix E.

2.4 Checking for nonlinearity graphically

This section illustrates graphical approaches for checking for nonlinearity in the relationship between a predictor and outcome variable. These approaches include 1) examining scatterplots of the predictor and outcome, 2) examining residual-versus-fitted plots, 3) creating plots based on locally weighted smoothers, and 4) plotting the mean of the outcome for each level of the predictor. Each of these approaches is considered in turn, beginning with the use of scatterplots.

2.4.1 Using scatterplots to check for nonlinearity

Let's use a subset of the auto dataset called `autosubset.dta`. It is the same as the auto dataset (`auto.dta`) that comes with Stata except that I have omitted three cars to make the nonlinear relationships a bit more clear.

```
. use autosubset
(1978 Automobile Data)
```

Let's look at a scatterplot of the size of the engine (`displacement`) by length of the car (`length`) with a line showing the linear fit, as shown in figure 2.5.

```
. graph twoway (scatter displacement length) (lfit displacement length),
> ytitle("Engine displacement (cu in.)") legend(off)
```

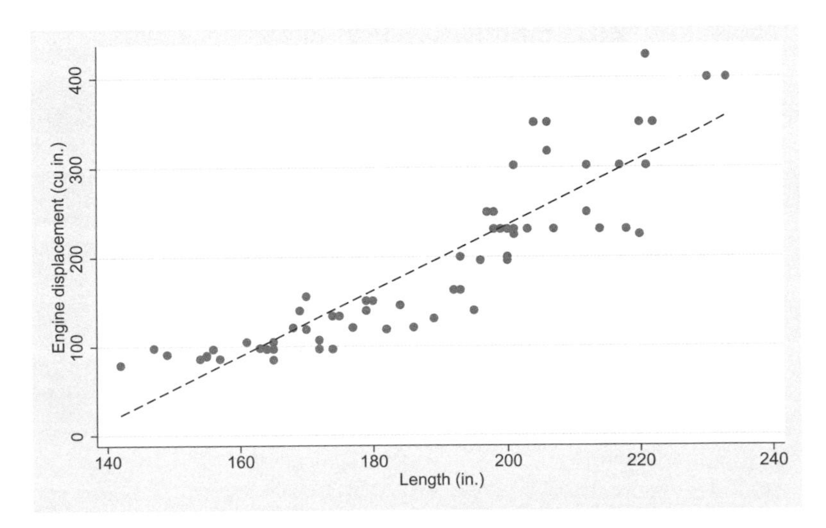

Figure 2.5. Scatterplot of engine displacement by car length with linear fit line

The relationship between these two variables looks fairly linear, but the addition of the linear fit line helps us to see the nonlinearity. Note how for short cars (when length is below 160) the fit line underpredicts and for longer cars (when length is above 210) the fit line also underpredicts.

Using a scatterplot like this can be a simple means of looking at the linearity of the simple relationship between a predictor and outcome variable. However, this does not account for other predictors that you might want to include in a model. To this end, let's next look at how we can use the residuals for checking linearity.

2.4.2 Checking for nonlinearity using residuals

We can check for nonlinearity by looking at the relationship between the residuals and predicted values, after accounting for other variables in the model. For example, let's run a regression predicting `displacement` from `length`, `trunk`, and `weight`, as shown below.

```
. regress displacement length trunk weight
      Source |       SS           df       MS      Number of obs   =        71
-------------+----------------------------------   F(3, 67)        =    189.26
       Model |  535697.677         3  178565.892   Prob > F        =    0.0000
    Residual |  63213.9848        67   943.49231   R-squared       =    0.8945
-------------+----------------------------------   Adj R-squared   =    0.8897
       Total |  598911.662        70  8555.88089   Root MSE        =    30.716

------------------------------------------------------------------------------
displacement |      Coef.   Std. Err.      t    P>|t|     [95% Conf. Interval]
-------------+----------------------------------------------------------------
      length |  -.7790159   .5796436    -1.34   0.183    -1.935989    .3779577
       trunk |   1.539695   1.230105     1.25   0.215    -.9156062    3.994995
      weight |   .1276394   .0155864     8.19   0.000     .0965288      .15875
       _cons |  -67.78624   61.93516    -1.09   0.278    -191.4093    55.83686
------------------------------------------------------------------------------
```

We can then look at the residuals versus the fitted values, as shown in figure 2.6. Note the U-shaped pattern of the residuals. This pattern suggests that the relationship between the predictors and outcome is not linear.

```
. rvfplot
```

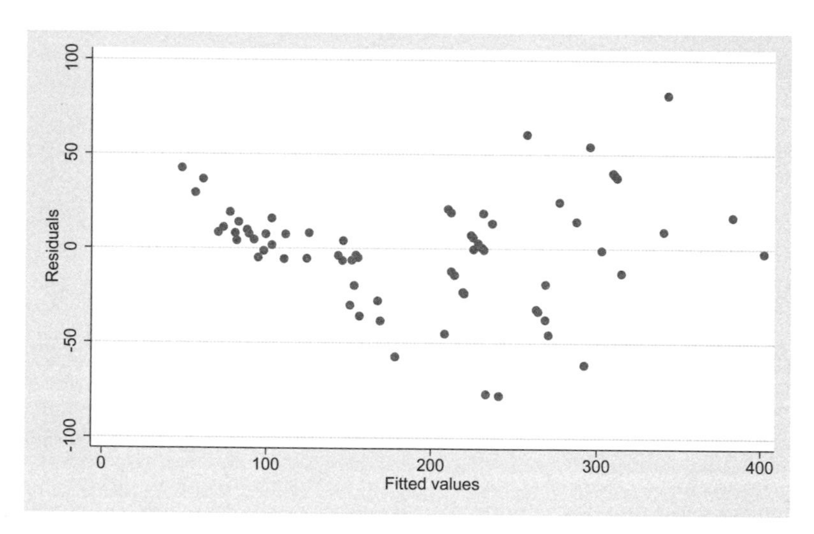

Figure 2.6. Residual-versus-fitted plot of engine displacement by car length, trunk size, and weight

There are many excellent resources that illustrate Stata's regression diagnostic tools, including the manual entry for [R] **regress postestimation**. You can also see the help entries for `avplot`, `rvfplot`, and `rvpplot`.

2.4.3 Checking for nonlinearity using locally weighted smoother

With larger datasets, it can be harder to visualize nonlinearity using scatterplots or residual-versus-fitted plots. To illustrate this, let's use an example from the dataset `gss_ivrm.dta`, which has many observations.

```
. use gss_ivrm
```

Suppose we want to determine the nature of the relationship between the year that the respondent was born (`yrborn`) and education level (`educ`). The `scatter` command can be used (as shown below) to create a scatterplot of these two variables. The resulting scatterplot is shown in figure 2.7.

```
. scatter educ yrborn, msymbol(oh)
```

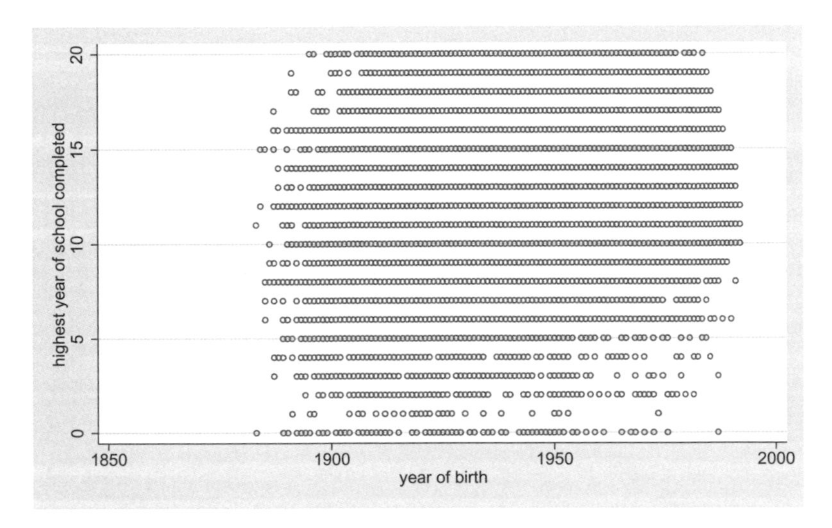

Figure 2.7. Scatterplot of education by year of birth

It is hard to discern the nature of the relationship between year of birth and education using this scatterplot. With so many observations, the scatterplot is saturated with data points creating one big blotch that tells us little about the shape of the relationship between the predictor and outcome.

Let's try a different strategy where we create a graph that shows the relationship between education and year of birth using a locally weighted regression (also called a locally weighted smoother or lowess). The `lowess` command below creates a graph showing the locally weighted regression of education on year of birth, as shown in figure 2.8. The `msymbol(p)` option displays the markers using tiny points.

```
. lowess educ yrborn, msymbol(p)
```

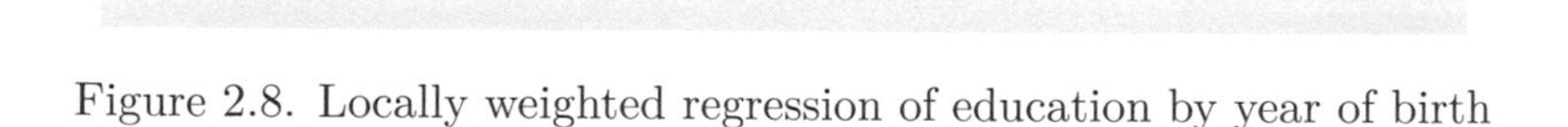

Figure 2.8. Locally weighted regression of education by year of birth

The lowess graph suggests that there is nonlinearity in the relationship between year of birth and education. Education increases with year of birth until the 1950s, at which point the smoothed education values level out and then start to decline[2] The graph produced by the `lowess` command is much more informative than the scatterplot alone.

2.4.4 Graphing outcome mean at each level of predictor

Another way to visualize the relationship between the predictor and outcome is to create a graph showing the mean of the outcome at each level of the predictor. Using the example predicting education from year of birth means creating a graph of the average of education at each level of year of birth. Although the variable `yrborn` can assume many values (from 1883 to 1990), the variable is composed of discrete integers with reasonably many observations (usually more than 100) for each value. In such a case, we can explore the nature of the relationship between the predictor and outcome by creating a graph of the mean of the outcome variable (education level) for each level of the predictor (year of birth). This kind of graph imposes no structure on the shape

2. We should note that people who were born in 1992 would have been 18 at the time of the interview (the last interview year in this dataset is 2010). This creates a hard upper limit to the years of education for those born in the late 1980s up to 1992. For this sake of this example, we will overlook this, but if we were publishing this finding, we would investigate whether the nonlinearity persists after accounting for this hard upper limit in education.

of the relationship between year of birth and education and allows us to observe the nature of the relationship between the predictor and the outcome.[3]

One simple way to create such a graph is to fit a regression model predicting the outcome treating the predictor variable as a factor variable. Following that, the **margins** command is used to obtain the predicted mean of the outcome for each level of the predictor. In the **regress** command below, specifying **i.yrborn** indicates that the variable **yrborn** should be treated as a factor variable. The following **margins** command computes the predicted mean of the outcome (education) at each year of birth.

```
. regress educ i.yrborn
  (output omitted)
. margins yrborn
  (output omitted)
```

Now, we can use the **marginsplot** command (shown below) to graph the average level of education as a function of the year of birth. The resulting graph is shown in figure 2.9.

```
. marginsplot
  Variables that uniquely identify margins: yrborn
```

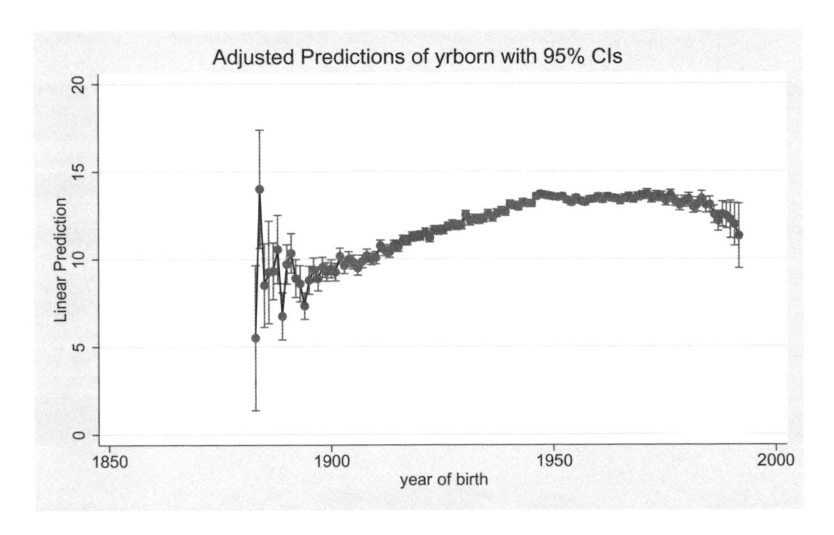

Figure 2.9. Mean education by year of birth

The graph in figure 2.9 shows that education level generally increases with year of birth until the 1950s where the level of education plateaus and then starts to decline. This is consistent with the graph created by the **lowess** command shown in figure 2.8.

3. This technique requires that the predictor have integer values. If the predictor were income measured to the penny, you could create income rounded to the nearest dollar. However, to have many observations per income category, you might be even better off using income rounded to the nearest \$1,000 or nearest \$10,000.

The predicted means vary erratically for the years before 1900 because of the few observations per year during those years. For these years, the confidence intervals are much wider compared with later years, reflecting greater uncertainty of the estimates because of fewer observations.

If all the years had such few observations, then the entire graph might be dominated by wild swings in the means and show little about the nature of the relationship between the predictor and outcome. For such a case, we could try grouping the predictor into larger bins such as creating a variable that contains the decade of birth, based on the year of birth. The `gss_ivrm.dta` dataset has a variable called `yrborndec`. The tabulation of this variable is shown below.

```
. tabulate yrborndec

  Year born |
(as decade) |
  (recoded) |      Freq.     Percent        Cum.
------------+-----------------------------------
      1880s |         60        0.11        0.11
      1890s |        608        1.11        1.22
      1900s |      2,051        3.74        4.95
      1910s |      4,135        7.53       12.49
      1920s |      5,892       10.73       23.22
      1930s |      6,401       11.66       34.88
      1940s |      9,505       17.32       52.20
      1950s |     11,631       21.19       73.39
      1960s |      8,265       15.06       88.45
      1970s |      4,570        8.33       96.77
      1980s |      1,707        3.11       99.88
      1990s |         64        0.12      100.00
------------+-----------------------------------
      Total |     54,889      100.00
```

Let's graph the mean of education by decade of birth but omit the 1880s and 1990s because of their very small-sample sizes. This graph is shown in figure 2.10.

```
. drop if yrborndec==1 | yrborndec==12
(124 observations deleted)
. regress educ i.yrborndec
  (output omitted)
. margins yrborndec
  (output omitted)
. marginsplot
  Variables that uniquely identify margins: yrborndec
```

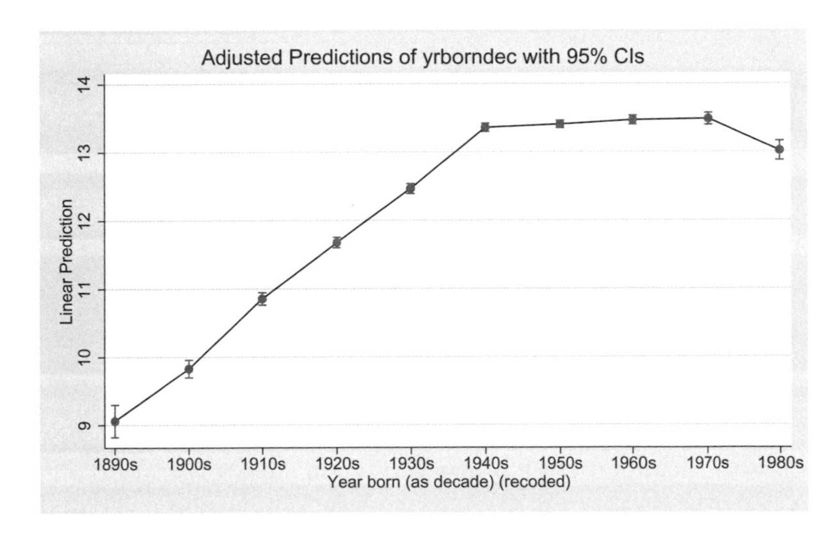

Figure 2.10. Mean education by decade of birth

For each decade of year of birth, the estimate of the mean of education is much more stable because of the larger sample size in each decade (compared with each year). However, if there were important changes in trend within a decade, this graph would conceal that information.

2.4.5 Summary

The previous sections have illustrated several ways to check for nonlinearity for a continuous predictor using visual approaches: creating scatterplots, examining residuals, creating plots based on locally weighted smoothers, and examining the predicted means for each level of the predictor by treating the predictor as a categorical variable.

2.5 Checking for nonlinearity analytically

This section shows how to check for nonlinearity using analytic approaches, including adding power terms and using factor variables.

2.5.1 Adding power terms

Another way to check for nonlinearity of a continuous variable is to add power terms (for example, quadratic or cubic). Let's continue using the example with education as a function of year of birth, beginning by showing the linear model predicting education from year of birth.

```
. use gss_ivrm
. regress educ yrborn
      Source |       SS           df       MS      Number of obs   =    54,745
-------------+----------------------------------   F(1, 54743)     =   5291.46
       Model |  48751.0917         1  48751.0917   Prob > F        =    0.0000
    Residual |  504356.171    54,743   9.2131628   R-squared       =    0.0881
-------------+----------------------------------   Adj R-squared   =    0.0881
       Total |  553107.263    54,744   10.103523   Root MSE        =    3.0353

------------------------------------------------------------------------------
        educ |      Coef.   Std. Err.      t    P>|t|     [95% Conf. Interval]
-------------+----------------------------------------------------------------
      yrborn |   .0464875   .0006391    72.74   0.000     .0452349    .0477401
       _cons |  -77.72922   1.243564   -62.51   0.000    -80.16662   -75.29183
------------------------------------------------------------------------------
```

Clearly, this model has a strong linear component, indicating that education tends to increase as year of birth increases. However, there may be nonlinearity in this relationship as well. Let's introduce a quadratic term (in addition to the linear term) by adding `c.yrborn#c.yrborn` to the model. This introduces a quadratic effect that would account for one bend in the line relating year of birth to education. We would expect the quadratic term to be significant based on the graphs we saw in figures 2.8 and 2.9 that showed that there was a bend in the relationship between year of birth and education. The model with the linear and quadratic terms is shown below.[4]

```
. regress educ yrborn c.yrborn#c.yrborn, noci
      Source |       SS           df       MS      Number of obs   =    54,745
-------------+----------------------------------   F(2, 54742)     =   3567.49
       Model |  63778.2779         2  31889.1389   Prob > F        =    0.0000
    Residual |  489328.985    54,742  8.93882184   R-squared       =    0.1153
-------------+----------------------------------   Adj R-squared   =    0.1153
       Total |  553107.263    54,744   10.103523   Root MSE        =    2.9898

-------------------------------------------------------------
               educ |      Coef.   Std. Err.      t    P>|t|
--------------------+----------------------------------------
             yrborn |   4.218252   .1017488    41.46   0.000
                    |
  c.yrborn#c.yrborn |  -.0010738   .0000262   -41.00   0.000
                    |
              _cons |      -4129   98.81569   -41.78   0.000
-------------------------------------------------------------
```

4. This model can also be specified as `regress educ c.yrborn##c.yrborn`.

Note! Using the noci option for clearer output

The statistical output in this book is limited to 80 columns. As a result, estimation commands (for example, the **regress** command) have a limited amount of space in the first column for labeling the coefficients. When models include interactions, those labels can get rather wide, and Stata is forced to display those labels across multiple lines, which can make the output confusing and hard to read. I think output is clearest when each term in the model is displayed, and labeled, in a single row. To achieve this, I sometimes include the **noci** option, as I did in the example above. This omits the display of the confidence intervals, which makes enough room to display the label for every term in the model in a single line. I will use this trick, as necessary, to make the output more readable in the book. I fret that this might be misconstrued as implying that confidence intervals are not important (they are!) or that confidence intervals are not important for a particular example. To avoid that connotation, I will often include a parenthetical notation or a footnote to explain why I am including the **noci** option and refer back to this callout. Your computer screen is probably wide enough to avoid this problem, so you probably will get readable output without adding the **noci** option.

Indeed, the quadratic term is significant in this model, confirming what we graphically saw in figures 2.8 and 2.9. Furthermore, the R^2 for this model increased to 0.1153 compared with 0.0881 for the linear model. This provides analytic support for including a quadratic term for year of birth when predicting education.

A cubic term would imply that the line fitting year of birth and education has a tendency to have two bends in it. We can test for a cubic trend by specifying the model as shown below. Including **c.yrborn##c.yrborn##c.yrborn** as a predictor is a shorthand for including the linear, quadratic, and cubic terms for year of birth (**yrborn**).

```
. regress educ c.yrborn##c.yrborn##c.yrborn, noci
note: c.yrborn#c.yrborn#c.yrborn omitted because of collinearity

      Source |       SS           df       MS      Number of obs   =    54,745
-------------+----------------------------------   F(2, 54742)     =   3567.49
       Model |  63778.2779         2  31889.1389   Prob > F        =    0.0000
    Residual |  489328.985    54,742  8.93882184   R-squared       =    0.1153
-------------+----------------------------------   Adj R-squared   =    0.1153
       Total |  553107.263    54,744   10.103523   Root MSE        =    2.9898

-----------------------------------------------------------------------
                          educ |      Coef.   Std. Err.      t    P>|t|
-------------------------------+---------------------------------------
                        yrborn |   4.218252    .1017488    41.46   0.000
                               |
             c.yrborn#c.yrborn |  -.0010738    .0000262   -41.00   0.000
                               |
    c.yrborn#c.yrborn#c.yrborn |          0  (omitted)
                               |
                         _cons |      -4129    98.81569   -41.78   0.000
-----------------------------------------------------------------------
```

There was a problem running this model. There is a note saying that the cubic term was omitted because of collinearity. This is a common problem when entering cubic terms, which can be solved by centering `yrborn`. The dataset includes a variable called `yrborn40`, which is the variable `yrborn` centered around the year 1940 (that is, 1940 is subtracted from each value of `yrborn`). Let's try fitting the above model again, but instead using the variable `yrborn40`.

```
. regress educ c.yrborn40##c.yrborn40##c.yrborn40, noci

      Source |       SS           df       MS      Number of obs   =    54,745
-------------+----------------------------------   F(3, 54741)     =   2399.43
       Model |  64279.431         3  21426.477     Prob > F        =    0.0000
    Residual |  488827.832    54,741  8.92983015   R-squared       =    0.1162
-------------+----------------------------------   Adj R-squared   =    0.1162
       Total |  553107.263    54,744  10.103523    Root MSE        =    2.9883

------------------------------------------------------------------------------
                              educ |      Coef.   Std. Err.      t    P>|t|
-----------------------------------+------------------------------------------
                          yrborn40 |   .0592234   .0011862    49.93   0.000
                                   |
              c.yrborn40#c.yrborn40 |  -.0010261   .0000269   -38.08   0.000
                                   |
   c.yrborn40#c.yrborn40#c.yrborn40 |  -7.56e-06   1.01e-06    -7.49   0.000
                                   |
                             _cons |   12.87476    .017648   729.53   0.000
------------------------------------------------------------------------------
```

Using the centered value `yrborn40` solved the problem of collinearity. The results show, as expected, that the cubic term is significant. However, this dataset has many observations, so the model has the statistical power to detect very small effects. Note how the R^2 is 0.1162 for the cubic model compared with 0.1153 for the quadratic model. This is a trivial increase, suggesting that the cubic trend is not really an important term to include in this model.

We could continue this exploration process by adding power terms, searching for additional nonlinear components, but for this example it seems that the major nonlinear component is strictly quadratic. See chapter 3 for more details about how to analyze and interpret such models.

2.5.2 Using factor variables

The strategy of using factor variables to check for nonlinearity makes sense only when you have a relatively limited number of levels of the predictor that are coded as whole numbers. We have such an example in the predictor variable named `agedec`, which is the age of the respondent expressed as a decade of life (for example, 30s, 40s, or 50s). Let's explore the relationship between the decade of life and the health status rating given by the respondent, with health status rated as: 4 = excellent, 3 = good, 2 = fair, and 1 = poor. We expect that as people age, their health status will decline, but does it decline linearly, or is there a nonlinearity to the relationship between age and health status? Let's investigate this using the variable `agedec` as the age of the person (as a decade), tabulated below.

```
. use gss_ivrm
. tabulate agedec
```

Age (as decade) of respondent	Freq.	Percent	Cum.
18-19	951	1.73	1.73
20s	10,849	19.76	21.50
30s	11,875	21.63	43.13
40s	10,082	18.37	61.50
50s	8,123	14.80	76.30
60s	6,539	11.91	88.21
70s	4,502	8.20	96.41
80s	1,969	3.59	100.00
Total	54,890	100.00	

Let's begin by graphing the mean health rating by each decade of life (using the strategy that was illustrated in section 2.4.4). The graph of the relationship between the mean of health at each level of age as a decade is shown in figure 2.11.

```
. regress health i.agedec
  (output omitted)
. margins agedec
  (output omitted)
. marginsplot
  Variables that uniquely identify margins: agedec
```

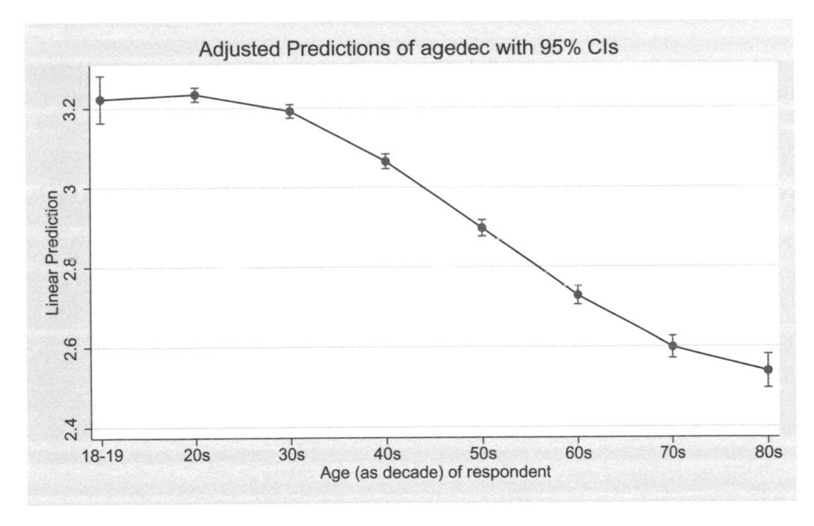

Figure 2.11. Average health by age (as a decade)

This graph shows nonlinearity in this relationship, but let's perform an analysis to detect any kind of nonlinearity in the relationship between decade of age and health status. One approach is to formulate a regression model as shown below, which includes both `c.agedec` and `i.agedec` as predictors in the same model.

```
. regress health c.agedec i.agedec
note: 8.agedec omitted because of collinearity
      Source |       SS           df       MS      Number of obs   =    40,984
-------------+----------------------------------   F(7, 40976)     =    451.59
       Model |  2111.10121         7  301.585887   Prob > F        =    0.0000
    Residual |  27364.9308    40,976  .667828261   R-squared       =    0.0716
-------------+----------------------------------   Adj R-squared   =    0.0715
       Total |   29476.032    40,983  .719225826   Root MSE        =    .81721

------------------------------------------------------------------------------
      health |      Coef.   Std. Err.      t    P>|t|     [95% Conf. Interval]
-------------+----------------------------------------------------------------
      agedec |  -.0977603   .0052951   -18.46   0.000    -.1081388   -.0873819
             |
      agedec |
      18-19  |          0  (base)
        20s  |   .1095554   .0275956     3.97   0.000     .0554674    .1636433
        30s  |   .1648267   .0240665     6.85   0.000     .1176558    .2119976
        40s  |   .1354884   .0217134     6.24   0.000     .0929297    .1780471
        50s  |   .0652255   .0206744     3.15   0.002     .0247032    .1057478
        60s  |  -.0058219    .021133    -0.28   0.783     -.047243    .0355992
        70s  |  -.0377352   .0236222    -1.60   0.110    -.0840352    .0085648
        80s  |          0  (omitted)
             |
       _cons |   3.320438    .034656    95.81   0.000     3.252511    3.388364
------------------------------------------------------------------------------
```

This unconventional-looking model divides the relationship between the age decade and the outcome into two pieces: the linear relationship, which is accounted for by `c.agedec`, and any remaining nonlinear components, which are explained by the indicator variables specified by `i.agedec`. Because of the inclusion of `c.agedec`, one of those indicators was redundant and Stata excluded it. (That is why the category for those born in their it 80s is labeled as `(omitted)` in the output.) Let's now use the `testparm` command to perform a test of the indicator variables, giving us an overall test of the nonlinearity in the relationship between age decade and health status.

```
. testparm i.agedec
 ( 1)  2.agedec = 0
 ( 2)  3.agedec = 0
 ( 3)  4.agedec = 0
 ( 4)  5.agedec = 0
 ( 5)  6.agedec = 0
 ( 6)  7.agedec = 0
       F(  6, 40976) =   20.61
            Prob > F =    0.0000
```

This test is significant, indicating that overall there is a significant contribution of nonlinear terms in the relationship between the age decade and health status. This general strategy tells us that there is nonlinearity between the age decade and health status but does not pinpoint the exact nature of the nonlinearity. Let's try another strategy that will pinpoint the nature of the nonlinearity. We begin by running a model predicting the variable **health** from **i.agedec**.

```
. regress health i.agedec
  (output omitted)
```

We now can use the **contrast** command with the **p.** contrast operator to obtain a detailed breakdown of possible nonlinear trends in the relationship between the age decade and health status.

```
. contrast p.agedec
Contrasts of marginal linear predictions
Margins      : asbalanced
```

	df	F	P>F
agedec			
(linear)	1	1187.58	0.0000
(quadratic)	1	26.65	0.0000
(cubic)	1	52.99	0.0000
(quartic)	1	1.18	0.2778
(quintic)	1	1.00	0.3179
(sextic)	1	0.15	0.6959
(septic)	1	0.02	0.8748
Joint	7	451.59	0.0000
Denominator	40976		

	Contrast	Std. Err.	[95% Conf.	Interval]
agedec				
(linear)	-.2599665	.0075437	-.2747524	-.2451806
(quadratic)	-.0384272	.0074435	-.0530165	-.0238378
(cubic)	.0466045	.0064025	.0340554	.0591536
(quartic)	.0056375	.005194	-.0045429	.0158178
(quintic)	-.0042917	.0042968	-.0127134	.00413
(sextic)	.0014878	.0038067	-.0059734	.0089489
(septic)	-.0005636	.003578	-.0075765	.0064494

These results show that in the relationship between the age decade and health status, there are significant linear, quadratic, and cubic effects. If we wished, we could limit our examination to only the nonlinear effects, as shown below.

```
. contrast p(2/7).agedec
Contrasts of marginal linear predictions
Margins      : asbalanced

             |         df           F        P>F
-------------+----------------------------------
      agedec |
 (quadratic) |          1       26.65     0.0000
     (cubic) |          1       52.99     0.0000
   (quartic) |          1        1.18     0.2778
   (quintic) |          1        1.00     0.3179
    (sextic) |          1        0.15     0.6959
    (septic) |          1        0.02     0.8748
       Joint |          6       20.61     0.0000
             |
 Denominator |      40976
------------------------------------------------

--------------------------------------------------------------
             |   Contrast   Std. Err.     [95% Conf. Interval]
-------------+------------------------------------------------
      agedec |
 (quadratic) |  -.0384272    .0074435     -.0530165   -.0238378
     (cubic) |   .0466045    .0064025      .0340554    .0591536
   (quartic) |   .0056375     .005194     -.0045429    .0158178
   (quintic) |  -.0042917    .0042968     -.0127134      .00413
    (sextic) |   .0014878    .0038067     -.0059734    .0089489
    (septic) |  -.0005636     .003578     -.0075765    .0064494
--------------------------------------------------------------
```

The overall (`Joint`) test matches the test we performed earlier looking at the contribution of the indicator variables (`i.agedec`) in the presence of the continuous version (`c.agedec`). Both of these tests ask the same question regarding the nonlinear contributions.

These tests suggest that linear, quadratic, and cubic terms may be appropriate to use when modeling the relationship predicting `health` from `agedec`. That kind of modeling strategy is illustrated in section 3.3.

2.6 Summary

This chapter has illustrated fitting linear models with a continuous predictor. These models have included simple regression models and multiple regression models. We have seen how to compute adjusted means using the `margins` command and how to create graphs of the adjusted means using the `marginsplot` command. We also saw how to assess the linearity assumption, illustrating both graphical (section 2.4) and analytic techniques (section 2.5), including the use of the `p.` contrast operator. You can find more information about the use of the `contrast` command with the `p.` contrast operator for performing polynomial contrasts in section 7.9.

If the relationship between a predictor and outcome is not linear, there are several ways you can account for the nonlinearity. Chapter 3 illustrates the use of polynomial terms to account for nonlinearity in the relationship between the predictor and outcome. Such approaches include the addition of quadratic (squared) terms (see section 3.2), the addition of cubic terms (see section 3.3), and fractional polynomial regression (see section 3.4). You could also account for nonlinearity using piecewise modeling, as illustrated in chapter 4. For example, you can fit a piecewise regression model with one or two known knots (see sections 4.2–4.6), a piecewise regression model with one unknown knot (see section 4.7), or a piecewise regression model with multiple unknown knots (see section 4.8).

For more background about the use of linear regression, I recommend Fox (2016), Hamilton (1992), Chatterjee and Hadi (2012), Cohen et al. (2003), or your favorite regression book. I also recommend Baum (2006), Kohler and Kreuter (2009), and Juul and Frydenberg (2014) as books that illustrate the use of regression methods using Stata. For more information about the difference between the average marginal effect and the marginal effect at the mean, see Cameron and Trivedi (2010).

3 Continuous predictors: Polynomials

3.1 Chapter overview

This chapter focuses on the use of polynomial terms to account for nonlinearity in the relationship between a continuous predictor and a continuous outcome. This chapter illustrates the use of quadratic (squared) terms in section 3.2, the use of cubic (third order) terms in section 3.3, and fractional polynomial models in section 3.4. The chapter concludes with a discussion of the interpretation of main effects in the presence of polynomial terms (see section 3.5).

3.2 Quadratic (squared) terms

3.2.1 Overview

A quadratic (squared) term can be used to model curved relationships, accounting for one bend in the relationship between the predictor and outcome. Such curves come in two general types: U-shaped (convex) and inverted U-shaped (concave).

Consider the graph shown in figure 3.1 illustrating a hypothetical relationship between age on the x axis and income on the y axis. The curve shows an inverted U-shape with incomes rising with increasing age until income reaches a peak at around 50 years of age, after which incomes decline with increasing age. Let's relate the nature of the curve to the regression equation predicting income (realrinc) from age (age), shown below.

$$\widehat{\texttt{realrinc}} = -30000 + 2500\texttt{age} + (-25\texttt{age}^2)$$

The curve has an inverted U-shape because the coefficient for age squared is negative. If the coefficient for the squared term was positive, the curve would have a U-shape.

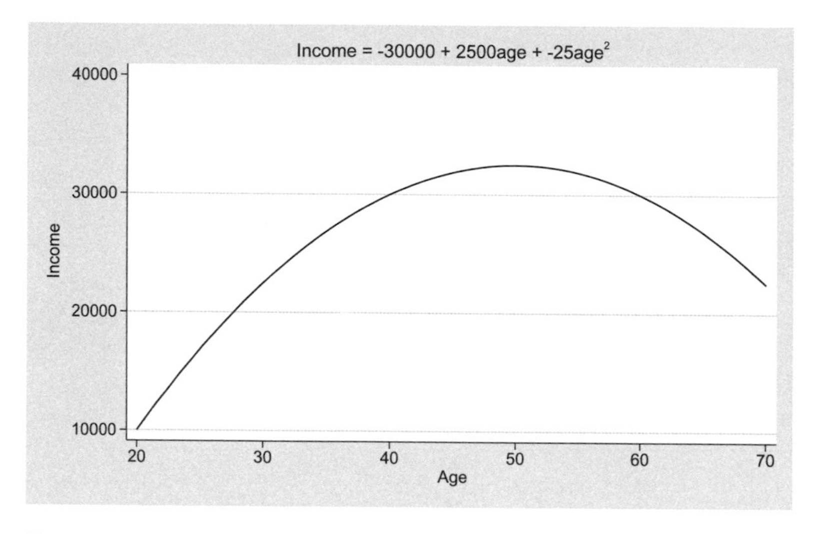

Figure 3.1. Quadratic regression with quadratic coefficient of −25

The magnitude of the squared term determines the degree of curvature. The squared term in figure 3.1 is −25. Contrast this with the left panel of figure 3.2, where the squared term is less negative (−22), and the right panel of figure 3.2, where the squared term is more negative (−28).

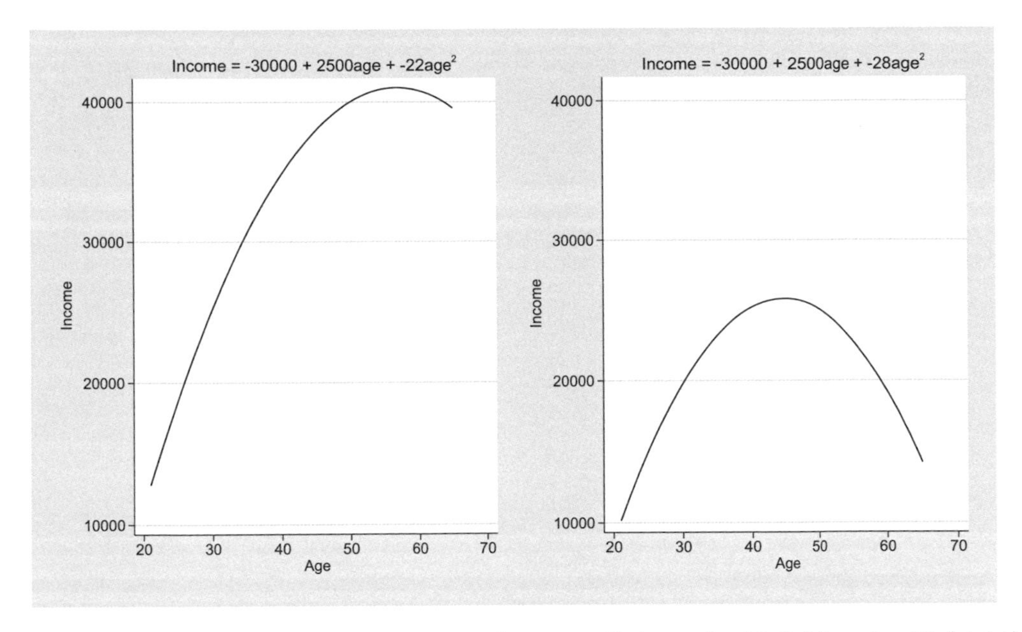

Figure 3.2. Quadratic regression with quadratic coefficient of −22 (left) and −28 (right)

The left panel shows less curvature in the relationship between age and income (due to the squared term being less negative), and the right panel shows more curvature (due to the squared term being more negative). If the quadratic coefficient is more negative, the curve will show a more strongly inverted U-shape.

The linear coefficient for age represents the `age` slope when age equals zero (which obviously is a preposterous value). To picture this, we need to extend the x axis of age further to the left to include zero. The linear coefficient determines the slope of the line when age is held constant at zero.

Terminology! Slopes

The above paragraph uses the term "`age` slope". This refers to the slope of the relationship between income and age or to the slope in the direction of age. In this chapter and through the book, I often use the shorthand term `age` slope in place of these longer descriptions.

In figure 3.1, the linear term is 2,500. Contrast this with the left panel of figure 3.3, where the linear term is lower (2,300), and the right panel of figure 3.3, where the linear term is higher (2,700).

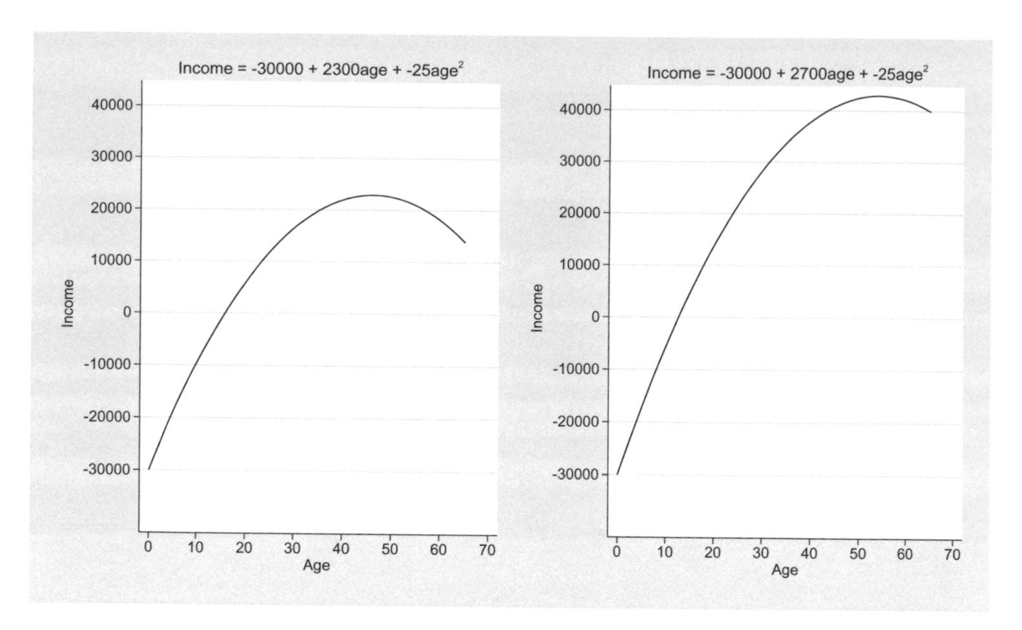

Figure 3.3. Quadratic regression with linear coefficient of 2,300 (left) and 2,700 (right)

A larger coefficient for the linear term increases the slope at the point where age is zero. Comparing the left and right panels of figure 3.3, we can see that in this example a higher linear term yields a steeper relationship between age and income before the bend in the curve.

For an inverted U-shaped curve, we can compute the value of x that corresponds to the maximum value of y. Referring back to figure 3.1, we can see that the maximum income corresponds to an age of around 50 years old. If we call the linear coefficient of age $b1$ and the quadratic coefficient of age $b2$, then the value of age that yields the maximum income is given by $-b1/(2 \times b2)$. Substituting 2,500 for $b1$ and -25 for $b2$ yields a value of 50; therefore, the maximum income occurs when someone is 50 years old. This also corresponds to the age where the slope of the relationship between age and income is flat. When age is less than 50, the slope is positive (and becomes progressively more positive as age further decreases below 50). Likewise, when age is greater than 50, the slope is negative (and becomes further negative as age increases beyond 50).

We can take this idea one step further and not only describe the place where the fit line is at its maximum (and the slope is zero) but also describe the slope for any given level of age. Figure 3.4 contains the same graph as figure 3.1 and displays the slope at three different values, 30, 50, and 60. At age 30, the slope is positive and has a value of 1,000. At age 50, the curve is at its maximum and the slope is zero. At age 60, the slope is negative and has a value of -500. This underscores that the slope changes every time age changes.

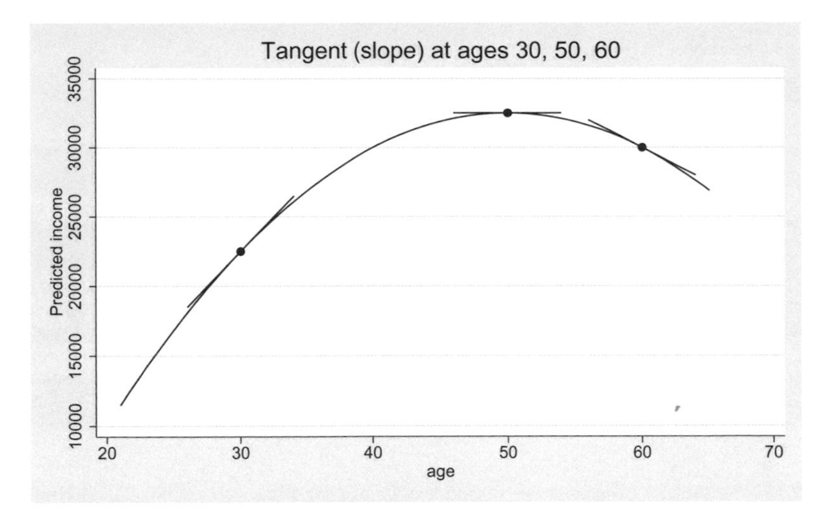

Figure 3.4. Linear slopes for quadratic regression with age at 30, 50, and 60

All of these examples have depicted a relationship where the quadratic coefficient is negative, yielding an inverted U-shaped relationship between x and y. If the quadratic coefficient were positive, the relationship between x and y would show a U-shaped relationship, but the other principles illustrated so far would remain the same. The degree of curvature (U-shape) would increase as the quadratic coefficient increases. The linear coefficient would still determine the slope when the predictor is at zero. Because the curve is U-shaped, the minimum of that curve would be represented by $-b1/(2 \times b2)$, where $b1$ is the linear coefficient and $b2$ is the quadratic coefficient.

3.2.2 Examples

Let's use the GSS dataset to illustrate a quadratic relationship between the age of the respondent (`age`) and income (`realrinc`). For this analysis, let's focus only on those who are 18 to 80 years old.

```
. use gss_ivrm
. keep if (age<=80)
(1,892 observations deleted)
```

Before fitting a quadratic model relating income to age, let's assess the shape of the relationship between these variables using a locally weighted smoother. The `lowess` command is used to create the variable `yhatlowess`, which is the predicted value based on the locally weighted regression predicting income from age. The `line` command then draws the graph showing the relationship between the fitted value of income and age. This graph is shown in figure 3.5.

```
. lowess realrinc age, nograph gen(yhatlowess)
. line yhatlowess age, sort
```

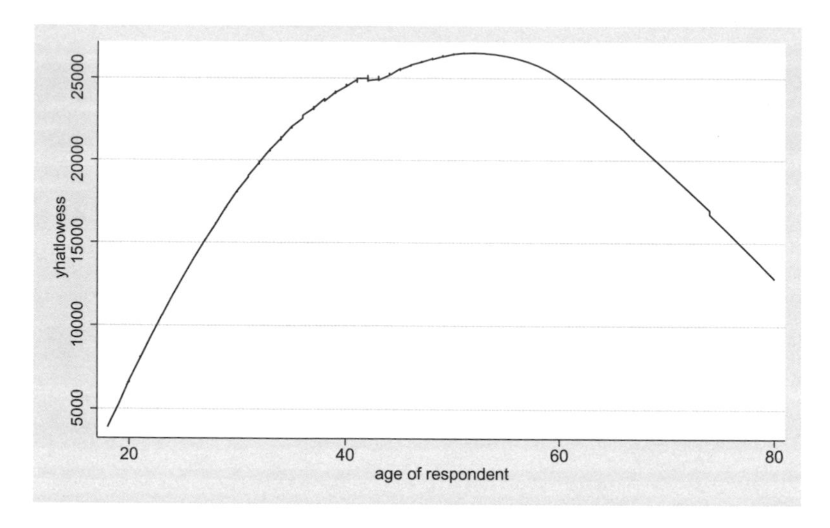

Figure 3.5. Lowess-smoothed values of income by age

The graph shows a curved relationship between the predictor and outcome. Income rises as a function of age until it reaches a peak at about 50 years of age, and then income declines thereafter. A model with only a linear term would not be able to account for this bend in the relationship between age and income. Introducing a quadratic term can account for such a bend.

Note! Checking for nonlinearity

Chapter 2 describes methods for checking for nonlinearity using graphical methods (see section 2.4). That chapter also describes methods for checking for nonlinearity using analytical methods (see section 2.5).

Let's fit a model with a quadratic (squared) term to account for the bend in the relationship between age and income. We do this using the interaction operator `#` (as shown below), which creates a term that multiplies `age` by `age`. Specifying `c.age` indicates to Stata that `age` should be treated as a continuous variable (instead of treating it as factor variable). By default, variables connected with the interaction operator (`#`) are assumed to be factor variables. This model also includes `female` as a covariate, specifying it as a dummy variable.

```
. regress realrinc age c.age#c.age female, vce(robust)
Linear regression                               Number of obs     =     32,100
                                                F(3, 32096)       =    1252.67
                                                Prob > F          =     0.0000
                                                R-squared         =     0.1089
                                                Root MSE          =      25407
```

realrinc	Coef.	Robust Std. Err.	t	P>|t|	[95% Conf.	Interval]
age	2412.339	58.05774	41.55	0.000	2298.544	2526.135
c.age#c.age	-24.20196	.6958905	-34.78	0.000	-25.56593	-22.83799
female	-12419.24	280.4746	-44.28	0.000	-12968.98	-11869.49
_cons	-25679.77	1038.412	-24.73	0.000	-27715.1	-23644.44

We could simplify this command by using the ## operator, which creates both main effects and products (interactions) of the connected variables:

```
. regress realrinc c.age##c.age female, vce(robust)
  (output omitted)
```

Note! Why not square age?

You might be wondering why we do not instead use the generate command to create a new variable, say, age2, that contains the age squared. We could then enter age and age2 into the model. If we do this, the results of the regress command would be the same, but this would confuse the margins command. The margins command would think that age and age2 are two completely different variables. So when estimating the adjusted means as a function of age, the margins command would do so, adjusting for age2. By specifying c.age##c.age, the margins command understands that both c.age and c.age#c.age are derived from the variable age and accordingly computes appropriate adjusted means.

Before interpreting the coefficients from this regression model, let's first create a graph showing the adjusted means of income as a function of age. The first step is to use the margins command to compute the adjusted means across the spectrum of age. The at() option is used to compute the adjusted means for ages ranging from 18 to 80 in one-year increments. Using one-year increments will make a smoother curve than using 5- or 10-year increments.

```
. margins, at(age=(18(1)80))
  (output omitted)
```

The second step is to use the marginsplot command, which creates the graph shown in figure 3.6.

```
. marginsplot, noci
  Variables that uniquely identify margins: age
```

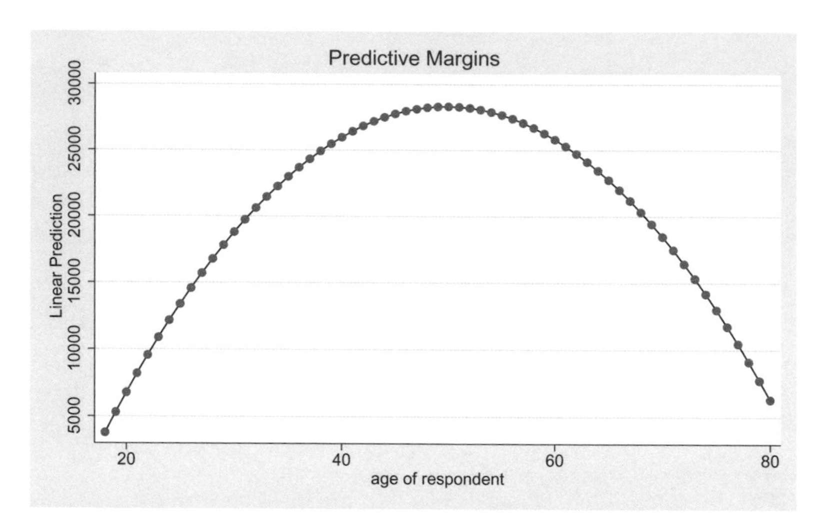

Figure 3.6. Adjusted mean of income by age

Interpreting the relationship between age and income

Now, let's interpret the coefficients from the regression output using the graph shown in figure 3.6. First, the `c.age#c.age` coefficient is negative, which is why the curve has an inverted U-shape. The coefficient for `age` is positive. This is the slope at the point where age equals zero. If we imagine the graph in figure 3.6 extending age to zero, the slope would be 2,412.34.

We can use the formula $-b1/(2 \times b2)$ (see section 3.2.1) to compute the age at which income is at its maximum. This yields $-2412.34/(2 \times -24.20)$, which equals 49.84. The adjusted mean of income is highest for those who are 49.84 years old. This age also corresponds to the point at which the slope of the fitted line is zero. Based on these computations, and by looking at figure 3.6, we see that the slope is positive when age equals zero (2,412.34) and becomes smaller with increasing age until the slope is zero at age 49.84. The slope is negative afterward.

Suppose we wanted to estimate the `age` slope for any given value of `age`. We can do so using the `margins` command combined with the `dydx(age)` option. For example, below we compute the `age` slope at 40 years of age. At 40 years of age, the slope is 476.18.

```
. margins, at(age=40) dydx(age)
Average marginal effects                        Number of obs     =     32,100
Model VCE    : Robust

Expression   : Linear prediction, predict()
dy/dx w.r.t. : age
at           : age             =          40

------------------------------------------------------------------------------
             |            Delta-method
             |      dy/dx   Std. Err.      t    P>|t|     [95% Conf. Interval]
-------------+----------------------------------------------------------------
         age |   476.1827   9.823473    48.47   0.000     456.9283    495.4371
------------------------------------------------------------------------------
```

To help you visualize this, I have created figure 3.7, which shows the adjusted means of income as a function of age. This includes a tangent line showing the age slope when age is 40. At that point, the age slope is 476.18.

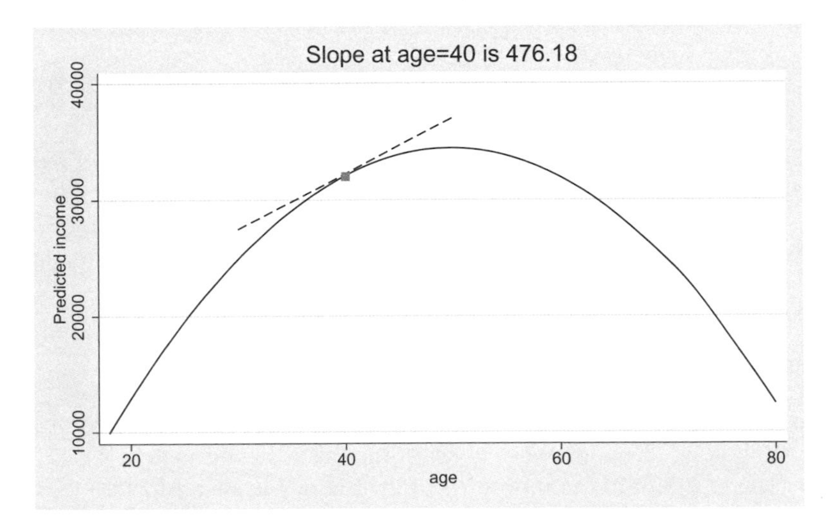

Figure 3.7. Slope when age is held constant at 40

In fact, we can use the margins command to estimate the age slope at any value of age. Below, we obtain the age slope for ages ranging from 30 to 70 in 10-year increments.

```
. margins, at(age=(30(10)70)) dydx(age) vsquish
Average marginal effects                        Number of obs     =     32,100
Model VCE      : Robust

Expression     : Linear prediction, predict()
dy/dx w.r.t. : age
1._at          : age           =          30
2._at          : age           =          40
3._at          : age           =          50
4._at          : age           =          60
5._at          : age           =          70

------------------------------------------------------------------------------
             |            Delta-method
             |      dy/dx   Std. Err.      t    P>|t|     [95% Conf. Interval]
-------------+----------------------------------------------------------------
age          |
         _at |
          1  |   960.2218   18.27387    52.55   0.000     924.4044    996.0393
          2  |   476.1827   9.823473    48.47   0.000     456.9283    495.4371
          3  |  -7.856494   15.69961    -0.50   0.617    -38.62833    22.91534
          4  |  -491.8957    27.9976   -17.57   0.000     -546.772   -437.0193
          5  |  -975.9348     41.336   -23.61   0.000    -1056.955   -894.9147
------------------------------------------------------------------------------
```

This output shows how the age slope changes as a function of age. For example, the age slope is 960.22 at 30 years of age. At 40 years of age, the slope is smaller (476.18), and by the age of 50, the slope is negative (−7.86). In a quadratic model, there is no single value that we can use for describing the age slope because the age slope is different at every age.

Graphing adjusted means with confidence intervals

The marginsplot command can be used to create a graph of the adjusted means with a shaded region showing the confidence interval for the adjusted means. We first need to run the margins command to compute the adjusted means as a function of age (shown below). We can then run the marginsplot command, adding the recast(line) and recastci(rarea) options to display the adjusted means as a line and the confidence interval as a shaded region. The resulting graph is shown in figure 3.8.

```
. margins, at(age=(18(1)80))
  (output omitted)
. marginsplot, recast(line) recastci(rarea)
  Variables that uniquely identify margins: age
```

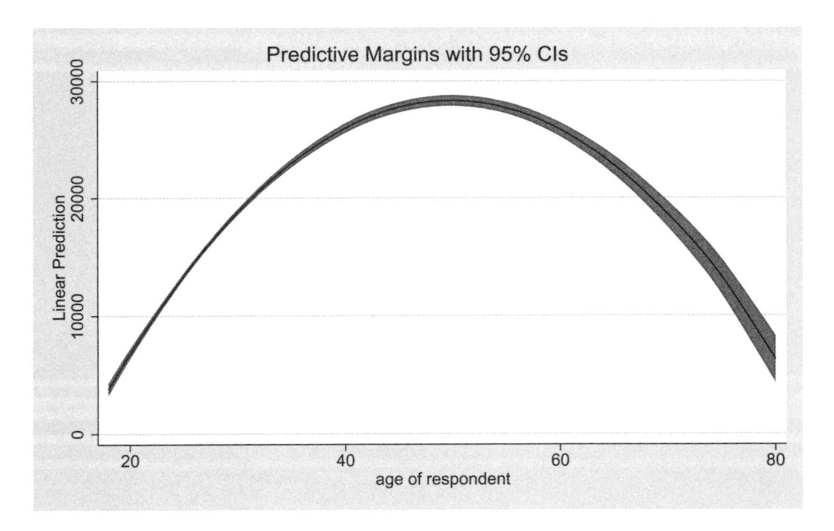

Figure 3.8. Adjusted means with shaded confidence region from quadratic regression

> **Note! Including linear and quadratic effects**
>
> Consider the model we ran earlier:
>
> ```
> . regress realrinc c.age##c.age female, vce(robust)
> ```
>
> This includes both the linear and quadratic effects of age. Even if the linear effect is not significant, it is important to still include it in the model.

3.3 Cubic (third power) terms

3.3.1 Overview

Previously, we saw that a quadratic term can account for one bend in the relationship between the predictor and outcome. Adding a cubic term can account for two bends in the relationship between the predictor and the outcome. This section illustrates the use of cubic terms for fitting a relationship between a predictor and outcome that includes two bends.

An example of such a cubic relationship is shown in figure 3.9. This shows a hypothetical relationship between the year of a woman's birth and her number of children.

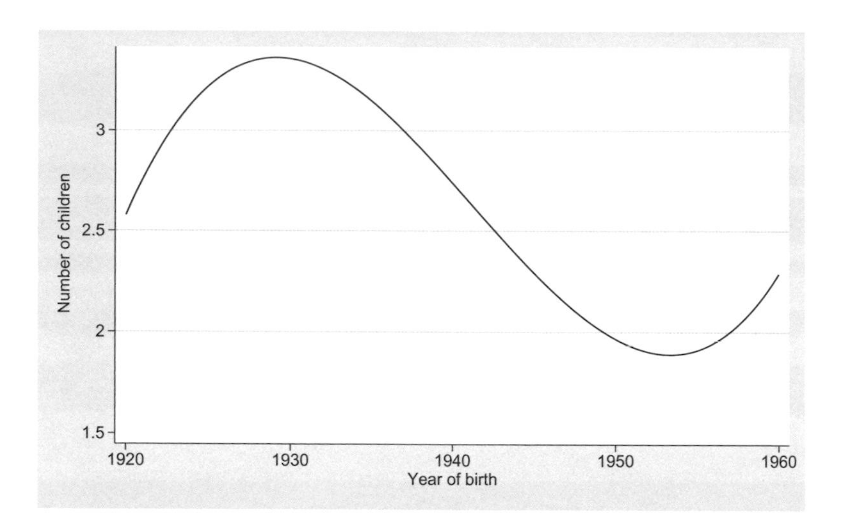

Figure 3.9. Number of children by year of birth

The number of children initially increases until it reaches a peak for those born around 1930. Then, the number of children declines until hitting a minimum around 1952, and then, the number of children rises again. A cubic term can be appropriate for modeling a relationship between a predictor and outcome that has two bends, like the relationship shown in figure 3.9.

3.3.2 Examples

Let's examine the relationship between the year of birth and number of children a woman has using the gss_ivrm.dta dataset, focusing on women aged 45 to 55 born between 1920 and 1960.

```
. use gss_ivrm
. keep if (age>=45 & age<=55) & (yrborn>=1920 & yrborn<=1960) & female==1
(50,021 observations deleted)
```

Let's get a sense of the relationship between year of birth (yrborn) and the average number of children a woman has (children) by looking at a lowess-smoothed regression relating year of birth to number of children. The lowess command is used with the generate() option to create the variable yhatlowess, which is the smoothed value of the average number of children.

```
. lowess children yrborn, generate(yhatlowess) nograph
```

The **graph twoway** command is then used to show the smoothed values as a function of the year of birth in figure 3.10.

```
. graph twoway line yhatlowess yrborn, sort
```

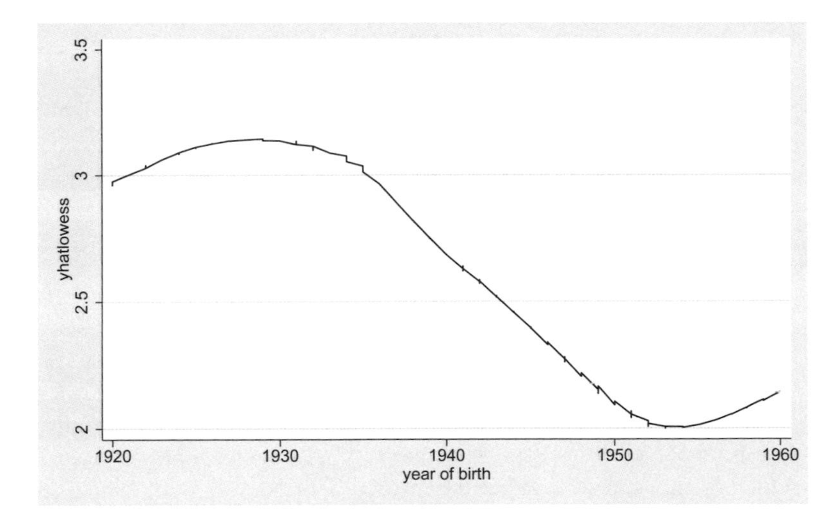

Figure 3.10. Lowess-smoothed fit of number of children by year of birth

The lowess-smoothed values show that there are two bends in the relationship between year of birth and number of children. The number of children rises initially, then declines, and then rises again. This suggests that including year of birth as a cubic term may account for this nonlinearity.

Below, we fit a model predicting `children` from `yrborn` fit using a cubic term. The model specifies `c.yrborn##c.yrborn##c.yrborn`, which includes the cubic term for `yrborn`, the quadratic term for `yrborn`, and the linear term for `yrborn`.[1]

1. The `noci` option is included to make the output more readable for this example. See the callout titled *Using the* `noci` *option for clearer output* in section 2.5.1 for more details.

```
. regress children c.yrborn##c.yrborn##c.yrborn, noci
note: c.yrborn#c.yrborn#c.yrborn omitted because of collinearity
      Source |       SS           df       MS      Number of obs   =     5,049
-------------+----------------------------------   F(2, 5046)      =    193.19
       Model |  1163.13728         2  581.56864   Prob > F        =    0.0000
    Residual |   15189.893     5,046  3.01028399   R-squared       =    0.0711
-------------+----------------------------------   Adj R-squared   =    0.0708
       Total |  16353.0303     5,048  3.2395068   Root MSE        =     1.735

------------------------------------------------------------------------
                     children |      Coef.   Std. Err.      t    P>|t|
------------------------------+-----------------------------------------
                       yrborn |   1.430931   .8156673     1.75   0.079
                              |
            c.yrborn#c.yrborn |  -.0003795   .0002101    -1.81   0.071
                              |
   c.yrborn#c.yrborn#c.yrborn |          0  (omitted)
                              |
                        _cons |  -1344.881   791.4741    -1.70   0.089
------------------------------------------------------------------------
```

There was a problem running this model. There is a note saying that the cubic term was omitted because of collinearity. This is a common problem when entering cubic terms, which can be solved by centering `yrborn`. The dataset includes a variable called `yrborn40`, which is the variable `yrborn` centered around the year 1940 (that is, 1940 is subtracted from each value of `yrborn`). Let's try fitting the above model again but instead using the variable `yrborn40`.

```
. regress children c.yrborn40##c.yrborn40##c.yrborn40, noci
      Source |       SS           df       MS      Number of obs   =     5,049
-------------+----------------------------------   F(3, 5045)      =    166.84
       Model |  1475.96126         3  491.987087   Prob > F        =    0.0000
    Residual |   14877.069     5,045  2.94887394   R-squared       =    0.0903
-------------+----------------------------------   Adj R-squared   =    0.0897
       Total |  16353.0303     5,048  3.2395068   Root MSE        =    1.7172

------------------------------------------------------------------------------
                           children |      Coef.   Std. Err.      t    P>|t|
------------------------------------+-----------------------------------------
                           yrborn40 |  -.0908745   .0052395   -17.34   0.000
                                    |
              c.yrborn40#c.yrborn40 |  -.0007751   .0002115    -3.66   0.000
                                    |
   c.yrborn40#c.yrborn40#c.yrborn40 |   .0002091   .0000203    10.30   0.000
                                    |
                              _cons |   2.741425   .0372098    73.67   0.000
------------------------------------------------------------------------------
```

Using the centered value `yrborn40` solved the problem of collinearity. The results show, as expected, that the cubic term is significant. This is consistent with the two bends that we saw in the lowess-smoothed relationship between `yrborn` and `children` in figure 3.10.

> **Note! Including linear, quadratic, and cubic terms**
>
> When fitting this kind of a cubic model, you might find that the linear or quadratic coefficients are not significant. For the sake of parsimony, you might be tempted to omit those variables because they are not significant. However, it is essential that these terms be included in the model (even if not significant) to preserve the interpretation of the cubic term.

Let's show a graph of the relationship between year of birth (`yrborn40`) and the predicted number of children (`children`). The first step is to use the `margins` command to compute the predicted means for the spectrum of `yrborn40` values with the `at()` option. Specifying `at(yrborn40=(-20(1)20))` yields predicted means for years of birth ranging from 1920 to 1960 in one-year increments.

```
. margins, at(yrborn40=(-20(1)20))
  (output omitted)
```

The `marginsplot` command is then used to create the graph shown in figure 3.11. The `recast(line)` and `recastci(rarea)` options display the predicted means as a solid line and the confidence interval as a shaded area. The graph shows that the predicted mean of `children` reaches a high at approximately −10 (those born about 1930) and then reaches a low at approximately 13 (those born about 1953).

```
. marginsplot, recast(line) recastci(rarea)
  Variables that uniquely identify margins: yrborn40
```

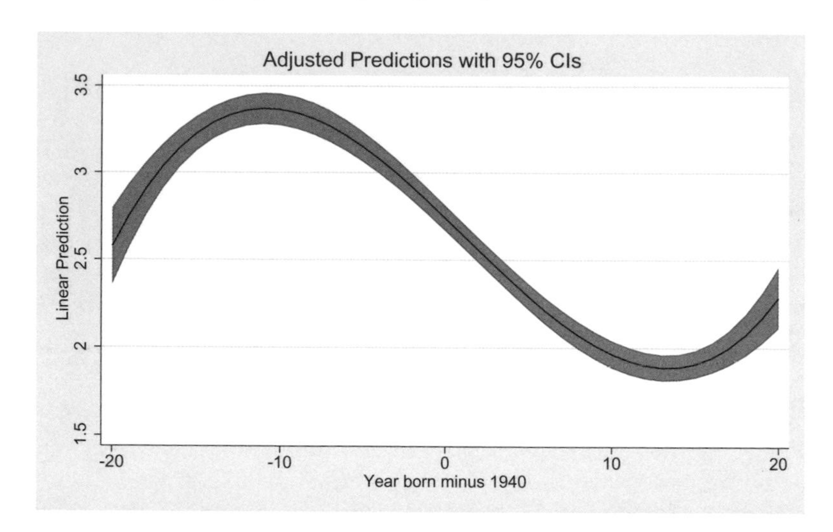

Figure 3.11. Predicted means from cubic regression with shaded confidence region

Based on this model, we might want to obtain predicted means for specified years of birth. However, we need to specify the year of birth with the variable `yrborn40`. For example, the `margins` command below obtains the predicted number of children for those born in 1950. This was accomplished by specifying the value of `yrborn40` as 10 (1950 − 1940).

```
. margins, at(yrborn40=10) vsquish
Adjusted predictions                            Number of obs     =      5,049
Model VCE    : OLS
Expression   : Linear prediction, predict()
at           : yrborn40        =          10

------------------------------------------------------------------------------
             |            Delta-method
             |     Margin   Std. Err.      t    P>|t|     [95% Conf. Interval]
-------------+----------------------------------------------------------------
       _cons |   1.964318   .0394603    49.78   0.000     1.886959    2.041678
------------------------------------------------------------------------------
```

The `margins` command below obtains the predicted number of children for the years 1920 to 1960 incrementing by 10.

```
. margins, at(yrborn40=(-20(10)20)) vsquish
Adjusted predictions                            Number of obs     =      5,049
Model VCE    : OLS

Expression   : Linear prediction, predict()
1._at        : yrborn40        =         -20
2._at        : yrborn40        =         -10
3._at        : yrborn40        =           0
4._at        : yrborn40        =          10
5._at        : yrborn40        =          20

------------------------------------------------------------------------------
             |            Delta-method
             |     Margin   Std. Err.      t    P>|t|     [95% Conf. Interval]
-------------+----------------------------------------------------------------
         _at |
          1  |   2.575693   .1118582    23.03   0.000     2.356402    2.794984
          2  |   3.363512   .0468221    71.84   0.000     3.271721    3.455304
          3  |   2.741425   .0372098    73.67   0.000     2.668477    2.814372
          4  |   1.964318   .0394603    49.78   0.000     1.886959    2.041678
          5  |   2.287081   .0886407    25.80   0.000     2.113307    2.460856
------------------------------------------------------------------------------
```

The **margins** command can also be used to estimate the **yrborn40** slope. For example, we might want to know the **yrborn40** slope when the year of birth is 1950 (that is, **yrborn40=10**). The **margins** command can compute this using **dydx()** option, as shown below.

```
. margins, at(yrborn40=10) dydx(yrborn40) vsquish
Conditional marginal effects                    Number of obs     =      5,049
Model VCE    : OLS

Expression   : Linear prediction, predict()
dy/dx w.r.t. : yrborn40
at           : yrborn40        =          10

------------------------------------------------------------------------------
             |            Delta-method
             |      dy/dx   Std. Err.      t    P>|t|     [95% Conf. Interval]
-------------+----------------------------------------------------------------
    yrborn40 |   -.043632    .004444    -9.82   0.000    -.0523442   -.0349198
------------------------------------------------------------------------------
```

You can further explore how the **yrborn40** slope varies as a function of year of birth by specifying multiple values within the **at()** option, as shown below.

```
. margins, at(yrborn40=(-20(5)20)) dydx(yrborn40) vsquish
Conditional marginal effects                    Number of obs     =      5,049
Model VCE    : OLS

Expression   : Linear prediction, predict()
dy/dx w.r.t. : yrborn40
1._at        : yrborn40        =         -20
2._at        : yrborn40        =         -15
3._at        : yrborn40        =         -10
4._at        : yrborn40        =          -5
5._at        : yrborn40        =           0
6._at        : yrborn40        =           5
7._at        : yrborn40        =          10
8._at        : yrborn40        =          15
9._at        : yrborn40        =          20

------------------------------------------------------------------------------
             |            Delta-method
             |      dy/dx   Std. Err.      t    P>|t|     [95% Conf. Interval]
-------------+----------------------------------------------------------------
yrborn40     |
         _at |
          1  |   .1911069   .0229201     8.34   0.000     .1461735    .2360404
          2  |   .0735533   .0122085     6.02   0.000     .0496193    .0974872
          3  |  -.0126282   .0053683    -2.35   0.019    -.0231524   -.0021041
          4  |  -.0674375   .0042965   -15.70   0.000    -.0758604   -.0590145
          5  |  -.0908745   .0052395   -17.34   0.000    -.1011462   -.0806028
          6  |  -.0829394   .0045764   -18.12   0.000    -.0919112   -.0739675
          7  |   -.043632    .004444    -9.82   0.000    -.0523442   -.0349198
          8  |   .0270476   .0100175     2.70   0.007      .007409    .0466863
          9  |   .1290994    .019872     6.50   0.000     .0901417    .1680572
------------------------------------------------------------------------------
```

The output from this `margins` command shows that the slope of `yrborn40` is positive when `yrborn40` is −20 and −15, is negative when `yrborn40` is between −10 and 10, and is positive when `yrborn40` is 15 and 20.

3.4 Fractional polynomial regression

3.4.1 Overview

The previous sections showed polynomial regressions, illustrating regression models with quadratic terms and cubic terms. Each of these terms is a positive integer, but we do not need to restrict ourselves to only positive integers. Fractional polynomial regression is more flexible by considering other kinds of power terms as well, including negative powers and fractional powers.

Stata has a convenience command for fitting fractional polynomial models called `fp`. The `fp` prefix automates the process of selecting the best fitting fractional polynomial model. It fits a variety of polynomial terms (and combinations of polynomial terms) and shows you the best fitting model. The default set of power terms the `fp` prefix will try includes −2, −1, −0.5, 0, 0.5, 1, 2, and 3 (where 0 indicates that the natural log of the predictor is used). This yields a large canvas of possible shapes.

Let's look at this canvas of shapes, beginning with the negative powers (−2, −1, and −0.5). These models take the form of $y = b * x^{power}$, where b could be positive or negative, and power could be −2, −1, or −0.5. Figure 3.12 shows some possible shapes of these models, showing the powers −2, −1, and −0.5 in columns 1, 2, and 3. The first row shows a positive value of b (specifically, 0.3), and the second row shows a negative value of b (specifically, −0.3).

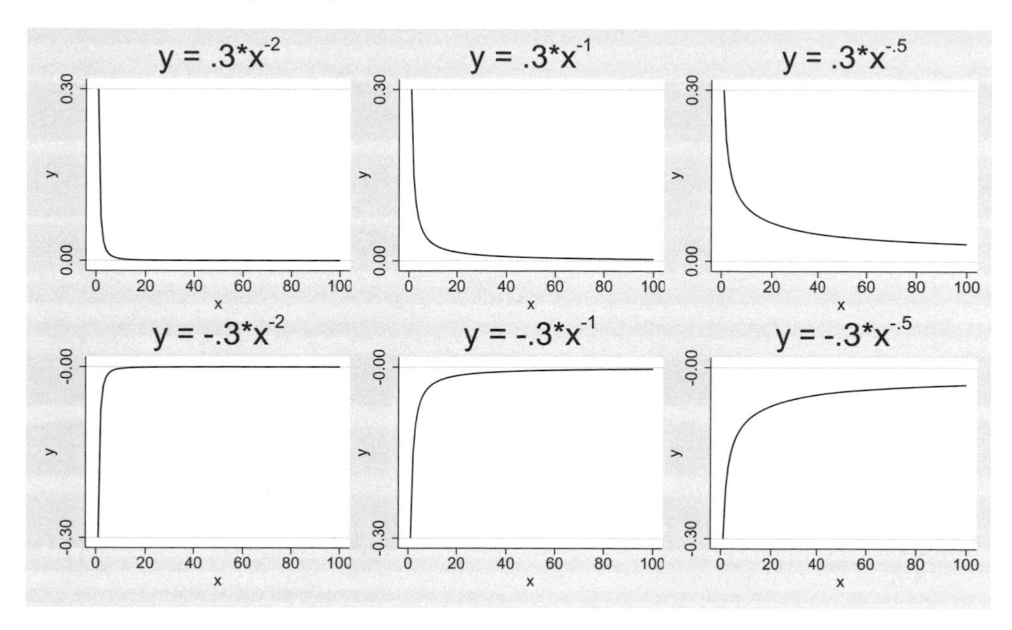

Figure 3.12. Fractional polynomials, powers = −2, −1, and −0.5 (columns) for $b = 0.3$ (top row) and $b = -0.3$ (bottom row)

The graph in the first row of column 1 shows the equation $y = 0.3 \times x^{-2}$. This shows that y drops rapidly with increasing values of x until y hits a floor. The equation in the second row of column 1 is the same, except that the coefficient for b is negative (that is, $y = -0.3 \times x^{-2}$). This figure shows that y rises rapidly with increasing values of x until y hits a ceiling.

The graphs in the first row (with the positive coefficients) are all typified by a steep descent and then reaching a floor. The more strongly negative the power term, the stronger the descent. The second row is a vertical mirror image of the first. When the coefficient is negative, there is a sharp ascent and then a ceiling is reached. The more negative power terms are associated with a sharper ascent.

The shapes of the relationship between x and y for the powers 1, 2, and 3 are shown in figure 3.13. The three columns correspond to the powers 1, 2, and 3; the first row represents a positive regression coefficient (of +0.3); and the second row illustrates a negative regression coefficient (of −0.3). The first column shows a linear relationship between x and y. The second and third columns show the second and third power, with the top row showing a U-shaped bend and the bottom row showing an inverted U-shaped bend. The higher power terms are associated with a more rapid change in the outcome for a unit change in the predictor.

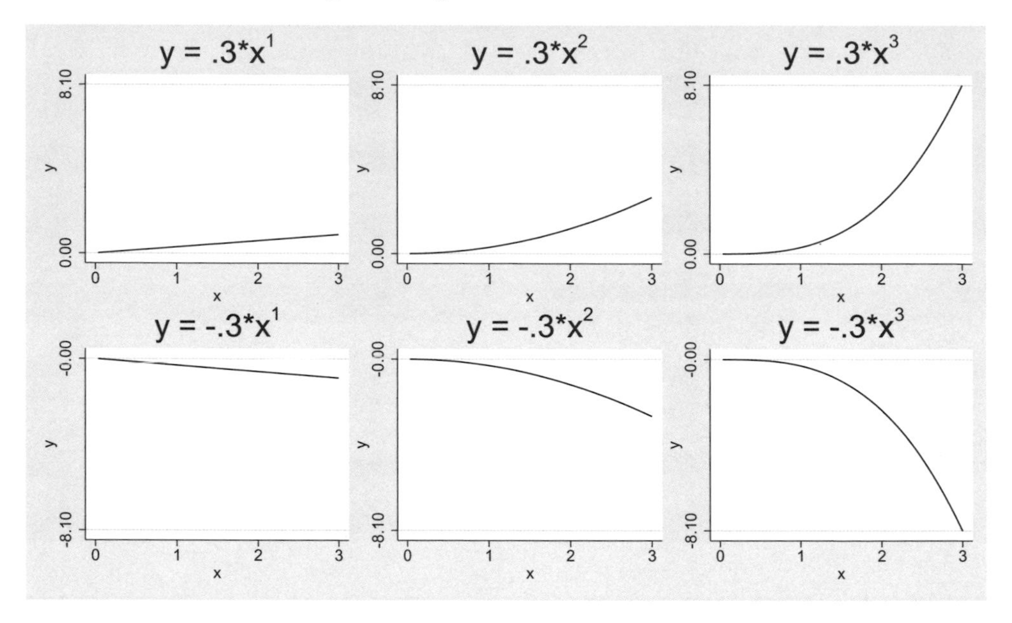

Figure 3.13. Fractional polynomials, powers = 1, 2, and 3 (columns) for $b = 0.3$ (top row) and $b = -0.3$ (bottom row)

Note! Cubic shapes

It might appear that the graphs for the cubic terms shown in the third column of figure 3.13 (with only one bend) are at odds with the graphs of cubic models from section 3.3, which showed two bends. The key is that the third column of figure 3.13 includes only a cubic term, whereas the examples from section 3.3 included linear, quadratic, and cubic terms all in the same model.

Figure 3.14 shows the remaining two of the default powers used by the `fp` prefix. The power of 0 (which it uses to indicate the natural log of the predictor) is shown in the left column and the power of 0.5 is shown in the right column. The top row shows a positive coefficient of 0.3 and the bottom row shows a negative coefficient of −0.3. The graph in the first row of column 1 shows an initial steep ascent in y followed by gradual growth in y thereafter. The second row shows the mirror image, an initial steep descent followed by a gradual decrease thereafter. The top row of column two shows a gradually curved ascent in y as a function of x, whereas the bottom row shows a gradual descent in y as a function of x.

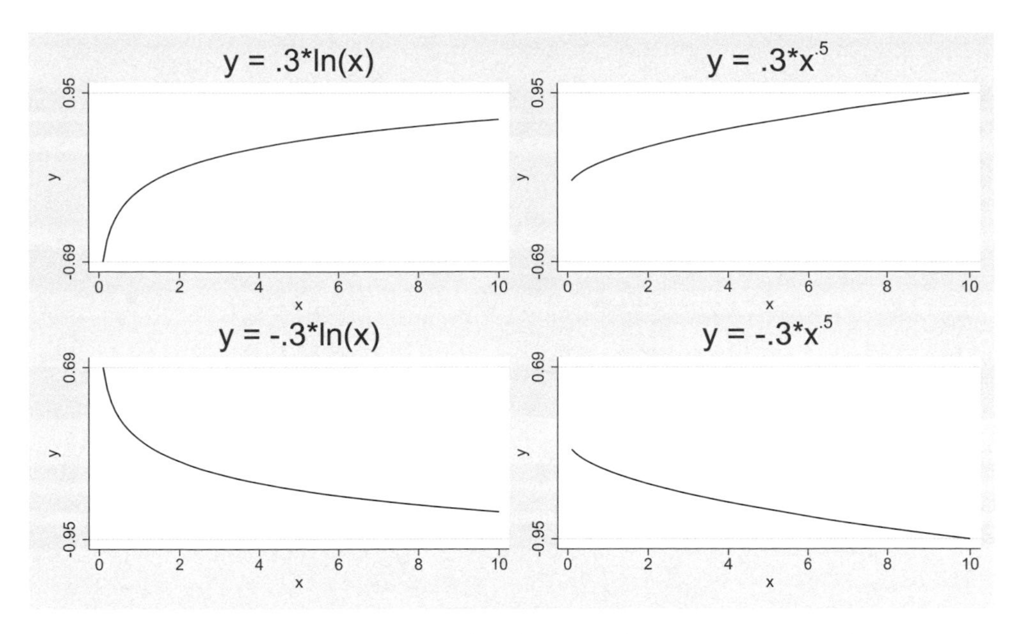

Figure 3.14. Fractional polynomials, $\ln(x)$ (column 1) and x to the 0.5 (column 2) for $B = 0.3$ (top row) and $B = -0.3$ (bottom row)

The graphs in figures 3.12, 3.13, and 3.14 cover eight different powers of the predictor (-2, -1, -0.5, 0, 0.5, 1, 2, and 3) crossed with one example of a positive coefficient (0.3) and one example of a negative coefficient (-0.3). This creates a wide variety of shapes for modeling the relationship between x and y. Furthermore, the `fp` prefix (by default) will not only fit each of these eight powers alone, but also includes all two-way combinations of these powers.

Consider the curves shown in figure 3.15. The left panel shows the curve given by $y = -10 \times x^{-1}$. In that curve, you can see y rises rapidly as x increases and then quickly reaches a ceiling. The middle curve is given by the formula $y = -1 \times x^2$, and in this curve the values of y show an inverted U-shape. The right panel shows the formula combining these two curves, $y = -10 \times x^{-1} + -1 \times x^2$. Note how this curve combines the rapid rise of the left panel with the gradual inverted U-shape of the middle panel. As you can imagine, being able to combine two different fractional polynomials can yield a flexible set of possible curve shapes for modeling your data.

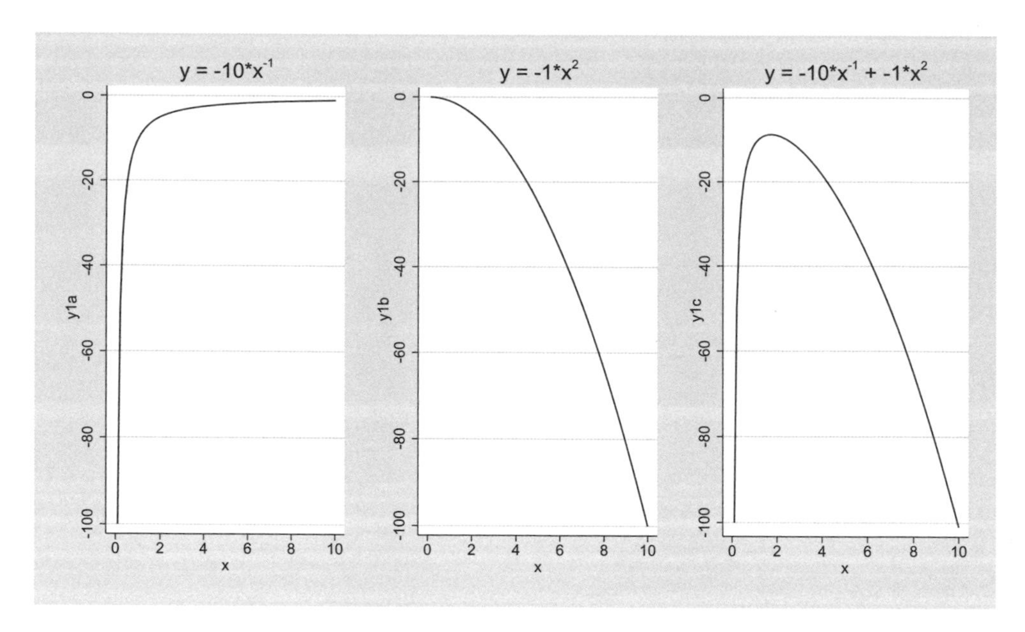

Figure 3.15. Combined fractional polynomials

3.4.2 Example using fractional polynomial regression

Let's now consider an example of fractional polynomial regression using the GSS dataset. For this example, let's look at the relationship between the respondent's age and education level. This example is a bit odd because the age of the person reflects both their chronological age and the era of their birth. That is, older people tended to be born in an earlier era when it was more common for people to have less education. This example is not ideal in this respect. However, it shows an excellent contrast between the fit of a quadratic model with a fractional polynomial model, so I will ask for your indulgence as I present this example.

This example uses the `gss_ivrm.dta` dataset, keeping only those who are 18 to 80 years old.

```
. use gss_ivrm
. keep if (age<=80)
(1,892 observations deleted)
```

To show the relationship between age and education, let's graph the average of education by each level of age. First, we will run a regression predicting `educ` from `age`, treating `age` as a categorical variable (the output of this is omitted to save space). Then, we obtain the predicted values using the `predict` command, calling the variable `yhatmean`. This variable contains the mean of education at each level of age.[2] The `graph` command then shows the relationship between the predicted mean of education and age. The graph is shown in figure 3.16.

```
. regress educ i.age
  (output omitted)
. predict yhatmean
(option xb assumed; fitted values)
. graph twoway line yhatmean age, sort xlabel(20(5)80)
```

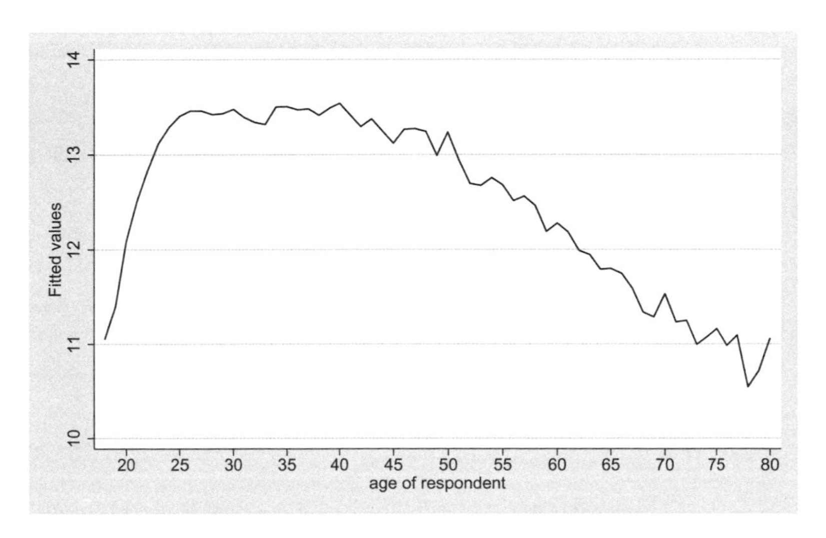

Figure 3.16. Average education at each level of age

Note how there is a curvilinear aspect to the relationship between age and education, suggesting that we might try including a quadratic term for `educ`. We fit such a model below (for more on quadratic models; see section 3.2).

2. I could have used the `margins` and `marginsplot` commands, but this strategy will allow me to overlay graphs comparing different fit methods.

```
. regress educ c.age##c.age

      Source |       SS           df       MS      Number of obs   =    53,070
-------------+----------------------------------   F(2, 53067)     =   1616.60
       Model |  29981.1031         2  14990.5515   Prob > F        =    0.0000
    Residual |  492083.501    53,067  9.27287205   R-squared       =    0.0574
-------------+----------------------------------   Adj R-squared   =    0.0574
       Total |  522064.604    53,069  9.83746828   Root MSE        =    3.0451

------------------------------------------------------------------------------
        educ |      Coef.   Std. Err.      t    P>|t|     [95% Conf. Interval]
-------------+----------------------------------------------------------------
         age |    .138446   .0049059    28.22   0.000     .1288304    .1480615
             |
 c.age#c.age |  -.0018433   .0000512   -36.02   0.000    -.0019436    -.001743
             |
       _cons |   10.76335   .1074823   100.14   0.000     10.55269    10.97402
------------------------------------------------------------------------------
```

This seems promising, with the quadratic term for `age` being significant. Let's use the `predict` command to compute the fitted values from the quadratic model, naming the variable `yhatq`. Then, let's graph the fitted values from the quadratic model and the mean of education by age, as shown below. The resulting graph is shown in figure 3.17.

```
. predict yhatq
(option xb assumed; fitted values)
. graph twoway line yhatmean yhatq age, sort xlabel(20(5)85)
> legend(label(1 "Mean at each age") label(2 "Quadratic fit"))
```

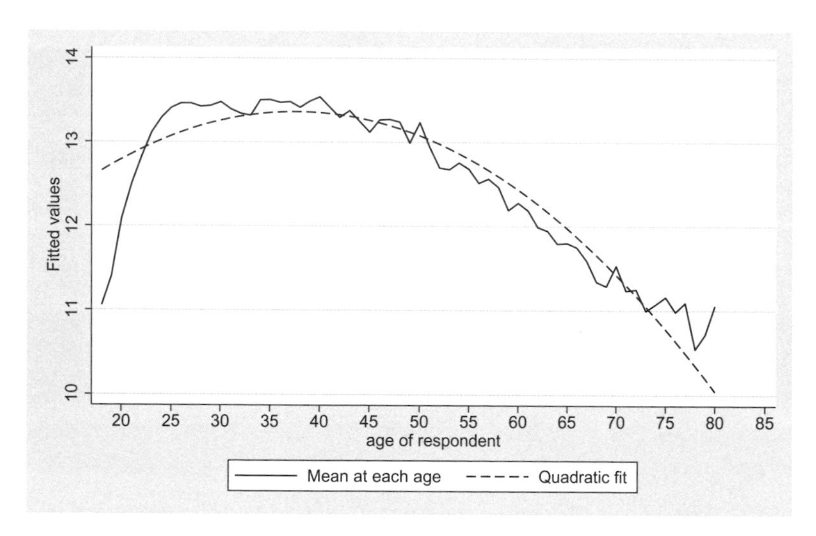

Figure 3.17. Fitted values of quadratic model compared with observed means

Note how the quadratic fit line yields predicted values that are too high in the younger years and too low in the older years. Another way of putting this is that the quadratic line does not account for the rapid rise in education in the late teens and early 20s, nor does it account for the slow decline of education in later years.

A fractional polynomial model, with its increased flexibility, could provide a more appropriate fit. We fit such a model by adding the `fp` prefix to the `regress` command, as shown below.

```
. fp <age>: regress educ <age>
(fitting 44 models)
(....10%....20%....30%....40%....50%....60%....70%....80%....90%....100%)
Fractional polynomial comparisons:

         age     df     Deviance   Res. s.d.    Dev. dif.    P(*)    Powers

     omitted      0   271933.68       3.136     3507.042    0.000
      linear      1   270076.53       3.082     1649.885    0.000    1
       m = 1      2   269297.58       3.060      870.937    0.000    3
       m = 2      4   268426.64       3.035        0.000       --    -2 1

(*) P = sig. level of model with m = 2 based on F with 53065
    denominator dof.

      Source          SS           df       MS      Number of obs   =    53,070
                                                    F(2, 53067)     =   1812.66
       Model   33384.5311          2  16692.2655    Prob > F        =    0.0000
    Residual   488680.073     53,067   9.2087375    R-squared       =    0.0639
                                                    Adj R-squared   =    0.0639
       Total   522064.604     53,069  9.83746828    Root MSE        =    3.0346

        educ        Coef.   Std. Err.       t    P>|t|     [95% Conf. Interval]

       age_1    -1655.009   40.42981   -40.94    0.000    -1734.252   -1575.766
       age_2    -.0902469   .0015578   -57.93    0.000    -.0933002   -.0871937
       _cons     18.08954   .0984116   183.82    0.000     17.89665    18.28243
```

The `fp` prefix tried 44 different models involving the polynomials -2, -1, -0.5, 0, 0.5, 1, 2, and 3 selected alone and in pairs (where 0 indicates that the natural log of the predictor). The `fp` results show that it selected the model using the powers -2 and 1 as the best fitting combination of polynomials.

The variable `age_1` reflects the term associated with $\texttt{age}^{-2}$, and the coefficient for this variable is negative. Referring to the bottom left panel of figure 3.12, we see the general shape of this kind of relationship. As x increases, there is a sharp rise in y followed by a plateau.

The variable `age_2` reflects the term associated with `age`, and the coefficient for this variable is negative. This is a linear effect with a negative coefficient, as illustrated in the bottom left panel of figure 3.13. This shows a linear decrease in y with increasing values of x.

Tip! Automatic modeling

Any time we let the computer perform automatic modeling for us, we want to be at least a little bit skeptical of the results. We want to ask if the results make sense, in both a practical sense and a theoretical sense. Also, we want to be sure that the model does not overfit our sample data. For example, we can explore the issue of overfitting with the use of cross validation.

The `fp` prefix selected two curves that, when blended, address the unique features of the relationship between age and education, where there is a rapid rise in education for the early values of age combined with a roughly linear decline in education for the later ages. The variable `age_1` helps to account for the sharp rise in education in the early years, and the variable `age_2` helps to account for the linear decline in education in the later years. Let's generate the predicted values based on this model and see if they conform to these expectations. The `predict` command is used to create the predicted values from the fractional polynomial model, calling the variable `yhatfp`.

```
. predict yhatfp
(option xb assumed; fitted values)
(125 missing values generated)
```

Let's now graph the predicted values from the fractional polynomial model (`yhatfp`) against age. Let's also include the predicted values from the quadratic model (`yhatq`) and the average education at each age (`yhatmean`).

```
. graph twoway line yhatmean yhatq yhatfp age, sort xlabel(20(5)85)
> legend(label(1 "Mean at each age") label(2 "Quadratic fit"))
> legend(label(3 "Fractional polynomial fit"))
```

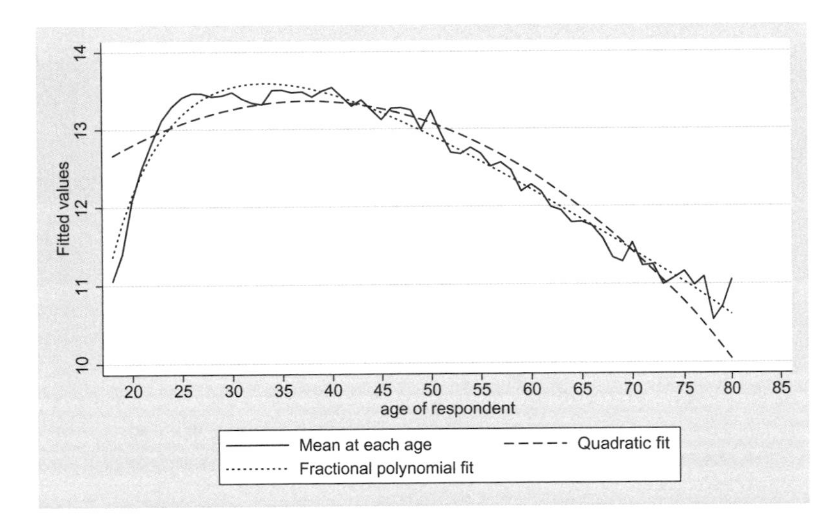

Figure 3.18. Fitted values of quadratic and fractional polynomial models compared with observed means

As figure 3.18 shows, the fitted values from the fractional polynomial model are more closely aligned with the mean at each age than the quadratic model. The fractional polynomial model fits the rapid rise in education in the late teens and early 20s, and it fits the steady decline in education for those who are older.

The graphs that we have created so far have used variables created by the `predict` command and could correctly do so because we had only one predictor in the model, `age`. However, if we had other predictors in the model, the `predict` command will not adjust for the other predictors in the model. That is a good job for the `margins` command.

In using the `margins` command, we normally use the `at()` option to compute adjusted means for specific values of the predictor, such as `at(age=20)`. After the fractional polynomial model, we need to express age in terms of the variables `age_1` and `age_2`. As we noted earlier, the output shows the best fitting model used the powers `-2 1`, meaning that `age_1` is `age` raised to the -2 power (that is, $\texttt{age}^{-2}$), and `age_2` is `age` raised to the 1 power (that is, just `age`). Let's select an age of 20. The value for `age_2` is 20, and the value for `age_1` is shown using the `display` command below.

```
. display "Age=20 converted into age_1 is " (20)^-2
Age=20 converted into age_1 is .0025
```

We now use these values in the margins command below. This shows that the adjusted mean for someone who is 20 years old is 12.15.[3]

```
. margins, at(age_1= .0025 age_2=20)
Adjusted predictions                            Number of obs     =     53,070
Model VCE    : OLS

Expression   : Linear prediction, predict()
at           : age_1           =       .0025
               age_2           =          20

------------------------------------------------------------------------------
             |            Delta-method
             |     Margin   Std. Err.      t    P>|t|     [95% Conf. Interval]
-------------+----------------------------------------------------------------
       _cons |   12.14708   .0439773   276.21   0.000     12.06088    12.23327
------------------------------------------------------------------------------
```

Suppose that we wanted to compute the adjusted mean for multiple ages, say, for the ages 20 to 80 in 10-year increments. We can use the forvalues command to loop across such a range of ages, as shown below. Within the loop, the local command is used to compute the local macros age1 and age2, using the powers associated with age_1 and age_2. The margins command is then used to compute the adjusted mean based on the values of age1 and age2. The matrix and local commands extract the adjusted mean, resulting in the local macro named adjmean. Finally, the display command displays the adjusted mean.

```
. forvalues age = 20(10)80 {
  2.    local age1 = `age'^-2
  3.    local age2 = `age'
  4.    margins, at(age_1=`age1' age_2=`age2')
  (output omitted)
  5.    matrix am = r(b)
  6.    local adjmean = am[1,1]
  7.    display "Adjusted mean for age = `age' is " `adjmean'
  8. }
Adjusted mean for age = 20 is 12.147077
Adjusted mean for age = 30 is 13.543231
Adjusted mean for age = 40 is 13.44528
Adjusted mean for age = 50 is 12.915188
Adjusted mean for age = 60 is 12.214998
Adjusted mean for age = 70 is 11.434496
Adjusted mean for age = 80 is 10.611189
```

We can see that adjusted mean at age 20 computed via the forvalues loop matches what we obtained when typing the margins command. This forvalues loop helps to automate the process of computing the adjusted means as a function of age. We can input these adjusted means into a dataset and graph them as a function of age, as shown below. The resulting graph is shown in figure 3.19. If we included more levels of age, we would obtain a smoother graph.

3. I am referring to this as an adjusted mean even though no additional predictors are present. This emphasizes that this technique would properly adjust for any additional predictors if they were present.

```
. preserve
. clear
. input age yhat
           age       yhat
  1. 20 12.147077
  2. 30 13.543231
  3. 40 13.44528
  4. 50 12.915188
  5. 60 12.214998
  6. 70 11.434496
  7. 80 10.611189
  8. end
. graph twoway line yhat age, sort
. restore
```

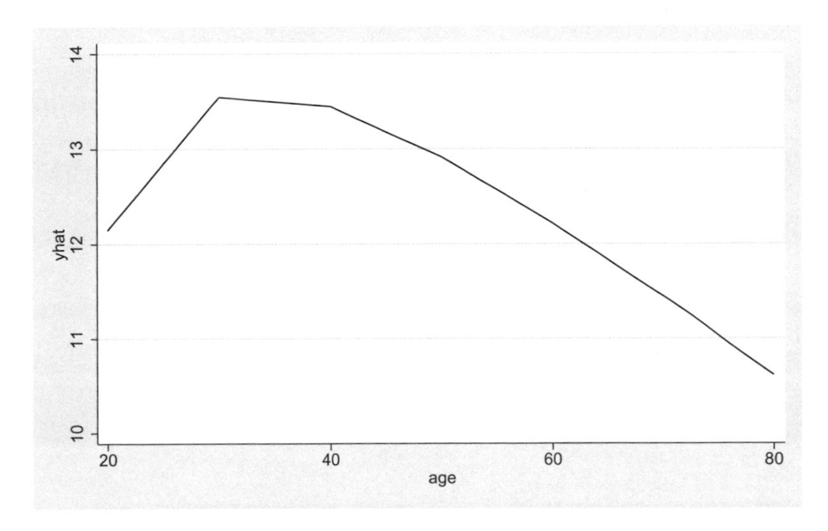

Figure 3.19. Adjusted means values from fractional polynomial model, ages 20 to 80 incrementing by 10

To avoid the process of typing the values into a dataset, we can automate the process of saving the adjusted means to a dataset using the `postfile` and `post` commands, as illustrated below. (See [P] **postfile** for more information on these commands.)

```
. * New: Open up a postfile named -adjmeans- that will
. *      contain the variables age and yhat
. postutil clear // Close any open postfiles
. postfile adjmeans age yhat using adjmeans, replace
(note: file adjmeans.dta not found)
. forvalues age = 18(1)80 {
  2.    local age1 = `age'^-2
  3.    local age2 = `age'
  4.    margins, at(age_1=`age1' age_2=`age2')
  (output omitted)
  5.    matrix mm = r(b)
  6.    local yhat = mm[1,1]
  7.    * New: Save local macros `age' and `yhat' to the postfile named adjmeans
.    post adjmeans (`age') (`yhat')
  8. }
. * New: Close the postfile named -adjmeans-
. postclose adjmeans
```

This creates a dataset named `adjmeans.dta` that contains the adjusted means. Let's use this dataset and list the first 10 observations.

```
. use adjmeans
. list in 1/10
```

	age	yhat
1.	18	11.35704
2.	19	11.79033
3.	20	12.14708
4.	21	12.4415
5.	22	12.68467
6.	23	12.8853
7.	24	13.05033
8.	25	13.18535
9.	26	13.29488
10.	27	13.38263

We can then graph the adjusted means using the **graph** command below, which creates the graph displayed in figure 3.20. When automating the process of storing the adjusted means, we computed adjusted means in one-year increments of age, yielding a much smoother graph than the manually created graph from figure 3.19.

```
. graph twoway line yhat age
```

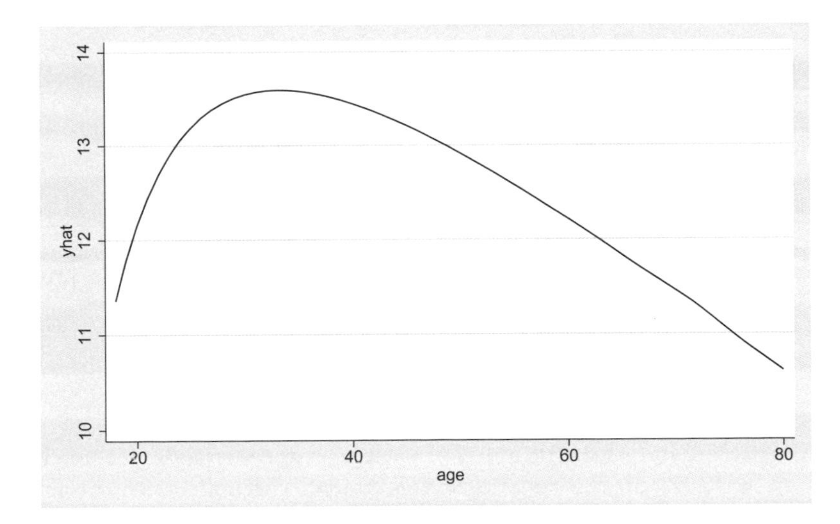

Figure 3.20. Adjusted means from fractional polynomial model, ages 18 to 80 incrementing by 1

> **Tip! graph command shortcuts**
>
> The previous example uses the `graph twoway line yhat age` command, which emphasizes that the underlying command is `graph twoway`. For your convenience, Stata will permit you to instead specify `twoway line yhat age` or even `line yhat age`.

3.5 Main effects with polynomial terms

The meaning of main effects changes when polynomial terms are included in the model. In fact, the inclusion of polynomial terms can substantially change the coefficient for the main effect when compared with the main effect–only model. Many find this distressing, wondering why the main effect changed after including a nonlinear term. Let's briefly explore why this can happen and what it means using the example we saw from section 3.2.2. That example predicted `realrinc` from `age` and `c.age#c.age`, including `female` as a control variable, specified as a dummy variable.

Let's begin by fitting the linear model shown below, predicting `realrinc` from `age` and `female`. As we did in section 3.2.2, we keep people who are at most 80 years old.

```
. use gss_ivrm
. keep if (age<=80)
(1,892 observations deleted)
. regress realrinc age female, vce(robust)
Linear regression                               Number of obs     =     32,100
                                                F(2, 32097)       =    1170.51
                                                Prob > F          =     0.0000
                                                R-squared         =     0.0787
                                                Root MSE          =      25833

------------------------------------------------------------------------------
             |               Robust
    realrinc |      Coef.   Std. Err.      t    P>|t|     [95% Conf. Interval]
-------------+----------------------------------------------------------------
         age |   320.0676   10.69559    29.93   0.000     299.1039    341.0314
      female |  -12373.25   285.0254   -43.41   0.000    -12931.91   -11814.59
       _cons |   15106.17   403.0443    37.48   0.000     14316.19    15896.16
------------------------------------------------------------------------------
```

The main effect of age is significant. For each year age increases, income is predicted to increase by $320.07. In the context of this model, the main effect of age describes the general trend in the relationship between income and age.

However, there is a problem. As we saw in section 3.2.2, the relationship between income and age is not linear. As we did in that section, let's add a quadratic term for age, as shown below.

```
. regress realrinc c.age##c.age female, vce(robust)
Linear regression                               Number of obs     =     32,100
                                                F(3, 32096)       =    1252.67
                                                Prob > F          =     0.0000
                                                R-squared         =     0.1089
                                                Root MSE          =      25407

------------------------------------------------------------------------------
             |               Robust
    realrinc |      Coef.   Std. Err.      t    P>|t|     [95% Conf. Interval]
-------------+----------------------------------------------------------------
         age |   2412.339   58.05774    41.55   0.000     2298.544    2526.135
             |
 c.age#c.age |  -24.20196   .6958905   -34.78   0.000    -25.56593   -22.83799
             |
      female |  -12419.24   280.4746   -44.28   0.000    -12968.98   -11869.49
       _cons |  -25679.77   1038.412   -24.73   0.000     -27715.1   -23644.44
------------------------------------------------------------------------------
```

The quadratic term is significant, but we might suddenly become concerned that the main effect of age has skyrocketed from 320.07 in the linear model to 2,412.34 in the quadratic model. Why did the main effect change so much? Did we do something wrong? The key is that the term "main effect" is really a misnomer, because we expect this term to describe the general trend of the relationship between income and age.

In the presence of the quadratic term, the coefficient for age represents the age slope when age is held constant at zero. At that point, the slope is 2,412.34. This value is meaningless for two reasons. First, nobody has income when they are zero years old.

Second, this term no longer reflects the general trend. As we saw in section 3.2.2, the age slope changes for every level of age, so there is no such thing as a measure of general trend in this kind of model.

This underscores the importance of remembering that the main effect in the presence of a quadratic term (or any other nonlinear term) really is no longer a main effect. It simply represents the linear slope relating the predictor and outcome when the predictor is held constant at zero.

3.6 Summary

This chapter has illustrated the use of polynomials for fitting nonlinear relationships between a predictor and outcome. Quadratic terms were illustrated for fitting a relationship between a predictor and outcome that has one bend. For relationships that have two bends, cubic models were introduced. Fractional polynomial models can account for polynomial trends that go beyond simple quadratic and cubic terms. The next chapter considers a different method for fitting nonlinear relationships between the predictor and outcome through the use of piecewise regression models.

For more information about fitting curvilinear relationships using polynomial terms, I recommend Cohen et al. (2003). Royston and Sauerbrei (2008) provide excellent and thorough coverage of fractional polynomial models.

4 Continuous predictors: Piecewise models

4.1 Chapter overview

There is a wide variety of modeling techniques that can be used to account for nonlinearity in the relationship between a predictor and outcome. This chapter illustrates the use of piecewise regression. This involves fitting separate line segments, demarcated by knots, that account for the nonlinearity between the predictor and outcome. This chapter illustrates piecewise regression models with one knot (see section 4.3), with two knots (see section 4.4), one knot and one jump (see section 4.5), and two knots and two jumps (see section 4.6). This chapter also covers piecewise regression models with one unknown knot (see section 4.7) and piecewise regression models with multiple unknown knots (see section 4.8). Piecewise regression models pose special difficulties in graphing the adjusted means as a function of the predictor, as described in section 4.9. This is followed by section 4.10, which illustrates ways to automate graphing the adjusted means from piecewise models. This chapter begins with an overview of piecewise models.

4.2 Introduction to piecewise regression models

A piecewise regression goes by several names, including spline regression, broken line regression, broken stick regression, and even hockey stick models. Consider the example, predicting annual income from education, shown in the left panel of figure 4.1.

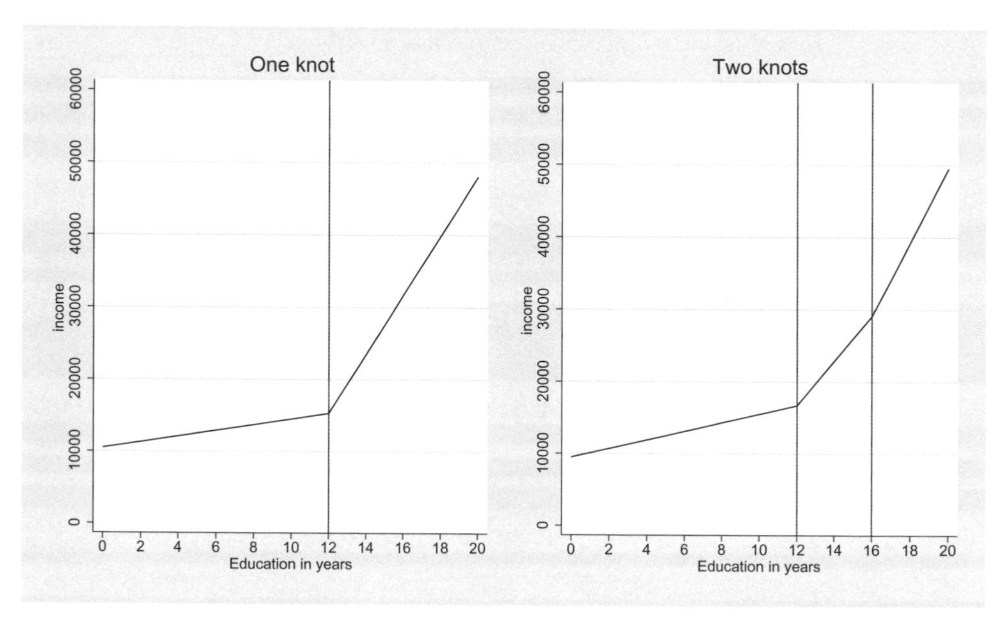

Figure 4.1. Piecewise regression with one knot (left) and two knots (right)

In this hypothetical example, the slope of the relationship between education and income (that is, the `educ` slope) for those who have fewer than 12 years of education is 400. The slope is different for those who have 12 or more years of education. For those people, the slope is 4,100. For those with fewer than 12 years of education, income is predicted to increase by $400 for each additional year of education. Starting at the 12th year of education, each additional year of education is associated with a $4,100 increase in income. When education reaches 12 years, the slope increases by 3,700 (to 4,100). The place where the slope changes is called the knot. In the left panel of figure 4.1, there is one knot at 12 years of education.

In the right panel of figure 4.1, there are two knots: one at 12 years of education, and another at 16 years of education. For those with fewer than 12 years of education, the slope of the relationship between income and education is 600. For those with between 12 and 16 years of education, the slope increases by 2,500 (to 3,100), and for those with 16 or more years of education the slope increases by an additional 2,000 (to 5,100).

Models with one knot are illustrated in section 4.3, and models with two knots are illustrated in section 4.4.

A knot can signify a change of slope and a change of intercept, yielding an increase (or decrease) in the outcome upon attaining a particular milestone. For example, in the left panel of figure 4.2, achieving 12 years of education results in not only a change of slope, but also a jump in income.[1] The slope for those with fewer than 12 years of education is 200, and that slope increases by 3,500 to 3,700 for those with 12 years of education. The predicted income also jumps at 12 years of education. At 12 years of education, predicted income increases by 2,200.

The right panel of figure 4.2 shows a model with two knots, both associated with a change of slope and change of intercept. The slope for those with fewer than 12 years of education is 100. At 12 years of education, the predicted income increases (jumps) by 3,500. Plus, the slope of the relationship between income and education also increases by 1,700 (to 1,800). At 16 years of education, the predicted income jumps by 5,500, and the slope increases by 2,700 (to 4,500).

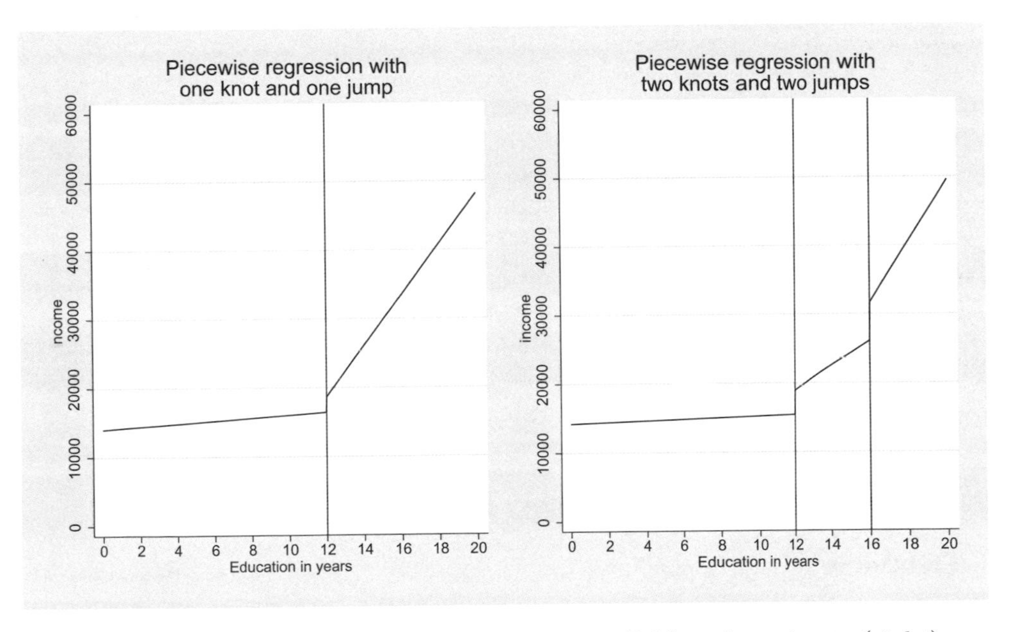

Figure 4.2. Piecewise regression with one knot (left) and two knots (right)

1. Although the term *jump* is somewhat informal, I feel it is simpler than the more technically correct term, change in intercept. I will use the term *jump* in this chapter and subsequent chapters to refer to a sudden increase in intercept and use the term *drop* to refer to a sudden decrease in intercept.

Models with one knot with a change in slope and intercept are illustrated in section 4.5, and models with two knots with a change of slope and intercept are illustrated in section 4.6.

Note! Instantaneous jumps?

In the example shown in the left panel of figure 4.2, achieving 12 years of education yields a sudden increase in income of $2,200. Is this plausible? As a thought experiment, imagine someone being one day short of graduating high school and the wages they would obtain as they seek a job. Compare this person with an identical job seeker who has one more day of education (that is, they graduated high school). These two people are identical except that one crossed the threshold of getting a diploma. It is indeed plausible that the second person would be offered an annual income $2,200 more than the first person. Another application of such a model is a regression discontinuity design, where the application of a treatment is based on reaching a threshold of a covariate (predictor). The change in intercept (jump) at that threshold is the estimate of the effect of the treatment.

Suppose you have a predictor that shows a nonlinear relationship with the outcome, and you believe that a piecewise model with one knot signifying a change in slope would fit your data well. However, unlike the previous examples, you do not have a theoretical or practical basis for selecting the placement of the knot. You could haphazardly try a variety of placements for the knot, trying to find a placement that results in the best fitting model. Alternatively, you can let Stata do the work for you by using a least-squares procedure for selecting a location for the knot that produces the lowest residual sum of squares. This process is described in section 4.7.

It is possible that you might have multiple unknown knots for a continuous variable, but you do not know how many knots should be selected, or where the knots should be placed. As illustrated in section 4.8, you can fit such a model by intentionally selecting too many knots and then progressively eliminating superfluous knots until you have a parsimonious (and hopefully sensible) model.

4.3 Piecewise with one known knot

4.3.1 Overview

This section illustrates piecewise regression with one knot. Such a model was visualized in the left panel of figure 4.1. That figure is reproduced in greater detail and shown in figure 4.3. This model predicts the income from education and includes one knot at 12 years of education. The slope of the relationship between income and education for those who have fewer than 12 years of education is 400, but for the person who has 12 years of education, the slope increases to 4,100.

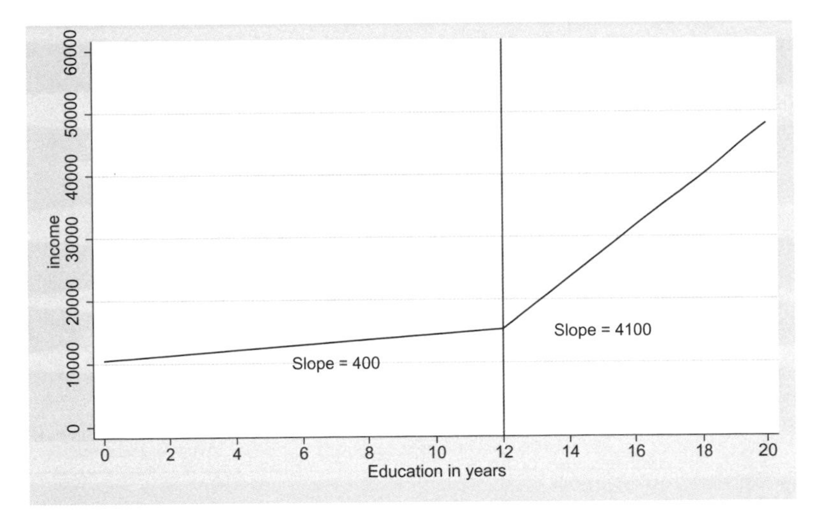

Figure 4.3. Hypothetical piecewise regression with one knot

4.3.2 Examples using the GSS

Let's use the GSS dataset to illustrate a piecewise model with one knot, focusing on the relationship between income and education.

```
. use gss_ivrm
```

Let's create a graph that shows the mean of income (`realrinc`) for each level of education (`educ`). (Section 2.4.4 provides more details about how to create such a graph.) First, we fit a model that predicts `realrinc` from `i.educ`. We then use the `margins` command followed by the `marginsplot` command to show a graph of the mean of income at each level of education. This graph is shown in figure 4.4. This allows us to see the shape of the relationship between education and income.

```
. regress realrinc i.educ, vce(robust)
  (output omitted)
. margins educ
  (output omitted)
. marginsplot, noci
  Variables that uniquely identify margins: educ
```

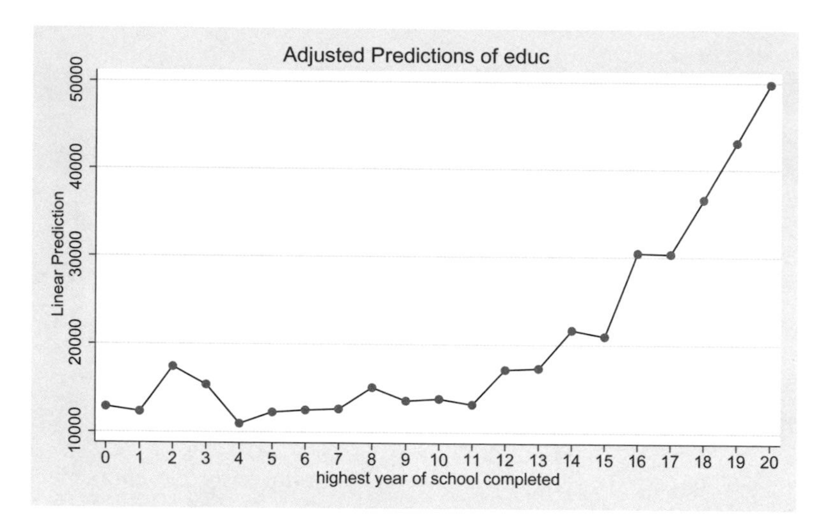

Figure 4.4. Mean of income at each level of education

It looks like the relationship between education and income could be fit well with a piecewise model with a knot at 12 (corresponding to graduating high school). The following subsections illustrate two different ways to fit this kind of piecewise model: using individual slope coding and using change in slope coding.

Individual slope coding

The individual slope coding scheme estimates the slope of each line segment of the piecewise model. In this example, the individual slope coding scheme would estimate the `educ` slope for those with fewer than 12 years of education (non–high school graduates) and estimate the slope for those with 12 or more years of education (high school graduates).

The first step is to create two new variables that are coded to represent the `educ` slope before and after graduating high school. The `mkspline` command creates the variables `ed1` and `ed2` based on the original variable `educ`. The value of 12 is inserted between `ed1` and `ed2`, indicating that this is the knot.

```
. mkspline ed1 12 ed2 = educ
```

The **showcoding** command, which you can download (see section 1.1), shows the correspondence of the values of educ with ed1 and ed2. The command shows that the variable ed1 corresponds to the variable educ for 12 or fewer years of education and contains the value of 12 for more than 12 years of education. The variable ed2 contains 0 for 12 or fewer years of education and contains educ minus 12 for more than 12 years of education.

```
. showcoding educ ed1 ed2
```

educ	ed1	ed2
0	0	0
1	1	0
2	2	0
3	3	0
4	4	0
5	5	0
6	6	0
7	7	0
8	8	0
9	9	0
10	10	0
11	11	0
12	12	0
13	12	1
14	12	2
15	12	3
16	12	4
17	12	5
18	12	6
19	12	7
20	12	8

The next step is to use the **regress** command to predict **realrinc** from **ed1** and **ed2**. This will yield a piecewise model with 12 years of education as the knot. The variable **female** is also included in the model as an additional predictor (covariate), specified as a dummy variable.

```
. regress realrinc ed1 ed2 female, vce(robust)
Linear regression                               Number of obs     =     32,183
                                                F(3, 32179)       =    1030.24
                                                Prob > F          =     0.0000
                                                R-squared         =     0.1420
                                                Root MSE          =      25045
```

realrinc	Coef.	Robust Std. Err.	t	P>|t|	[95% Conf.	Interval]
ed1	832.2703	72.37999	11.50	0.000	690.4028	974.1378
ed2	3441.326	93.42434	36.84	0.000	3258.211	3624.441
female	-12372.38	276.3531	-44.77	0.000	-12914.05	-11830.72
_cons	12052.18	793.4194	15.19	0.000	10497.04	13607.31

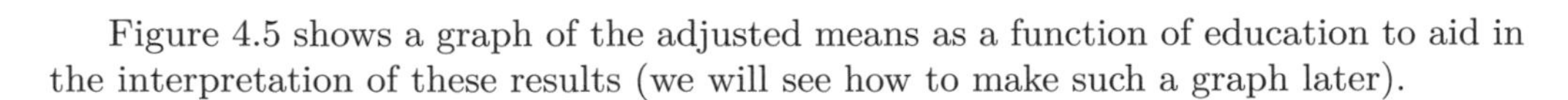

Figure 4.5 shows a graph of the adjusted means as a function of education to aid in the interpretation of these results (we will see how to make such a graph later).

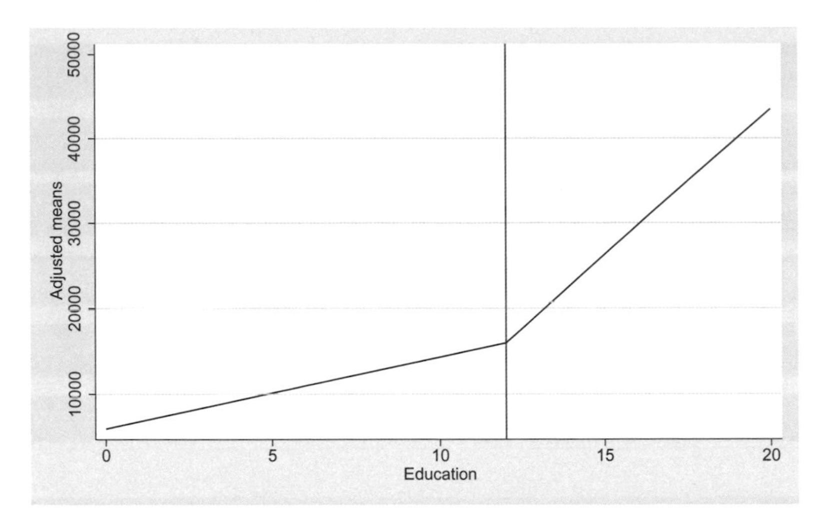

Figure 4.5. Adjusted means for income from a piecewise model with one knot at `educ` = 12

The coefficient for `ed1` is the slope of the relationship between income and education for those with fewer than 12 years of education (non–high school graduates). Among non–high school graduates, each additional year of education is predicted to increase income by $832.27. The coefficient for `ed2` is the slope for those with 12 or more years of education (high school graduates). Income increases by $3,441.33 for each additional year of education beyond 12 years of education. Each of these slopes is significantly different from 0.

We might ask whether the difference in these slopes is significant. In other words, is the slope for high school graduates steeper than the slope for non–high school graduates? We can test this using the `lincom` command, as shown below.

```
. lincom ed2 - ed1
 ( 1)  - ed1 + ed2 = 0
```

realrinc	Coef.	Std. Err.	t	P>|t|	[95% Conf.	Interval]
(1)	2609.055	134.397	19.41	0.000	2345.632	2872.479

The slope of the relationship between income and education is significantly higher for high school graduates than for non–high school graduates (t = 19.41, p = 0.000). High school graduates earn $2,609.06 more for each additional year of education than non–high school graduates.

Say that we want to compute the adjusted mean given that a person has eight years of education. We can compute this adjusted mean using the **margins** command; however, we cannot specify the variable **educ** in the **at()** option (because **educ** does not appear in the regression model). Instead, we express education in terms of the variables **ed1** and **ed2**. Referring to the **showcoding** results above, we see that for eight years of education, the corresponding value of **ed1** is 8 and **ed2** is 0. We specify these values in the **at()** option below. The adjusted mean (adjusting for gender) given eight years of education is $12,623.14.

```
. margins, at(ed1=8 ed2=0)
Predictive margins                          Number of obs     =     32,183
Model VCE    : Robust

Expression   : Linear prediction, predict()
at           : ed1             =           8
               ed2             =           0

------------------------------------------------------------------------------
             |            Delta-method
             |     Margin   Std. Err.      t    P>|t|     [95% Conf. Interval]
-------------+----------------------------------------------------------------
       _cons |   12623.14   244.9678    51.53   0.000     12142.99    13103.28
------------------------------------------------------------------------------
```

We can compute the adjusted mean given 16 years of education by referring again to the output of the **showcoding** command and specifying a value of 12 for **ed1** and 4 for **ed2**, as shown below. At 16 years of education, the adjusted mean of income is $29,717.52.

```
. margins, at(ed1=12 ed2=4)
Predictive margins                          Number of obs     =     32,183
Model VCE    : Robust

Expression   : Linear prediction, predict()
at           : ed1             =          12
               ed2             =           4

------------------------------------------------------------------------------
             |            Delta-method
             |     Margin   Std. Err.      t    P>|t|     [95% Conf. Interval]
-------------+----------------------------------------------------------------
       _cons |   29717.52   299.4035    99.26   0.000     29130.68    30304.36
------------------------------------------------------------------------------
```

To graph the adjusted means as a function of education, we need to compute the adjusted means when education is 0, 12, and 20. The **margins** command below computes these three adjusted means by using the **at()** option three times.

```
. margins, at(ed1=0 ed2=0) at(ed1=12 ed2=0) at(ed1=12 ed2=8) vsquish

Predictive margins                              Number of obs     =     32,183
Model VCE    : Robust

Expression   : Linear prediction, predict()
1._at        : ed1             =           0
               ed2             =           0
2._at        : ed1             =          12
               ed2             =           0
3._at        : ed1             =          12
               ed2             =           8

------------------------------------------------------------------------------
             |            Delta-method
             |     Margin   Std. Err.      t    P>|t|     [95% Conf. Interval]
-------------+----------------------------------------------------------------
         _at |
          1  |   5964.977   794.2416     7.51   0.000     4408.233     7521.72
          2  |   15952.22     161.46    98.80   0.000     15635.75    16268.69
          3  |   43482.83   657.6587    66.12   0.000     42193.79    44771.86
------------------------------------------------------------------------------
```

We can then manually input the adjusted means into a dataset. The `graph` command creates a graph showing the adjusted means as a function of income.[2] This creates the graph that we saw in figure 4.5.

```
. preserve
. clear
. input educ yhat
          educ        yhat
  1. 0  5964.977
  2. 12 15952.22
  3. 20 43482.83
  4. end
. graph twoway line yhat educ, xlabel(0(4)20) xline(12)
. restore
```

Change in slope coding

Let's fit this model again, but this time use change in slope coding. This coding scheme estimates the slope for the line segment before the first knot (for example, for non–high school graduates) and then estimates the difference in the slopes of adjacent line segments (for example, for high school graduates versus non–high school graduates). This strategy emphasizes the change in slope that occurs at each knot.

The process of fitting this kind of model is similar to fitting a individual slope model (as we saw in section 4.3.2). The key difference is that we add the `marginal` option[3] to the `mkspline` command, as shown below. I name the variables `ed1m` and `ed2m`, adding the `m` to emphasize that these variables were created using the `marginal` option.

2. This process of graphing the adjusted means can be automated, as shown in section 4.10.
3. Not to be confused with the `margins` command.

```
. mkspline ed1m 12 ed2m = educ, marginal
```

Let's use the showcoding command, which you can download (see section 1.1), to see the correspondence between educ and ed1m and between educ and ed2m, showing the impact of the addition of the marginal option on the mkspline command.

```
. showcoding educ ed1m ed2m
```

educ	ed1m	ed2m
0	0	0
1	1	0
2	2	0
3	3	0
4	4	0
5	5	0
6	6	0
7	7	0
8	8	0
9	9	0
10	10	0
11	11	0
12	12	0
13	13	1
14	14	2
15	15	3
16	16	4
17	17	5
18	18	6
19	19	7
20	20	8

The variable ed1m corresponds to educ. The variable ed2m contains 0 for 12 or fewer years of education and contains the value of educ minus 12 for more than 12 years of education. (If you are curious, you can compare the coding of ed1m and ed2m with ed1 and ed2 from the previous section.)

The next step is to enter ed1m and ed2m as predictors of income, as shown below.

```
. regress realrinc ed1m ed2m female, vce(robust)
Linear regression                               Number of obs     =     32,183
                                                F(3, 32179)       =    1030.24
                                                Prob > F          =     0.0000
                                                R-squared         =     0.1420
                                                Root MSE          =      25045
```

realrinc	Coef.	Robust Std. Err.	t	P>|t|	[95% Conf.	Interval]
ed1m	832.2703	72.37999	11.50	0.000	690.4028	974.1378
ed2m	2609.055	134.397	19.41	0.000	2345.632	2872.479
female	-12372.38	276.3531	-44.77	0.000	-12914.05	-11830.72
_cons	12052.18	793.4194	15.19	0.000	10497.04	13607.31

The coefficient for ed1m is the slope for non–high school graduates. The coefficient for ed2m is the change in the slope after the knot compared with before the knot (that is, for high school graduates compared with non–high school graduates). For those without a high school diploma, the slope is 832.27. The change in the slope upon graduating high school is 2,609.06 (and this is significantly different from 0).

If we want to estimate the slope for those with 12 or more years of education, we can use the following lincom command. Each additional year of education beyond 12 years is associated with a $3,441.33 increase in income.

```
. lincom ed1m + ed2m
 ( 1)  ed1m + ed2m = 0
```

realrinc	Coef.	Std. Err.	t	P>|t|	[95% Conf.	Interval]
(1)	3441.326	93.42434	36.84	0.000	3258.211	3624.441

The process of computing and graphing adjusted means for the change in slope model is the same as we saw for the individual slope model. The key difference is that ed1m and ed2m would be used to express the level of education. To save space, I refer you to section 4.3.2 for a description of how to compute and graph adjusted means.

Summary

In this section, we have seen how to fit a piecewise model with one known knot. In fact, we have seen two ways we can fit such a model. The individual slope coding method omitted the marginal option on the mkspline command and yielded estimates of the educ slope before and after the knot. We also saw the change in slope coding method, which included the marginal option on the mkspline command, and yielded an estimate of the educ slope before the knot and the change in the educ slope before and after the knot.

> **Note! Coding of remaining examples**
>
> The remaining examples in this chapter will use change in slope coding (that is, including the marginal option on the mkspline command). If you prefer, you can fit these models using individual slope coding by omitting the marginal option.

4.4 Piecewise with two known knots

4.4.1 Overview

We can extend the previous example to include two knots. We saw such an example in the right panel of figure 4.1. That graph is shown in more detail in figure 4.6, which shows the fitted values from a hypothetical piecewise model predicting income from education with a knot at 12 and 16 years of education. Because this model has two knots, there are three slopes. The slope of the relationship between income and education is 600 for those who have fewer than 12 years of education. The slope increases by 2,500 (from 600 to 3,100) for those who have 12 or more years of education, and the slope increases again by 2,000 (from 3,100 to 5,100) for those who have 16 or more years of education. In other words, the slope for non–high school graduates is 600, the slope for high school graduates is 3,100, and the slope for college graduates is 5,100.

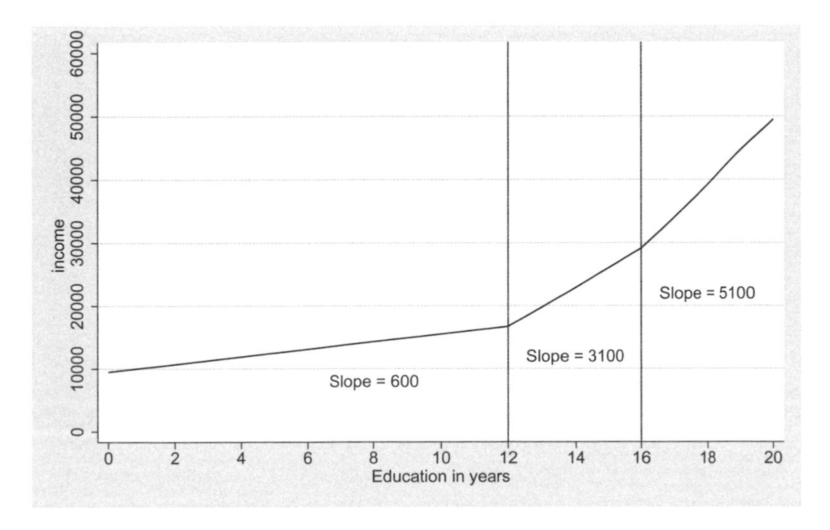

Figure 4.6. Hypothetical piecewise regression with two knots

4.4.2 Examples using the GSS

Let's fit a model with 2 knots corresponding to 12 years of education (graduating high school) and 16 years of education (graduating college) using the `gss_ivrm.dta` dataset. As shown in section 4.3, the `mkspline` command is used, but now we have knots at 12 and 16. The variables `ed1m`, `ed2m`, and `ed3m` are created using the `mkspline` command, and the `marginal` option is used so `ed2m` and `ed3m` will be coded to represent the change in slope from the previous piece of the regression line. The coefficient for `ed2m` will represent the change in slope for high school graduates versus non–high school graduates. The coefficient for `ed3m` will represent the change in slope comparing college graduates with high school graduates.

```
. use gss_ivrm
. mkspline ed1m 12 ed2m 16 ed3m = educ, marginal
```

Let's use the `showcoding` command (see section 1.1) to show how `ed1m`, `ed2m`, and `ed3m` are coded.

```
. showcoding educ ed1m ed2m ed3m
```

educ	ed1m	ed2m	ed3m
0	0	0	0
1	1	0	0
2	2	0	0
3	3	0	0
4	4	0	0
5	5	0	0
6	6	0	0
7	7	0	0
8	8	0	0
9	9	0	0
10	10	0	0
11	11	0	0
12	12	0	0
13	13	1	0
14	14	2	0
15	15	3	0
16	16	4	0
17	17	5	1
18	18	6	2
19	19	7	3
20	20	8	4

The variable `ed1m` is coded as the value of `educ`. The variable `ed2m` is coded as 0 for those with fewer than 12 years of education, and `educ` minus 12 for those with 12 or more years of education. Likewise, `ed3m` is coded as 0 for those with fewer than 16 years of education, and `educ` minus 16 for those with 16 or more years of education. We will use this information later to compute adjusted means for any given value of education.

We can now run the piecewise regression as shown below. Note the `female` variable is included as a covariate.

```
. regress realrinc ed1m ed2m ed3m female, vce(robust)
Linear regression                               Number of obs     =     32,183
                                                F(4, 32178)       =     784.42
                                                Prob > F          =     0.0000
                                                R-squared         =     0.1432
                                                Root MSE          =      25029

------------------------------------------------------------------------------
             |               Robust
    realrinc |      Coef.   Std. Err.      t    P>|t|     [95% Conf. Interval]
-------------+----------------------------------------------------------------
        ed1m |   939.6728   70.54666    13.32   0.000     801.3987    1077.947
        ed2m |   2022.422   144.2374    14.02   0.000     1739.711    2305.132
        ed3m |   1657.452   412.1507     4.02   0.000     849.6215    2465.283
      female |  -12336.74   277.0882   -44.52   0.000    -12879.84   -11793.64
       _cons |   11215.32    787.534    14.24   0.000     9671.725    12758.92
------------------------------------------------------------------------------
```

Figure 4.7 shows the adjusted means of income as a function of education for this model. Later, we will see how to create this graph. Let's refer to figure 4.7 to help interpret the coefficients. The slope for those without a high school degree is 939.67. Upon attaining a high school degree (when `educ` = 12), the slope increases by 2,022.42 and that increase is significant. Upon receiving a college degree (when `educ` = 16), the slope increases by 1,657.45, and that increase is also significant.

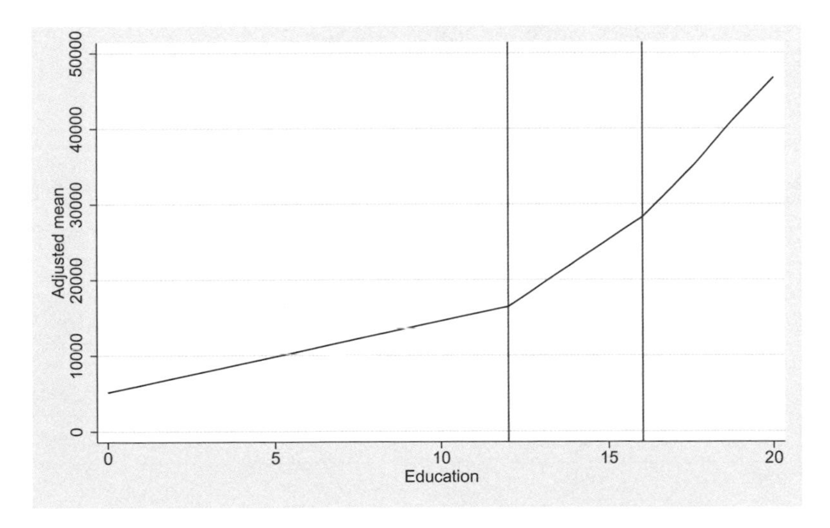

Figure 4.7. Adjusted means from piecewise model with knots at `educ` = 12 and `educ` = 16

We can use the `lincom` command to compute the slope for high school graduates by adding together the coefficients of `ed1m` and `ed2m`, as shown below. The slope for high school graduates is 2,962.09, and that slope is significantly different from 0.

```
. lincom ed1m + ed2m
 ( 1)  ed1m + ed2m = 0

   realrinc |      Coef.   Std. Err.      t    P>|t|     [95% Conf. Interval]
        (1) |   2962.094   110.5663    26.79   0.000      2745.38    3178.809
```

Likewise, we can compute the slope for college graduates by adding the slopes of ed1m, ed2m, and ed3m. The slope for college graduates is 4,619.55 and is significantly different from 0.

```
. lincom ed1m + ed2m + ed3m
 ( 1)  ed1m + ed2m + ed3m = 0

   realrinc |      Coef.   Std. Err.      t    P>|t|     [95% Conf. Interval]
        (1) |   4619.547   347.5461    13.29   0.000     3938.343     5300.75
```

We can use the **margins** command to compute adjusted means for any given value of education by expressing education in terms of the variables **ed1m**, **ed2m**, and **ed3m**. Referring to the output of the **showcoding** command, we can compute the adjusted mean for a person with 14 years of education by specifying that **ed1m** equals 14, **ed2m** equals 2, and **ed3m** equals 0. After adjusting for gender, a person with 14 years of education is predicted to have $22,346 of income.

```
. margins, at(ed1m=14 ed2m=2 ed3m=0)
Predictive margins                              Number of obs     =     32,183
Model VCE    : Robust

Expression   : Linear prediction, predict()
at           : ed1m            =          14
               ed2m            =           2
               ed3m            =           0

             |            Delta-method
             |     Margin   Std. Err.      t    P>|t|     [95% Conf. Interval]
       _cons |   22345.92   187.3352   119.28   0.000     21978.74    22713.11
```

To graph the entire range of education values, we need to compute the adjusted means for the minimum of education (0), for each of the knots (12 and 16), and for the maximum of education (20). The **margins** command below computes these adjusted means using the **at()** option once for each of these four levels of education.

```
. margins, at(ed1m=0  ed2m=0 ed3m=0)
>          at(ed1m=12 ed2m=0 ed3m=0)
>          at(ed1m=16 ed2m=4 ed3m=0)
>          at(ed1m=20 ed2m=8 ed3m=4) vsquish

Predictive margins                              Number of obs     =     32,183
Model VCE    : Robust

Expression   : Linear prediction, predict()
1._at        : ed1m            =           0
               ed2m            =           0
               ed3m            =           0
2._at        : ed1m            =          12
               ed2m            =           0
               ed3m            =           0
3._at        : ed1m            =          16
               ed2m            =           4
               ed3m            =           0
4._at        : ed1m            =          20
               ed2m            =           8
               ed3m            =           4

------------------------------------------------------------------------------
             |            Delta-method
             |     Margin   Std. Err.      t    P>|t|     [95% Conf. Interval]
-------------+----------------------------------------------------------------
         _at |
          1  |    5145.66    785.1576     6.55   0.000     3606.721    6684.599
          2  |   16421.73    145.5957   112.79   0.000     16136.36    16707.11
          3  |   28270.11    383.1319    73.79   0.000     27519.16    29021.06
          4  |    46748.3    1242.792    37.62   0.000     44312.38    49184.22
------------------------------------------------------------------------------
```

We can then manually input the adjusted means from the **margins** command into a dataset as shown below.[4] The **graph** command is then used to graph the relationship between education and income. The resulting graph matches the one we saw in figure 4.7 (hence, it is not repeated).

```
. preserve
. clear
. input educ yhat
          educ        yhat
  1. 0 5145.66
  2. 12 16421.73
  3. 16 28270.11
  4. 20 46748.3
  5. end
. graph twoway line yhat educ, xlabel(0(4)20) xline(12 16)
> xtitle(Education) ytitle(Adjusted mean)
. restore
```

4. This process of graphing the adjusted means can be automated, as shown in section 4.10.

Note! Individual slope coding

You could fit this model by omitting the `marginal` option on the `mkspline` command. The slope terms would then correspond to the three line segments of the piecewise model: for those with fewer than 12 years of education, for those with 12 up to (but not including) 16 years of education, and for those with 16 or more years of education.

4.5 Piecewise with one knot and one jump

4.5.1 Overview

In section 4.3, we saw a piecewise model with one knot where there was a change in the slope at the knot. It is possible that there can also be a jump (or fall) that coincides with the knot as well. We saw such an example in the left panel of figure 4.2. That graph is shown in more detail in figure 4.8. This model has one knot at 12 years of education that signifies a change in both slope and intercept. For those with fewer than 12 years of education, the slope is 200. Upon reaching 12 years of education, there is a change of both the intercept and the slope. The intercept increases (jumps) by 2,200 and the slope increases by 3,500 (from 200 to 3,700).

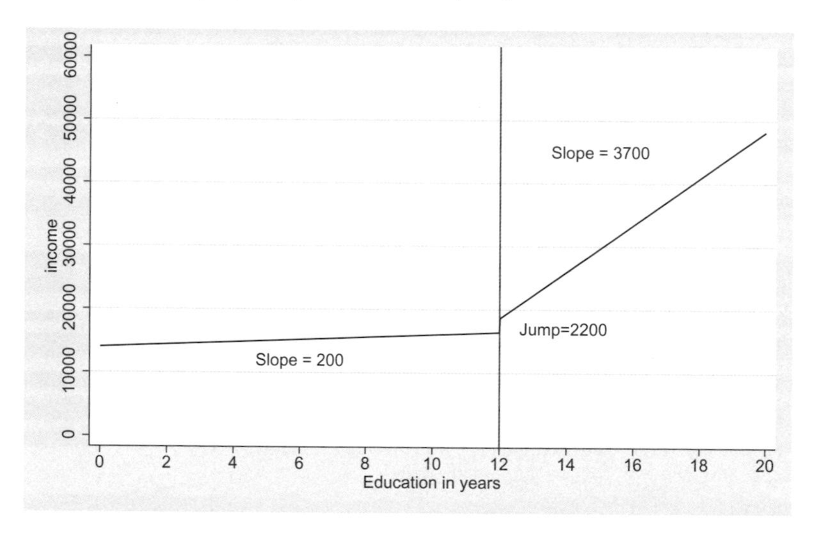

Figure 4.8. Hypothetical piecewise regression with one knot and one jump

4.5.2 Examples using the GSS

Let's illustrate a model with one knot and one jump using the GSS dataset.

```
. use gss_ivrm
```

As shown in section 4.3.2, the `mkspline` command is used with the `marginal` option to create the variables `ed1m` and `ed2m`. `ed1m` represents the slope for non–high school graduates, and `ed2m` represents the change in slope for high school graduates compared with non–high school graduates.

```
. mkspline ed1m 12 ed2m = educ, marginal
```

To fit the jump (change in intercept) associated with having 12 or more years of education, we need a dummy variable that reflects having 12 or more years of education. The variable `hsgrad` is coded 1 if someone has 12 or more years of education, and 0 otherwise.

We can fit a piecewise model that includes a change in both slope and intercept (jump) by including `ed1m`, `ed2m`, and `hsgrad` as predictors in the model. The variable `female` is also included in the model as a covariate.

```
. regress realrinc ed1m ed2m hsgrad female, vce(robust)
Linear regression                               Number of obs     =     32,183
                                                F(4, 32178)       =     803.67
                                                Prob > F          =     0.0000
                                                R-squared         =     0.1425
                                                Root MSE          =      25039

------------------------------------------------------------------------------
             |               Robust
    realrinc |      Coef.   Std. Err.      t    P>|t|     [95% Conf. Interval]
-------------+----------------------------------------------------------------
        ed1m |   273.1002   105.8843     2.58   0.010     65.56293    480.6374
        ed2m |   3102.954   141.8274    21.88   0.000     2824.967    3380.941
      hsgrad |   2721.536    412.816     6.59   0.000     1912.402    3530.671
      female |  -12391.31   276.3032   -44.85   0.000    -12932.87   -11849.74
       _cons |   16345.24   988.3754    16.54   0.000     14407.98    18282.49
------------------------------------------------------------------------------
```

To help interpret the results, a graph of the adjusted means is shown in figure 4.9. This shows the adjusted mean of income at specific levels of education, for those having 0, 1, 12, 13, and 20 years of education. At 12 years of education, the adjusted mean is computed twice, once assuming the absence of a high school degree and once assuming a high school degree, illustrating the jump in income due to graduating high school. (We will see how to make this kind of graph later.)

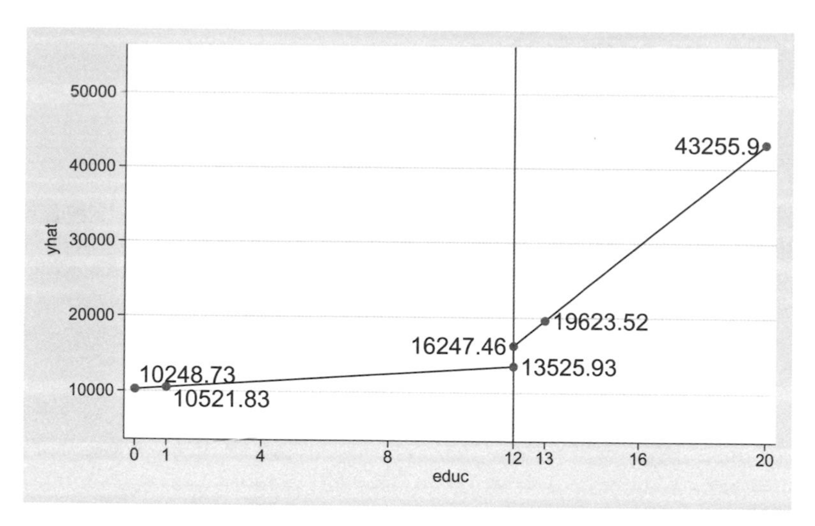

Figure 4.9. Adjusted means from piecewise model with one knot and one jump at `educ` = 12

Let's interpret the regression coefficients by relating the coefficients to the adjusted means shown in figure 4.9. The coefficient for `ed1m` is 273.10, representing the slope for those who have fewer than 12 years of education. For each additional year of education (up to 12), income increases by \$273.10. For example, figure 4.9 shows the adjusted means given zero and one year of education. This difference in these adjusted means equals 273.10 (10521.83 − 10248.73).

The coefficient of `ed2m` is 3,102.95, which is the change (increase) in slope for high school graduates compared with non–high school graduates. The slope of the relationship between income and education for high school graduates is significantly different from the slope for non–high school graduates ($t = 21.88$, $p = 0.000$). We can estimate the slope for high school graduates using the `lincom` command below. That slope is 3,376.05 and is significant ($t = 35.29$, $p = 0.000$). Note how this slope corresponds to the change in the adjusted means for 13 years of education versus 12 years of education (19623.52 − 16247.46 = 3376.05). For 12 or more years of education, a one-unit increase in education is associated with a \$3,376.05 increase in income.

```
. lincom ed1m + ed2m
 ( 1)  ed1m + ed2m = 0
```

realrinc	Coef.	Std. Err.	t	P>|t|	[95% Conf.	Interval]
(1)	3376.054	95.65935	35.29	0.000	3188.558	3563.55

Finally, the `hsgrad` coefficient (2,721.54) represents the predicted jump (increase in income) due to graduating high school. This is shown in figure 4.9 by showing the adjusted mean of income at 12 years of education omitting the benefit of a high school diploma ($13,525.93) compared with 12 years of education, including the benefit of a high school diploma ($16,247.46). The difference in these adjusted means is $2,721.54, showing the jump in income due to graduating high school (which corresponds to the `hsgrad` coefficient).

We can use the `margins` command to compute adjusted means for any given value of education. However, we need to express education in terms of the variables `ed1m`, `ed2m`, and `hsgrad`. The coding of these variables is illustrated using the `showcoding` command. (Section 1.1 shows how to download this command.)

```
. showcoding educ hsgrad ed1m ed2m
```

educ	hsgrad	ed1m	ed2m
0	0	0	0
1	0	1	0
2	0	2	0
3	0	3	0
4	0	4	0
5	0	5	0
6	0	6	0
7	0	7	0
8	0	8	0
9	0	9	0
10	0	10	0
11	0	11	0
12	1	12	0
13	1	13	1
14	1	14	2
15	1	15	3
16	1	16	4
17	1	17	5
18	1	18	6
19	1	19	7
20	1	20	8

As you can see from the output above, `ed1m` is coded as the value of `educ`. The variable `ed2m` is coded as 0 for those with fewer than 12 years of education, and `educ` minus 12 for those with 12 or more years of education. The variable `hsgrad` is coded as 0 for those with fewer than 12 years of education and 1 for those with 12 or more years of education. We can use this information to compute adjusted means of the outcome variable (income) for any given value of education.

Below, we compute the adjusted mean of income for a person with 14 years of education (corresponding to `ed1m` equaling 14, `ed2m` equaling 2, and `hsgrad` equaling 1). At 14 years of education, the adjusted mean of income is $22,999.57.

```
. margins, at(ed1m=14 ed2m=2 hsgrad=1)
Predictive margins                              Number of obs     =     32,183
Model VCE    : Robust

Expression   : Linear prediction, predict()
at           : ed1m            =          14
               ed2m            =           2
               hsgrad          =           1

------------------------------------------------------------------------------
             |            Delta-method
             |     Margin   Std. Err.      t    P>|t|     [95% Conf. Interval]
-------------+----------------------------------------------------------------
       _cons |   22999.57   152.7437   150.58   0.000     22700.19    23298.96
------------------------------------------------------------------------------
```

To graph the entire range of education values, we need to compute the adjusted means when education is zero, at the value of the knot (that is, when education is 12) with and without a high school diploma,[5] and at the maximum of education (that is, when education is 20). The `margins` command below computes these adjusted means.

```
. margins, at(ed1m=0  ed2m=0 hsgrad=0)
>          at(ed1m=12 ed2m=0 hsgrad=0)
>          at(ed1m=12 ed2m=0 hsgrad=1)
>          at(ed1m=20 ed2m=8 hsgrad=1) vsquish
Predictive margins                              Number of obs     =     32,183
Model VCE    : Robust

Expression   : Linear prediction, predict()
1._at        : ed1m            =           0
               ed2m            =           0
               hsgrad          =           0
2._at        : ed1m            =          12
               ed2m            =           0
               hsgrad          =           0
3._at        : ed1m            =          12
               ed2m            =           0
               hsgrad          =           1
4._at        : ed1m            =          20
               ed2m            =           8
               hsgrad          =           1

------------------------------------------------------------------------------
             |            Delta-method
             |     Margin   Std. Err.      t    P>|t|     [95% Conf. Interval]
-------------+----------------------------------------------------------------
         _at |
          1  |   10248.73   987.4159    10.38   0.000     8313.353     12184.1
          2  |   13525.93   374.9727    36.07   0.000     12790.97    14260.89
          3  |   16247.46   174.7942    92.95   0.000     15904.86    16590.07
          4  |    43255.9   664.0012    65.14   0.000     41954.43    44557.37
------------------------------------------------------------------------------
```

We can then manually input these values into a dataset and graph them as shown in the commands below.[6] The resulting graph is shown in figure 4.10.

5. This illustrates the jump in income due to graduating high school.
6. This process of graphing the adjusted means can be automated, as shown in section 4.10.

```
. preserve
. clear
. input educ yhat
            educ        yhat
  1. 0 10248.73
  2. 12 13525.93
  3. 12 16247.46
  4. 20 43255.9
  5. end
. graph twoway line yhat educ, xlabel(0(4)20) xline(12)
. restore
```

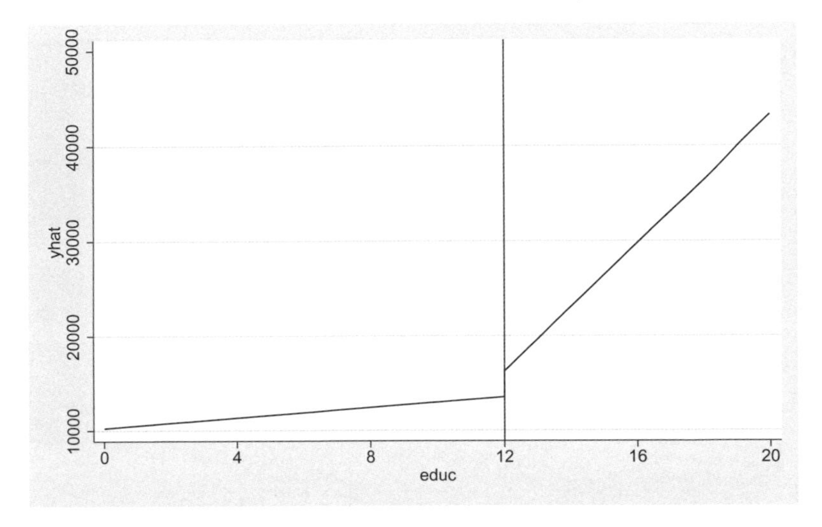

Figure 4.10. Adjusted means from piecewise model with one knot and one jump at `educ` = 12

This shows how you can create a visual representation of the adjusted means. It illustrates the slope when education is fewer than 12 years, the slope (or change in slope) when education is at least 12 years, and the jump in income upon achieving 12 years of education.

Note! Individual slope coding

This model was fit using the change in slope coding method (that is, with the `marginal` option on the `mkspline` command). If you wished, you could fit this model using individual slope coding by omitting the `marginal` option on the `mkspline` command. This would yield estimates of the `educ` slope before having 12 years of education and the `educ` slope after reaching 12 years of education.

4.6 Piecewise with two knots and two jumps

4.6.1 Overview

In section 4.4, we saw a piecewise regression model that contained two knots with a change of slope (but not a change of intercept). Section 4.5 showed a model with one knot that signified a change of slope and intercept. This section will illustrate a model with two knots signifying a change of slope and intercept. We saw such an example in the right panel of figure 4.2. That graph is shown in more detail in figure 4.11. It shows the results of a hypothetical piecewise model predicting income from education with knots at 12 and 16 years of education, each representing a change in slope and change in intercept.

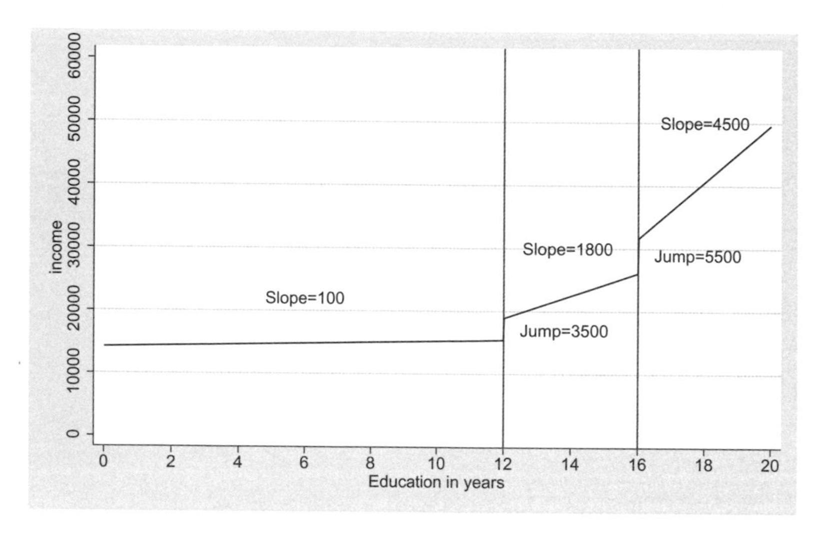

Figure 4.11. Hypothetical piecewise regression with two knots and two jumps

In figure 4.11, the slope for those with fewer than 12 years of education is 100. At 12 years of education, the intercept increases (jumps) by 3,500 and the slope increases by 1,700 (from 100 to 1,800). At 16 years of education, the intercept jumps by 5,500 and the slope increases by 2,700 (from 1,800 to 4,500).

4.6.2 Examples using the GSS

Let's use the GSS dataset to fit a piecewise model predicting income from education, including two knots at 12 and 16 years of education. Each knot will include a change in slope and a change in intercept.

Looking back to section 4.4.2, we saw how to use the mkspline command to create the variables corresponding to knots at 12 and 16 years of education. We use the same mkspline command below. This creates the variables ed1m, ed2m, and ed3m, which will include changes in the slope (knots) at 12 and 16 years of education.

```
. use gss_ivrm
. mkspline ed1m 12 ed2m 16 ed3m = educ, marginal
```

In section 4.5.2, we saw how to include a change of intercept at 12 years of education by including a dummy variable for achieving 12 years of education (hsgrad). The same strategy is used for this model, also including a dummy variable for achieving 16 years of education (cograd). These two dummy variables will fit the change in intercept at 12 and 16 years of education.

We can now fit a piecewise model with knots at 12 and 16 years of education, allowing for a change in slope and change in intercept at each knot using the regress command below. This model also includes gender as a covariate.

```
. regress realrinc ed1m ed2m ed3m hsgrad cograd female, vce(robust)
Linear regression                               Number of obs     =     32,183
                                                F(6, 32176)       =     544.02
                                                Prob > F          =     0.0000
                                                R-squared         =     0.1458
                                                Root MSE          =      24991

------------------------------------------------------------------------------
             |               Robust
    realrinc |      Coef.   Std. Err.      t    P>|t|     [95% Conf. Interval]
-------------+----------------------------------------------------------------
        ed1m |   272.2715   105.8731     2.57   0.010      64.7562    479.7868
        ed2m |   1329.815   189.3145     7.02   0.000     958.7514    1700.879
        ed3m |   2540.166   398.7068     6.37   0.000     1758.686    3321.647
      hsgrad |   3925.583   405.8082     9.67   0.000     3130.184    4720.982
      cograd |   5740.762    733.167     7.83   0.000     4303.727    7177.797
      female |  -12358.72   276.6883   -44.67   0.000    -12901.04    -11816.4
       _cons |    16339.1   988.2776    16.53   0.000     14402.04    18276.16
------------------------------------------------------------------------------
```

To help us interpret the coefficients from this model, let's first visualize the adjusted means using the graph shown in figure 4.12. This shows the adjusted means as education ranges from 0 to 20, along with the adjusted means for selected values of education, namely, 0, 1, 12, 13, 16, 17, and 20 years of education.[7] (We will see how to create a line graph like this one shortly.)

7. The adjusted means are computed at 12 years of education twice, once assuming the absence of a high school degree and again assuming a high school degree. The difference in these adjusted means illustrates the predicted jump in income due to graduating high school. Likewise, at 16 years of education the adjusted means are computed once assuming no college degree and again assuming a college degree. The difference in these predicted incomes illustrates the jump in income due to graduating college.

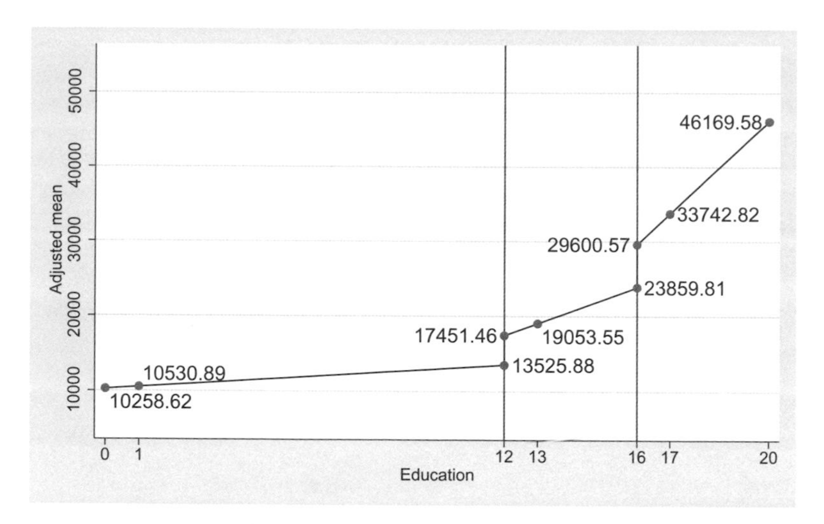

Figure 4.12. Adjusted means from piecewise model with knots and jumps at `educ` = 12 and `educ` = 16

Let's begin by interpreting the coefficients for `hsgrad` and `cograd`. The coefficient for `hsgrad` is 3,925.58, which represents the jump in income due to graduating high school. This can be seen in figure 4.12 by the jump from $13,525.88 to $17,451.46, which is an increase in $3,925.58. The coefficient for `cograd` is 5,740.76, which represents the predicted jump in income at 16 years of education. This is represented in figure 4.12 by the jump from $23,859.81 to $29,600.57 that occurs due to graduating college.

Now, let's interpret the coefficients for `ed1m`, `ed2m`, and `ed3m`. Note that we used the `marginal` option when creating these terms. The coefficient for `ed1m` is 272.27, which is the slope for those with fewer than 12 years of education. We can see this in the change from zero to one year of education where the adjusted mean of income increases from $10,258.62 to $10,530.89, an increase of $272.27.

The coefficient for `ed2m` is the difference between the slope for high school graduates and the slope for non–high school graduates. The difference in these slopes is significant ($t = 7.02$, $p = 0.000$). To estimate the slope for high school graduates, we can use the `lincom` command, as shown below. The slope for high school graduates is 1,602.09. Note how this corresponds to the increase in income as education increases from 12 to 13 years ($19053.55 - 17451.46 = 1602.09$).

```
. lincom ed1m + ed2m
 ( 1)  ed1m + ed2m = 0
```

realrinc	Coef.	Std. Err.	t	P>|t|	[95% Conf.	Interval]
(1)	1602.086	156.9098	10.21	0.000	1294.537	1909.636

Finally, the coefficient for ed3m is the change in the slope for college graduates compared with high school graduates. This change in slope is significant ($t = 6.37$, $p = 0.000$). We can estimate the slope for college graduates by adding the coefficients for ed1m, ed2m, and ed3m using the lincom command below. The slope for college graduates is 4,142.25. We can see this reflected in figure 4.12 by the change in income as education increases from 16 to 17 years ($33742.82 - 29600.57 = 4142.25$).

```
. lincom ed1m + ed2m + ed3m
 ( 1)  ed1m + ed2m + ed3m = 0
```

realrinc	Coef.	Std. Err.	t	P>|t|	[95% Conf.	Interval]
(1)	4142.253	366.6663	11.30	0.000	3423.573	4860.933

Now that we understand how to interpret the coefficients from this model, let's show how to use the margins command to compute adjusted means for any given value of education. To use the margins command to compute adjusted means, we need to understand the coding of ed1m, ed2m, ed3m, hsgrad, and cograd. Let's show this below using the showcoding command.

```
. showcoding educ ed1m ed2m ed3m hsgrad cograd
```

educ	ed1m	ed2m	ed3m	hsgrad	cograd
0	0	0	0	0	0
1	1	0	0	0	0
2	2	0	0	0	0
3	3	0	0	0	0
4	4	0	0	0	0
5	5	0	0	0	0
6	6	0	0	0	0
7	7	0	0	0	0
8	8	0	0	0	0
9	9	0	0	0	0
10	10	0	0	0	0
11	11	0	0	0	0
12	12	0	0	1	0
13	13	1	0	1	0
14	14	2	0	1	0
15	15	3	0	1	0
16	16	4	0	1	1
17	17	5	1	1	1
18	18	6	2	1	1
19	19	7	3	1	1
20	20	8	4	1	1

As you can see from the output above, ed1m is coded as the value of educ. The variable ed2m is coded as 0 for those with fewer than 12 years of education, and educ minus 12 for those with 12 or more years of education. The variable ed3m is coded as 0 for those with fewer than 16 years of education, and educ minus 16 for those with 16 or more years of education. The variables hsgrad and cograd are dummy coded.

The variable **hsgrad** is coded 0 for those with fewer than 12 years of education and 1 for those with 12 or more years of education. The variable **cograd** is coded as 0 for those with fewer than 16 years of education and 1 for those with 16 or more years of education.

Using this information, we can compute adjusted means as a function of education. The **margins** command below computes the adjusted mean for a person with 14 years of education (corresponding to **ed1m** equaling 14, **ed2m** equaling 2, **ed3m** equaling 0, **hsgrad** equaling 1, and **cograd** equaling 0). Adjusting for gender, the predicted mean of income at 14 years of education is $20,655.64.

```
. margins, at(ed1m=14 ed2m=2 ed3m=0 hsgrad=1 cograd=0)
Predictive margins                              Number of obs     =     32,183
Model VCE    : Robust

Expression   : Linear prediction, predict()
at           : ed1m            =          14
               ed2m            =           2
               ed3m            =           0
               hsgrad          =           1
               cograd          =           0

------------------------------------------------------------------------------
             |            Delta-method
             |     Margin   Std. Err.      t    P>|t|     [95% Conf. Interval]
-------------+----------------------------------------------------------------
       _cons |   20655.64   264.3417    78.14   0.000     20137.52    21173.76
------------------------------------------------------------------------------
```

To graph the adjusted means as a function of education, we need to compute the adjusted mean when education equals zero, the value of the first knot (that is, 12) with and without a high school diploma,[8] the value of the second knot (that is, 16) with and without a college degree,[9] and at the maximum of education (that is, 20). The **margins** command below computes these adjusted means. (The **noatlegend** option suppresses the display of the legend of the covariate values to save space.)

8. This illustrates the jump in income due to graduating high school.
9. This illustrates the jump in income due to graduating college.

```
. margins, at(ed1m=0  ed2m=0 ed3m=0 hsgrad=0 cograd=0)
>          at(ed1m=12 ed2m=0 ed3m=0 hsgrad=0 cograd=0)
>          at(ed1m=12 ed2m=0 ed3m=0 hsgrad=1 cograd=0)
>          at(ed1m=16 ed2m=4 ed3m=0 hsgrad=1 cograd=0)
>          at(ed1m=16 ed2m=4 ed3m=0 hsgrad=1 cograd=1)
>          at(ed1m=20 ed2m=8 ed3m=4 hsgrad=1 cograd=1) vsquish noatlegend
Predictive margins                              Number of obs     =     32,183
Model VCE    : Robust

Expression   : Linear prediction, predict()

------------------------------------------------------------------------------
             |            Delta-method
             |     Margin   Std. Err.      t    P>|t|     [95% Conf. Interval]
-------------+----------------------------------------------------------------
         _at |
          1  |   10258.62   987.3208    10.39   0.000     8323.437    12193.81
          2  |   13525.88   374.9279    36.08   0.000     12791.01    14260.75
          3  |   17451.46   157.3674   110.90   0.000     17143.02    17759.91
          4  |   23859.81   558.5285    42.72   0.000     22765.07    24954.55
          5  |   29600.57   475.6116    62.24   0.000     28668.35    30532.79
          6  |   46169.58   1255.734    36.77   0.000      43708.3    48630.87
------------------------------------------------------------------------------
```

We can then manually input the adjusted means from the **margins** command into a dataset and graph the adjusted means by education.[10] The resulting graph is shown in figure 4.13.

10. This process of graphing the adjusted means can be automated, as shown in section 4.10.

```
. preserve
. clear
. input educ yhat
          educ        yhat
  1. 0  10258.62
  2. 12 13525.88
  3. 12 17451.46
  4. 16 23859.81
  5. 16 29600.57
  6. 20 46169.58
  7. end
. graph twoway line yhat educ, xlabel(0(4)20) xline(12 16)
> xtitle(Education) ytitle(Adjusted means)
. restore
```

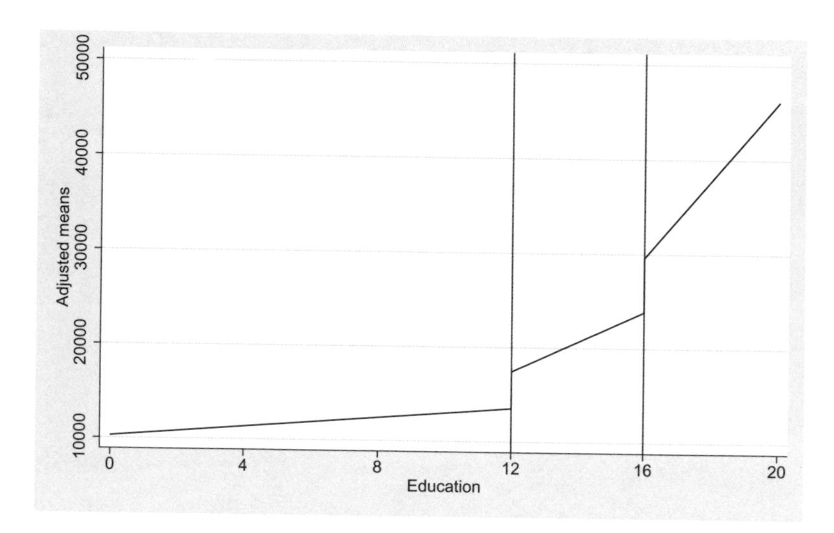

Figure 4.13. Adjusted means from piecewise model with knots and jumps at `educ` = 12 and `educ` = 16

The graph in figure 4.13 shows the relationship between education and the adjusted mean of income. This illustrates the change of intercept and change of slope that occur at 12 and 16 years of education.

Note! Individual slope coding

If you prefer, you can omit the `marginal` option on the `mkspline` command. This would estimate the `educ` slopes for each of the three line segments of the piecewise model: the slope for those with fewer than 12 years of education, for those with between 12 and 16 years of education, and for those with 16 or more years of education.

4.7 Piecewise with an unknown knot

This section explores the use of a piecewise model with one knot where the placement of the knot is unknown. Contrast this with the model illustrated in section 4.3, which used a piecewise model relating income to education. That model had one knot corresponding to 12 years of education because receiving a high school diploma is an important milestone with respect to one's income. Sometimes, however, we may want to fit a piecewise model with one knot without a predefined location for the knot. This section illustrates a method for fitting a piecewise model with one knot where the location of the knot is uncertain.

The `gss_ivrm.dta` dataset is used to illustrate this, focusing on people who were born between 1905 and 1985 with nonmissing data on `educ`.

```
. use gss_ivrm
. keep if (yrborn>=1905 & yrborn<=1985) & !missing(educ)
(2,214 observations deleted)
```

Consider the relationship between year of birth and level of education. In this dataset, the year of birth is recorded in the variable `yrborn` and education level is recorded in the variable `educ`. To visualize the relationship between year of birth and education, let's make a graph showing the mean of `educ` at each level of `yrborn`. (This technique is illustrated in more detail in section 2.4.4.) The `regress` command is used to predict `educ`[11] from the variable `yrborn`, treating it as a categorical (factor) variable. The `margins` command is then used to compute the adjusted means, and the `marginsplot` command graphs the means as a function of `yrborn`. The resulting graph is shown in figure 4.14.

11. The variable `educ` is not skewed like `realrinc`, so the `vce(robust)` option is not used for analyses where `educ` is the outcome.

```
. regress educ i.yrborn
  (output omitted)
. margins yrborn
  (output omitted)
. marginsplot, noci
  Variables that uniquely identify margins: yrborn
```

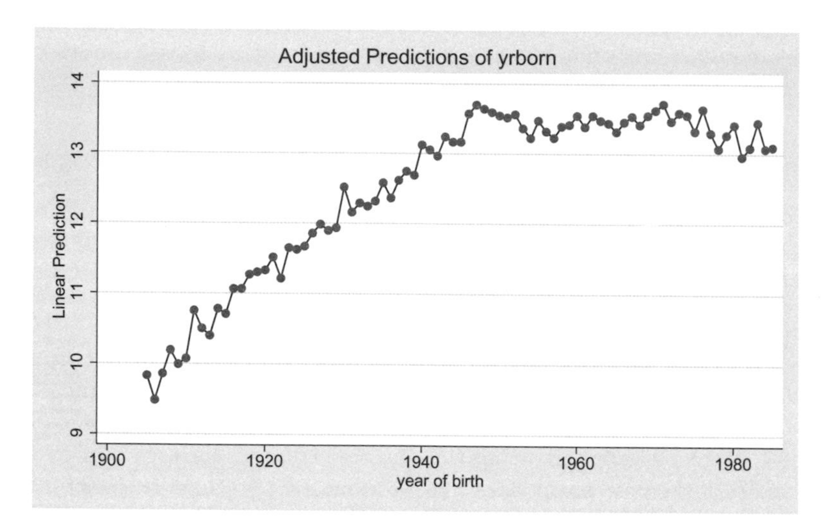

Figure 4.14. Average education at each level of year of birth

Picture fitting the observed values shown in figure 4.14 using a piecewise model with one knot. Imagine two line segments, where the first line segment would begin in 1905 and extend to a period somewhere in the 1940s. The second line segment would begin where the first line segment left off (in the 1940s) and extend until 1985. This model is shown in figure 4.15, where I have overlaid a hand-drawn fitted line spanning from 1905 to 1945 (labeled "Piece 1") and a second such fitted line spanning from 1945 to 1985 (labeled "Piece 2").

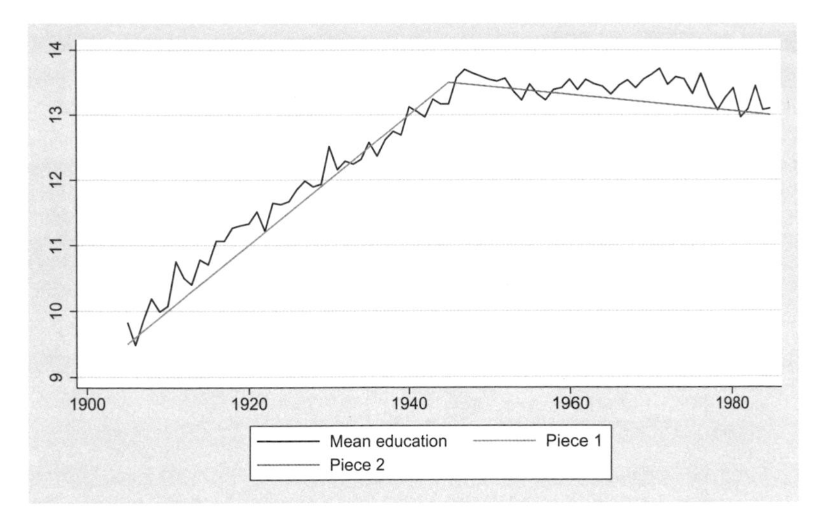

Figure 4.15. Average education with hand-drawn fitted lines

Unfortunately, I have no basis for selecting the year 1945 as the knot. We could haphazardly select different years for the placement of the knot and select the knot that yields the best fitting model. Rather than manually doing this selection process, we can use the `nl` command to automate the process of selecting the optimal knot, by selecting a knot location that yields the lowest residual sum of squares. The code for fitting such a model is shown below. You can use this as a template for fitting your own model.[12]

12. In place of `educ`, you would insert your outcome variable, and in place of `yrborn`, you would place your predictor variable. Then for the `initial()` option, you would insert plausible values for `knot`, `b1`, `b2`, and `cons`. These correspond to the placement of the knot, the slope before the knot, the slope after the knot, and the constant, respectively. In this example, I used 1945 as the knot. To estimate `b1`, I looked at the graph and saw that the mean education rose about 4 units from 1905 to 1945. Taking a change of 4 units divided by 40 gave me an estimate of 0.1 for `b1`. Likewise, I estimated that education declined by about 0.5 units from 1945 to 1985 and divided -0.5 units by 40 to yield -0.0125 as an estimate of `b2`. Finally, to estimate `cons` I estimated the average education to be 9.5 units at 1905 and then used the estimated slope of 0.1 to estimate the education at year 0 would be $9.5 - 1905 \times 0.1 = -181$.

```
. nl (educ = ({cons} + {b1}*yrborn)*(yrborn < {knot}) +
>     ({cons} + {b1}*{knot} + {b2}*(yrborn-{knot}))*(yrborn >= {knot})),
>     initial(knot 1945 b1 .1  b2 -.0125 cons -181)
(obs = 52,873)

Iteration 0:  residual SS =    464140
Iteration 1:  residual SS =  464092.3
Iteration 2:  residual SS =  464092.3

      Source |       SS            df       MS
-------------+----------------------------------    Number of obs =     52,873
       Model |  48574.186           3  16191.3952    R-squared     =     0.0947
    Residual |  464092.28       52869  8.77815515    Adj R-squared =     0.0947
-------------+----------------------------------    Root MSE      =   2.962795
       Total |  512666.47       52872  9.69636992    Res. dev.     =   264897.3

------------------------------------------------------------------------------
        educ |      Coef.   Std. Err.      t    P>|t|     [95% Conf. Interval]
-------------+----------------------------------------------------------------
       /cons |  -152.0708   3.243928   -46.88   0.000    -158.4289   -145.7127
         /b1 |   .0850539   .0016812    50.59   0.000     .0817588    .0883491
       /knot |   1946.965   .5123086  3800.38   0.000     1945.961    1947.969
         /b2 |  -.0053593   .0017983    -2.98   0.003    -.008884   -.0018347
------------------------------------------------------------------------------
  Parameter cons taken as constant term in model & ANOVA table
```

The coefficient labeled `cons` is the constant for the model (the predicted value of `educ` when `yrborn` is 0). This value is generally uninteresting. The coefficient labeled `knot` is the location of the knot, selected as 1946.97 (which we can round to 1947). The coefficient labeled `b1` is the slope of the relationship between education and year of birth for those born before the knot (before 1947). The coefficient labeled `b2` is the slope for those born in or after 1947. The key is that 1947 was selected as the optimal location for the knot.

Now that we have identified the knot as 1947, we can fit a piecewise model with one knot (as illustrated in section 4.3). The `mkspline` command is used to create the variables `yrborn1` and `yrborn2`, which are separated by a knot at 1947. The `marginal` option is omitted, so the coefficient for `yrborn1` will be the slope for the years before 1947, and `yrborn2` will be the slope for the years after 1947. As I would expect, the results of the `regress` command below match closely to the results using the `nl` command above.

```
. mkspline yrborn1 1947 yrborn2 = yrborn
. regress educ yrborn1 yrborn2
      Source |       SS           df       MS      Number of obs   =    52,873
-------------+----------------------------------   F(2, 52870)     =   2766.81
       Model |   48574.145         2  24287.0725   Prob > F        =    0.0000
    Residual |  464092.325    52,870  8.77798989   R-squared       =    0.0947
-------------+----------------------------------   Adj R-squared   =    0.0947
       Total |   512666.47    52,872  9.69636992   Root MSE        =    2.9628

------------------------------------------------------------------------------
        educ |      Coef.   Std. Err.      t    P>|t|     [95% Conf. Interval]
-------------+----------------------------------------------------------------
     yrborn1 |   .0849812   .0012992    65.41   0.000     .0824348    .0875276
     yrborn2 |  -.0054247   .0015211    -3.57   0.000    -.0084061   -.0024432
       _cons |  -151.9311   2.513393   -60.45   0.000    -156.8573   -147.0048
------------------------------------------------------------------------------
```

I will not belabor the explanation of these results because this kind of model was explained in section 4.3. Briefly, the `educ` slope for those born before 1947 is 0.085 and for those born in 1947 or later the slope is −0.005. The `lincom` command below estimates the difference in the slopes (for 1947 and later compared with before 1947).

```
. lincom yrborn2 - yrborn1
 ( 1)  - yrborn1 + yrborn2 = 0

------------------------------------------------------------------------------
        educ |      Coef.   Std. Err.      t    P>|t|     [95% Conf. Interval]
-------------+----------------------------------------------------------------
         (1) |  -.0904059   .0024594   -36.76   0.000    -.0952263   -.0855855
------------------------------------------------------------------------------
```

The slope for those born in 1947 and later shows a significant decrease compared with the slope for those born before 1947.

4.8 Piecewise model with multiple unknown knots

This section illustrates the use of piecewise regression models where there could be multiple knots in the relationship between the predictor and outcome, and you have no basis for the placement of the knots. Contrast this with earlier sections of this chapter (namely, sections 4.3, 4.4, 4.5, and 4.6), where we fit models with one or two knots where the selection of the knot was known, or section 4.7, where there was one knot with an unknown value.

Consider the relationship between age and income. I would suspect that income would initially rise rapidly with increasing age, hit a peak, and then decline with increasing age. Even though I have some general prediction about the pattern of the relationship between age and income, I do not have a strong basis for the placement of the knots for a piecewise model. In such a case, we can intentionally fit a model with too many knots and then progressively remove the superfluous knots. For this example, we will use the `gss_ivrm.dta` dataset and will keep only those who are 80 years old or younger.

```
. use gss_ivrm
. keep if age<=80
(1,892 observations deleted)
```

I like to start this process by fitting an indicator model in which the predictor (for example, age) is treated as a categorical variable. Such a model makes no assumption about the relationship between the predictor and outcome (in this case, between `age` and `realrinc`). Let's call this model 0. This model is run below (the output is suppressed to save space).

```
. * Model 0
. regress realrinc i.age, vce(robust)
  (output omitted)
```

This model includes 62 predictors (corresponding to the indicators for ages 19 to 80). The test of these indicators is significant, and they have an R^2 value of 0.0581. This R^2 value is the highest amount of variance we could hope to explain using age as a predictor. This is because this set of indicators accounts for every tiny bump and drop in the relationship between age and income. Any simpler model (for example, linear, quadratic, or piecewise) will not account for all the bumps and drops and thus will have a lower R^2. However, a good model will account for the major bumps and drops and have an R^2 that is not too much smaller than the R^2 from the indicator model. This is our goal in fitting the piecewise model with multiple knots.

Let's visualize the relationship between age and income. We can do this by using the `predict` command to create predicted values based on the indicator model, in this case named `yhatind`. These predicted values contain the mean of income at each level of age (as described in section 2.4.4). The `graph` command shows the mean of income at each level of age, as shown in figure 4.16.

```
. predict yhatind
(option xb assumed; fitted values)
. graph twoway line yhatind age, sort xlabel(18 20(5)80)
```

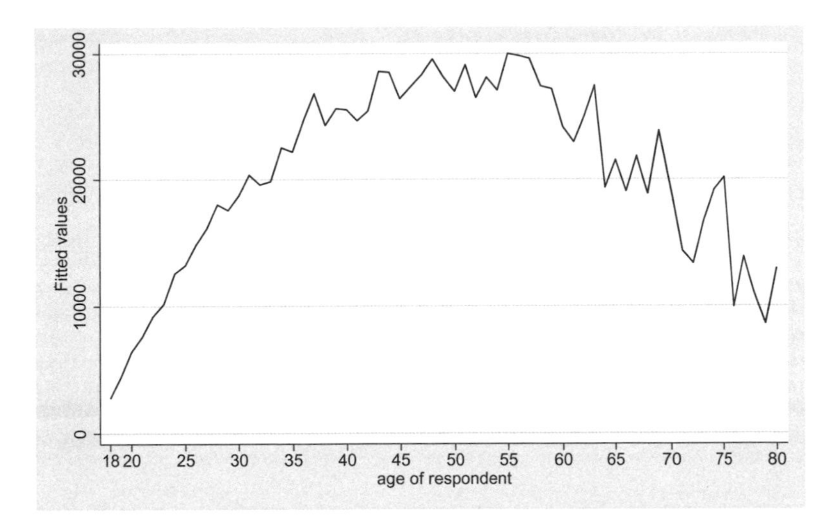

Figure 4.16. Income predicted from age using indicator model

This figure is consistent with my intuition about the relationship between age and income. Incomes rise rapidly from ages 18 to about 35. The incomes continue to rise from ages 35 to 55 but not as rapidly as in the younger years. Then around age 55, the incomes start to decline. Using these visual observations, we can see that it might be to our advantage to include more knots in the early years because of the greater degree of change in the slope of the relationship between income and age during that time period. Likewise, fewer knots seem to be needed for later ages.

Let's begin with a model with knots at ages 25, 30, 35, 45, 55, and 65. This provides knots closer together at the earlier years and provides knots at milestone years of 55 (when people might consider early retirement) and 65 (a standard retirement age). Let's first visualize such a model and see whether this makes sense by repeating figure 4.16, but let's include lines on the x axis showing the proposed placement of these knots. This is shown in figure 4.17.

```
. graph twoway line yhatind age, xlabel(25 30 35 45 55 65)
> xline(25 30 35 45 55 65) sort
```

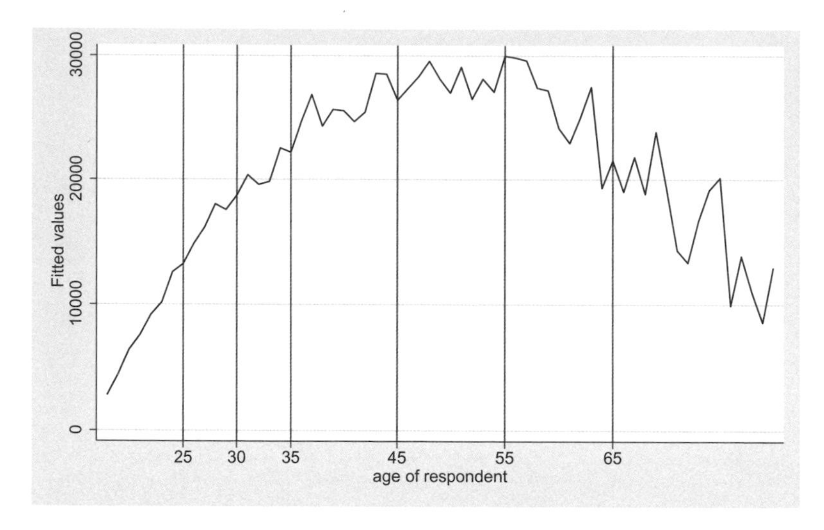

Figure 4.17. Income predicted from age using indicator model with lines at ages 25, 30, 35, 45, 55, and 65

As you look at figure 4.17, imagine imposing a linear relationship between each segment created by the knots. The piecewise model can account for nonlinearities that occur at the knots but cannot account for nonlinearities between any pair of knots. As I look at this, I can picture the overall curve broken into a series of straight lines connected at each knot. In fact, I could picture eliminating some of the knots and still having a series of straight lines. But before I leapfrog ahead, let's first fit a model specified by the knots at ages 25, 30, 35, 45, 55, and 65, which we will call model 1. Note how the `mkspline` command is used to create knots at each of the ages, and this creates the variables `age18to24m` to `age65to80m`.[13] These seven predictors are then used in the `regress` command to predict `realrinc`.

13. I included the `m` suffix to emphasize that these variables were created using the `marginal` option.

```
. * model 1: full model
. mkspline age18to24m 25 age25to29m 30 age30to34m 35 age35to44m 45
> age45to54m 55 age55to64m 65 age65to80m=age, marginal
. regress realrinc age18to24m-age65to80m, vce(robust)
Linear regression                               Number of obs     =     32,100
                                                F(7, 32092)       =     712.84
                                                Prob > F          =     0.0000
                                                R-squared         =     0.0559
                                                Root MSE          =      26153

------------------------------------------------------------------------------
             |               Robust
    realrinc |      Coef.   Std. Err.      t    P>|t|     [95% Conf. Interval]
-------------+----------------------------------------------------------------
  age18to24m |   1559.142   66.34466    23.50   0.000     1429.104     1689.18
  age25to29m |  -582.0045   131.6503    -4.42   0.000     -840.044   -323.9649
  age30to34m |  -45.67108   188.3367    -0.24   0.808    -414.8181    323.4759
  age35to44m |  -494.4584   184.9306    -2.67   0.008    -856.9294   -131.9874
  age45to54m |  -310.1838   154.2959    -2.01   0.044    -612.6096   -7.758053
  age55to64m |  -827.8346   190.3534    -4.35   0.000    -1200.934   -454.7348
  age65to80m |  -4.333834   214.9239    -0.02   0.984    -425.5929    416.9252
       _cons |   -25134.6   1502.493   -16.73   0.000    -28079.54   -22189.66
------------------------------------------------------------------------------
```

Recall (from section 4.4) the interpretation of the coefficients when using the option `marginal`. The coefficient for `age18to24` is the slope for the line segment before the first knot (that is, the slope of those aged 18 to 24). Then, `age25to29` is the change in the slope comparing the second line segment with the first line segment (that is, the slope of those aged 25 to 29 compared with those aged 18 to 24). The coefficient for `age30to34` is the change in the slope comparing the slope of those aged 30 to 34 compared with those aged 25 to 29, and so forth. When a coefficient regarding the change in the slope in the line segments is not significant, it indicates a knot that can be removed because the slopes before and after the knot are not significantly different. We can use this as a guide for removing superfluous knots.

The knot for `age65to80` (corresponding to change in the slope for ages 65 to 80 compared with ages 55 to 64) is not significant. Referring to figure 4.17, we can see that the slope for those who are 55 to 64 years old is similar to the slope for those who are 65 years old and older. Thus, the knot at age 65 is not really needed and we can assume one slope from ages 55 to 80. Let's refit the model omitting the knot at age 65, creating a model we will call model 2.

First, the variables `age18to24m` to `age65to80m` are dropped so we can then use the `mkspline` command to make the new set of predictors. The `mkspline` command is used to create the predictors `age18to24m` to `age55to80m`, and then the `regress` command is run using the new set of predictors. Although the names of some of the predictors are the same in model 2 and model 1, the meaning changes based on knots specified in the new `mkspline` command.

```
. * model 2: drop knot at age 65
. drop age18to24m-age65to80m
. mkspline age18to24m 25 age25to29m 30 age30to34m 35 age35to44m 45
> age45to54m 55 age55to80m=age, marginal
. regress realrinc age18to24m-age55to80m, vce(robust)
Linear regression                               Number of obs     =     32,100
                                                F(6, 32093)       =     830.57
                                                Prob > F          =     0.0000
                                                R-squared         =     0.0559
                                                Root MSE          =      26152

------------------------------------------------------------------------------
             |               Robust
    realrinc |      Coef.   Std. Err.      t    P>|t|     [95% Conf. Interval]
-------------+----------------------------------------------------------------
  age18to24m |   1559.147    66.34308    23.50   0.000     1429.113    1689.182
  age25to29m |  -582.0308    131.6404    -4.42   0.000     -840.051   -324.0107
  age30to34m |  -45.56634    188.2265    -0.24   0.809    -414.4975    323.3648
  age35to44m |    -494.68    184.3103    -2.68   0.007    -855.9352   -133.4248
  age45to54m |   -309.493    146.2357    -2.12   0.034    -596.1205   -22.86545
  age55to80m |  -829.9867    133.1483    -6.23   0.000    -1090.962   -569.0111
       _cons |  -25134.71    1502.459   -16.73   0.000    -28079.59   -22189.83
------------------------------------------------------------------------------
```

In model 2, the coefficient for age30to34m is not significant. This suggests that the slope for those aged 30 to 34 are not different from those aged 25 to 29. Let's consolidate the slope for the ages 25 to 35 and refit the model, calling it model 3.

```
. * model 3: drop knot at age 30
. drop age18to24m-age55to80m
. mkspline age18to24m 25 age25to34m 35 age35to44m 45
> age45to54m 55 age55to80m=age, marginal
. regress realrinc age18to24m-age55to80m, vce(robust)
Linear regression                               Number of obs     =     32,100
                                                F(5, 32094)       =     961.72
                                                Prob > F          =     0.0000
                                                R-squared         =     0.0559
                                                Root MSE          =      26152

------------------------------------------------------------------------------
             |               Robust
    realrinc |      Coef.   Std. Err.      t    P>|t|     [95% Conf. Interval]
-------------+----------------------------------------------------------------
  age18to24m |   1570.773    66.98834    23.45   0.000     1439.473    1702.073
  age25to34m |  -617.8846    105.5122    -5.86   0.000    -824.6925   -411.0767
  age35to44m |  -520.3487    108.2807    -4.81   0.000    -732.5829   -308.1145
  age45to54m |   -303.907    139.2214    -2.18   0.029    -576.7863    -31.0278
  age55to80m |  -831.4752    132.5477    -6.27   0.000    -1091.274   -571.6766
       _cons |   -25377.5    1512.811   -16.78   0.000    -28342.66   -22412.33
------------------------------------------------------------------------------
```

All terms in model 3 are significant, but we should be mindful that we have a very large-sample size with considerable statistical power to detect minor changes in slopes. If we focus solely on statistical significance as our guide for removing knots, we could overfit the model. Let's visualize the fit of this model against the mean of income at each level of age to see how well the model fits. This is shown in figure 4.18.

```
. predict yhat
(option xb assumed; fitted values)
. line yhat yhatind age, sort xline(25 35 45 55) xlabel(25 35 45 55)
> legend(label(1 "Piecewise model") label(2 "Indicator model"))
```

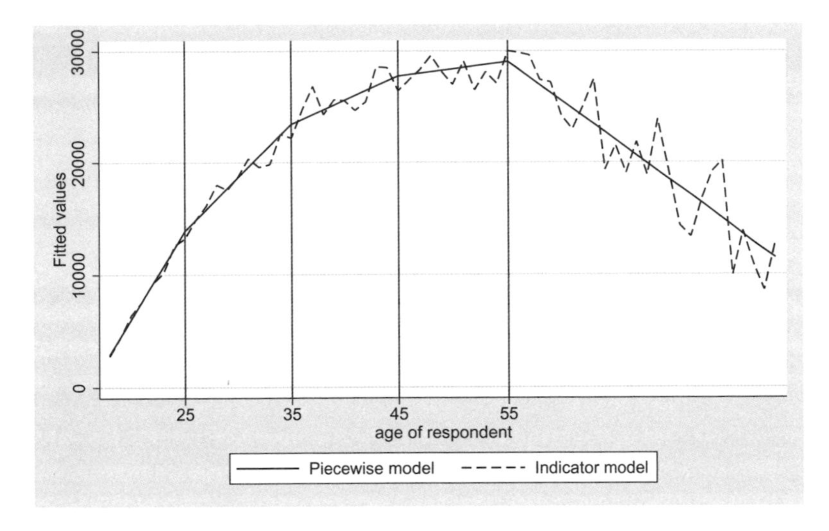

Figure 4.18. Income predicted from age using indicator model and piecewise model 3 with four knots at 25, 35, 45, and 55

The change in slope at age 45 does not seem as necessary as some of the other knots. The slope for ages 35 and 44 is not visually much different from the slope for ages 45 to 54. This is also reflected in the coefficient for `age45to54m`, which, although significant, shows the smallest change of all the coefficients. For the sake of creating a parsimonious model, let's try removing the knot at age 45. This fits one slope for those who are 35 to 54 years old. Let's call this model 4.

```
. * model 4: drop knot at age 45
. drop age18to24m-age55to80m
. mkspline age18to24m 25 age25to34m 35 age35to54m 55
> age55to80m=age, marginal
. regress realrinc age18to24m-age55to80m, vce(robust)
Linear regression                               Number of obs     =     32,100
                                                F(4, 32095)       =    1201.52
                                                Prob > F          =     0.0000
                                                R-squared         =     0.0557
                                                Root MSE          =      26154

------------------------------------------------------------------------------
             |               Robust
    realrinc |      Coef.   Std. Err.      t    P>|t|     [95% Conf. Interval]
-------------+----------------------------------------------------------------
  age18to24m |   1526.973   66.44659    22.98   0.000     1396.735     1657.21
  age25to34m |  -503.6532   101.8505    -4.95   0.000    -703.2841   -304.0223
  age35to54m |  -736.2145   72.34209   -10.18   0.000    -878.0077   -594.4212
  age55to80m |  -1047.584   78.89915   -13.28   0.000    -1202.229   -892.9382
       _cons |  -24462.77   1502.233   -16.28   0.000    -27407.21   -21518.34
------------------------------------------------------------------------------
```

Let's compare the R^2 for the current model (model 4, without the knot at age 45) with the previous model (model 3, with the knot at age 45). The R^2 for model 3 is 0.0559 and the R^2 for model 4 is 0.0557. This is a trivial drop in the R^2.

There are two issues that we have not covered with respect to this kind of model: how to compute adjusted means with the `margins` command and how to graph the adjusted means with the `marginsplot` command. The techniques illustrated in section 4.4 (which showed piecewise regression with two knots) can be applied to this kind of model. I refer you to that section for more information about how to compute adjusted means with the `margins` command and how to graph the adjusted means using the `marginsplot` command.

4.9 Piecewise models and the marginsplot command

The examples in this chapter have used a manual method for graphing the results of the `margins` command instead of using the `marginsplot` command. You might be rightly asking yourself why have I not been using the `marginsplot` command instead. Although the `marginsplot` command is extremely smart in terms of creating useful graphs following the `margins` command, the graphs it creates in the context of a piecewise regression model are not very useful. Let's see why by using the example of a piecewise model with one knot and one jump from section 4.5.2. As we did in that section, we use the `mkspline` command to create the variables `ed1m` and `ed2m` and use those variables, as well as `hsgrad`, in the `regress` command to predict `realrinc`. The variable `female` is included as a covariate.

```
. use gss_ivrm
. mkspline ed1m 12 ed2m = educ, marginal
. regress realrinc ed1m ed2m hsgrad female, vce(robust)
Linear regression                               Number of obs     =     32,183
                                                F(4, 32178)       =     803.67
                                                Prob > F          =     0.0000
                                                R-squared         =     0.1425
                                                Root MSE          =      25039

------------------------------------------------------------------------------
             |               Robust
    realrinc |      Coef.   Std. Err.      t    P>|t|     [95% Conf. Interval]
-------------+----------------------------------------------------------------
        ed1m |   273.1002   105.8843     2.58   0.010     65.56293    480.6374
        ed2m |   3102.954   141.8274    21.88   0.000     2824.967    3380.941
      hsgrad |   2721.536    412.816     6.59   0.000     1912.402    3530.671
      female |  -12391.31   276.3032   -44.85   0.000    -12932.87   -11849.74
       _cons |   16345.24   988.3754    16.54   0.000     14407.98    18282.49
------------------------------------------------------------------------------
```

Now, I would like to compute and graph the adjusted means as a function of education. But the variable **educ** is not in the model. Instead, the model represents education using **ed1m**, **ed2m**, and **hsgrad**. As we did in section 4.5.2, let's use the **margins** command to compute the adjusted means as a function of **ed1m**, **ed2m**, and **hsgrad**.

```
. margins, at(ed1m=0 ed2m=0 hsgrad=0)
>          at(ed1m=12 ed2m=0 hsgrad=0)
>          at(ed1m=12 ed2m=0 hsgrad=1)
>          at(ed1m=20 ed2m=8 hsgrad=1) vsquish
Predictive margins                              Number of obs     =     32,183
Model VCE    : Robust

Expression   : Linear prediction, predict()
1._at        : ed1m            =           0
               ed2m            =           0
               hsgrad          =           0
2._at        : ed1m            =          12
               ed2m            =           0
               hsgrad          =           0
3._at        : ed1m            =          12
               ed2m            =           0
               hsgrad          =           1
4._at        : ed1m            =          20
               ed2m            =           8
               hsgrad          =           1

------------------------------------------------------------------------------
             |            Delta-method
             |     Margin   Std. Err.      t    P>|t|     [95% Conf. Interval]
-------------+----------------------------------------------------------------
         _at |
           1 |   10248.73   987.4159    10.38   0.000     8313.353     12184.1
           2 |   13525.93   374.9727    36.07   0.000     12790.97    14260.89
           3 |   16247.46   174.7942    92.95   0.000     15904.86    16590.07
           4 |    43255.9   664.0012    65.14   0.000     41954.43    44557.37
------------------------------------------------------------------------------
```

Now, let's try using the `marginsplot` command to graph the adjusted means. The resulting graph is shown in figure 4.19.

```
. marginsplot
  Variables that uniquely identify margins: _atopt
  Multiple at() options specified:
      _atoption=1: ed1m=0 ed2m=0 hsgrad=0
      _atoption=2: ed1m=12 ed2m=0 hsgrad=0
      _atoption=3: ed1m=12 ed2m=0 hsgrad=1
      _atoption=4: ed1m=20 ed2m=8 hsgrad=1
```

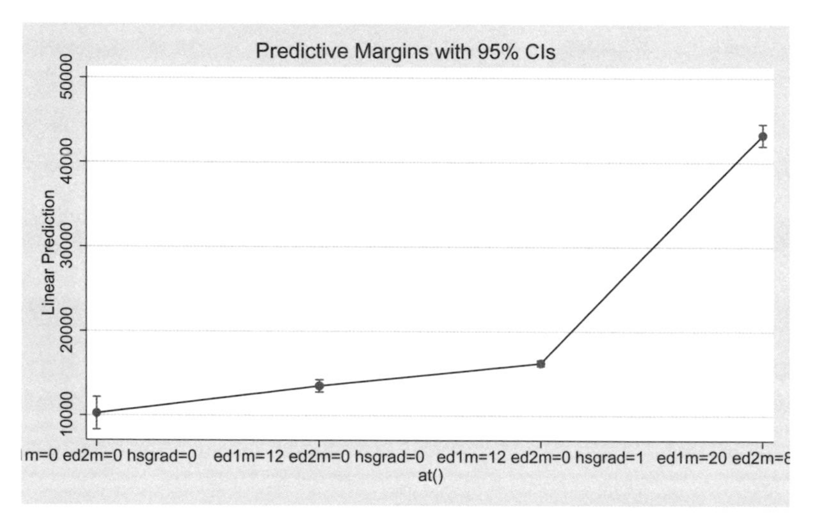

Figure 4.19. Graph of adjusted means from a piecewise regression using the `marginsplot` command

The `marginsplot` command graphs each of the four adjusted means in the order in which they were computed by the `margins` command. Although it is hard to read, the x axis is labeled with the values of `ed1m`, `ed2m`, and `hsgrad` for each adjusted mean. Although this is logical behavior for the `marginsplot` command, it does not produce the graph that we wanted. The problem is that the specification of the `margins` command obscures the true meaning of what we want.

Unfortunately, this is a rare case where the `marginsplot` command does not create the graph we want; thus, we need to manually create the graph ourselves. The process of creating the graph for this example was illustrated in section 4.5.2. The steps illustrated in that section involved manually retyping the adjusted means, which was simple to understand but not efficient. The following section shows a more complex but more efficient method for graphing adjusted means from a piecewise model.

4.10 Automating graphs of piecewise models

So far, I have illustrated how to graph adjusted means by using the `margins` command and then retyping those values into a dataset. This is simple to understand, but if you change your data (for example, fix incorrect values), the adjusted means will not be automatically updated to reflect the new data. This section illustrates a more efficient, but trickier, way of creating such graphs that does not require retyping the data.

First, let's use the `gss_ivrm.dta` dataset and run the analysis that was shown in section 4.6 in which we fit a piecewise model with two knots that specified a change in both slope and intercept. This is followed by the `margins` command that computes adjusted means at key values of education. The output of these commands is omitted to save space.

```
. use gss_ivrm
. mkspline ed1m 12 ed2m 16 ed3m = educ, marginal
. regress realrinc ed1m ed2m ed3m hsgrad cograd female, vce(robust)
  (output omitted)
. margins, at(ed1m=0  ed2m=0 ed3m=0 hsgrad=0 cograd=0)
>          at(ed1m=1  ed2m=0 ed3m=0 hsgrad=0 cograd=0)
>          at(ed1m=12 ed2m=0 ed3m=0 hsgrad=0 cograd=0)
>          at(ed1m=12 ed2m=0 ed3m=0 hsgrad=1 cograd=0)
>          at(ed1m=13 ed2m=1 ed3m=0 hsgrad=1 cograd=0)
>          at(ed1m=16 ed2m=4 ed3m=0 hsgrad=1 cograd=0)
>          at(ed1m=16 ed2m=4 ed3m=0 hsgrad=1 cograd=1)
>          at(ed1m=17 ed2m=5 ed3m=1 hsgrad=1 cograd=1)
>          at(ed1m=20 ed2m=8 ed3m=4 hsgrad=1 cograd=1)
  (output omitted)
```

The following steps save the adjusted means from the `margins` command, as well as the corresponding values of education into the active dataset. The `graph` command is then used to graph the adjusted means by education, creating the graph shown in figure 4.20.

```
. matrix yhat = r(b)'
. svmat yhat
. matrix educ = (0 \ 1 \ 12 \ 12 \ 13 \ 16 \ 16 \ 17 \ 20)
. svmat educ
. graph twoway line yhat1 educ1, xline(12 16)
> xtitle(Education) ytitle(Adjusted mean)
```

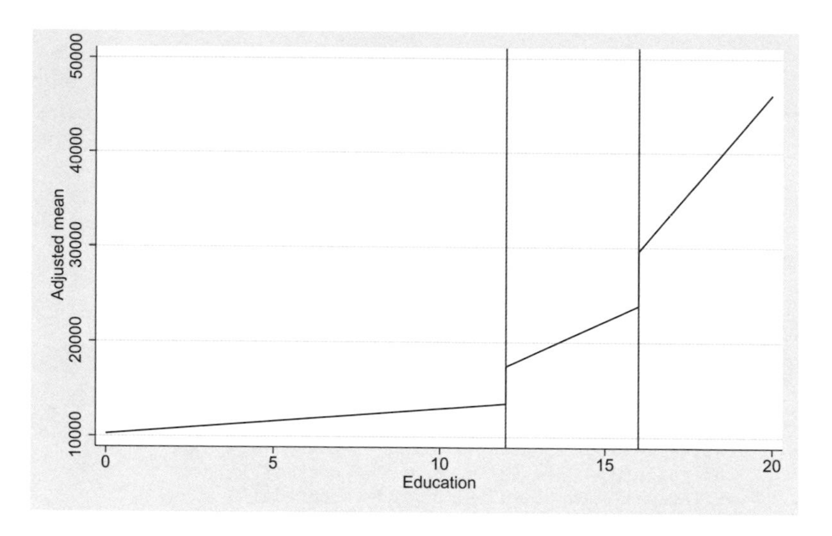

Figure 4.20. Adjusted means from piecewise regression with two knots and two jumps

Let's walk through this process again, but do so more slowly. First, repeat the `use gss_ivrm` command, as well as the `mkspline`, `regress`, and `margins` commands from above. Now, the adjusted means computed by the `margins` command are stored in a matrix named `r(b)` with one row and nine columns, corresponding to the nine values specified with the `at()` option. We can list this matrix, as shown below.

```
. matrix list r(b)
r(b)[1,9]
             1.          2.          3.          4.          5.          6.          7.
            _at         _at         _at         _at         _at         _at         _at
r1   10258.623   10530.894   13525.881   17451.464    19053.55    23859.81   29600.571
             8.          9.
            _at         _at
r1   33742.824   46169.582
```

Let's store these adjusted means as a matrix called `yhat` and in the process transpose the matrix (converting the columns to rows). The `matrix list` command shows the adjusted means from the matrix named `yhat`. This matrix has nine rows and one column.

```
. matrix yhat = r(b)'
. matrix list yhat
yhat[9,1]
                r1
1._at  10258.623
2._at  10530.894
3._at  13525.881
4._at  17451.464
5._at   19053.55
6._at   23859.81
7._at  29600.571
8._at  33742.824
9._at  46169.582
```

We can then save `yhat` into the current dataset with the `svmat` command. The result is that the first nine observations contain the values from `yhat`, stored in the variable named `yhat1`. The remaining observations in the dataset contain missing values. We can see this below, listing the first 10 observations.

```
. svmat yhat
. list yhat1 in 1/10

     +----------+
     |    yhat1 |
     |----------|
  1. | 10258.62 |
  2. | 10530.89 |
  3. | 13525.88 |
  4. | 17451.46 |
  5. | 19053.55 |
     |----------|
  6. | 23859.81 |
  7. | 29600.57 |
  8. | 33742.82 |
  9. | 46169.58 |
 10. |        . |
     +----------+
```

Now, let's make a matrix containing the values of education. This is stored in the matrix named `educ`. We then save this matrix into the dataset, as shown below.

```
. matrix educ = (0 \ 1 \ 12 \ 12 \ 13 \ 16 \ 16 \ 17 \ 20)
. svmat educ
. list yhat1 educ1 in 1/10

     +-------------------+
     |    yhat1    educ1 |
     |-------------------|
  1. | 10258.62        0 |
  2. | 10530.89        1 |
  3. | 13525.88       12 |
  4. | 17451.46       12 |
  5. | 19053.55       13 |
     |-------------------|
  6. | 23859.81       16 |
  7. | 29600.57       16 |
  8. | 33742.82       17 |
  9. | 46169.58       20 |
 10. |        .        . |
     +-------------------+
```

Now, we can graph the adjusted means, called `yhat1`, by the levels of education, called `educ1`, as shown below.

```
. graph twoway line yhat1 educ1, xline(12 16)
> xtitle(Education) ytitle(Adjusted mean)
  (output omitted)
```

Although the process of creating this graph is more complicated, the benefit is that it will automatically be updated if the dataset changes. This can be a little more work in the short run but saves us time in the long run.

4.11 Summary

This chapter has covered a variety of piecewise models you can use for fitting a nonlinear relationship between a continuous predictor and continuous outcome. This included piecewise models permitting a change of slope (with one and with multiple knots), piecewise models permitting a change of slope and intercept (with one and with two knots), and piecewise models with one unknown knot. This chapter also illustrated how to fit a piecewise model with multiple unknown knots. The chapter concluded with an example showing how the process of graphing piecewise models can be automated.

For more information about piecewise regression models, I recommend Marsh and Cormier (2002). You can also see Panis (1994) for a piecewise regression example using logistic regression.

5 Continuous by continuous interactions

5.1 Chapter overview

This chapter illustrates how to interpret interactions of two continuous predictors. Section 5.2 covers models involving the interaction of two linear continuous predictors. Then, section 5.3 covers models involving a linear continuous predictor interacted with a quadratic continuous predictor.

5.2 Linear by linear interactions

5.2.1 Overview

A linear by linear interaction implies that the slope of the relationship between one of the predictors and the outcome changes as a linear function of the other predictor. Suppose that `realrinc` is the outcome, and `age` and `educ` are the predictors of interest. An interaction of `age` and `educ` describes the degree to which the slope of one predictor changes as a linear function of the other predictor. The interaction can be described as the degree to which the `age` slope changes as a function of education. It can also be described as the degree to which the `educ` slope changes as a function of age. I think this is difficult to explain in the abstract, so I will illustrate this using a hypothetical example. I will begin with a hypothetical example in which there is no linear by linear interaction, and then contrast that with an example that includes such an interaction.

Let's begin with a hypothetical example in which we predict income (`realrinc`) from age (`age`) and education (`educ`) without an interaction. The regression equation for this hypothetical example is shown in (5.1).

$$\widehat{\texttt{realrinc}} = -41300 + 600\texttt{age} + 3000\texttt{educ} \tag{5.1}$$

The fitted values are shown using a three-dimensional graph in figure 5.1.[1]

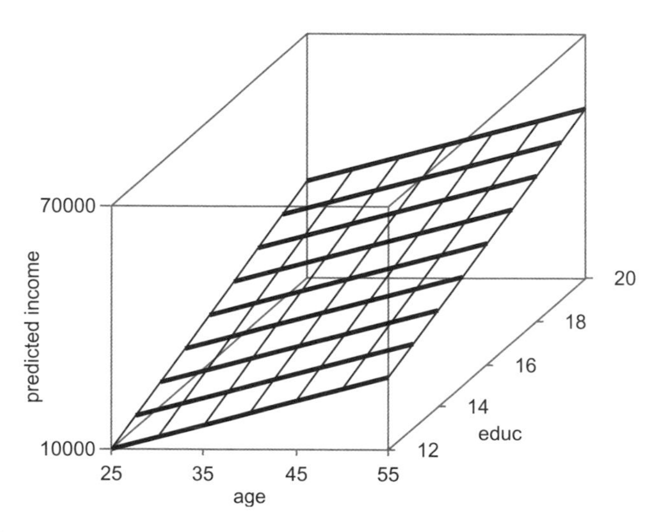

Figure 5.1. Three-dimensional graph of fitted values from model without an interaction

The fitted values in figure 5.1 form a flat regression plane. The coefficients from (5.1) describe the slope of the plane with respect to age and education. The slope of the plane in the direction of age is 600 (regardless of the value of education), and the slope of the plane in the direction of education is 3,000 (regardless of the value of age).

Let's now consider a second hypothetical regression model that contains an interaction of age (`age`) and education (`educ`). The regression equation for this second model is shown in (5.2).

$$\widehat{\texttt{realrinc}} = 24000 + -1100\texttt{age} - 1600\texttt{educ} + 120\texttt{age*educ} \tag{5.2}$$

1. These three-dimensional graphs were created using a modified version of the `surface` command written by Adrian Mander (2011).

Let's ignore the coefficients associated with age and education and focus on the coefficient for the interaction. The interaction term is 120. To help understand this interaction term, let's visualize the fitted values using a three-dimensional graph (see figure 5.2). Note how the regression plane in figure 5.2 is no longer flat due to the inclusion of the interaction term. The plane looks like a piece of paper that has a twist.

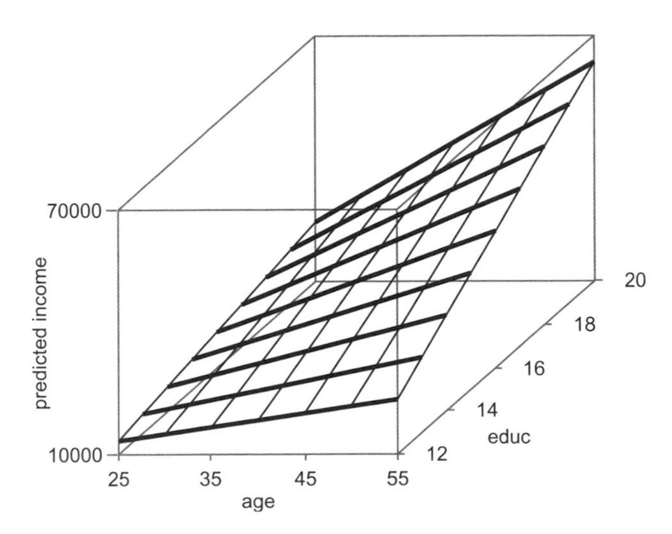

Figure 5.2. Three-dimensional graph of fitted values from model with an interaction

Three-dimensional graphs provide the most comprehensive visualization of these regression models, but I feel that two-dimensional graphs can be easier to understand, especially for precisely visualizing how the interaction changes the slope associated with each predictor. Let's transition to the use of two-dimensional graphs for illustrating the precise role of linear by linear interactions. I have created a two-dimensional version of figures 5.1 and 5.2, shown in figure 5.3. The left panel shows a two-dimensional representation of the model without an interaction, and the right panel shows a two-dimensional representation of the model with an interaction.

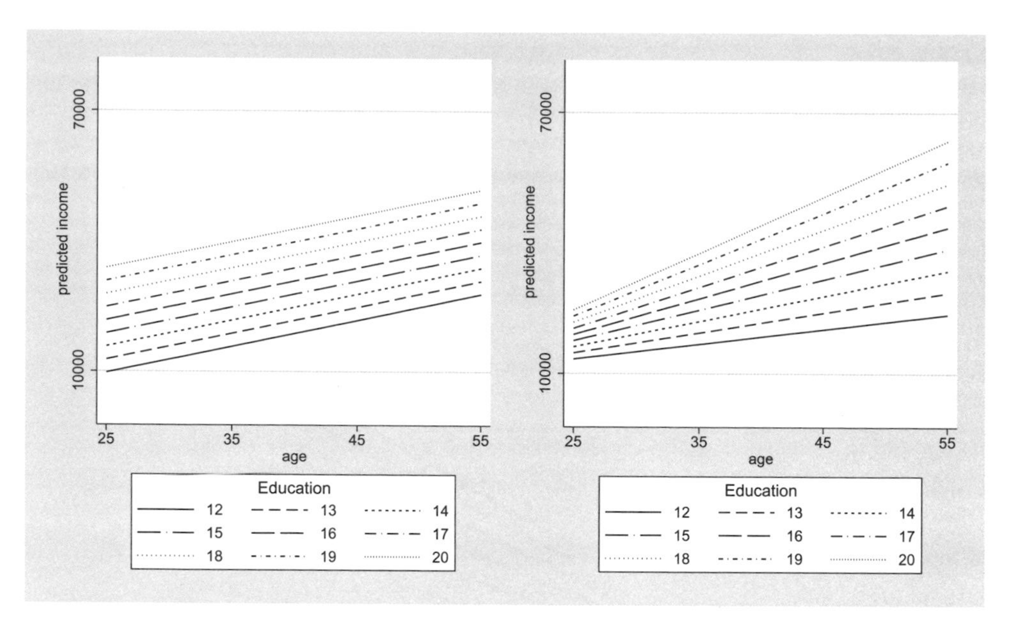

Figure 5.3. Two-dimensional graph of fitted values from model without an interaction (left panel) and with an interaction (right panel)

The left panel of figure 5.3 clearly shows that the slope of the relationship between income and age is the same at each level of education. From (5.1), we can see that the `age` slope is 600. By contrast, the right panel of figure 5.3 clearly shows how, in the presence of an interaction, the `age` slope grows with increasing values of education.[2] The `age` slope for each of the lines depicted in the right panel of figure 5.3 is shown in table 5.1.

Note! Describing slopes

The graph in the right panel of figure 5.3 shows that the slope of the relationship between income and age changes as a function of education. Instead of referring to the slope of the relationship between income and age, I will refer to this as the `age` slope. Others might refer to this as the *marginal effect of age* or the *age effect*.

2. For a real analysis with real data, the `margins` command could easily estimate the `age` slope at each level of education. This will be illustrated later in this chapter.

Table 5.1. The `age` slope at each level of education

Education	12	13	14	15	16	17	18	19	20
`age` slope	340	460	580	700	820	940	1060	1180	1300

For those with 12 years of education, the `age` slope is 340. The `age` slope increases to 460 for those with 13 years of education, and the `age` slope increases to 580 for those with 14 years of education. Note how the `age` slope increases by 120 units for every one-unit increase in education. This is due to the age by education interaction term. As shown in (5.2), the age by education interaction is 120. This interaction can be described as the degree to which the `age` slope increases for every one-unit increase in education.

The interaction can also be described as the degree to which the `educ` slope changes as a linear function of age. We can visualize the interaction focusing on the slope of the relationship between income and education, as shown in figure 5.4. This graph shows the fitted values of predicted income on the y axis and education on the x axis, using separate lines for ages 25, 35, 45, and 55.

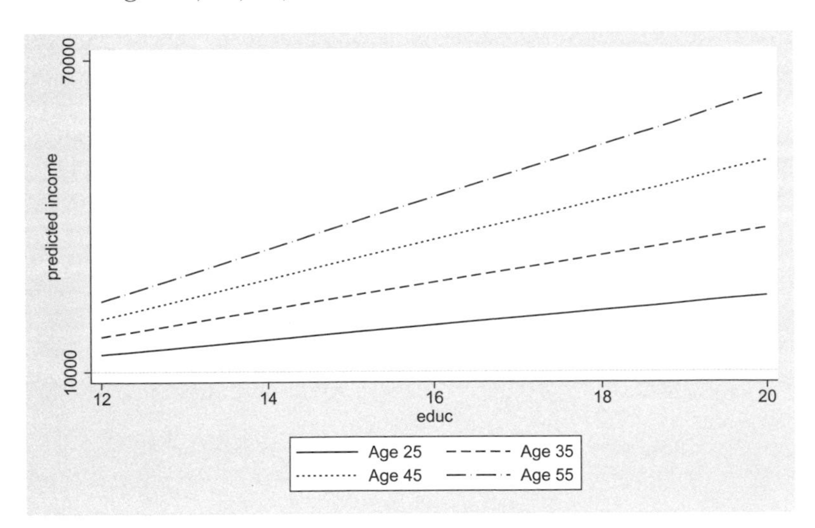

Figure 5.4. Two-dimensional graph of an interaction, focusing on `educ` slope

Note how the `educ` slope increases as a function of age. For this hypothetical example, the `educ` slope for each of the lines from figure 5.4 is shown in table 5.2.

Table 5.2. The educ slope at 25, 35, 45, and 55 years of age

Age	25	35	45	55
educ slope	1400	2600	3800	5000

The educ slope is 1,400 for 25-year-olds, 2,600 for 35-year-olds, and 3,800 for 45-year-olds. Note how the educ slope increases by 1,200 units for every 10-year increase in age. This corresponds to a 120-unit increase for every one-year increase in age, which matches the coefficient for the age by education interaction. This shows another way to interpret the age by education interaction. It shows the degree to which the educ slope changes for a one-year increase in age.

5.2.2 Example using GSS data

Let's now consider an example that includes two continuous predictors, as well as the interaction of the two continuous predictors. This will be illustrated using the GSS dataset predicting realrinc from the continuous variables age and educ, as well as the interaction of these two variables. The model illustrated in this section assumes that the relationship between the predictor and outcome is linear. Actually, both age and educ exhibit considerable nonlinearity in the prediction of realrinc. However, if we focus on those aged 22 to 55, the relationship between age and income is reasonably linear. Likewise, for those with 12 or more years of education, educ is linearly related to realrinc. So for the examples in this section, we will use the gss_ivrm.dta dataset and focus on those who are 22 to 55 years old and have 12 or more years of education.

```
. use gss_ivrm
. keep if (age>=22 & age<=55) & (educ>=12)
(25,024 observations deleted)
```

Let's fit a model where we predict realrinc from educ, age, and the interaction of these two variables.

```
. regress realrinc c.educ##c.age female, vce(robust)
Linear regression                               Number of obs     =     22,367
                                                F(4, 22362)       =     664.04
                                                Prob > F          =     0.0000
                                                R-squared         =     0.1712
                                                Root MSE          =      24677

------------------------------------------------------------------------------
             |               Robust
    realrinc |      Coef.   Std. Err.      t    P>|t|     [95% Conf. Interval]
-------------+----------------------------------------------------------------
        educ |  -1487.578   339.5978    -4.38   0.000    -2153.214   -821.9428
         age |  -1036.792   126.2278    -8.21   0.000    -1284.207   -789.3764
             |
c.educ#c.age |   114.9126   9.382337    12.25   0.000     96.52259    133.3027
             |
      female |  -12553.79    328.284   -38.24   0.000    -13197.25   -11910.33
       _cons |   28738.43   4550.477     6.32   0.000     19819.18    37657.69
------------------------------------------------------------------------------
```

The `c.educ#c.age` interaction is significant. This interaction can be interpreted by focusing on the `age` slope or by focusing on the `educ` slope. The next section, section 5.2.3, interprets the interaction by focusing on the `age` slope, followed by section 5.2.4, which interprets the interaction by focusing on the `educ` slope.

5.2.3 Interpreting the interaction in terms of age

We can use the `margins` and `marginsplot` commands to visualize this interaction. Let's visualize this interaction by graphing the adjusted means with `age` on the x axis and with separate lines for each level of `educ`. The `margins` command is used with the `at()` option to compute the adjusted means for ages 22 and 55 and educations ranging from 12 to 20 in two-year increments.

```
. margins, at(age=(22 55) educ=(12(2)20))
  (output omitted)
```

Then, the `marginsplot` command is used to graph the adjusted means with `age` on the x axis and with separate lines for each level of `educ`. The `legend()` option is included to customize the display of the graph legend. The resulting graph is shown in figure 5.5.

```
. marginsplot, noci legend(rows(2) title(Education))
  Variables that uniquely identify margins: age educ
```

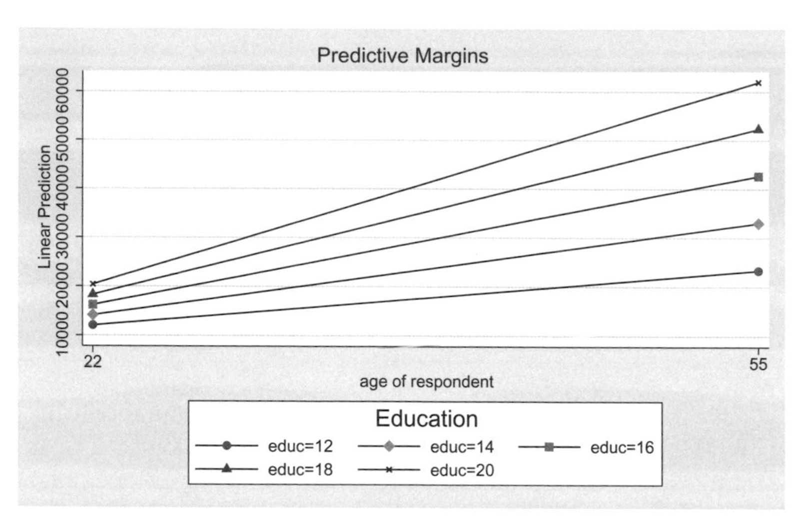

Figure 5.5. Adjusted means for a linear by linear interaction with age on the x axis

Note! Choosing the x-axis dimension using the marginsplot command

The `marginsplot` command chooses the variable that is placed on the x axis and the variable that is represented using separate lines. In figure 5.5, the variable `age` was placed on the x axis and the variable `educ` was represented using separate lines. Had the `marginsplot` command placed `educ` onto the x axis, you could specify the `xdimension(age)` option, which can be abbreviated as `x(age)`, to place `age` on the x axis. Appendix C provides more details about options you can use with the `marginsplot` command.

Note how the `age` slope increases with increasing values of `educ`. The `age` slope for those with 14 years of education is greater than the `age` slope for those with 12 years of education. In fact, we can use the `margins` command combined with the `dydx(age)` option to estimate the slope for each of the lines displayed in figure 5.5.

```
. margins, at(educ=(12(2)20)) dydx(age) vsquish
Average marginal effects                        Number of obs     =     22,367
Model VCE    : Robust

Expression   : Linear prediction, predict()
dy/dx w.r.t. : age
1._at        : educ            =          12
2._at        : educ            =          14
3._at        : educ            =          16
4._at        : educ            =          18
5._at        : educ            =          20

------------------------------------------------------------------------------
             |            Delta-method
             |      dy/dx   Std. Err.      t    P>|t|     [95% Conf. Interval]
-------------+----------------------------------------------------------------
age          |
         _at |
          1  |   342.1599   19.78151    17.30   0.000     303.3868    380.9331
          2  |   571.9852   16.30163    35.09   0.000     540.0329    603.9375
          3  |   801.8104   29.05863    27.59   0.000     744.8535    858.7674
          4  |   1031.636   46.12256    22.37   0.000     941.2322    1122.039
          5  |   1261.461   64.14361    19.67   0.000     1135.735    1387.187
------------------------------------------------------------------------------
```

The estimates of the **age** slope increase as a function of **educ**. For example, at 12 years of education, the **age** slope is 342.16, and at 14 years of education, the **age** slope is 571.99. For a two-unit increase in education, the **age** slope increases by 229.83 (571.99−342.16). We can relate this to the coefficient for the `c.educ#c.age` interaction, which is the amount by which the **age** slope changes for every one-year increase in **educ**. For every one-year increase in **educ**, the **age** slope increases by 114.91.

5.2.4 Interpreting the interaction in terms of education

Now, let's explore the meaning of the interaction by focusing on the **educ** slope. Let's visualize this by creating a graph of the adjusted means showing **educ** on the x axis and separate lines for **age**. First, the `margins` command is used to create adjusted means for 12 and 20 years of education and ages ranging from 25 to 55 in 10-year increments.

```
. margins, at(educ=(12 20) age=(25(10)55))
  (output omitted)
```

Second, the `marginsplot` command is used to graph these adjusted means with **educ** on the x axis and with separate lines for **age**. The resulting graph is shown in figure 5.6.

```
. marginsplot, noci legend(title(Age))
  Variables that uniquely identify margins: educ age
```

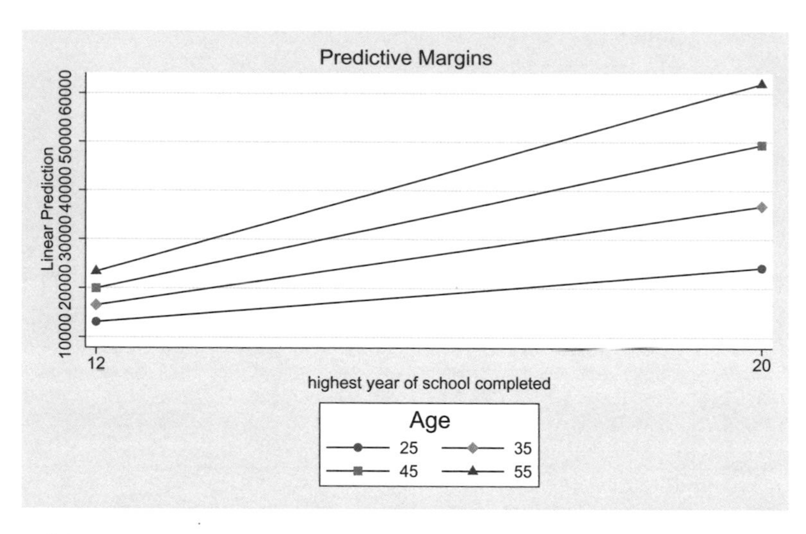

Figure 5.6. Adjusted means for a linear by linear interaction with education on the x axis

Note how the slope of the lines increase with increasing values of age. In fact, we can estimate the slope of each line (that is, the educ slope) using the margins command, as shown below.

```
. margins, at(age=(25(10)55)) dydx(educ) vsquish
Average marginal effects                        Number of obs     =     22,367
Model VCE      : Robust

Expression     : Linear prediction, predict()
dy/dx w.r.t.   : educ
1._at          : age             =          25
2._at          : age             =          35
3._at          : age             =          45
4._at          : age             =          55

------------------------------------------------------------------------------
             |            Delta-method
             |      dy/dx   Std. Err.      t    P>|t|     [95% Conf. Interval]
-------------+----------------------------------------------------------------
educ         |
         _at |
          1  |   1385.237    129.777    10.67   0.000     1130.865     1639.61
          2  |   2534.364   90.87424    27.89   0.000     2356.244    2712.484
          3  |    3683.49   131.4528    28.02   0.000     3425.833    3941.147
          4  |   4832.616   209.5404    23.06   0.000     4421.902     5243.33
------------------------------------------------------------------------------
```

The educ slope for a 25-year-old is 1,385.24. This corresponds to the slope of the line for 25-year-olds shown in figure 5.6. This means that for someone who is 25 years old, their income is expected to increase by $1,385.24 for every year of additional education. As age increases, so does the educ slope. For every one-unit increase in age, the educ slope increases by 114.91, the estimate of the age#educ interaction. For a 10-unit increase in age, the educ slope would be expected to increase by 1,149.12. We see this when comparing the age slope for a 35-year-old with a 25-year-old. The educ slope for a 35-year-old is 2,534.36 compared with 1,385.24 for a 25-year-old. The difference in these slopes is 1,149.12.

5.2.5 Interpreting the interaction in terms of age slope

We can visualize the age by educ interaction by illustrating the way that the age slope changes as a function of education. The margins command below includes the dydx(age) option to estimate the age slope at each level of educ.

```
. margins, dydx(age) at(educ=(12(1)20)) vsquish
Average marginal effects                        Number of obs     =     22,367
Model VCE      : Robust

Expression     : Linear prediction, predict()
dy/dx w.r.t.  : age
1._at          : educ            =          12
2._at          : educ            =          13
3._at          : educ            =          14
4._at          : educ            =          15
5._at          : educ            =          16
6._at          : educ            =          17
7._at          : educ            =          18
8._at          : educ            =          19
9._at          : educ            =          20

------------------------------------------------------------------------------
             |            Delta-method
             |      dy/dx   Std. Err.      t    P>|t|     [95% Conf. Interval]
-------------+----------------------------------------------------------------
age          |
         _at |
          1  |   342.1599   19.78151    17.30   0.000     303.3868    380.9331
          2  |   457.0726   15.50798    29.47   0.000     426.6758    487.4693
          3  |   571.9852   16.30163    35.09   0.000     540.0329    603.9375
          4  |   686.8978   21.61123    31.78   0.000     644.5383    729.2573
          5  |   801.8104   29.05863    27.59   0.000     744.8535    858.7674
          6  |   916.7231   37.38742    24.52   0.000     843.4411     990.005
          7  |   1031.636   46.12256    22.37   0.000     941.2322    1122.039
          8  |   1146.548   55.07103    20.82   0.000     1038.605    1254.491
          9  |   1261.461   64.14361    19.67   0.000     1135.735    1387.187
------------------------------------------------------------------------------
```

This shows that the age slope increases as a function of education. In fact, the age slope increases by 114.91 units for every one-unit increase in educ. We can visualize these age slopes as a function of education using the marginsplot command (shown below). This creates the graph shown in figure 5.7, which visually depicts the age slope increasing linearly as a function of education. The graph also includes the confidence interval with respect to each age slope, illustrating that the age slope is significant at every level of education.[3]

```
. marginsplot
Variables that uniquely identify margins: educ
```

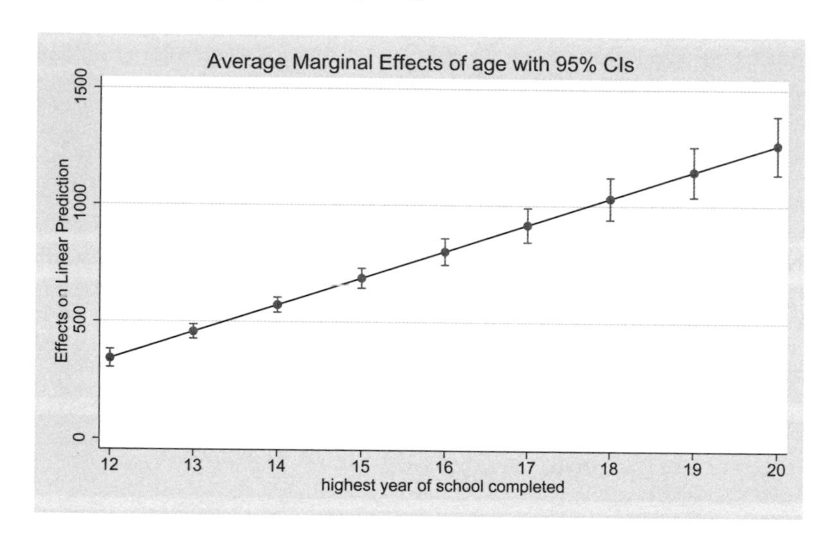

Figure 5.7. age slope as a function of education

5.2.6 Interpreting the interaction in terms of the educ slope

We can visualize the age by educ interaction by focusing on the way that the educ slope changes as a function of age. The margins command below estimates the educ slope for ages ranging from 25 to 55 in five-year increments.

3. Each age slope is significant because each confidence interval excludes zero.

```
. margins, dydx(educ) at(age=(25(5)55)) vsquish
Average marginal effects                    Number of obs     =     22,367
Model VCE    : Robust

Expression   : Linear prediction, predict()
dy/dx w.r.t. : educ
1._at        : age             =          25
2._at        : age             =          30
3._at        : age             =          35
4._at        : age             =          40
5._at        : age             =          45
6._at        : age             =          50
7._at        : age             =          55

------------------------------------------------------------------------------
             |            Delta-method
             |      dy/dx   Std. Err.      t    P>|t|     [95% Conf. Interval]
-------------+----------------------------------------------------------------
educ         |
         _at |
          1  |   1385.237    129.777    10.67   0.000     1130.865     1639.61
          2  |   1959.801    101.732    19.26   0.000     1760.399    2159.202
          3  |   2534.364   90.87424    27.89   0.000     2356.244    2712.484
          4  |   3108.927   102.8021    30.24   0.000     2907.428    3310.426
          5  |    3683.49   131.4528    28.02   0.000     3425.833    3941.147
          6  |   4258.053   168.5017    25.27   0.000     3927.778    4588.328
          7  |   4832.616   209.5404    23.06   0.000     4421.902     5243.33
------------------------------------------------------------------------------
```

This shows that the `educ` slope increases as a function of age. For each five-year increase in `age`, the `educ` slope increases by 574.56 (that is, the `age#educ` coefficient multiplied by five, 114.91×5). For example, as `age` increases from 25 to 30, the `educ` slope increases by 574.56 (1959.80−1385.24). We can visualize the `educ` slopes estimated by the `margins` command using the `marginsplot` command, shown in figure 5.8.

```
. marginsplot, noci
  Variables that uniquely identify margins: age
```

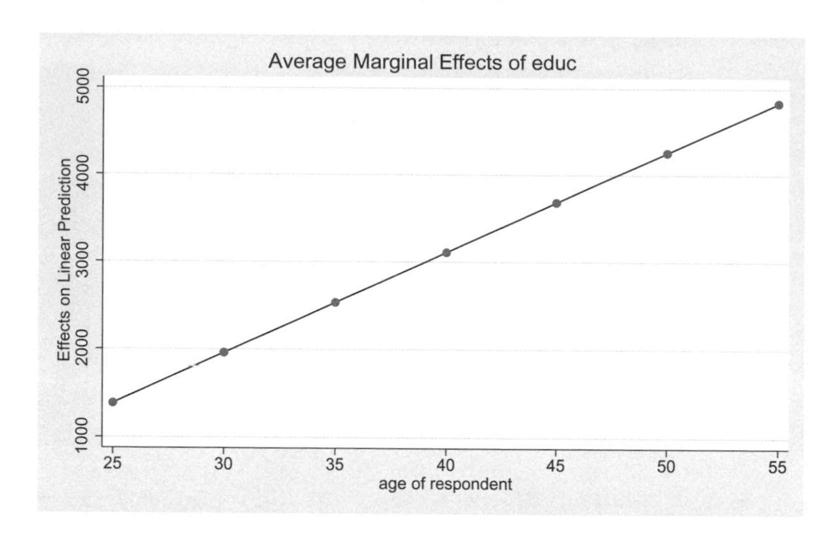

Figure 5.8. `educ` slope as a function of age

Figure 5.8 shows that the `educ` slopes increases linearly as a function of age.

5.3 Linear by quadratic interactions

This section considers models with an interaction of two continuous variables, where one of the variables is fit linearly and the other is fit using a quadratic term. This combines ideas found in the previous section on interaction of two continuous variables with section 3.2 on quadratic predictors.

5.3.1 Overview

A quadratic term introduces curvature in the fitted relationship between the predictor and outcome. As we saw in section 3.2.1, this curvature can be convex (U-shape) or concave (inverted U-shape). By interacting a linear variable with a quadratic variable, the quadratic term can change linearly as a function of the linear term.

Let's consider a hypothetical example that illustrates a greater degree of curvature with increasing values of the linear predictor. Let's first illustrate a model that contains a quadratic predictor and a linear predictor but no interaction between these predictors. This hypothetical example uses `realrinc` as the outcome variable, `age` as the quadratic predictor, and `educ` as the linear predictor. The regression equation for this hypothetical example is shown in (5.3).

$$\widehat{\texttt{realrinc}} = -70000 + 2100\texttt{age} + -20\texttt{age}^2 + 3000\texttt{educ} \tag{5.3}$$

The fitted values from this model are visualized using a three-dimensional graph in figure 5.9.

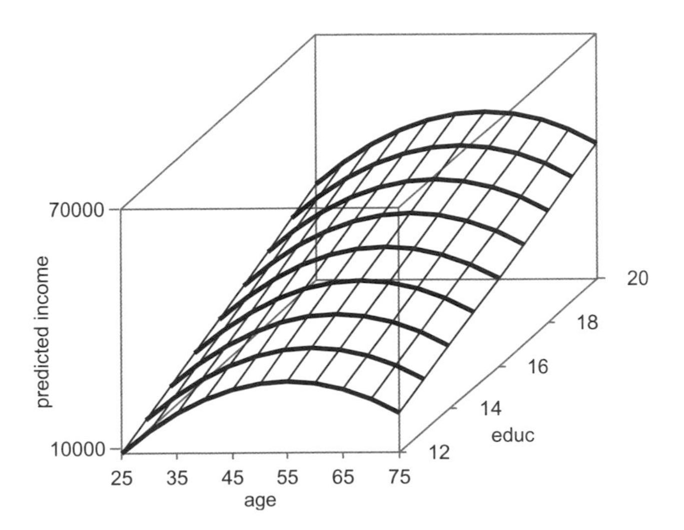

Figure 5.9. Three-dimensional graph of fitted values for linear and quadratic models without an interaction

Focusing on age in figure 5.9, we can see how incomes rise until about age 50 and then diminish thereafter. As we saw in section 3.2.1, such inverted U-shaped relationships arise from a negative coefficient for the quadratic term. The quadratic term from (5.3) is indeed negative. Note how the degree of curvature in age is constant across the levels of education.

Let's now consider a model that includes an interaction between `age` (as a quadratic term) and `educ`. This regression equation is shown in (5.4).

$$\begin{aligned}\widehat{\texttt{realrinc}} =& 165000 + -8600\texttt{age} + 95\texttt{age}^2 + -14000\texttt{educ} \\ &+ 780\texttt{age*educ} + -8\texttt{age}^2\texttt{*educ}\end{aligned} \tag{5.4}$$

This model includes age, age squared, education, age interacted with education, and age squared interacted with education. The key term is the age squared by education interaction. This governs the degree of the curvature in age as a function of education. The fitted values from this regression equation are graphed in figure 5.10.

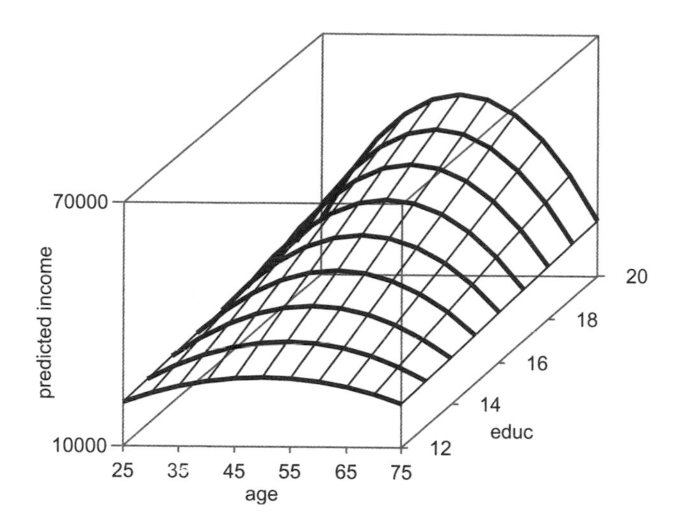

Figure 5.10. Three-dimensional graph of fitted values for linear and quadratic models with an interaction

Overall, this figure looks a little bit like a hill. Let's focus on the `age` slope at each level of education. At 12 years of education, the relationship between age and income looks almost linear. At 20 years of education, the relationship between age and income has a considerable inverted U-shape. As education increases from 12 to 20 years, the degree of the inverted U-shape grows as a function of education. This is due to the interaction of education with age squared. Note that this term is -8. For every one-unit increase in education, the quadratic term for age changes by -8. When quadratic terms become more negative, the inverted U-shape becomes more pronounced. Thus for every one-unit increase in education, the relationship between income and age shows more of an inverted U-shape.

It can be difficult to visualize these relationships using three-dimensional graphs, so let's show how to visualize them using two dimensions.

The left panel of figure 5.11 shows a two-dimensional representation of figure 5.9, where there was no interaction. We can clearly see how the degree of the curvature in relationship between income and age is the same across the levels of education. The right panel of figure 5.11 shows a two-dimensional representation of figure 5.10, where there was an interaction of education with the quadratic term for age. We can see how the relationship between income and age is nearly linear for 12 years of education and grows more concave with increasing levels of education.

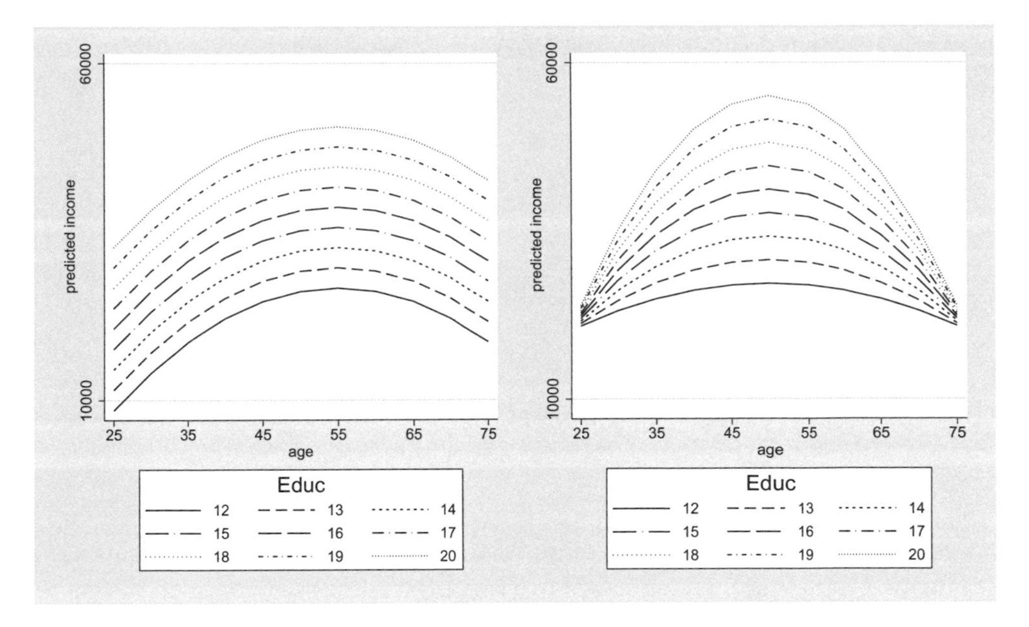

Figure 5.11. Two-dimensional graph of fitted values for linear and quadratic models without an interaction (left panel) and with linear by quadratic interaction (right panel)

5.3.2 Example using GSS data

Let's use the GSS dataset to fit a model that predicts `realrinc` from `educ` (treated as a continuous linear variable) interacted with `age` (treated as a continuous variable fit using a quadratic term). The examples in this section will focus on people with 12 to 20 years of education who are 22 to 80 years old.[4] Let's use the `gss_ivrm.dta` dataset, focusing only on those aged 22 to 80 with 12 or more years of education.

```
. use gss_ivrm
. keep if (age>=22 & age<=80) & (educ>=12)
(15,934 observations deleted)
```

Let's begin by fitting a model that predicts `realrinc` using `educ` as a linear term and `age` as a quadratic term but no interaction between education and age. This variable also includes `female` as an additional predictor, entered as a dummy variable.

4. As described in section 1.2.3, the relationship between income and education is linear for 12 to 20 years of education, and the relationship between income and age is quadratic for 18 to 80 years of age. Because we are interacting `age` with `educ`, the sample is further restricted to those aged 22 to 80 to avoid observations in the 18 to 21 year age range, because people at such a young age would not have had much of a chance to have any higher levels of education.

```
. regress realrinc c.educ c.age##c.age female, vce(robust)
Linear regression                               Number of obs     =     25,964
                                                F(4, 25959)       =     785.49
                                                Prob > F          =     0.0000
                                                R-squared         =     0.1658
                                                Root MSE          =      26078

------------------------------------------------------------------------------
             |               Robust
    realrinc |      Coef.   Std. Err.      t    P>|t|     [95% Conf. Interval]
-------------+----------------------------------------------------------------
        educ |   3071.265   93.45769    32.86   0.000     2888.083    3254.447
         age |   2245.623   79.05411    28.41   0.000     2090.673    2400.573
             |
 c.age#c.age |  -21.76865   .9428415   -23.09   0.000    -23.61667   -19.92063
             |
      female |  -13344.78    319.659   -41.75   0.000    -13971.33   -12718.23
       _cons |  -64839.08   2060.085   -31.47   0.000    -68876.96    -60801.2
------------------------------------------------------------------------------
```

The quadratic term is significant and is negative. This negative coefficient indicates that the relationship between age and income has an inverted U-shape. Let's use the **margins** and **marginsplot** commands to visualize the adjusted means of income as a function of age while holding education constant at 12 to 20 years of education (in two-year increments). We first compute the adjusted means as a function of age and education using the **margins** command. Then, the **marginsplot** command graphs the adjusted means on the y axis and **age** on the x axis, with separate lines for each level of education. This graph, shown in figure 5.12, shows that the degree of the curvature in the relationship between income and age is the same across levels of education.

```
. margins, at(age=(22(1)80) educ=(12(2)20))
  (output omitted)
. marginsplot, noci legend(rows(2)) recast(line) scheme(s1mono)
  Variables that uniquely identify margins: age educ
```

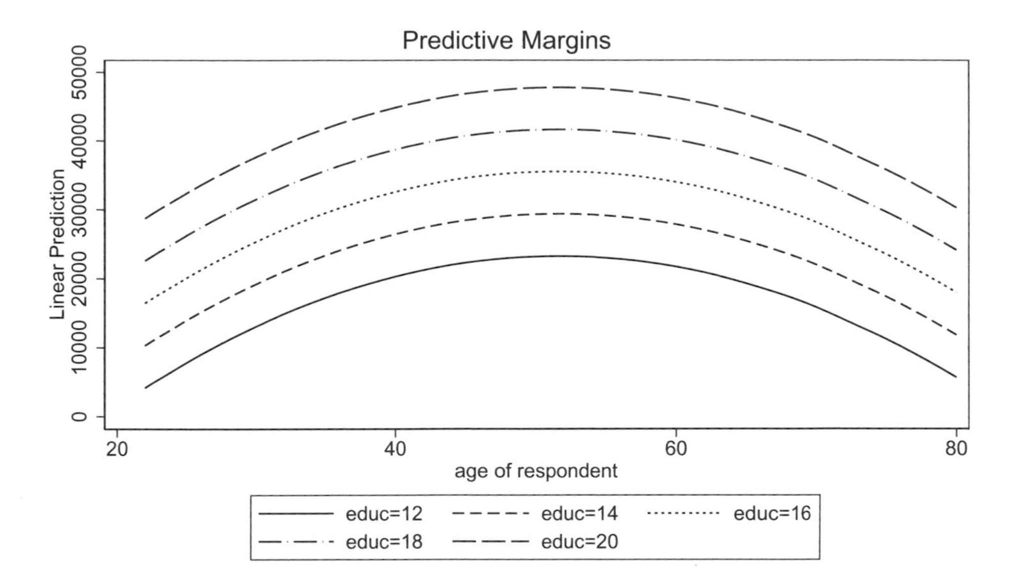

Figure 5.12. Adjusted means at 12, 14, 16, 18, and 20 years of education

Let's now fit a model that includes an interaction of `educ` with the quadratic term for `age`. This model is shown below. Note this model also includes `female` as a covariate.[5]

```
. regress realrinc c.educ##c.age##c.age female, vce(robust) noci
Linear regression                               Number of obs     =     25,964
                                                F(6, 25957)       =     558.27
                                                Prob > F          =     0.0000
                                                R-squared         =     0.1757
                                                Root MSE          =      25924

------------------------------------------------------------------------
                     |               Robust
            realrinc |      Coef.   Std. Err.      t    P>|t|
---------------------+--------------------------------------------------
                educ |  -9317.781   817.8192   -11.39   0.000
                 age |  -5009.255   551.9681    -9.08   0.000
                     |
        c.educ#c.age |   517.9477   41.56362    12.46   0.000
                     |
         c.age#c.age |   46.88394   6.386088     7.34   0.000
                     |
  c.educ#c.age#c.age |  -4.905396   .4819612   -10.18   0.000
                     |
              female |  -13160.33   317.9152   -41.40   0.000
               _cons |   108559.9   10897.04     9.96   0.000
------------------------------------------------------------------------
```

The interaction of `c.educ#c.age#c.age` is significant. Let's interpret this as a function of age by graphing the adjusted means on the y axis and `age` on the x axis, with separate lines to indicate the levels of education. The graph is shown in figure 5.13.

5. The `noci` option is included to make the output more readable for this example. See the callout titled *Using the* `noci` *option for clearer output* in section 2.5.1 for more details.

```
. margins, at(age=(22 25(5)80) educ=(12(2)20))
  (output omitted)
. marginsplot, plotdimension(, allsimple) legend(subtitle(Education)
> rows(2)) noci recast(line) scheme(s1mono)
  Variables that uniquely identify margins: age educ
```

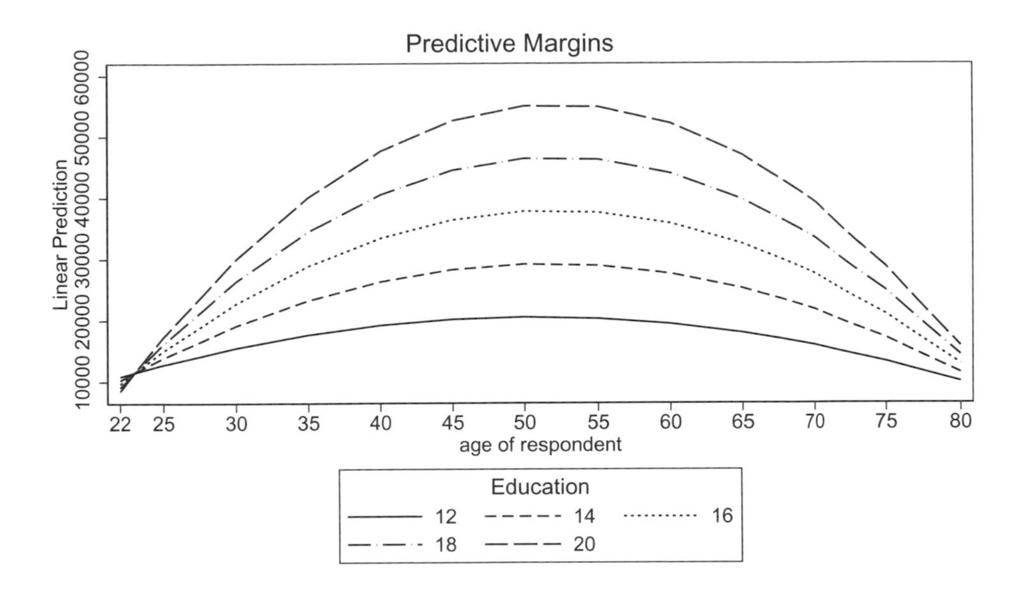

Figure 5.13. Adjusted means from linear by quadratic model

Figure 5.13 shows that the relationship between income and age has an inverted U-shape, but the degree of the inverted U-shape varies by the levels of education. As education increases, so does the degree of the inverted U-shape. People with 12 years of education show a mild inverted U-shape, and people with 20 years of education show a more inverted U-shape. In fact, the `c.educ#c.age#c.age` coefficient describes the degree to which the inverted U-shape changes as a function of education. This coefficient, which is −4.9, represents the change in the quadratic term for `age` for a one-unit increase in `educ`. As education increases by one unit, the quadratic term for `age` decreases by 4.9, creating a more inverted U-shape. The more negative a quadratic term is, the more it exhibits an inverted U-shape. Thus, increasing values of `educ` are associated with a greater inverted U-shape in the relationship between age and income. For those with lower educations, the relationship between age and income tends to be more linear, and for those with higher levels of education, the relationship between age and income is more curved, showing a greater rise and fall across ages.

5.4 Summary

This chapter illustrated the application of models involving interactions of two continuous predictors. Section 5.2 investigated the interaction of `educ` and `age`, finding an interaction between these two variables. This interaction was interpreted by focusing on how the `age` slope changes as a function of education. It was also interpreted by focusing on how the `educ` slope changes with increasing `age`. Section 5.3 illustrated a model involving a linear continuous predictor interacted with a quadratic continuous predictor. We saw that `age` (when treated as a quadratic term) interacted with `educ` in the prediction of income. This interaction showed that increasing education was associated with a greater curvature in the relationship between age and income.

For more information about continuous by continuous interactions, I recommend Aiken and West (1991) and Cohen et al. (2003). Also, West, Aiken, and Krull (1996) provide an excellent introduction and example of the use of such models.

6 Continuous by continuous by continuous interactions

6.1 Chapter overview

This chapter explores models that involve the interaction of three continuous linear predictors. This builds upon the foundation of what we learned about the interaction of two linear predictors in section 5.2. That section illustrated the prediction of `realrinc` from `age`, `educ`, and the interaction of those two variables. This chapter extends that example by adding a third continuous predictor, `yrborn` (the year the respondent was born). This chapter will illustrate the prediction of `realrinc` from the three-way interaction of `age`, `educ`, and `yrborn`. In exploring the three-way interaction, we will focus on the `age` by `educ` interaction, illustrating how the size of this interaction linearly increases as a function of `yrborn`.

6.2 Overview

Let's first consider a hypothetical example looking at income (adjusted for inflation) as a function of age, education, and year born. To illustrate the meaning of an interaction of these three continuous predictor, let's first consider a model that does not include a three-way interaction. We can then contrast such a model with one that includes a three-way interaction. Consider the regression equation shown in (6.1). This regression equation predicts `realrinc` from `age`, `educ`, `yrborn`, and the two-way interactions of these predictors (`age*educ`, `age*yrborn`, and `educ*yrborn`). Note this equation does not include a three-way interaction of `age*educ*yrborn`.

$$\widehat{\texttt{realrinc}} = 3365394 + -55747\texttt{age} + -66053\texttt{educ} + -1712\texttt{yrborn} \\ + 109\texttt{age*educ} + 28\texttt{age*yrborn} + 33.2\texttt{educ*yrborn} \quad (6.1)$$

The predicted values from this equation are visualized using a three-dimensional graph as shown in figure 6.1. The variable `age` is plotted on the x (horizontal) axis, `educ` is on the z axis (representing depth), and the predicted value of `realrinc` is on the y (vertical) axis. Separate graphs are used to represent `yrborn`. Reading from left to right and then top to bottom, the graphs represent the years of birth 1930, 1940, 1950, and 1960. The lines with respect to `age` are drawn thicker to help accentuate changes in the slope of the relationship between `age` and the outcome as a function of `educ` and `yrborn`.

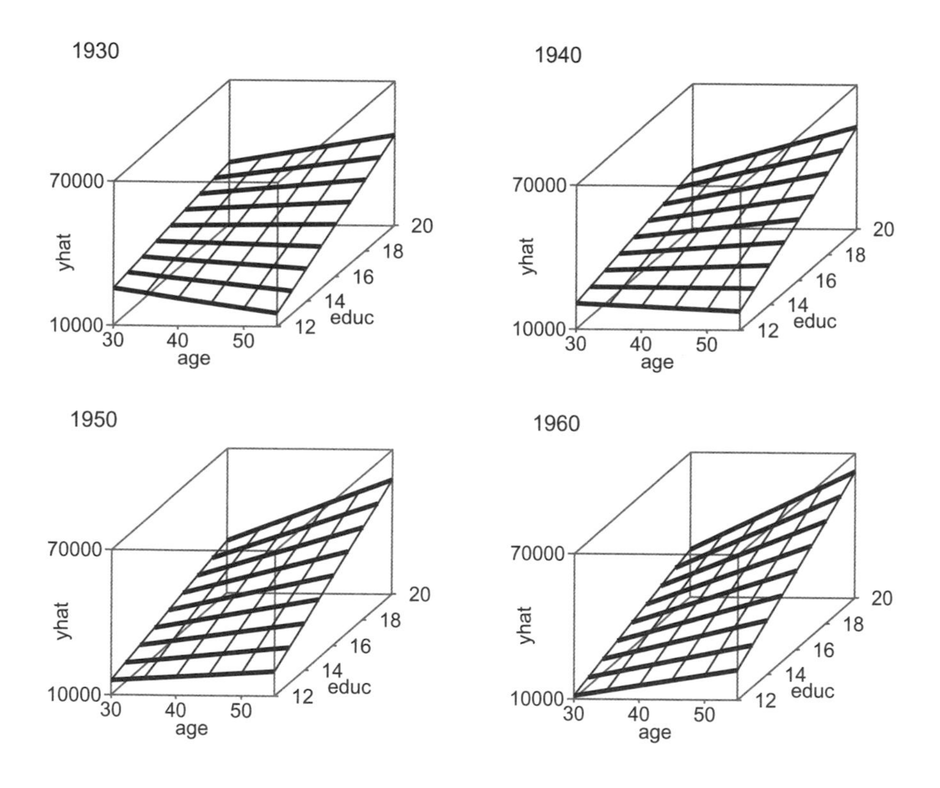

Figure 6.1. Three-dimensional graph of predicted values from model without the three-way interaction

Focusing on the top left graph, for those born in 1930, the `age` slope is moderately negative for those with 12 years of education, and it increases to being mildly positive for those with 20 years of education. For those born 10 years later (in 1940, in the top right graph), the age slope is mildly negative for those with 12 years of education and moderately positive for those with 20 years of education. Note the pattern that is emerging. The age slope grows more positive with increasing education. The same pattern is illustrated for those born in 1950 and 1960 (in the bottom left and bottom right panels, respectively). Although it might be hard to judge visually, all four graphs exhibit the same degree in the growth of the age slope as a function of educ. In other words, the size of the age*educ interaction is the same across the four graphs (across the levels of yrborn). This is because of the absence of a three-way interaction term. The three-way interaction term would be the factor that would permit the age*educ interaction to vary across the levels of yrborn.

So let's now consider a model that includes a three-way interaction. Consider the regression equation shown in (6.2). This regression model predicts realrinc from age, educ, yrborn, the two-way interactions of these predictors (age*educ, age*yrborn, and educ*yrborn), and the three-way interaction (age*educ*yrborn).

$$\begin{aligned}\widehat{\texttt{realrinc}} = &-2004225 + 66407\texttt{age} + 313720\texttt{educ} + 1042\texttt{yrborn}\\ &+ -8524\texttt{age*educ} + -34.7\texttt{age*yrborn}\\ &+ -161.62\texttt{educ*yrborn} + 4.43\texttt{age*educ*yrborn} \qquad (6.2)\end{aligned}$$

The predicted values from this model are visualized in figure 6.2, with age on the x axis, educ on the z axis, the predicted value of realrinc on the y axis. Separate graphs represent yrborn for the years of birth 1930, 1940, 1950, and 1960 shown from left to right and then top to bottom.

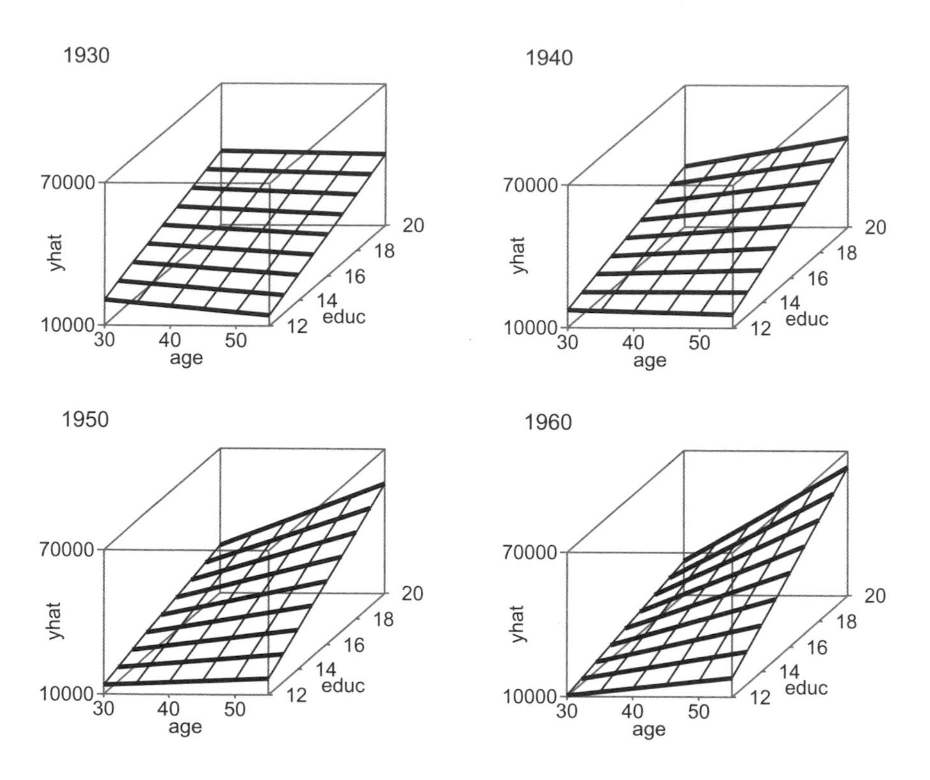

Figure 6.2. Three-dimensional graph of predicted values from model with the three-way interaction

Comparing figure 6.1 (without the three-way interaction) with figure 6.2 (with the three-way interaction), we can get an overall sense of the contribution of the three-way interaction. Let's begin to understand the three-way interaction by focusing on figure 6.2. For those born in 1930 (the top left panel), the `age` slope is mildly negative for those with 12 years of education and remains mildly negative across the different levels of education. Skipping forward to those born in 1960 (the bottom right panel), the `age` slope is mildly positive for those with 12 years of education and is sharply positive for those with 20 years of education. Looking across the panels from those born in 1930 (top left) to those born in 1960 (bottom right), we can see how the increase in the `age` slope due to higher levels of education grows as the year of birth increases from 1930 to 1960. For those born in 1930, the `age` slope remains largely the same for all education levels. By contrast, for those born in 1960, the `age` slope increases considerably with increasing levels of `educ`. In other words, the size of the `age*educ` interaction changes as a function of `yrborn`. This is a result of, and a way to describe, the interaction of `age*educ*yrborn`.

Three-dimensional graphs provide a useful mental model for visualizing the results. However, such graphs can be difficult to read in terms of clearly seeing the exact pattern of results or being able to ascertain the predicted values for a given set of predictors. Let's see how these models can be visualized using two-dimensional graphs. Let's visualize figure 6.2 (in which there is a three-way interaction) using a two-dimensional graph, as shown in figure 6.3. The variable `age` is plotted on the x axis and `yrborn` is shown as separate panels. The variable `educ` is depicted using separate lines ranging from 12 (shown at the bottom as a solid line) to 20 (shown at the top as dots). I could have included a legend for `educ` but omitted it to save space.

Focusing on those born in 1930, the slope of the relationship between income and age (that is, the `age` slope) is mildly negative for all levels of education. Jumping forward to those born in 1960, the `age` slope is mildly positive for those with 12 years of education (the solid line) and is sharply positive for those with 20 years of education (the dotted line at the top). In 1960, the `age` slope increases considerably with increasing education. We can compare how the `age` slope changes as a function of education for those born in 1930 with those born in 1960. For those born in 1930, the `age` slope remains mildly negative for all levels of education. By contrast, for those born in 1960, the `age` slope grows increasingly positive with increasing levels of education. As we saw in the three-dimensional graph, this two-dimensional graph illustrates how the `age*educ` interaction changes as a function of `yrborn`, which is another way of saying that there is an `age*educ*yrborn` interaction.

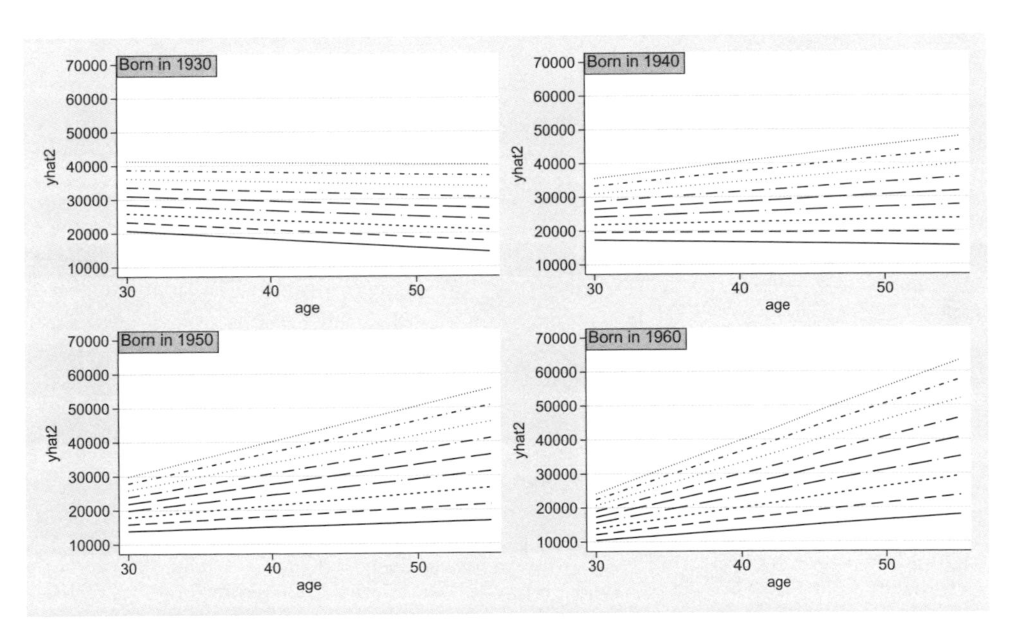

Figure 6.3. Two-dimensional graph of adjusted means for model with the three-way interaction

6.3 Examples using the GSS data

Let's now illustrate a model with an interaction of three continuous variables using the GSS dataset. This example predicts `realrinc` from `age`, `educ`, and `yrborn`. This example will build upon the example from section 5.2.2 that predicted `realrinc` from `age` and `educ`. That example focused on those aged 22 to 55 with 12 or more years of education. In this example, we are further introducing the year of birth (`yrborn`). Let's focus on those who were born from 1930 to 1960 and further restrict age to be between 30 and 55[1]

```
. use gss_ivrm
. keep if (age>=30 & age<=55) & (educ>=12) & (yrborn>=1930 & yrborn<=1960)
(39,526 observations deleted)
```

6.3.1 A model without a three-way interaction

As we did with the hypothetical example, let's begin with a regression model that predicts `realrinc` from `age`, `educ`, `yrborn`, and the two-way interactions of these terms, but not the three-way interaction.[2] This regression model is shown below and includes the variable `i.race` as a covariate.[3]

1. As you might recall, this version of the GSS dataset reflects interviews conducted from 1972 to 2010. So, actually, for those born in 1930, the lowest age is 42; similarly, for those born in 1960, the highest age is 50. In other words, the examples illustrated in this chapter involve values extrapolated outside of the actual sample, a practice that I would not recommend in a real world analysis. However, I hope you will forgive this statistical transgression for the sake of having an example that strives to make the illustration of three way interactions as simple as possible.
2. We will see a model in the next section that includes a three-way interaction so we can see the effect of including the three-way interaction term.
3. The `noci` option is included to make the output more readable for this example. See the callout titled *Using the* `noci` *option for clearer output* in section 2.5.1 for more details.

```
. regress realrinc c.age c.educ c.yrborn
> c.age#c.educ c.age#c.yrborn c.educ#c.yrborn i.race, vce(robust) noci
Linear regression                               Number of obs     =     11,765
                                                F(8, 11756)       =     138.89
                                                Prob > F          =     0.0000
                                                R-squared         =     0.1142
                                                Root MSE          =      24595

-----------------------------------------------------------------
                 |               Robust
        realrinc |      Coef.   Std. Err.      t    P>|t|
-----------------+-----------------------------------------------
             age |  -48960.89   8316.819    -5.89   0.000
            educ |  -61659.03   34481.49    -1.79   0.074
          yrborn |  -1529.751   290.8206    -5.26   0.000
                 |
    c.age#c.educ |   109.4449   18.74366     5.84   0.000
                 |
  c.age#c.yrborn |   24.52853   4.280298     5.73   0.000
                 |
 c.educ#c.yrborn |   30.92122    17.4986     1.77   0.077
                 |
            race |
          white  |          0  (base)
          black  |  -5274.182   464.2166   -11.36   0.000
          other  |  -4616.737   1145.586    -4.03   0.000
                 |
           _cons |    3011547   570438.5     5.28   0.000
-----------------------------------------------------------------
```

We can visualize these results using the **margins** and the **marginsplot** commands. The **margins** command is used to obtain the fitted value for ages 30 and 55, for educations of 12 to 20 years (in two-year increments), and for the years of birth 1930, 1940, 1950, and 1960. The **marginsplot** command is then used to graph these values, placing **age** on the x axis, using separate panels for **yrborn**, and using separate lines for **educ**. (Appendix C includes more details about the options you can use to customize the look of graphs using the **marginsplot** command.) The resulting graph is shown in figure 6.4.

```
. margins, at(age=(30 55) educ=(12(2)20) yrborn=(1930(10)1960))
  (output omitted)
. marginsplot, xdimension(age) bydimension(yrborn)
> plotdimension(educ, allsimple) legend(row(2) subtitle(Education))
> recast(line) scheme(s1mono) noci ylabel(, angle(0))
  Variables that uniquely identify margins: age educ yrborn
```

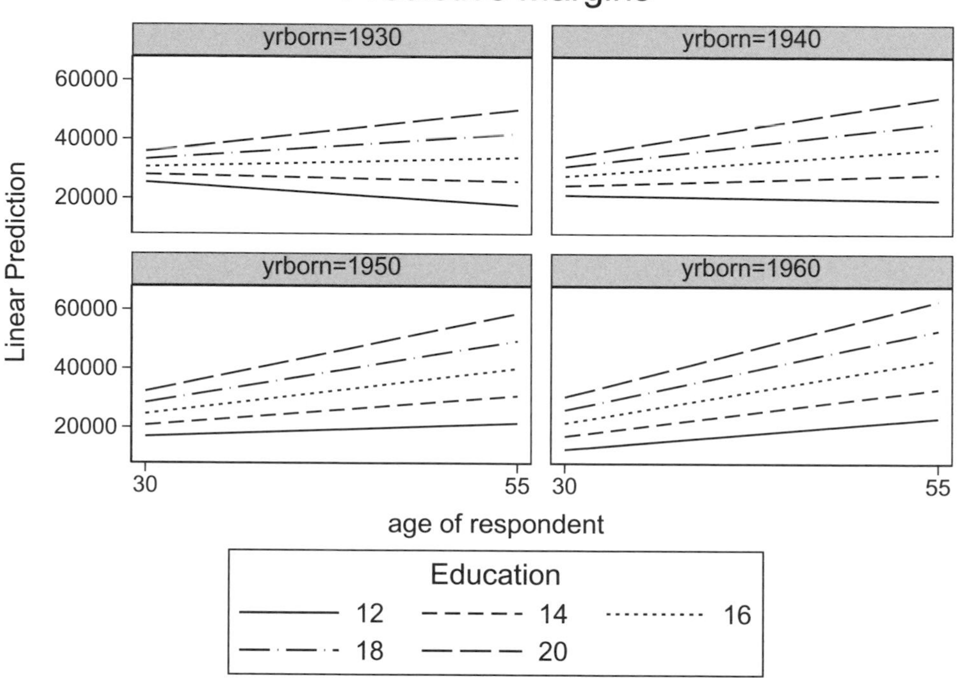

Figure 6.4. Adjusted means from model without the three-way interaction, showing age on the x axis, separate panels for year of birth, and separate lines for education

Let's first focus on the age slope for those with 12 years of education across each of the years of birth. For those born in 1930, the age slope is mildly negative; in 1940, the age slope approximately zero; in 1950, the age slope grows to be mildly positive; and in 1960, the age slope is moderately positive. That is, greater age was associated with less income for people with 12 years of education for those born in 1930; compared with those born in 1960 greater age was associated with higher income for people with 12 years of education. The relationship between age and income (for those with 12 years of education) gradually became more positive as the year of birth increased from 1930 to 1960.

These age slopes (for those with 12 years of education) form a kind of baseline for each of the years of birth. Let's now consider how the age slope changes with respect to these baseline values. For those born in 1930, the age slope gently increases as a function of educ. The same pattern is revealed for the other years of birth as well. Although the people born in different years might show a different age slope at 12 years of education, the increase in the age slope as a function of education is the same across the years of birth. In fact, the coefficient for c.age#c.educ describes the exact degree of this increase. For every unit increase in educ, the age slope increases by 109.44. This is true for all years of birth.

Let's see this for ourselves using the margins command. The margins command below estimates the age slope for those with 14 and 15 years of education who were born in 1930.

```
. margins, dydx(age) at(educ=(14 15) yrborn=1930) vsquish
Average marginal effects                        Number of obs     =     11,765
Model VCE    : Robust

Expression   : Linear prediction, predict()
dy/dx w.r.t. : age
1._at        : educ            =          14
               yrborn          =        1930
2._at        : educ            =          15
               yrborn          =        1930

------------------------------------------------------------------------------
             |            Delta-method
             |      dy/dx   Std. Err.      t    P>|t|     [95% Conf. Interval]
-------------+----------------------------------------------------------------
age          |
         _at |
          1  |  -88.59318   81.20867    -1.09   0.275    -247.7756    70.58928
          2  |   20.85175   86.72145     0.24   0.810    -149.1367    190.8402
------------------------------------------------------------------------------
```

The age slope is 20.85 for someone with 15 years of education born in 1930 and is −88.59 for someone with 14 years of education born in the same year. The difference in these slopes is 109.44. Note how this corresponds to the c.age#c.educ coefficient. This coefficient describes the increase in the age slope for a one-unit increase in educ. Let's make the same comparison, except for those born in 1940.

```
. margins, dydx(age) at(educ=(14 15) yrborn=1940) vsquish
Average marginal effects                        Number of obs     =     11,765
Model VCE    : Robust

Expression   : Linear prediction, predict()
dy/dx w.r.t. : age
1._at        : educ            =          14
               yrborn          =        1940
2._at        : educ            =          15
               yrborn          =        1940

------------------------------------------------------------------------------
             |            Delta-method
             |      dy/dx   Std. Err.      t    P>|t|     [95% Conf. Interval]
-------------+----------------------------------------------------------------
age          |
         _at |
          1  |   156.6922   46.03623     3.40   0.001     66.45352    246.9308
          2  |   266.1371   53.96508     4.93   0.000     160.3566    371.9176
------------------------------------------------------------------------------
```

The difference in the age slopes is the same for someone born in 1940, 266.1371 − 156.6922 = 109.44. You would find that this difference is the same for those born in 1950 and 1960 as well. Another way to say this is that the size of the age#educ interaction is the same across levels of yrborn. This is because this model does not include a three-way interaction of age#educ#yrborn. So let's now consider a model with such a three-way interaction.

6.3.2 A three-way interaction model

Let's now consider a model that includes a three-way interaction of age, educ, and yrborn when predicting realrinc. Such models can have a high degree of multicollinearity that can lead to numerical instability of the estimates and inflated standard errors. To avoid these problems, Aiken and West (1991) recommend centering the predictors prior to the analysis.[4] Let's adopt this advice and center age, educ, and yrborn before attempting an analysis that includes the interaction of these three variables. The following generate commands create centered versions of these variables, centering them around values that I chose to ease the interpretation of the centered values.

```
. generate yrborn30 = yrborn - 1930
. generate age40 = age - 40
. generate educ16 = educ - 16
(39 missing values generated)
```

The variable yrborn is centered by subtracting 1930, creating a new variable called yrborn30. This centered variable will range from 0 to 30, corresponding to the years of birth ranging from 1930 to 1960. The centered version of age is called age40 and contains the value of age minus 40. An age of 30 would be represented as −10 using the

4. Centering means to subtract a constant from the variable. One form of centering is called mean centering, in which the mean is subtracted from the variable.

centered version of age (age40). Finally, the centered version of educ is called educ16, which contains educ minus 16. We could refer to 20 years of education via the centered variable by specifying that educ16 equals 4. The rest of this section will use these centered variables for the analysis.[5]

Let's now form a model that predicts realrinc from age40, educ16, and yrborn30, the two-way interactions among these variables, and the three-way interaction of these variables. We could manually specify the main effect and interaction terms, or we can specify c.age40##c.educ16##c.yrborn30, which is a Stata shorthand for the main effects, the two-way interactions, and the three-way interaction of age40, educ16, and yrborn30. The variable i.race is also included as a covariate.[6]

```
. regress realrinc c.age40##c.educ16##c.yrborn30 i.race, vce(robust) noci
Linear regression                               Number of obs     =     11,765
                                                F(9, 11755)       =     126.24
                                                Prob > F          =     0.0000
                                                R-squared         =     0.1145
                                                Root MSE          =      24591

------------------------------------------------------------------------------
                             |               Robust
                    realrinc |      Coef.   Std. Err.      t    P>|t|
-----------------------------+------------------------------------------------
                       age40 |  -27.89286   135.3081    -0.21   0.837
                      educ16 |   2800.943   366.4629     7.64   0.000
                             |
             c.age40#c.educ16 |   26.63371   42.54171     0.63   0.531
                             |
                    yrborn30 |  -88.92108     55.933    -1.59   0.112
                             |
           c.age40#c.yrborn30 |   32.55721   6.689896     4.87   0.000
                             |
          c.educ16#c.yrborn30 |   12.89238   17.90104     0.72   0.471
                             |
c.age40#c.educ16#c.yrborn30 |   4.319807    2.19088     1.97   0.049
                             |
                        race |
                       white |          0  (base)
                       black |  -5279.075   463.4662   -11.39   0.000
                       other |  -4647.999   1142.813    -4.07   0.000
                             |
                       _cons |   33441.15   1161.797    28.78   0.000
------------------------------------------------------------------------------
```

The coefficient for the three-way interaction is significant ($p = 0.049$). We can visualize the c.age40#c.educ16#c.yrborn30 interaction in a variety of ways. We can

5. You might wonder if the results of the following analysis would change if these variables were centered around different values. For example, suppose age was centered around 45 instead of 40. The estimates regarding the three-way interaction would remain the same, whether age was centered around 40 or 45. The estimates of the adjusted means would remain the same as well. However, there would be differences in the main effects and two-way interactions. These differences are not consequential because these lower order effects are not important in the presence of the three-way interaction term.
6. The noci option is included to make the output more readable for this example. See the callout titled *Using the noci option for clearer output* in section 2.5.1 for more details.

focus on the age40 slope, the educ16 slope, or the yrborn30 slope. As we have done throughout this chapter, let's focus our attention on the age slope (via the centered variable age40).

> **Note! Retaining main effects and two-way interactions**
>
> Looking at the estimates from the regress command, you might notice that some of the main effects and two-way interaction terms are not significant. Even though they are not significant, it is important to retain these effects to preserve the interpretation of the three-way interaction term.

Visualizing the three-way interaction

To visualize the three-way interaction, the margins command is first used to compute the adjusted means as a function of age40, educ16, and yrborn30. Then, the marginsplot command is used to graph the adjusted means.

The margins command is used to compute the adjusted means at different values of the centered variable using the at() option. In terms of the uncentered variables, the at() option specifies ages 30 and 55, educations from 12 to 20 in 2-unit increments, and years of birth from 1930 to 1960 in 10-unit increments. Then, the marginsplot command is used to graph the fitted values, placing age40 on the x axis, with separate lines for educ16 and with separate panels for yrborn30. The graph is shown in figure 6.5.

```
. margins, at(age40=(-10 15) educ16=(-4(2)4) yrborn30=(0(10)30))
  (output omitted)
. marginsplot, xdimension(age40) bydimension(yrborn30)
> plotdimension(educ16, allsimple) legend(row(2) subtitle(Education))
>   recast(line) scheme(s1mono) noci ylabel(, angle(0))
  Variables that uniquely identify margins: age40 educ16 yrborn30
```

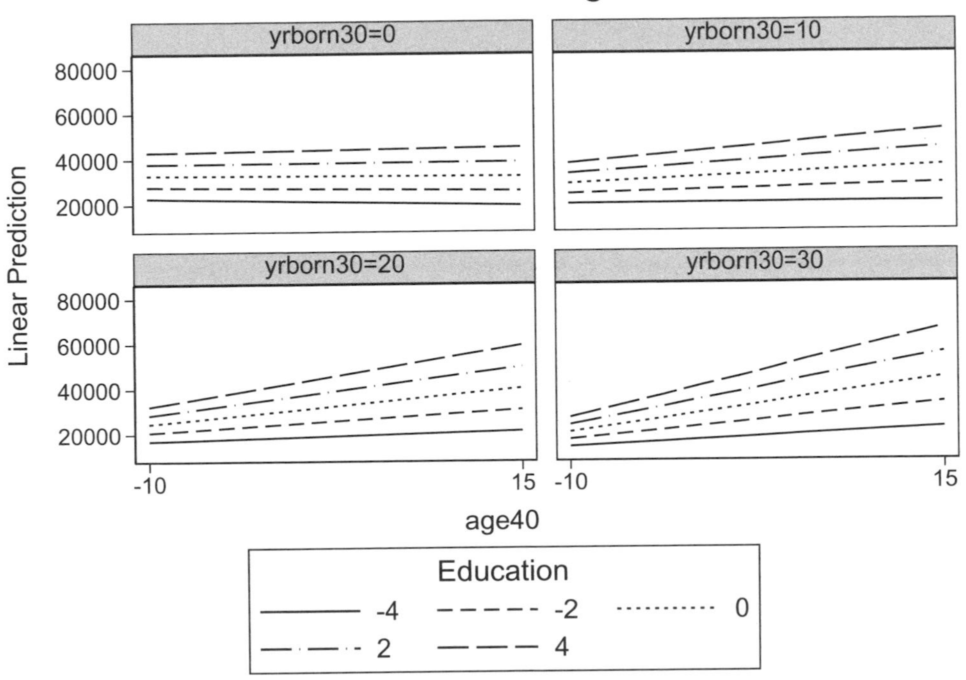

Figure 6.5. Adjusted means from model with the three-way interaction as a function of age (x axis), year of birth (separate panels), and education (separate lines)

Let's focus on the age40 slope, beginning with those born in 1930 (that is, yrborn30 = 0). The age40 slope is mildly negative for those born in 1930, across all levels of education. Jumping forward to 1960 (that is, yrborn30 = 30), the age40 slope is mildly positive for those with 12 years of education and is sharply positive for those with 20 years of education. Looking across the years of birth, we can see that increasing education is more highly related to an increase in the age40 slope as the year of birth increases. Let's use the margins command to obtain exact estimates of the age40 slope for specific values of educ16 and yrborn30 to further quantify what we see in figure 6.5.

Let's begin by estimating the age40 slope when yrborn30 is 0 and educ16 is −2 and −1. This corresponds to those born in 1930 with 14 and 15 years of education.

```
. margins, dydx(age40) at(yrborn30=0 educ16=(-2 -1)) vsquish
Average marginal effects                        Number of obs     =     11,765
Model VCE        : Robust

Expression       : Linear prediction, predict()
dy/dx w.r.t.     : age40
1._at            : educ16          =          -2
                   yrborn30        =           0
2._at            : educ16          =          -1
                   yrborn30        =           0

------------------------------------------------------------------------------
             |            Delta-method
             |      dy/dx   Std. Err.      t    P>|t|     [95% Conf. Interval]
-------------+----------------------------------------------------------------
age40        |
         _at |
          1  |  -81.16028   80.00272    -1.01   0.310    -237.9789    75.65832
          2  |  -54.52657   102.6867    -0.53   0.595    -255.8095    146.7564
------------------------------------------------------------------------------
```

The age40 slope is −81.16 for those with 14 years of education compared with −54.53 for those with 15 years of education, a gain of 26.63 units. This illustrates that a one-unit increase in education leads to a 26.63-unit increase in the age40 slope for those born in 1930.

Let's now obtain the same estimates for those born one year later, born in 1931. This will allow us to assess the impact of a one-unit increase in yrborn30.

```
. margins, dydx(age40) at(yrborn30=1 educ16=(-2 -1)) vsquish
Average marginal effects                        Number of obs     =     11,765
Model VCE    : Robust
Expression   : Linear prediction, predict()
dy/dx w.r.t. : age40
1._at        : educ16          =          -2
               yrborn30        =           1
2._at        : educ16          =          -1
               yrborn30        =           1

------------------------------------------------------------------------------
             |            Delta-method
             |      dy/dx   Std. Err.      t    P>|t|     [95% Conf. Interval]
-------------+----------------------------------------------------------------
age40        |
         _at |
          1  |  -57.24268   76.23312    -0.75   0.453    -206.6722    92.18688
          2  |  -26.28917   98.05195    -0.27   0.789    -218.4872    165.9089
------------------------------------------------------------------------------
```

For those born in 1931, the change in the `age40` slope due to a one-unit increase in education is 30.95 (−26.29 − −57.24). This is larger than the value we found for 1930. A one-unit increase in `educ16` yields a larger increase in the `age40` slope for those born in 1931 (30.95) than for those born in 1930 (26.63). This is due to the inclusion of the three-way interaction.

Consider the value we obtain if we take 30.95 minus 26.63. We obtain 4.32, which corresponds to the coefficient for the `c.age40#c.educ16#c.yrborn30` interaction. This interaction can be interpreted as the degree to which the `age40` slope increases as a function of `educ16` for every one-unit increase in `yrborn30`. For every one-unit increase in `yrborn30`, the `age40` slope increases by 4.32 units for every one-unit increase in `educ16`.

Visualizing the age slope

We can visualize the contribution that `educ16` and `yrborn30` make with respect to the `age40` slope using the `margins` and `marginsplot` commands. First, the `margins` command is used with the `dydx(age40)` option to obtain estimates of the `age40` slope for those with 12 to 20 years of education (in one-year increments) and those born in 1930 to 1960 (in 10-year increments). The `marginsplot` command is then used to graph the `age40` slope as a function of `educ16` and `yrborn30`, as shown in figure 6.6.

```
. margins, dydx(age40) at(educ16=(-4(1)4) yrborn30=(0(10)30))
  (output omitted)
. marginsplot, noci
  Variables that uniquely identify margins: educ16 yrborn30
```

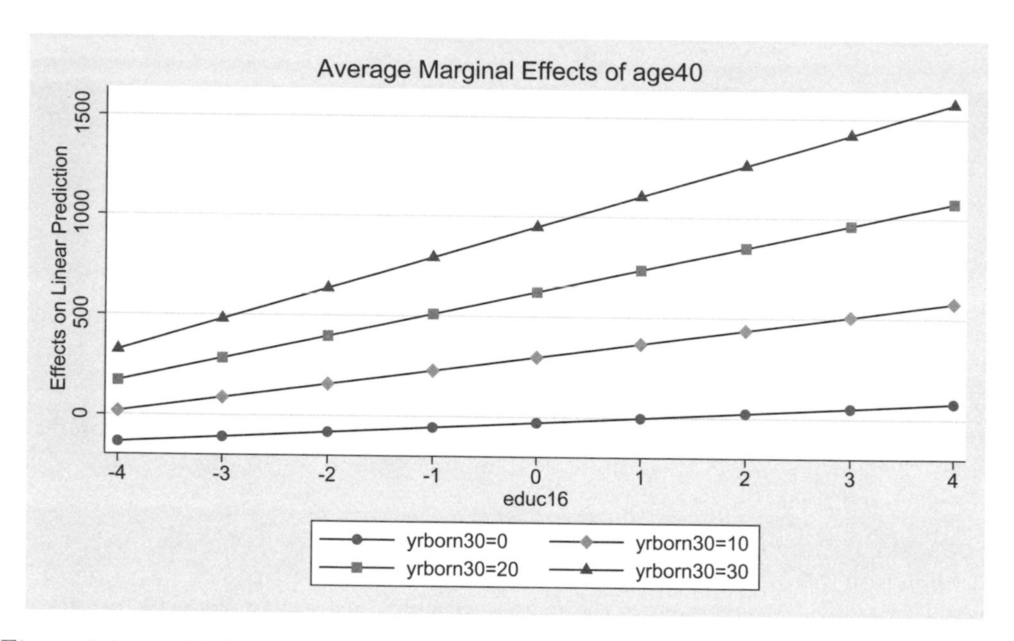

Figure 6.6. age40 slope as a function of education (x axis) and year of birth (separate lines)

Focusing on those born in 1930 (that is, when yrborn30 equals 0), figure 6.6 shows that the age40 slope increases only slightly as educ16 increases. Contrast this with those born in 1960 (that is, when yrborn30 equals 30), where increasing levels of educ16 are associated with a considerable increase in the age40 slope.

Note! Interpreting main effects and two-way interactions

In a model that includes a three-way interaction, it might be tempting to draw conclusions regarding the main effects (for example, effect of age40, educ16, or yrborn30) or the two-way interactions (for example, age40#educ). However, in the presence of the three-way interaction, these main effects and two-way interactions are no longer meaningful and it is not fruitful to dwell on the interpretation of such effects. Instead, as illustrated in this chapter, the margins command can be used to further understand the three-way interaction.

6.4 Summary

This chapter has illustrated the application of models that involve the interaction of three continuous predictors. The visualization of these models using three dimensions was illustrated, as well as their two-dimensional counterparts. In these examples, we saw how the `age` slope changes little as a function of `educ` for those born in the 1930s. By contrast, those born in the 1960s showed a much more rapid increase in the `age` slope as a function of `educ`. We saw how to visualize this by showing the adjusted means on the y axis or by showing the `age` slope on the y axis.

For more information about the interaction of three continuous predictors, see Aiken and West (1991).

Part II

Categorical predictors

This part of the book focuses on the interpretation of categorical predictors. This includes the interpretation of one categorical predictor, the interaction of two categorical predictors, and the interaction of three categorical predictors.

Chapter 7 covers models involving one categorical predictor. This chapter introduces the `contrast` command, a powerful tool for forming contrasts among the levels of categorical predictors. Such contrasts are formed using contrast operators, which are described and illustrated in this chapter. The chapter also illustrates the use of the `margins` and `marginsplot` commands for computing and visualizing the mean of the outcome as a function of the categorical variable. The chapter also describes the use of the `pwcompare` command for making pairwise comparisons among groups.

Chapter 8 covers models involving the interaction of two categorical predictors. The emphasis of this chapter is how to visualize and interpret the interactions. This chapter illustrates how to dissect two-way interactions using simple effects, simple contrasts, partial interactions, and interaction contrasts. The `contrast` command is illustrated for dissecting the interactions, and the `margins` and `marginsplot` commands are used to display and graph the means associated with the interactions.

Chapter 9 covers models involving the interaction of three categorical predictors. This chapter, like the previous chapter, emphasizes the visualization and interpretation of the interactions, in this case focusing on the three-way interaction. Figures are used to visually understand the three-way interactions and a variety of analytic and graphical methods are illustrated for understanding three-way interactions.

7 Categorical predictors

7.1 Chapter overview

This chapter covers models that involve categorical predictors. The emphasis is on how to make contrasts among levels of the categorical predictor to answer interesting questions regarding the differences among the categories. The `contrast` command is used to illustrate how you can easily form contrasts of your choosing among the levels of a categorical predictor.

The chapter begins with a simple comparison of two groups using a *t* test (see section 7.2), followed by an example that includes a multigroup categorical predictor along with additional predictors in section 7.3. That example introduces the use of the `contrast` command and the concept of a contrast operator. Section 7.4 lists and describes the contrast operators.

The following sections illustrate how contrast operators can be used to perform reference group contrasts (see section 7.5), perform contrasts against the grand mean (see section 7.6), compare adjacent means (see section 7.7), compare each mean with subsequent or previous levels (see section 7.8), and make polynomial contrasts (see section 7.9). The use of custom contrasts, for making any contrast of your choosing, is illustrated in section 7.10 followed by a discussion of weighted contrasts (see section 7.11).

The `pwcompare` command, for performing pairwise comparisons, is covered in section 7.12. Section 7.13 provides some caveats about the interpretation of confidence intervals produced by the `marginsplot` command. The chapter concludes with a comparison of the `anova` and `regress` commands for the analysis of categorical predictors (see section 7.14).

7.2 Comparing two groups using a t test

The simplest kind of categorical predictor has two levels. Examples of such two-level predictors include gender (male versus female), treatment assignment (treatment group versus control group), or whether one is married (married versus not married). Suppose we are interested in comparing the happiness rating of people who are married with those who are not married. We might hypothesize that those who are married would be happier than those who are unmarried. Let's explore this question using the GSS dataset.

```
. use gss_ivrm, clear
```

The variable `happy7` indicates the happiness rating of the respondent on a 1 to 7 scale, where 7 is completely happy and 1 is completely unhappy. To compare the average happiness between those who are married and unmarried, we can perform an independent groups *t* test, as shown below.[1]

1. Some might be bothered by analyzing a Likert scale like `happy7` as though it were measured on an interval scale. For the sake of these examples, let's assume that `happy7` is measured on an interval scale.

```
. ttest happy7, by(married)
Two-sample t test with equal variances

   Group │     Obs        Mean    Std. Err.   Std. Dev.   [95% Conf. Interval]
─────────┼──────────────────────────────────────────────────────────────────
Unmarrie │     604     5.35596    .0425197    1.044982    5.272456    5.439465
 Married │     556    5.705036    .0368763    .8695302    5.632602     5.77747
─────────┼──────────────────────────────────────────────────────────────────
combined │   1,160    5.523276    .0287773    .9801179    5.466815    5.579737
─────────┼──────────────────────────────────────────────────────────────────
    diff │           -.3490757    .0567084               -.4603384   -.237813

    diff = mean(Unmarrie) - mean(Married)                         t =  -6.1556
Ho: diff = 0                                     degrees of freedom =     1158

    Ha: diff < 0                 Ha: diff != 0                 Ha: diff > 0
 Pr(T < t) = 0.0000         Pr(|T| > |t|) = 0.0000          Pr(T > t) = 1.0000
```

The ttest command shows that the average happiness is 5.356 for those who are unmarried and 5.705 for those who are married. The difference between these means is −0.349, and that difference is significantly different from 0 (with a two-tailed *p*-value of 0.0000). The difference between these means is negative. We can interpret this result to say that those who are unmarried are significantly less happy than those who are married. We could also say that those who are married are significantly happier than those who are unmarried. I belabor this point here because this issue will arise many times throughout this chapter. The contrast of those who are unmarried versus married is negative. Thus, the mean of the first group (unmarried) is lower; the mean of the second group (married) is higher.

7.3 More groups and more predictors

We are seldom interested in simply comparing two groups in the absence of any additional predictors (covariates). Let's extend the previous example in two ways. First, let's use a five-level measure of marital status, which is coded: 1 = married, 2 = widowed, 3 = divorced, 4 = separated, and 5 = never married. Second, let's include additional predictors (covariates): **gender**,[2] **race**, and **age**. We begin by testing the overall null hypothesis that the average happiness is equal among the five marital status groups:

$$H_0: \mu_1 = \mu_2 = \mu_3 = \mu_4 = \mu_5$$

We can perform that test using the **anova** command (shown below) that predicts happiness from marital status, gender, race, and age. Note that I included the **i.** prefix before each of the categorical variables and the **c.** prefix in front of **age** to specify that it is a continuous variable.[3]

2. In this chapter, I use the variable **gender** (coded: 1 = Male and 2 = Female). This yields output that clearly distinguishes the variable name (that is, **gender**) and its values (that is, **Male** and **Female**).

3. I could have omitted the **i.** prefix in front of the categorical variables because the **anova** command assumes variables are categorical unless we specify otherwise. The inclusion of the **c.** prefix before **age** is mandatory; not including the **c.** prefix would cause **age** to be treated as a categorical variable.

```
. anova happy7 i.marital i.gender i.race c.age

                         Number of obs =      1,156    R-squared     =  0.0488
                         Root MSE      =    .957826    Adj R-squared =  0.0422

                  Source | Partial SS         df         MS        F    Prob>F
              -----------+----------------------------------------------------
                   Model |  53.979104          8    6.747388     7.35  0.0000
                         |
                 marital |  47.156057          4   11.789014    12.85  0.0000
                  gender |  3.8954776          1   3.8954776     4.25  0.0396
                    race |  .92449374          2   .46224687     0.50  0.6043
                     age |  5.3200497          1   5.3200497     5.80  0.0162
                         |
                Residual |  1052.2934      1,147   .91743103
              -----------+----------------------------------------------------
                   Total |  1106.2725      1,155   .95781168
```

Note! The anova and regress commands

In the `anova` output above, the output for categorical variables like `marital` and `race` provide the omnibus *F*-test of the null hypothesis that all the population means are equal. In this chapter (as well as chapters 8 and 9), my focus will be on starting the analysis with an assessment of the main effects (and interactions, if any) using the omnibus *F*-tests provided by the `anova` command. I would emphasize, however, that these same results could be obtained via the `regress` command. In fact, section 7.14 covers the use of the `regress` command for fitting the models involving categorical predictors, showing the equivalency of models fit using the `anova` and `regress` commands.

The overall test of `marital` is significant ($F = 12.85$, $p = 0.000$). After adjusting for gender, race, and age, we can reject the null hypothesis that the average happiness is equal among the five marital status groups.

Let's probe this finding in more detail. Suppose that our research hypothesis (prior to even seeing the data) was that those who are married will be happier than each of the four other marital status groups. We can frame this as four separate null hypotheses, shown below, in which the happiness of each of the other marital status groups (widowed [group 2], divorced [group 3], separated [group 4], and never married [group 5]) is compared with the happiness of those who are married (group 1).

$$H_0\#1\colon\ \mu_2 = \mu_1$$
$$H_0\#2\colon\ \mu_3 = \mu_1$$
$$H_0\#3\colon\ \mu_4 = \mu_1$$
$$H_0\#4\colon\ \mu_5 = \mu_1$$

Let's begin the exploration of these tests by using the `margins` command to compute the adjusted mean of happiness by marital status. This output shows, for example, that the adjusted mean of happiness (after adjusting for gender, race, and age) for those who are married is 5.70.

```
. margins marital
Predictive margins                              Number of obs     =      1,156
Expression   : Linear prediction, predict()

------------------------------------------------------------------------------
             |            Delta-method
             |     Margin   Std. Err.      t    P>|t|     [95% Conf. Interval]
-------------+----------------------------------------------------------------
     marital |
    married  |   5.698407   .0408098   139.63   0.000     5.618336    5.778477
    widowed  |   5.245245   .1178389    44.51   0.000     5.014041    5.476449
   divorced  |   5.199342   .0684853    75.92   0.000     5.064972    5.333713
  separated  |   5.148789   .1647873    31.25   0.000     4.825471    5.472107
never marr.. |   5.542625   .0630837    87.86   0.000     5.418853    5.666398
------------------------------------------------------------------------------
```

The **marginsplot** command can be used to make a graph of the adjusted means and confidence intervals that were computed by the **margins** command. This produces the graph shown in figure 7.1.

```
. marginsplot
  Variables that uniquely identify margins: marital
```

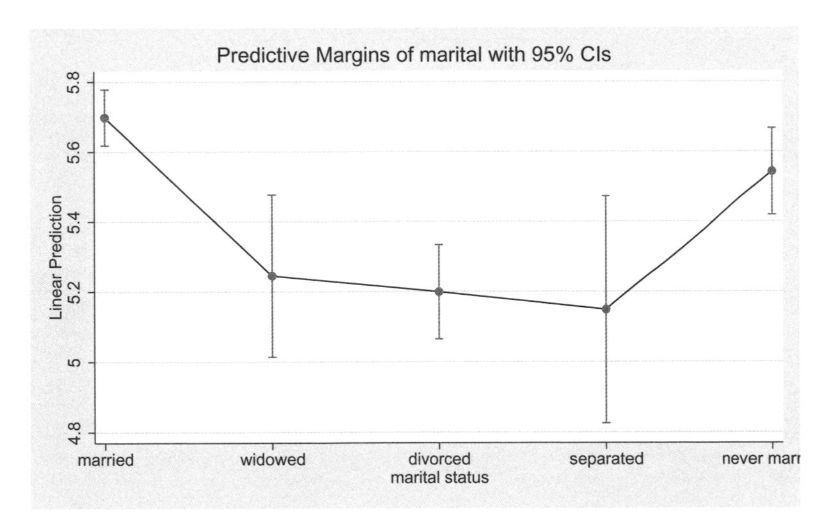

Figure 7.1. Adjusted mean of happiness by marital status

Note! Interpreting confidence intervals

Figure 7.1 shows the adjusted mean of happiness at each level of marital status along with a 95% confidence interval for each of the adjusted means. It is not appropriate to use these confidence intervals for making inferences about the differences between marital status groups. See section 7.13 for more details.

The results so far have given some indirect support for our research hypotheses. The overall test of `marital` is significant, and the pattern of means is consistent with our hypothesis—the adjusted means of those who are married appear greater than the means for the other groups. But let's directly test the four null hypotheses based on our research question—the happiness ratings for each group compared with those who are married to determine if these differences are statistically significant. We can perform such tests using the `contrast` command, as shown below. (I will discuss the syntax of the `contrast` command shortly.)

```
. contrast r.marital
Contrasts of marginal linear predictions
Margins      : asbalanced

------------------------------------------------------------
                            |         df           F        P>F
----------------------------+-------------------------------
                    marital |
      (widowed vs married)  |          1       13.44     0.0003
     (divorced vs married)  |          1       39.38     0.0000
    (separated vs married)  |          1       10.46     0.0013
(never married vs married)  |          1        4.18     0.0411
                      Joint |          4       12.85     0.0000
                            |
                Denominator |       1147
------------------------------------------------------------

---------------------------------------------------------------------------------
                            |   Contrast   Std. Err.     [95% Conf. Interval]
----------------------------+----------------------------------------------------
                    marital |
      (widowed vs married)  |  -.4531618    .123599     -.6956673   -.2106563
     (divorced vs married)  |   -.499064   .0795273     -.6550994   -.3430287
    (separated vs married)  |  -.5496177   .1699695     -.8831036   -.2161317
(never married vs married)  |   -.155781   .0761826     -.3052539   -.0063082
---------------------------------------------------------------------------------
```

This `contrast` command compares each marital status group with the first group (those who are married), adjusting for gender, race, and age. The first test compares group 2 versus 1, comparing those who are widowed versus married. The upper portion of the output shows that this difference is statistically significant ($F = 13.44$, $p = 0.0003$). The lower portion of the output shows the difference in the adjusted means for those who are widowed versus married is -0.45 with a 95% confidence interval of $[-0.70, -0.21]$. Those who are widowed are significantly less happy than those who are married. Put another way, those who are married are happier than those who are widowed. We can reject H_0#1 and say the results are consistent with our prediction that those who are married are significantly happier.

Let's now consider the output for the second, third, and fourth contrasts. These test the second, third, and fourth null hypotheses. The contrasts of groups 3 versus 1 (divorced versus married), 4 versus 1 (separated versus married), and 5 versus 1 (never married versus married) are each statistically significant (as shown in the upper portion of the output). Furthermore, the difference in the means (as shown in the lower portion of the output) is always negative, indicating that those who are married are happier than the group they are being compared with. We can reject the second, third, and fourth null hypotheses. Moreover, the results are in the predicted direction: those who are married are happier.

Now that we are ready to publish these excellent findings, we might want to include a graph that visually depicts these contrasts. We can create such a graph using the `margins` command followed by the `marginsplot` command,[4] creating the graph shown in figure 7.2.[5] This graph illustrates the differences in the adjusted means (comparing each group with those who are unmarried) along with the 95% confidence interval for each difference. For example, the estimate of the difference in the adjusted means for the widowed versus married groups is −0.45 and the 95% confidence interval is [−0.70, −0.21]. This confidence interval excludes 0, indicating that this difference is significant at the 5% level.

4. The `marginsplot` command cannot be used following the `contrast` command. So instead we use the `margins` command followed by the `marginsplot` command.
5. See appendix C for details about how to customize the graphs created by the `marginsplot` command, including tips for handling x-axis labels that are displayed beyond the edge of the graph.

```
. margins r.marital

Contrasts of predictive margins                 Number of obs     =      1,156
Expression   : Linear prediction, predict()

------------------------------------------------------------------
                             |         df           F        P>F
-----------------------------+------------------------------------
                     marital |
       (widowed vs married)  |          1       13.44     0.0003
      (divorced vs married)  |          1       39.38     0.0000
     (separated vs married)  |          1       10.46     0.0013
 (never married vs married)  |          1        4.18     0.0411
                      Joint  |          4       12.85     0.0000
                             |
                 Denominator |       1147
------------------------------------------------------------------

--------------------------------------------------------------------------------
                             |            Delta-method
                             |   Contrast   Std. Err.     [95% Conf. Interval]
-----------------------------+--------------------------------------------------
                     marital |
       (widowed vs married)  |  -.4531618    .123599     -.6956673   -.2106563
      (divorced vs married)  |   -.499064   .0795273     -.6550994   -.3430287
     (separated vs married)  |  -.5496177   .1699695     -.8831036   -.2161317
 (never married vs married)  |   -.155781   .0761826     -.3052539   -.0063082
--------------------------------------------------------------------------------

. marginsplot
  Variables that uniquely identify margins: marital
```

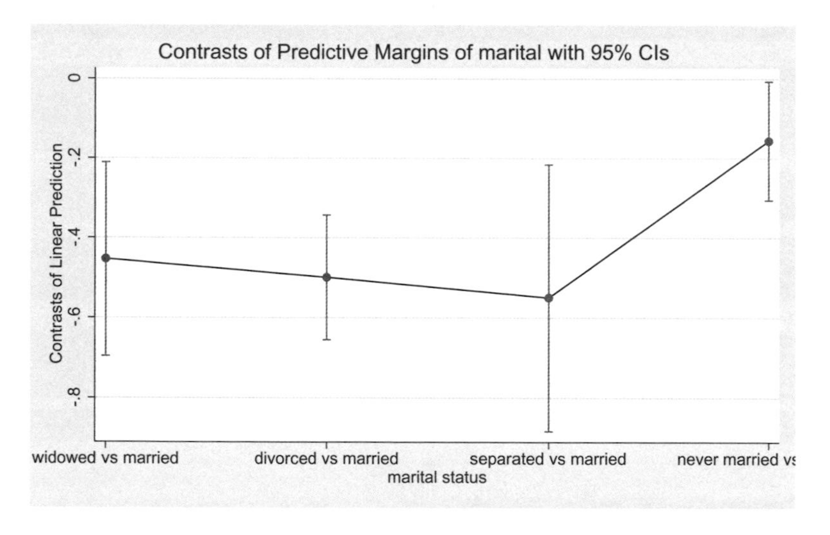

Figure 7.2. Contrasts of adjusted means to those who are married

In these examples, we specified `r.marital` after the `contrast` and `margins` commands. You might be rightly asking: what is this `r.` prefix and what does it mean? Stata calls this a *contrast operator*, and this is just one of many contrast operators that you can choose from. The `r.` contrast operator compares each group with a reference group (which, by default, is the first group, (that is, group 1, those who are married).

The `r.` contrast operator tested contrasts that corresponded to the null hypotheses based on the research questions for this study. It compared the happiness of each group with the reference group (married). There are many other kinds of questions that we could have asked, implying different kinds of contrasts among the groups. Fortunately, Stata offers a wide variety of contrast operators that answer many interesting research questions. These are described in the following section.

7.4 Overview of contrast operators

The examples I have shown so far have illustrated one of the contrast operators you can use with the `contrast` and `margins` commands, the `r.` contrast operator. Table 7.1 provides an overview of the contrast operators that can be used with the `contrast` and `margins` commands. This table provides a brief description of each contrast operator and shows the section of this chapter in which each contrast operator is covered.[6]

6. The custom contrast operator is left empty in table 7.1. It takes the form `{`*varname numlist*`}`, described in more detail in section 7.10.

Table 7.1. Summary of contrast operators

Operator	Section	Description
`r.`	7.5	differences from the reference (base) level; the default
`g.`	7.6	differences from the balanced grand mean
`a.`	7.7	differences from the next level (adjacent contrasts)
`ar.`	7.7	differences from the previous level (reverse adjacent contrasts)
`h.`	7.8	differences from the balanced mean of subsequent levels (Helmert contrasts)
`j.`	7.8	differences from the balanced mean of previous levels (reverse Helmert contrasts)
`p.`	7.9	orthogonal polynomial in the level values
`q.`	7.9	orthogonal polynomial in the level sequence
	7.10	custom contrasts
`gw.`	7.11	differences from the observation-weighted grand mean
`hw.`	7.11	differences from the observation-weighted mean of subsequent levels
`jw.`	7.11	differences from the observation-weighted mean of previous levels
`pw.`	7.11	observation-weighted orthogonal polynomial in the level values
`qw.`	7.11	observation-weighted orthogonal polynomial in the level sequence

The following sections cover each of these contrast operators in turn. Examples are provided illustrating their usage and how to interpret the results. This discussion begins with a more detailed discussion of the `r.` contrast operator in the following section.

Note! As-balanced and as-observed contrasts

In presenting the different contrasts, I first present the contrasts that provide *as-balanced* estimates (in sections 7.5 to 7.10). Then, in section 7.11, I discuss the meaning of *as-observed* contrasts and present the contrast operators that perform such contrasts.

7.5 Compare each group against a reference group

This section provides further examples illustrating the `r.` contrast operator for making reference group contrasts. Let's continue with the example from section 7.3 predicting happiness from marital status, adjusting for gender, race, and age. The `anova` command for that analysis is repeated below (the output is omitted to save space).

```
. use gss_ivrm
. anova happy7 i.marital i.gender i.race c.age
  (output omitted)
```

7.5.1 Selecting a specific contrast

Suppose we wanted to focus only on the contrast of those who have never been married to those who are married (group 5 versus 1). We can specify the `r5.` contrast operator and this yields a contrast of group 5 (the group we specified) compared with the reference group (group 1).

```
. contrast r5.marital, nowald pveffects
Contrasts of marginal linear predictions
Margins      : asbalanced

------------------------------------------------------------------------
                          |   Contrast   Std. Err.        t    P>|t|
--------------------------+---------------------------------------------
                  marital |
(never married vs married)|   -.155781   .0761826     -2.04    0.041
------------------------------------------------------------------------
```

Note! Options on the `contrast` command

The previous `contrast` command included two options: `nowald` and `pveffects`. This yields concise output that fits well on the pages of this book. I will frequently use these options in conjunction with the `contrast` command. Note, however, that this output omits the 95% confidence interval for the estimated difference. When you run such analyses yourself, I recommend specifying the options `nowald` and `effects`, which will display both significance tests and confidence intervals associated with each comparison. There are other options that can be used following the `contrast` command. Appendix D provides more details about these options, as well as other options you can use with the `contrast` command.

If we wanted only to compare those who are divorced with those who are married, we could have specified `r3.marital`. This would have shown only the contrast of group 3 to group 1.

Suppose you wanted to focus on the contrast of group 3 to group 1 (divorced versus married) and group 5 to group 1 (never married versus married). You can perform those two contrasts by specifying the `r(3 5).` contrast operator. This compares each of the groups within the parentheses with the reference group (group 1).

```
. contrast r(3 5).marital, pveffects
Contrasts of marginal linear predictions
Margins      : asbalanced
```

	df	F	P>F
marital			
(divorced vs married)	1	39.38	0.0000
(never married vs married)	1	4.18	0.0411
Joint	2	19.78	0.0000
Denominator	1147		

	Contrast	Std. Err.	t	P>|t|
marital				
(divorced vs married)	-.499064	.0795273	-6.28	0.000
(never married vs married)	-.155781	.0761826	-2.04	0.041

The output shows the test of the contrast of those who are divorced versus married (3 versus 1), and the contrast of those who have never been married versus those who are married (5 versus 1). The upper portion of the output includes the joint test of these two contrasts ($F = 19.78$, $p = 0.0000$). This jointly tests the contrast of groups 3 versus 1 and 5 versus 1. In other words, this tests the equality of the means for groups 1, 3, and 5 (that is, those who are married, divorced, and never married). This is a useful way to test the overall equality of the means for a subset of groups.

7.5.2 Selecting a different reference group

Suppose that instead we wanted to compare each group with a different reference group. For example, let's compare each group with those who have never been married, group 5. We can specify the `rb5.` contrast operator, which requests reference group contrasts using group 5 (never married) as the baseline (reference) group.

```
. contrast rb5.marital, nowald pveffects
Contrasts of marginal linear predictions
Margins      : asbalanced
```

	Contrast	Std. Err.	t	P>|t|
marital				
(married vs never married)	.155781	.0761826	2.04	0.041
(widowed vs never married)	-.2973808	.1446159	-2.06	0.040
(divorced vs never married)	-.343283	.0947271	-3.62	0.000
(separated vs never married)	-.3938366	.1761455	-2.24	0.026

Each of these contrasts is statistically significant. For example, the contrast of those who are separated versus those who have never been married is significant ($p = 0.026$). Those who are separated are significantly less happy than those who have never been married.

7.5.3 Selecting a contrast and reference group

You can both specify the reference group and specify the contrasts to be made at one time. For example, the `contrast` command below compares group 3 (divorced) with group 5 (never married). People who are divorced are significantly less happy than those who have never been married.

```
. contrast r3b5.marital, nowald pveffects
Contrasts of marginal linear predictions
Margins      : asbalanced
```

	Contrast	Std. Err.	t	P>|t|
marital				
(divorced vs never married)	-.343283	.0947271	-3.62	0.000

More than one specific contrast can be specified at once, through the use of parentheses. The `contrast` command below compares group 1 with group 5 and group 3 with group 5. This compares the happiness of those who are married versus never married (group 1 versus 5) and compares the happiness of those who are divorced versus never married (group 3 versus 5). People who are married are significantly happier than those who have never been married ($p = 0.041$), but those who are divorced are significantly less happy than those who have never been married ($p = 0.000$).

```
. contrast r(1 3)b5.marital, nowald pveffects
Contrasts of marginal linear predictions
Margins      : asbalanced
```

	Contrast	Std. Err.	t	P>|t|
marital				
(married vs never married)	.155781	.0761826	2.04	0.041
(divorced vs never married)	-.343283	.0947271	-3.62	0.000

7.6 Compare each group against the grand mean

This section illustrates the `g.` contrast operator that compares each group with the grand mean of all groups. This allows you to assess whether a particular group significantly differs from all groups combined. Continuing the example from the previous section regarding marital status and happiness, a researcher might be interested in comparing the mean happiness of each marital status group versus the grand mean of all groups. This allows you to assess if the mean of each group is significantly different from the grand mean of all of the groups.

Let's begin the analysis by using the GSS dataset and running the **anova** command below. The output of the **anova** command is the same as that from the previous section, so it is omitted.

```
. use gss_ivrm
. anova happy7 i.marital i.gender i.race c.age
  (output omitted)
```

The following **margins** command uses the **g.** contrast operator to compare the mean happiness of each marital status group with the grand mean.[7]

```
. margins g.marital, contrast(nowald pveffects)
Contrasts of predictive margins                   Number of obs     =      1,156
Expression   : Linear prediction, predict()
```

	Contrast	Delta-method Std. Err.	t	P>|t|
marital				
(married vs mean)	.3315249	.0541916	6.12	0.000
(widowed vs mean)	-.1216369	.1035071	-1.18	0.240
(divorced vs mean)	-.1675391	.0685289	-2.44	0.015
(separated vs mean)	-.2180927	.1350667	-1.61	0.107
(never married vs mean)	.1757439	.0708618	2.48	0.013

Note! The contrast() option

The previous **margins** command included the **contrast(nowald pveffects)** option. This yields concise output that is narrow enough to fit within the pages of this book. I will frequently use these options in conjunction with the **margins** command. Note, however, that this output omits the 95% confidence interval for the estimated difference. When you run such analyses yourself, I recommend specifying **contrast(nowald effects)**, which will display both significance tests and confidence intervals associated with each comparison. There are other options that can be used following the **margins** command. See appendix B for more examples of options that you can use with the **margins** command.

The adjusted mean of those who are married (group 1) is 0.33 units greater than the grand adjusted mean and this difference is significant. The contrasts of group 3

7. I chose the **margins** command (instead of the **contrast** command) so we can graph the results using the **marginsplot** command.

(divorced) versus the grand mean and group 5 (never married) versus the grand mean are also significant.

Let's now graph these differences, along with the confidence intervals, using the `marginsplot` command, shown in figure 7.3. When the confidence interval for a contrast excludes zero, the difference is significant at the 5% level.

```
. marginsplot, yline(0)
Variables that uniquely identify margins: marital
```

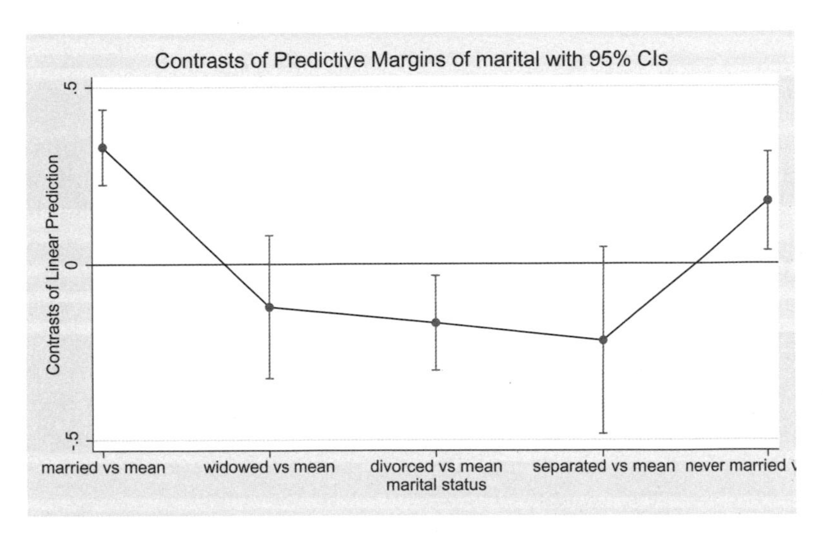

Figure 7.3. Contrasts comparing each group with the grand mean

7.6.1 Selecting a specific contrast

Suppose we only wanted to focus on the contrast of those who are married (group 1) compared with the grand mean. Using the `g1.` contrast operator, only the contrast of group 1 to the grand mean is performed.

```
. contrast g1.marital, nowald pveffects
Contrasts of marginal linear predictions
Margins      : asbalanced
```

	Contrast	Std. Err.	t	P>|t|
marital (married vs mean)	.3315249	.0541916	6.12	0.000

As we have seen previously, we can use parentheses to specify more than one contrast. In the example below, we compare the first group (married) with the grand mean, and we compare the third group (divorced) with the grand mean.

```
. contrast g(1 3).marital, nowald pveffects
Contrasts of marginal linear predictions
Margins      : asbalanced

------------------------------------------------------------------
                    |   Contrast   Std. Err.        t    P>|t|
--------------------+---------------------------------------------
            marital |
 (married vs mean)  |    .3315249   .0541916     6.12    0.000
(divorced vs mean)  |   -.1675391   .0685289    -2.44    0.015
------------------------------------------------------------------
```

7.7 Compare adjacent means

This section illustrates contrasts that compare the means of adjacent groups, for example, group 1 versus 2, group 2 versus 3, group 3 versus 4. These kinds of contrasts are especially useful for studies where you expect a nonlinear relationship between an ordinal or interval predictor and outcome. For example, consider a hypothetical study about the dosage of a new pain medication where the researchers expect that at a certain dosage level the effects of the medication will kick in and lead to a statistically significant reduction in pain. To study this, people in pain are given different dosages of the medication and their pain level is measured on a 100 point scale, where 100 is the worst pain and 0 is no pain. The medication dosages range from 0 mg to 250 mg incrementing by 50 mg, yielding six dosage groups: 1) 0 mg, 2) 50 mg, 3) 100 mg, 4) 150 mg, 5) 200 mg, and 6) 250 mg. The dataset for this example, `pain.dta`, is used below. The `codebook` command is then used to show the coding of `dosegrp`, confirming that `dosegrp` and is coded as 1 = 0 mg, 2 = 50 mg, 3 = 100 mg, 4 = 150 mg, 5 = 200 mg, and 6 = 250 mg.

```
. use pain
. codebook dosegrp
--------------------------------------------------------------------------------
dosegrp                                                  Medication dosage group
--------------------------------------------------------------------------------

                  type:  numeric (float)
                 label:  dosegrp

                 range:  [1,6]                        units:  1
         unique values:  6                        missing .:  0/180

            tabulation:  Freq.   Numeric  Label
                            30         1  0mg
                            30         2  50mg
                            30         3  100mg
                            30         4  150mg
                            30         5  200mg
                            30         6  250mg
```

Let's begin the analysis relating pain to medication dosage by testing the most general null hypothesis that could be tested—that the average pain is equal across all six dosage groups:

$H_0: \mu_1 = \mu_2 = \mu_3 = \mu_4 = \mu_5 = \mu_6$

This null hypothesis is tested using the **anova** command shown below.

```
. anova pain i.dosegrp
                         Number of obs =        180    R-squared     =  0.4602
                         Root MSE      =    10.4724    Adj R-squared =  0.4447

                  Source | Partial SS         df         MS        F    Prob>F
              -----------+----------------------------------------------------
                   Model |  16271.694          5   3254.3389     29.67  0.0000
                         |
                 dosegrp |  16271.694          5   3254.3389     29.67  0.0000
                         |
                Residual |  19082.633        174   109.67031
              -----------+----------------------------------------------------
                   Total |  35354.328        179   197.51021
```

The test associated with **dosegrp** tests the null hypothesis above. The F value is 29.67 and is significant. We can reject the overall null hypothesis. Let's use the **margins** command and the **marginsplot** command to display and graph the predicted mean of **pain** by **dosegrp**. The graph of the means is shown in figure 7.4.

```
. margins dosegrp
Adjusted predictions                              Number of obs     =        180
Expression   : Linear prediction, predict()

------------------------------------------------------------------------------
             |            Delta-method
             |     Margin   Std. Err.      t    P>|t|     [95% Conf. Interval]
-------------+----------------------------------------------------------------
     dosegrp |
         0mg |   71.83333   1.911982    37.57   0.000     68.05967      75.607
        50mg |       70.6   1.911982    36.93   0.000     66.82634    74.37366
       100mg |   72.13333   1.911982    37.73   0.000     68.35967      75.907
       150mg |       70.4   1.911982    36.82   0.000     66.62634    74.17366
       200mg |       54.7   1.911982    28.61   0.000     50.92634    58.47366
       250mg |       48.3   1.911982    25.26   0.000     44.52634    52.07366
------------------------------------------------------------------------------

. marginsplot
  Variables that uniquely identify margins: dosegrp
```

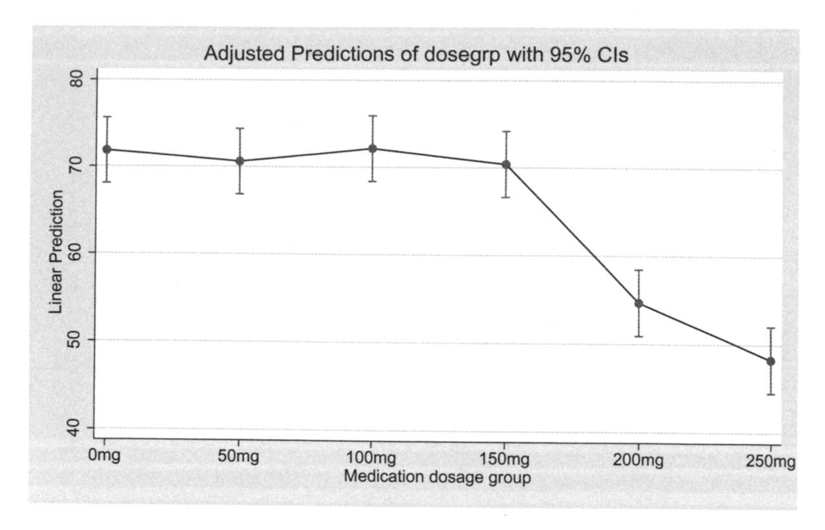

Figure 7.4. Mean pain rating by dosage group

For this study, the research question of interest focuses on the test of each dosage against the previous dosage to determine the dosage that leads to a statistically significant decrease in pain. This leads us to five specific null hypotheses.

$$H_0\#1: \ \mu_1 = \mu_2$$
$$H_0\#2: \ \mu_2 = \mu_3$$
$$H_0\#3: \ \mu_3 = \mu_4$$
$$H_0\#4: \ \mu_4 = \mu_5$$
$$H_0\#5: \ \mu_5 = \mu_6$$

Let's now test each of the hypotheses using the `contrast` command with the `a.` contrast operator to compare each dosage with the adjacent (subsequent) dosage.

```
. margins a.dosegrp, contrast(nowald pveffects)
Contrasts of adjusted predictions                Number of obs   =        180
Expression   : Linear prediction, predict()

------------------------------------------------------------------
                   |            Delta-method
                   |   Contrast   Std. Err.        t    P>|t|
-------------------+----------------------------------------------
           dosegrp |
     (0mg vs 50mg) |   1.233333   2.703952      0.46   0.649
   (50mg vs 100mg) |  -1.533333   2.703952     -0.57   0.571
  (100mg vs 150mg) |   1.733333   2.703952      0.64   0.522
  (150mg vs 200mg) |       15.7   2.703952      5.81   0.000
  (200mg vs 250mg) |        6.4   2.703952      2.37   0.019
------------------------------------------------------------------
```

The contrast of 0 mg versus 50 mg, 50 mg versus 100 mg, and 100 mg versus 150 mg are each not significant. The contrast of 150 mg versus 200 mg is significant ($p = 0.000$). Likewise, the contrast of 200 mg versus 250 mg is also significant ($p = 0.019$).

I am also interested in the 95% confidence interval of the estimate of each of these differences in means. To obtain this, I repeat the command from above but specify `cieffects` in lieu of `pveffects`.[8]

```
. margins a.dosegrp, contrast(nowald cieffects)
Contrasts of adjusted predictions                Number of obs     =        180
Expression   : Linear prediction, predict()
```

	Contrast	Delta-method Std. Err.	[95% Conf.	Interval]
dosegrp				
(0mg vs 50mg)	1.233333	2.703952	-4.103433	6.570099
(50mg vs 100mg)	-1.533333	2.703952	-6.870099	3.803433
(100mg vs 150mg)	1.733333	2.703952	-3.603433	7.070099
(150mg vs 200mg)	15.7	2.703952	10.36323	21.03677
(200mg vs 250mg)	6.4	2.703952	1.063234	11.73677

The output above shows the estimated difference in the means as well as the confidence interval for the difference. For example, the estimated difference in mean pain for those given 150 mg versus 200 mg is 15.7 with a 95% confidence interval of [10.36, 21.04]. Because we used the `margins` command to estimate these differences, we can graph the differences using the `marginsplot` command. The resulting graph is shown in figure 7.5.

8. On your computer, you could specify `contrast(nowald effects)` and your output would include both the significance tests and confidence intervals associated with each comparison.

```
. marginsplot, yline(0)
  Variables that uniquely identify margins: dosegrp
```

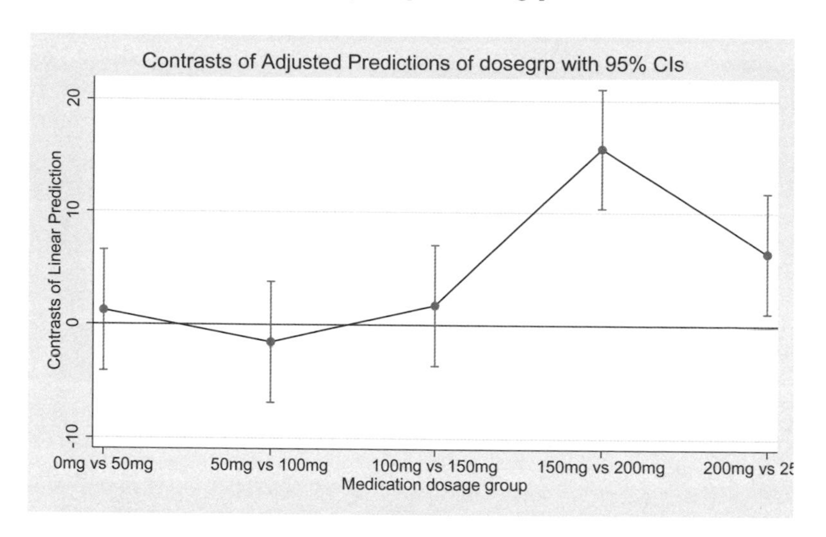

Figure 7.5. Contrasts of each dosage to the previous dosage

Figure 7.5 visually presents the results from the `margins` command. It shows that the adjacent group contrasts for groups 0 mg versus 50 mg, 50 mg versus 100 mg, and 100 mg versus 150 mg not significant (because each of these confidence intervals include zero). It also shows that the contrasts for 150 mg versus 200 mg and 200 mg versus 250 mg are significant. This pain medication appears to become effective at 200 mg, and the increase from 200 mg to 250 mg yields additional pain relief.

7.7.1 Reverse adjacent contrasts

In looking at the estimated differences in the means from the previous section, I notice that the contrasts focus on the contrast of a lower dose to a higher dose. For example, the contrast of group 1 to group 2 compares 0 mg with 50 mg. The results might be clearer if we reverse the order of the contrasts. The `ar.` contrast operator, shown below, provides adjacent group contrasts in reverse order.

```
. contrast ar.dosegrp, nowald pveffects
Contrasts of marginal linear predictions
Margins      : asbalanced
```

	Contrast	Std. Err.	t	P>|t|
dosegrp				
(50mg vs 0mg)	-1.233333	2.703952	-0.46	0.649
(100mg vs 50mg)	1.533333	2.703952	0.57	0.571
(150mg vs 100mg)	-1.733333	2.703952	-0.64	0.522
(200mg vs 150mg)	-15.7	2.703952	-5.81	0.000
(250mg vs 200mg)	-6.4	2.703952	-2.37	0.019

By supplying the `ar.` contrast operator, the contrasts are formed in reverse order, comparing the higher group (dosage) with the adjacent lower group (dosage). By comparing each group with the previous group, the difference in the means can be interpreted as the reduction in pain comparing the higher dosage with the lower dosage.

7.7.2 Selecting a specific contrast

When making adjacent group contrasts, you can select a specific contrast. For example, using the `a.` contrast operator, we can specify particular contrasts of interest. We can select the contrast of group 1 (0 mg) versus its adjacent category (group 2, 50 mg) as shown below. Note that the `a1` operator contrasts the first group to its adjacent group. If we had specified `a3`, this would contrast group 3 to the subsequent category (group 4).

```
. contrast a1.dosegrp, nowald pveffects
Contrasts of marginal linear predictions
Margins      : asbalanced
```

	Contrast	Std. Err.	t	P>|t|
dosegrp				
(0mg vs 50mg)	1.233333	2.703952	0.46	0.649

We can also select contrasts using the `ar.` contrast operator. The `ar.` contrast operator forms contrasts with the previous group, so specifying `ar3.` (as shown below) contrasts group 3 (100 mg) with the previous group (group 2, 50 mg).

```
. contrast ar3.dosegrp, nowald pveffects
Contrasts of marginal linear predictions
Margins      : asbalanced
```

	Contrast	Std. Err.	t	P>|t|
dosegrp				
(100mg vs 50mg)	1.533333	2.703952	0.57	0.571

We can combine selected contrasts as well. Suppose we wanted to test the equality of the mean pain ratings for the first four groups. The individual contrasts of the adjacent means suggest that the overall test of the equality of these means would be nonsignificant, but we have not formally performed such a test. Let's perform this test using the `contrast` command, as shown below.

```
. contrast a(1 2 3).dosegrp
Contrasts of marginal linear predictions
Margins      : asbalanced

------------------------------------------------
                    |         df           F        P>F
--------------------+---------------------------
           dosegrp  |
    (0mg vs 50mg)   |          1        0.21     0.6489
   (50mg vs 100mg)  |          1        0.32     0.5714
  (100mg vs 150mg)  |          1        0.41     0.5223
              Joint |          3        0.21     0.8918
                    |
        Denominator |        174
------------------------------------------------

--------------------------------------------------------------------------
                    |   Contrast   Std. Err.     [95% Conf. Interval]
--------------------+-----------------------------------------------------
           dosegrp  |
    (0mg vs 50mg)   |   1.233333   2.703952     -4.103433    6.570099
   (50mg vs 100mg)  |  -1.533333   2.703952     -6.870099    3.803433
  (100mg vs 150mg)  |   1.733333   2.703952     -3.603433    7.070099
--------------------------------------------------------------------------
```

This `contrast` command compares groups 1 versus 2 (0 mg versus 50 mg), 2 versus 3 (50 mg versus 100 mg), and 3 versus 4 (100 mg versus 150 mg). Each of these contrasts is individually nonsignificant, but let's focus on the joint test. That test is not significant ($p = 0.8918$). The joint test simultaneously tests all specified contrasts, providing a test of the null hypothesis of the equality of the means for groups 1, 2, 3, and 4 (that is, 0 mg, 50 mg, 100 mg, and 150 mg). We could cite this statistical test to indicate that the test of the equality of the pain ratings for the first four dosage groups (that is, 0 mg, 50 mg, 100 mg, and 150 mg) was not significant.

7.8 Comparing the mean of subsequent or previous levels

This section describes contrasts that compare each group mean with the mean of the subsequent groups (also known as *Helmert* contrasts). It also illustrates contrasts that compare each group with the mean of the previous groups (also known as *reverse Helmert* contrasts). For example, consider a follow-up to the pain study described in section 7.7. The previous study focused on the lowest dosages needed to achieve significant pain reduction. This hypothetical study focuses on determining the dosage at which no significant pain reductions are achieved. The participants in this study all suffer from pain and are given one of six different medication dosages: 300 mg, 400 mg, 500 mg, 600 mg, 800 mg, or 1000 mg. In this example, the variable `dosage` contains the actual size of the dosage (in milligrams): values 300, 400, 500, 600, 800, or 1000. The dataset for this example, `pain2.dta`, is used below, and the `tabulate` command shows the tabulation of the variable `dosage`.

```
. use pain2
. tabulate dosage

 Medication
  dosage in
         mg |      Freq.     Percent        Cum.
------------+-----------------------------------
        300 |         30       16.67       16.67
        400 |         30       16.67       33.33
        500 |         30       16.67       50.00
        600 |         30       16.67       66.67
        800 |         30       16.67       83.33
       1000 |         30       16.67      100.00
------------+-----------------------------------
      Total |        180      100.00
```

The overall null hypothesis regarding dosage is that the average pain is equal across all six dosage groups:

$H_0\colon\ \mu_{300} = \mu_{400} = \mu_{500} = \mu_{600} = \mu_{800} = \mu_{1000}$

Let's begin by testing this overall null hypothesis using the `anova` command below.

```
. use pain2
. anova pain i.dosage

                         Number of obs =        180    R-squared     =  0.2052
                         Root MSE      =    10.5056    Adj R-squared =  0.1824

                  Source | Partial SS         df         MS        F    Prob>F
              -----------+----------------------------------------------------
                   Model |  4958.8667          5   991.77333      8.99  0.0000
                         |
                  dosage |  4958.8667          5   991.77333      8.99  0.0000
                         |
                Residual |  19204.133        174   110.36858
              -----------+----------------------------------------------------
                   Total |      24163        179   134.98883
```

We can reject the overall null hypothesis of the equality of the six means ($F = 8.99$, $p = 0.000$).

Let's use the `margins` and `marginsplot` commands to show and graph the means by the six levels of `dosage`. The graph of the means is shown in figure 7.6. Note how the spacing of the dosages on the x axis reflect the actual dosage.

```
. margins dosage
Adjusted predictions                            Number of obs     =        180
Expression   : Linear prediction, predict()

------------------------------------------------------------------------------
             |            Delta-method
             |     Margin   Std. Err.      t    P>|t|     [95% Conf. Interval]
-------------+----------------------------------------------------------------
      dosage |
        300  |   43.83333    1.91806    22.85   0.000     40.04768    47.61899
        400  |       37.6    1.91806    19.60   0.000     33.81434    41.38566
        500  |   31.86667    1.91806    16.61   0.000     28.08101    35.65232
        600  |   29.63333    1.91806    15.45   0.000     25.84768    33.41899
        800  |   30.63333    1.91806    15.97   0.000     26.84768    34.41899
       1000  |   29.43333    1.91806    15.35   0.000     25.64768    33.21899
------------------------------------------------------------------------------

. marginsplot
  Variables that uniquely identify margins: dosage
```

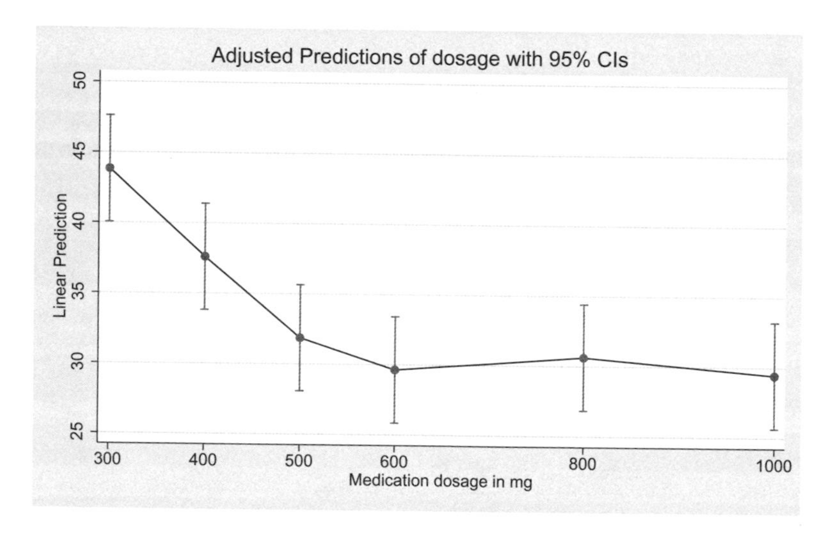

Figure 7.6. Mean pain rating by dosage

For this study, we want to establish the dosage threshold where no statistically significant pain reduction is achieved when compared with higher dosages. Thus, the mean pain rating at each dosage is compared with the mean pain rating for all the higher dosages. At some dosage, the pain rating will not be significantly different from those receiving higher dosages. We can express this using the following null hypotheses:

$$H_0\#1\colon\ \mu_{300} = \mu_{>300}$$
$$H_0\#2\colon\ \mu_{400} = \mu_{>400}$$
$$H_0\#3\colon\ \mu_{500} = \mu_{>500}$$
$$H_0\#4\colon\ \mu_{600} = \mu_{>600}$$
$$H_0\#5\colon\ \mu_{800} = \mu_{1000}$$

Let's now test each of the null hypotheses below using the `margins` command combined with the `h.` contrast operator.

```
. margins h.dosage, contrast(nowald pveffects)
Contrasts of adjusted predictions                Number of obs     =        180
Expression   : Linear prediction, predict()

------------------------------------------------------------
                 |            Delta-method
                 |   Contrast   Std. Err.        t    P>|t|
-----------------+------------------------------------------
          dosage |
 (300 vs > 300)  |         12   2.101129      5.71    0.000
 (400 vs > 400)  |   7.208333   2.144456      3.36    0.001
 (500 vs > 500)  |   1.966667   2.214784      0.89    0.376
 (600 vs > 600)  |        -.4   2.349134     -0.17    0.865
 (800 vs   1000) |        1.2   2.712546      0.44    0.659
------------------------------------------------------------
```

The `h.` contrast operator compares the mean of each group with the mean of the subsequent groups. The comparisons are specified in terms of the actual dosage, for example, `(300 vs > 300)`. The contrast of those receiving 300 mg with those receiving more than 300 mg is significant ($p = 0.000$), as is the contrast of those receiving 400 mg with those receiving more than 400 mg ($p = 0.001$). The contrast of those receiving 500 mg with those receiving more than 500 mg is not significant ($p = 0.375$). All subsequent contrasts are also not significant. Referring to the null hypotheses, we would reject the first and second null hypotheses, and we would not reject the third through fifth null hypotheses. In other words, those receiving 400 mg experience significantly more pain than those receiving 500 mg or more. Comparing those receiving 500 mg with those receiving 600 mg or more, there is no statistically significant difference in pain.

We can visualize these contrasts using the `marginsplot` command, as shown below. This creates the graph in figure 7.7, which shows each of the contrasts with a confidence interval. When the confidence interval for the contrast excludes zero, the difference is significant at the 5% level. (The `xlabel()` option is used to make the labels of the x axis more readable.)

```
. marginsplot, yline(0) xlabel(, angle(45))
  Variables that uniquely identify margins: dosage
```

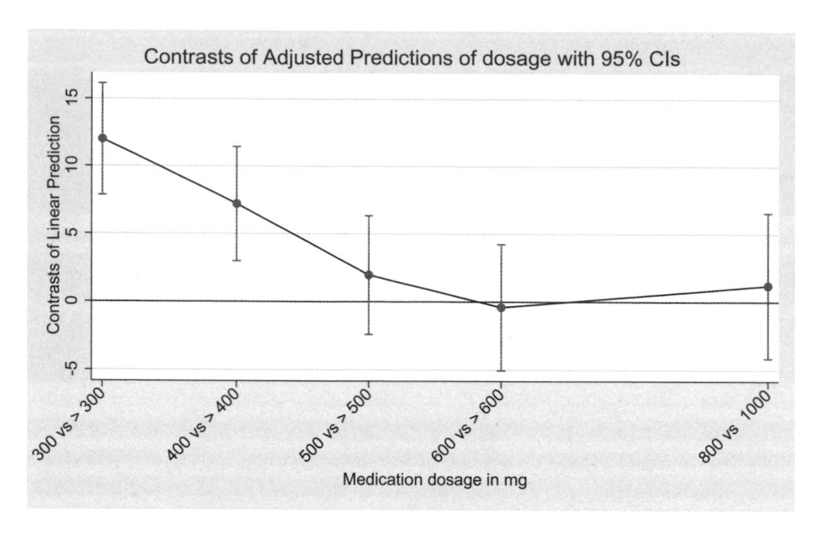

Figure 7.7. Mean pain rating by dosage

7.8.1 Comparing the mean of previous levels

We have seen how the `h.` operator compares the mean of each group with the mean of the subsequent groups. You might be interested in making the same kind of contrast, but in the reverse direction—comparing the mean of each group with the mean of the previous groups. The `j.` contrast operator is used for such contrasts.

Although it does not make much sense in the context of this example, the `contrast` command is shown below illustrating the use of the `j.` contrast operator.

```
. contrast j.dosage, nowald pveffects
Contrasts of marginal linear predictions
Margins      : asbalanced
```

	Contrast	Std. Err.	t	P>|t|
dosage				
(400 vs 300)	-6.233333	2.712546	-2.30	0.023
(500 vs < 500)	-8.85	2.349134	-3.77	0.000
(600 vs < 600)	-8.133333	2.214784	-3.67	0.000
(800 vs < 800)	-5.1	2.144456	-2.38	0.018
(1000 vs <1000)	-5.28	2.101129	-2.51	0.013

When using the `j.` contrast operator, the mean of each group is compared with the mean of the previous groups. In this case, all the contrasts are significant.

7.8.2 Selecting a specific contrast

Returning to the `h.` contrast operator, let's illustrate how you can select specific contrasts. Say that we wanted to focus only on the contrast of those whose value of `dosage` was 400 to those who have higher values of `dosage`. We can specify `h400.dosage`, as shown in the `contrast` command below.[9]

```
. contrast h400.dosage, nowald pveffects
Contrasts of marginal linear predictions
Margins      : asbalanced

                    Contrast   Std. Err.        t    P>|t|

          dosage
  (400 vs >400)     7.208333    2.144456     3.36    0.001
```

Alternatively, we might want to focus only on the contrasts of 400 mg versus the subsequent groups and 500 mg versus the subsequent groups. We can perform these contrasts, as shown below.

```
. contrast h(400 500).dosage, nowald pveffects
Contrasts of marginal linear predictions
Margins      : asbalanced

                    Contrast   Std. Err.        t    P>|t|

          dosage
  (400 vs >400)     7.208333    2.144456     3.36    0.001
  (500 vs >500)     1.966667    2.214784     0.89    0.376
```

This selection process can be used with the `j.` operator as well. If we wanted to select the contrast of 500 mg versus the mean of the previous groups, we could specify `j500.dosage`, as shown below.

```
. contrast j500.dosage, nowald pveffects
Contrasts of marginal linear predictions
Margins      : asbalanced

                    Contrast   Std. Err.        t    P>|t|

          dosage
  (500 vs <500)        -8.85    2.349134    -3.77    0.000
```

9. We specify `h400.` because that is the actual value stored in the variable `dosage`. Specifying the contrast in this way is related to the coding of the variable `dosage` and not because we are using the `h.` contrast operator.

7.9 Polynomial contrasts

Let's consider the use of polynomial contrasts for assessing nonlinear trends (for example, quadratic, cubic, or quartic). Let's refer back to the example from section 7.7 that looked at the relationship between the dosage of pain medication and pain ratings using the pain.dta dataset. In that dataset, the medication dosages ranged from 0 mg to 250 mg and were recorded in the variable dosegrp. This variable was coded: 1 = 0 mg, 2 = 50 mg, 3 = 100 mg, 4 = 150 mg, 5 = 200 mg, and 6 = 250 mg. Let's use the pain.dta dataset followed by the anova command to predict pain from i.dosage. The anova output is omitted to save space.

```
. use pain
. anova pain i.dosegrp
  (output omitted)
```

We can specify q.dosegrp on the contrast command to compute tests of polynomial trend with respect to dosegrp. (The noeffects option is used to save space and focus on the results of the Wald tests.)

```
. contrast q.dosegrp, noeffects
Contrasts of marginal linear predictions
Margins      : asbalanced
```

	df	F	P>F
dosegrp			
(linear)	1	109.12	0.0000
(quadratic)	1	29.25	0.0000
(cubic)	1	0.00	0.9824
(quartic)	1	8.39	0.0043
(quintic)	1	1.62	0.2048
Joint	5	29.67	0.0000
Denominator	174		

The test of linear trend is significant ($F = 109.12$, $p = 0.000$), and the test of quadratic trend is also significant ($F = 29.25$, $p = 0.000$). The test of cubic trend is not significant, but the test of quartic trend is significant ($F = 8.39$, $p = 0.0043$).

Suppose you wanted to fit the relationship between a predictor and outcome using a linear term and wanted to assess whether there are significant nonlinear trends in the relationship between the predictor and outcome. You can use the contrast command to test only the nonlinear terms, as shown below. The joint test of all the nonlinear terms (powers 2 through 6) is significant ($F = 9.81$, $p = 0.0000$). In such a case, it would be inadvisable to fit the relationship between the predictor and outcome using only a linear fit.

```
. contrast q(2/6).dosegrp, noeffects
Contrasts of marginal linear predictions
Margins      : asbalanced
```

	df	F	P>F
dosegrp			
(quadratic)	1	29.25	0.0000
(cubic)	1	0.00	0.9824
(quartic)	1	8.39	0.0043
(quintic)	1	1.62	0.2048
Joint	4	9.81	0.0000
Denominator	174		

The `q.` contrast operator assumes that the levels of `dosegrp` are equidistant from each other. Indeed, the levels of `dosegrp` range from 0 mg to 250 mg in 50 mg increments. Let's consider an example using the `pain2.dta` dataset (from section 7.8) where the levels of `dosage` were not equidistant. In the `pain2.dta` dataset, the dosages were recorded as the actual dosage in milligrams: 300, 400, 500, 600, 800, or 1,000. The `pain2.dta` dataset is used below, and the `anova` command is used to predict `pain` from the `dosage` groups. The output of the `anova` command is omitted to save space.

```
. use pain2
. anova pain i.dosage
  (output omitted)
```

The `contrast` command is used, applying the `p.` contrast operator to `dosage`. This tests the polynomial trends based on the actual dosage, accounting for the differing gaps among the levels of `dosage`.

```
. contrast p.dosage, noeffects
Contrasts of marginal linear predictions
Margins      : asbalanced
```

	df	F	P>F
dosage			
(linear)	1	28.37	0.0000
(quadratic)	1	13.44	0.0003
(cubic)	1	2.60	0.1084
(quartic)	1	0.47	0.4956
(quintic)	1	0.05	0.8202
Joint	5	8.99	0.0000
Denominator	174		

The test of linear and quadratic trends are significant, whereas the cubic, quartic, and quintic trends are each individually not significant. Let's form a test of only the cubic, quartic, and quintic trends using the `contrast` command below.

```
. contrast p(3/6).dosage, noeffects
Contrasts of marginal linear predictions
Margins      : asbalanced
```

	df	F	P>F
dosage			
(cubic)	1	2.60	0.1084
(quartic)	1	0.47	0.4956
(quintic)	1	0.05	0.8202
Joint	3	1.04	0.3759
Denominator	174		

The joint test of the cubic, quartic, and quintic trends is not significant ($F = 1.04$, $p = 0.3759$). These tests show that there are significant linear and quadratic trends in the relationship between `pain` and `dosage`. In addition, the joint test of the cubic, quartic, and quintic trends is not significant. Altogether, these tests suggest that we could be justified in modeling `pain` as a linear and quadratic function of `dosage`.

Tip! Minding your Ps and Qs

In this example, the levels of dosage are not equidistant, and we would have obtained different results by specifying `q.dosage` compared with specifying `p.dosage`. In the previous example, where the levels of `dosegrp` were equidistant, specifying `p.dosegrp` yields the same results as specifying `q.dosegrp`.

7.10 Custom contrasts

In sections 7.5 to 7.9, we have seen several contrast operators that can be used to form contrasts among the levels of a categorical variable. Specifically, we have seen the `r.`, `a.`, `ar.`, `g.`, `h.`, `j.`, `p.`, and `q.` contrast operators. These contrast operators automate the process of forming most of the kinds of contrasts you might make. For those times when you want to make another kind of contrast, you can specify a custom contrast. Let's illustrate the use of custom contrasts using the analysis we saw in section 7.3. Using the GSS dataset, let's predict happiness from marital status, gender, race, and age. The output from the `anova` command is the same as shown in section 7.3, so it is omitted to save space.

```
. use gss_ivrm, clear
. anova happy7 i.marital i.gender i.race c.age
  (output omitted)
```

Let's begin by illustrating how to perform custom contrasts using simple examples that compare one group with another group. For the first example, let's compare the mean of group 1 (married) with group 5 (not married). The custom contrast is enclosed within curly braces by specifying the variable name followed by the contrast coefficients. The contrast coefficients map to the levels (groups) of the variable. In this example, the contrast coefficient of 1 is applied to group 1, and −1 is applied to group 5. (A contrast coefficient of 0 is applied to groups 2, 3, and 4.) The result is a contrast of group 1 minus group 5. The `contrast` command computes difference in the adjusted means for group 1 versus group 5 as 0.16 and that difference is significant.

```
. contrast {marital 1 0 0 0 -1}, nowald pveffects
Contrasts of marginal linear predictions
Margins      : asbalanced

------------------------------------------------------
             |   Contrast   Std. Err.      t    P>|t|
-------------+----------------------------------------
     marital |
        (1)  |    .155781    .0761826    2.04   0.041
------------------------------------------------------
```

Let's switch the above contrast. Let's compare group 5 (not married) with group 1 (married), as shown below. Note how the results are the same as the previous results, except that the sign of the contrast changes from positive to negative.

```
. contrast {marital -1 0 0 0 1}, nowald pveffects
Contrasts of marginal linear predictions
Margins      : asbalanced

------------------------------------------------------
             |   Contrast   Std. Err.      t    P>|t|
-------------+----------------------------------------
     marital |
        (1)  |   -.155781    .0761826   -2.04   0.041
------------------------------------------------------
```

Now, let's compare those who are separated (group 4) with those who are divorced (group 3). We perform this contrast by using a contrast coefficient of 1 for group 4 and −1 for group 3. The difference in the adjusted means for group 4 versus group 3 is −0.05; this test is not significant ($p = 0.777$).

```
. contrast {marital 0 0 -1 1 0}, nowald pveffects
Contrasts of marginal linear predictions
Margins      : asbalanced

------------------------------------------------------
             |   Contrast   Std. Err.      t    P>|t|
-------------+----------------------------------------
     marital |
        (1)  |  -.0505536    .1784857   -0.28   0.777
------------------------------------------------------
```

All the above contrasts could have been performed using the `r.` contrast operator but were useful for getting us familiar with how to specify custom contrasts. Let's now consider a contrast that cannot be performed with one of the standard contrast operators. Say that we want to compare those who are married (group 1) with the average of those who are separated and divorced (groups 3 and 4). We can form that contrast as shown below. The contrast is statistically significant. The adjusted mean for group 1 is significantly different from the average of the adjusted means for groups 3 and 4. Put another way, the adjusted mean for group 1 minus the average of the adjusted means of groups 3 and 4 equals 0.52, and that contrast is significantly different from 0.

```
. contrast {marital 1 0 -.5 -.5 0}, nowald pveffects
Contrasts of marginal linear predictions
Margins      : asbalanced

             |   Contrast   Std. Err.      t    P>|t|
-------------+--------------------------------------
     marital |
        (1)  |   .5243409   .0981978    5.34   0.000
```

Suppose we want to compare those who are married (group 1) with the average of those who are widowed, divorced, and separated (groups 2, 3, and 4). We can perform that contrast as shown below. The adjusted mean of those who are married is 0.50 units greater than the average of the adjusted means for those are widowed, divorced, and separated. This contrast is significant.

```
. contrast {marital 1 -.33333333 -.33333333 -.33333333 0}, nowald pveffects
Contrasts of marginal linear predictions
Margins      : asbalanced

             |   Contrast   Std. Err.      t    P>|t|
-------------+--------------------------------------
     marital |
        (1)  |   .5006146   .0820748    6.10   0.000
```

Note! Contrasts must sum to zero

The contrast coefficients that we specify in a custom contrast must sum to zero. In the previous example, the contrast coefficients for groups 2, 3, and 4 are expressed as `-.33333333`, using eight digits after the decimal point. Although the sum of the coefficients for that custom contrast is not exactly zero, it is close enough to zero to satisfy the `margins` command. Had we used seven or fewer digits, the sum of the coefficients would be sufficiently different from 0 for the `margins` command to complain with the following error:

```
invalid contrast vector
r(198);
```

You can use *inline expansions* to directly specify the fraction 1/3, as shown in the `contrast` command below.

```
. contrast {marital 1 `=-1/3' `=-1/3' `=-1/3' 0}, nowald pveffects
```

The opening single quote is often found below the tilde key, and the close single quote is often found below the double quote key.

Let's form a contrast of the average of those who are married (group 1) and separated (group 4) to the average of those who are widowed (group 2) and divorced (group 3). Note how the coefficients for groups 1 and 4 are specified as 0.5 and the coefficients for groups 2 and 3 are specified as −0.5.

```
. contrast {marital .5 -.5 -.5 .5 0}, nowald pveffects
Contrasts of marginal linear predictions
Margins      : asbalanced
```

	Contrast	Std. Err.	t	P>|t|
marital				
(1)	.2013041	.1091831	1.84	0.065

To formulate the contrast coefficients, it can be helpful to write out the null hypothesis that you want to test. For example, let's write the null hypothesis corresponding to the contrast above:

$$H_0\colon\ (\mu_1 + \mu_4)/2 = (\mu_2 + \mu_3)/2$$

Let's now rewrite this showing the coefficient multiplied by each mean. That yields the following:

$$H_0\colon\ (1/2) * \mu_1 + (1/2) * \mu_4 = (1/2) * \mu_2 + (1/2) * \mu_3$$

Then, let's solve this for 0 by moving the right side of the equation to the left side of the equals sign and making those coefficients negative.

$$H_0: (1/2) * \mu_1 + (1/2) * \mu_4 + -(1/2) * \mu_2 + -(1/2) * \mu_3 = 0$$

We can then sort the groups to yield the following:

$$H_0: (1/2) * \mu_1 + -(1/2) * \mu_2 + -(1/2) * \mu_3 + (1/2) * \mu_4 = 0$$

We then use these contrast coefficients in the `contrast` command, repeated below. Note that a value of 0 is inserted with respect to group 5.

```
. contrast {marital .5 -.5 -.5 .5 0}, pveffects
  (output omitted)
```

7.11 Weighted contrasts

The examples I have presented so far have sidestepped the issue of how to account for unequal sample sizes. For example, say that we are comparing group 1 versus groups 2 and 3 combined. So far, the examples I have shown estimate the mean for groups 2 and 3 combined by obtaining the mean for group 2 and the mean for group 3, and then averaging those means. Stata calls this the *as-balanced* approach, because it gives equal weights to the groups even if their sample sizes are different. We could, instead, weight the means for groups 2 and 3 proportionate to their sample size. Stata calls this the *as-observed* approach, because the means are weighted in proportion to their observed sample size.

Let's illustrate this using the GSS dataset, predicting happiness from the three-level marital status coded: 1 = married, 2 = previously married, and 3 = never married. The `anova` command is used below to predict happiness from marital status (the output is omitted to save space).

```
. use gss_ivrm

. anova happy7 i.marital3
  (output omitted)
```

We can obtain the average happiness by marital status using the **margins** command, as shown below.

```
. margins marital3
Adjusted predictions                            Number of obs     =      1,160
Expression   : Linear prediction, predict()

------------------------------------------------------------------------------
             |            Delta-method
             |     Margin   Std. Err.      t    P>|t|     [95% Conf. Interval]
-------------+----------------------------------------------------------------
    marital3 |
     Married |   5.705036   .0408275   139.74   0.000     5.624932     5.78514
 Prevmarried |   5.263323   .0539007    97.65   0.000     5.157569    5.369077
Never marr.. |   5.459649   .0570253    95.74   0.000     5.347765    5.571534
------------------------------------------------------------------------------
```

Let's use the h. contrast operator to compare each group with the mean of the subsequent groups, using the as-balanced approach. Let's focus our attention on the first contrast, Married vs >Married, which compares those who are married (group 1) versus those who are previously married and never married (groups 2 and 3).

```
. margins h.marital3, contrast(nowald pveffects)
Contrasts of adjusted predictions               Number of obs     =      1,160
Expression   : Linear prediction, predict()

-----------------------------------------------------------------------
                                  |            Delta-method
                                  |   Contrast   Std. Err.      t    P>|t|
----------------------------------+------------------------------------
                         marital3 |
          (Married vs >Married)   |     .34355   .0566231     6.07   0.000
(Prevmarried vs  Never married)   |  -.1963262   .0784677    -2.50   0.012
-----------------------------------------------------------------------
```

The **margins** command with the h. contrast operator estimates the difference in the means for group 1 versus groups 2 and 3 as 0.34355. We can manually compute this estimate by taking the mean of group 1 (married) minus the average of the means from groups 2 and 3 (previously married and never married), displayed below.

```
. display 5.705036 - (5.263323 + 5.459649)/2
.34355
```

With a tiny bit of algebra, we can express this estimate a little bit differently, as shown below. The fractions (1/2) represent the fact that groups 2 and 3 (previously married and never married) are weighted equally.

```
. display 5.705036 - (5.263323*(1/2) + 5.459649*(1/2))
.34355
```

Let's compare this estimate with the as-observed approach, which weights the mean of groups 2 and 3 (previously married and never married) by their sample size. The hw. contrast operator is used to obtain the as-observed estimate.

```
. margins hw.marital3, contrast(nowald pveffects)
Contrasts of adjusted predictions              Number of obs     =      1,160
Expression   : Linear prediction, predict()

------------------------------------------------------------------------------
                                   |            Delta-method
                                   |   Contrast   Std. Err.      t    P>|t|
-----------------------------------+------------------------------------------
                          marital3 |
            (Married vs >Married)  |   .3490757     .05658     6.17   0.000
(Prevmarried vs  Never married)    |  -.1963262   .0784677    -2.50   0.012
------------------------------------------------------------------------------
```

Instead of weighting groups 2 and 3 equally—by (1/2)—groups 2 and 3 are weighted by their individual sample size divided by the combined sample size. The N for group 2 (previously married) is 319, and the N for group 3 (never married) is 285, and the combined N for the two groups is 604. The estimate of the difference using the as-observed approach is computed, as shown below, weighting the mean for group 2 (previously married) by (319/604), and weighting the mean for group 3 (never married) by (285/604).

```
. display 5.705036 - (5.263323*(319/604)+ 5.459649*(285/604))
.34907573
```

Although this example has focused on the `h.` contrast operator, this issue arises for all contrast operators that involve contrasts among more than two groups, namely, the `g.`, `h.`, `j.`, `p.`, and `q.` operators. Used without the `w`, these will provide as-balanced estimates. If you include the `w`, by specifying `gw.`, `hw.`, `jw.`, `pw.`, or `qw.`, then the as-observed estimates are computed. I think the as-balanced estimates are the most commonly used contrast method. However, if you want estimates that are weighted as a function of the proportion of observations in each group, then the as-observed estimates will provide you the kind of estimates that you desire.

7.12 Pairwise comparisons

Sometimes, you want to test all pairwise comparisons that can be formed for a factor variable. The `pwcompare` command can be used to form such comparisons. Let's illustrate the `pwcompare` command to form pairwise comparisons of the happiness ratings among the five marital status groups using the analysis from section 7.3. To begin, we use the GSS dataset and use the `anova` command to predict happiness from marital status, gender, race, and age. (The output from the `anova` command is omitted but is the same as shown in section 7.3.)

```
. use gss_ivrm, clear
. anova happy7 i.marital i.gender i.race c.age
  (output omitted)
```

We can use the `pwcompare` command to form all pairwise comparisons among the marital status groups. The `pveffects` option is specified to include t-values and p-values for each comparison.

```
. pwcompare marital, pveffects
Pairwise comparisons of marginal linear predictions
Margins      : asbalanced
```

	Contrast	Std. Err.	Unadjusted t	Unadjusted P>|t|
marital				
widowed vs married	-.4531618	.123599	-3.67	0.000
divorced vs married	-.499064	.0795273	-6.28	0.000
separated vs married	-.5496177	.1699695	-3.23	0.001
never married vs married	-.155781	.0761826	-2.04	0.041
divorced vs widowed	-.0459022	.1339294	-0.34	0.732
separated vs widowed	-.0964558	.202613	-0.48	0.634
never married vs widowed	.2973808	.1446159	2.06	0.040
separated vs divorced	-.0505536	.1784857	-0.28	0.777
never married vs divorced	.343283	.0947271	3.62	0.000
never married vs separated	.3938366	.1761455	2.24	0.026

Tip! More on the pwcompare command

More details about the `pwcompare` command are contained in appendix E.

Because of the many comparisons, we might want to make adjustments to the p-values to account for the multiple comparisons. For example, let's use Šidák's method for adjusting for multiple comparisons by adding the `mcompare(sidak)` option.

```
. pwcompare marital, pveffects mcompare(sidak)
Pairwise comparisons of marginal linear predictions
Margins      : asbalanced
```

	Number of Comparisons
marital	10

	Contrast	Std. Err.	Sidak t	Sidak P>|t|
marital				
widowed vs married	-.4531618	.123599	-3.67	0.003
divorced vs married	-.499064	.0795273	-6.28	0.000
separated vs married	-.5496177	.1699695	-3.23	0.013
never married vs married	-.155781	.0761826	-2.04	0.343
divorced vs widowed	-.0459022	.1339294	-0.34	1.000
separated vs widowed	-.0964558	.202613	-0.48	1.000
never married vs widowed	.2973808	.1446159	2.06	0.335
separated vs divorced	-.0505536	.1784857	-0.28	1.000
never married vs divorced	.343283	.0947271	3.62	0.003
never married vs separated	.3938366	.1761455	2.24	0.228

The **bonferroni** or **scheffe** method could have alternatively been specified within the **mcompare()** option. When you have balanced data, the **tukey**, **snk**, **duncan**, or **dunnett** method can be specified within **mcompare()** (see [R] **pwcompare** for more details).

7.13 Interpreting confidence intervals

The **marginsplot** command displays margins and confidence intervals that were computed from the most recent **margins** command. Sometimes, these confidence intervals might tempt you into falsely believing that they tell us about differences among groups. To illustrate this point, let's use the analysis from section 7.3 predicting happiness from marital status, gender, race, and age. The GSS dataset is used below, and the **anova** command is shown (the output is omitted to save space).

```
. use gss_ivrm, clear
. anova happy7 i.marital i.gender i.race c.age
  (output omitted)
```

The **margins** command is used to estimate the adjusted means of happiness by **marital**. The output also includes the 95% confidence interval for each adjusted mean. For example, among those who are married (group 1), we are 95% confident that the population mean of happiness (adjusting for gender, race, and age) is between 5.62 and 5.78.

```
. margins marital
Predictive margins                              Number of obs     =      1,156
Expression   : Linear prediction, predict()
```

	Margin	Delta-method Std. Err.	t	P>|t|	[95% Conf.	Interval]
marital						
married	5.698407	.0408098	139.63	0.000	5.618336	5.778477
widowed	5.245245	.1178389	44.51	0.000	5.014041	5.476449
divorced	5.199342	.0684853	75.92	0.000	5.064972	5.333713
separated	5.148789	.1647873	31.25	0.000	4.825471	5.472107
never marr..	5.542625	.0630837	87.86	0.000	5.418853	5.666398

We can graph the adjusted means and confidence intervals computed by the **margins** command using the **marginsplot** command, as shown below.

```
. marginsplot
  Variables that uniquely identify margins: marital
```

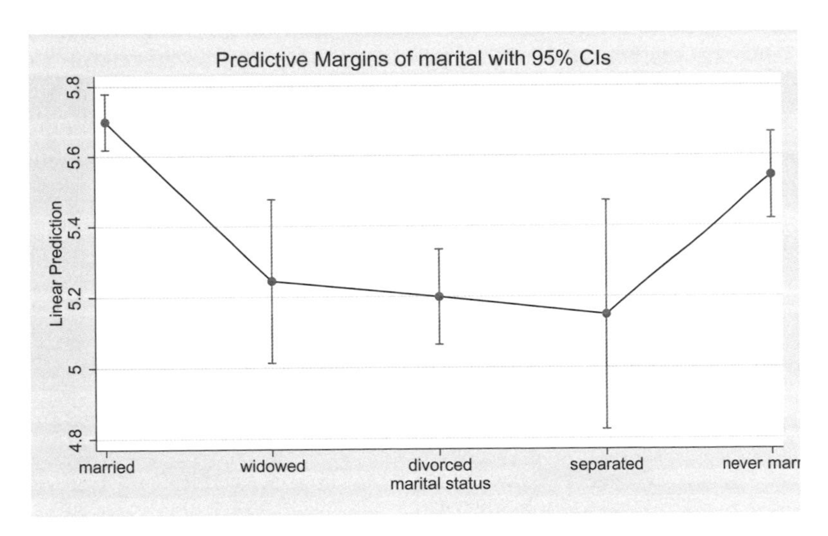

Figure 7.8. Adjusted means of happiness by marital status

Figure 7.8 is a graphical representation of the adjusted means and confidence intervals computed by the **margins** command. Our eye might be tempted to use the overlap (or lack of overlap) of confidence intervals between groups to draw conclusions about the significance of the differences between groups. However, such conclusions would not be appropriate. For example, although the confidence intervals for those who are separated and never married overlap, the output from the **contrast** command below shows that the difference in these means is statistically significant ($p = 0.026$).

```
. contrast ar.marital, nowald pveffects
Contrasts of marginal linear predictions
Margins      : asbalanced
```

	Contrast	Std. Err.	t	P>|t|
marital				
(widowed vs married)	-.4531618	.123599	-3.67	0.000
(divorced vs widowed)	-.0459022	.1339294	-0.34	0.732
(separated vs divorced)	-.0505536	.1784857	-0.28	0.777
(never married vs separated)	.3938366	.1761455	2.24	0.026

Using the **contrast** command, like the one shown above, is the appropriate way to assess the significance of group differences.

7.14 Testing categorical variables using regression

The analyses in this chapter have been conducted using the **anova** command because it produces concise printed output. However, this is not to imply that we cannot perform such tests using the **regress** command. All tests illustrated in this chapter could have been performed using the **regress** command and related linear modeling commands. (In fact, these tests could also have been performed using nonlinear modeling commands such as **logit**, as illustrated in chapter 18.) Consider the **anova** command we have frequently issued in this chapter, repeated below.

```
. use gss_ivrm, clear
. anova happy7 i.marital i.gender i.race c.age
                   Number of obs =      1,156    R-squared     =  0.0488
                   Root MSE      =    .957826    Adj R-squared =  0.0422

          Source | Partial SS         df         MS        F    Prob>F
      -----------+----------------------------------------------------
           Model |  53.979104          8    6.747388     7.35  0.0000
                 |
         marital |  47.156057          4   11.789014    12.85  0.0000
          gender |  3.8954776          1   3.8954776     4.25  0.0396
            race |  .92449374          2   .46224687     0.50  0.6043
             age |  5.3200497          1   5.3200497     5.80  0.0162
                 |
        Residual |  1052.2934      1,147   .91743103
      -----------+----------------------------------------------------
           Total |  1106.2725      1,155   .95781168
```

The output includes tests of the overall effect of each of the categorical variables: **marital**, **gender**, and **race**.[10] The output also includes the test of **age**, treated as a continuous variable.

10. The variable **gender** has only two levels, so there is only one test with respect to this variable.

Let's now use the **regress** command with the same set of predictors,[11]

```
. regress happy7 i.marital i.gender i.race c.age
      Source |       SS           df       MS      Number of obs   =     1,156
-------------+----------------------------------   F(8, 1147)      =      7.35
       Model |  53.9791041         8  6.74738801   Prob > F        =    0.0000
    Residual |  1052.29339     1,147  .917431026   R-squared       =    0.0488
-------------+----------------------------------   Adj R-squared   =    0.0422
       Total |  1106.27249     1,155  .957811681   Root MSE        =    .95783

------------------------------------------------------------------------------
      happy7 |      Coef.   Std. Err.      t    P>|t|     [95% Conf. Interval]
-------------+----------------------------------------------------------------
     marital |
     married |          0  (base)
     widowed |  -.4531618    .123599    -3.67   0.000    -.6956673   -.2106563
    divorced |   -.499064   .0795273    -6.28   0.000    -.6550994   -.3430287
   separated |  -.5496177   .1699695    -3.23   0.001    -.8831036   -.2161317
never marr.. |   -.155781   .0761826    -2.04   0.041    -.3052539   -.0063082
             |
      gender |
        Male |          0  (base)
      Female |   .1196351   .0580584     2.06   0.040     .0057225    .2335476
             |
        race |
       white |          0  (base)
       black |  -.0354866   .0850047    -0.42   0.676    -.2022688    .1312956
       other |  -.1123621   .1179656    -0.95   0.341    -.3438146    .1190905
             |
         age |   .0050064    .002079     2.41   0.016     .0009273    .0090854
       _cons |   5.415511   .1130652    47.90   0.000     5.193673    5.637349
------------------------------------------------------------------------------
```

The output of the **regress** command is lengthier because it uses dummy coding for each of the categorical variables and shows the effect for each of the dummy variables. To obtain the test of the overall effect of **marital** and the test of the overall effect of **race**, we can use the **contrast** command, as shown below. The overall test of **marital** is significant ($F = 12.85$, $p = 0.0000$). The overall test of **race** is not significant ($F = 0.50$, $p = 0.6043$). Note that these F tests match those produced by the **anova** command.

```
. contrast marital race
Contrasts of marginal linear predictions
Margins      : asbalanced

------------------------------------------------
             |         df           F        P>F
-------------+----------------------------------
     marital |          4       12.85     0.0000
             |
        race |          2        0.50     0.6043
             |
 Denominator |       1147
------------------------------------------------
```

11. As described in the callout in section 1.1, I have used the **set baselevels on** command to display the base level of factor variables by default.

The **contrast**, **margins**, **marginsplot**, and **pwcompare** commands work the same way after the **regress** command as they do after the **anova** command. For example, we can use the **margins** command to obtain the adjusted means broken down by **marital**.

```
. margins marital

Predictive margins                              Number of obs     =      1,156
Model VCE    : OLS

Expression   : Linear prediction, predict()

------------------------------------------------------------------------------
             |            Delta-method
             |     Margin   Std. Err.      t    P>|t|     [95% Conf. Interval]
-------------+----------------------------------------------------------------
     marital |
    married  |   5.698407   .0408098   139.63   0.000     5.618336    5.778477
    widowed  |   5.245245   .1178389    44.51   0.000     5.014041    5.476449
   divorced  |   5.199342   .0684853    75.92   0.000     5.064972    5.333713
  separated  |   5.148789   .1647873    31.25   0.000     4.825471    5.472107
never marr.. |   5.542625   .0630837    87.86   0.000     5.418853    5.666398
------------------------------------------------------------------------------
```

The **contrast** command can also be used to make contrasts among the adjusted means, as shown below.

```
. contrast r.marital, nowald pveffects

Contrasts of marginal linear predictions

Margins      : asbalanced

------------------------------------------------------------------------
                              |   Contrast   Std. Err.      t    P>|t|
------------------------------+-----------------------------------------
                      marital |
      (widowed vs married)    |  -.4531618    .123599    -3.67   0.000
     (divorced vs married)    |   -.499064   .0795273    -6.28   0.000
    (separated vs married)    |  -.5496177   .1699695    -3.23   0.001
(never married vs married)    |   -.155781   .0761826    -2.04   0.041
------------------------------------------------------------------------
```

The **margins** and **marginsplot** commands can also be used following the **regress** command to display and graph the adjusted means. The **pwcompare** command can also be used to obtain pairwise comparisons of means. (These commands are not illustrated to save space.)

The **anova** and **regress** commands provide the same results, but in a different format. The **anova** command provides the overall test of the equality of the means by default. When using the **regress** command, we need to use the **contrast** command to obtain such tests.

Note! Interaction terms with the anova and regress commands

If you include interaction terms, the estimates of the main effects will differ when using the **regress** command compared with the **anova** command. This is covered in more detail in section 8.6.

In the case of ordinary least-squares regression, you can use either the `anova` command or the `regress` command, depending on your taste. However, there are features provided by the `regress` command that are not available with the `anova` command. For example, the `regress` command permits estimation of robust standard errors with the `vce(robust)` option or the use of the `svy` prefix. These features are not available with the `anova` command. In such cases, the `regress` command can be used as illustrated in this section.

7.15 Summary

In this section, we have seen how you can interpret the effects of a categorical predictor. The `margins` and `marginsplot` commands make it easy to compute and graph the adjusted means of the outcome as a function of the categorical predictor. Furthermore, the `contrast` and `margins` commands can be used with contrast operators to form meaningful contrasts among the levels of the categorical variable. There are many built-in contrast operators that you can use; you can also specify custom contrast operators. Furthermore, if you use the `margins` command to form such contrasts, the `marginsplot` command can be used to graph the contrasts.

For more details about options you can use with the `contrast` command, see appendix D. For more details about options you can use with the `margins` command, see appendix B. Appendix C contains more details about customizing the appearance of the graphs created by the `marginsplot` command. For more information about the `pwcompare` command, see appendix E. For further help with these commands, you can also see [R] **contrast**, [R] **margins**, [R] **marginsplot**, and [R] **pwcompare**.

For more information about modeling categorical predictors in regression models, I recommend Keppel and Wickens (2004), Maxwell et al. (2018), and Pedhazur and Schmelkin (1991). For more information about different coding schemes, see Davis (2010) and Wendorf (2004).

8 Categorical by categorical interactions

8.1 Chapter overview

This chapter illustrates models that involve the interaction of two categorical variables. An example of such a model is a factorial ANOVA model. A key feature of such a model is the number of levels of each factor. Suppose our model has two factors, named A and B and that factor A has 2 levels and factor B has 3 levels. Such a model would frequently be referred to as a two by three factorial ANOVA.[1] The emphasis of this chapter is not only how to test for interactions between factor variables but also how to understand and dissect those interactions.

1. Using this nomenclature, factor A and factor B are categorical variables, in the abstract sense. This is different than a factor variable, such as specifying `i.marital` after the `regress` or `anova` command.

Note! Dissecting interactions

There are many ways that we strive to gain a greater understanding of something complicated. When we were young, perhaps we tore an old radio apart to try and understand how it worked. Likewise, we can tear apart statistical interactions to better understand them. This is called dissecting the interaction. In this and later chapters, I will illustrate a variety of techniques for dissecting interactions. For the sake of clarity (but at the risk of some repetition), I will consistently use this one term, *dissect*, to describe the general process of tearing an interaction apart to better understand it.

The way that you can dissect an interaction depends on the number of levels of each factor. This chapter focuses on three types of interactions: two by two interactions, two by three interactions, and three by three interactions. These three types of interactions cover the most common interactions that researchers will encounter. Furthermore, these interactions form the building blocks that can be used for analyzing larger interactions. The tools used to dissect two by three interactions generalize to two by X interactions. Likewise, the section on three by three interactions can be generalized to any kind of two-way interaction.

Four examples are presented in this chapter, one illustrating a two by two design (see section 8.2), two examples of a two by three design (see sections 8.3.1 and 8.3.2), and an example illustrating a three by three design (see section 8.4). For each design, I present each of the techniques that you could use for dissecting and understanding the interaction. You do not need to apply the techniques in the order in which they are presented, and you do not need to apply all the techniques that are illustrated. You can pick and choose the techniques that are most applicable to your study to dissect your interaction. Furthermore, I recommend that you create an analysis plan (prior to examining your data) in which you describe the predicted pattern of results and how you plan to dissect the interaction to test for the exact pattern of results that you anticipate.

Each example is based on a hypothetical research study conducted by a research psychologist. Rather than studying the reduction in depression, this psychologist is interested in increasing optimism. The research focuses on the effectiveness of different kinds of therapy for increasing optimism and determining how the effectiveness of the different kinds of therapy can depend on the person's depression status from before the beginning of the study.

The first example uses a two by two design with two levels of treatment (control group and happiness therapy) and two levels of depression status (nondepressed and depressed). The second example extends this example using a two by three design, using the same two levels of treatment (control group and happiness therapy) but including three levels of depression status (nondepressed, mildly depressed, and severely depressed). The third example also uses a two by three design, but instead focuses on three types of treatment (control group, traditional therapy, and happiness therapy) and two levels of depression (nondepressed and depressed). The fourth example illustrates a three by three design, including three levels of treatment (control group, traditional therapy, and happiness therapy), and three levels of depression status (nondepressed, mildly depressed, and severely depressed).

In each of these examples, the outcome variable is the optimism score of the person at the end of the study. Scores on this hypothetical optimism scale can theoretically range from 0 to 100, with a value of 50 representing the average optimism for people in general.

In discussing these designs, I often borrow terminology used regarding the analysis of variance. For example, as mentioned earlier, a categorical variable can be referred to as a factor. In describing a factor and its levels, it can be useful to refer to the factor using one letter, and to the levels using one letter followed by a number designating the level. Consider the variable depression status that has three levels: 1) nondepressed, 2) mildly depressed, and 3) severely depressed. As a shorthand, I might refer to this as factor D (for depression). Also, as a shorthand, instead of referring to the nondepressed group, I might simply refer to that group as *D1*.

8.2 Two by two models: Example 1

The field of psychology has a long history of studying depression and trying to find ways of lessening it. Imagine a hypothetical research psychologist who, instead, focuses her research on studying optimism and finding ways of increasing it. She developed a validated measure of optimism that, although correlated with conventional measures of depression, is conceptually and operationally distinct from depression. As part of her research, she has created a new kind of therapy, called happiness therapy, that she believes can be effective in increasing optimism. In this first hypothetical study, she seeks to determine the effectiveness of happiness therapy by comparing the optimism of people who have completed happiness therapy treatment with the optimism of people in a control group who received no treatment. The researcher is interested in not only assessing the effectiveness of happiness therapy but also assessing whether its effectiveness depends on whether the person has been diagnosed as clinically depressed. This yields a two by two research design, crossing treatment group assignment (control group versus happiness therapy) with depression status (nondepressed versus depressed).

The dataset for this example is used below, and the first five observations are displayed. The variable `treat` indicates the treatment assignment, coded: 1 = control group (`Con`) and 2 = happiness therapy (`HT`). The variable `depstat` reflects the person's depression status at the beginning of the study and is coded: 1 = nondepressed and 2 = depressed. The variable `opt` is the optimism score at the end of the study. In this dataset, `opt` has a mean of 44.5, a minimum of 16, and a maximum of 80.

```
. use opt-2by2, clear
. list in 1/5

     +-----------------------+
     | treat   depstat   opt |
     |-----------------------|
  1. |   Con       Non  41.0 |
  2. |    HT       Non  64.0 |
  3. |   Con       Dep  47.0 |
  4. |    HT       Dep  30.0 |
  5. |   Con       Non  27.0 |
     +-----------------------+
```

The `tabulate` command is used to show the frequencies broken down by `depstat` and `treat`. This illustrates the research design, showing that the study involved a total of 120 participants. Half of them (60) were nondepressed prior to the beginning of the study, and half were depressed. Each of these 60 people was randomly assigned (in equal numbers) to the treatment or control group.[2]

```
. tabulate depstat treat

Depression |    Treatment group
    status |       Con         HT |     Total
-----------+----------------------+----------
       Non |        30         30 |        60 
       Dep |        30         30 |        60 
-----------+----------------------+----------
     Total |        60         60 |       120 
```

The `table` command is used to display the mean optimism broken down by treatment and depression status. For example, the mean optimism for someone who is nondepressed in the control group was 44.9.

```
. table depstat treat, contents(mean opt)

----------------------
          | Treatment 
Depressio |   group   
n status  |  Con    HT
----------+-----------
      Non | 44.9  60.0
      Dep | 34.6  38.8
----------------------
```

The effect of happiness therapy (as compared with the control group) can be assessed by comparing the mean optimism of the happiness therapy group with the mean optimism of the control group. For example, among those who are nondepressed, the effect of happiness therapy is 60.0 minus 44.9, or 15.1. Among those who are depressed, the effect of happiness therapy is 38.8 minus 34.6, or 4.2.

By including both depressed and nondepressed people in this study, we can ask whether the effect of happiness therapy is the same for those who are depressed as those

2. Note that the analytic methods illustrated in this chapter do not depend on the data being balanced. These same methods work equally well if you have unbalanced data (which is more common than perfectly balanced data).

who are nondepressed.[3] In other words, we can ask if there is an interaction between treatment and depression status. We can get a sense of whether such an interaction might exist by graphing the mean optimism by treatment and depression status, as shown in figure 8.1. (We will see how to create such a graph later.)

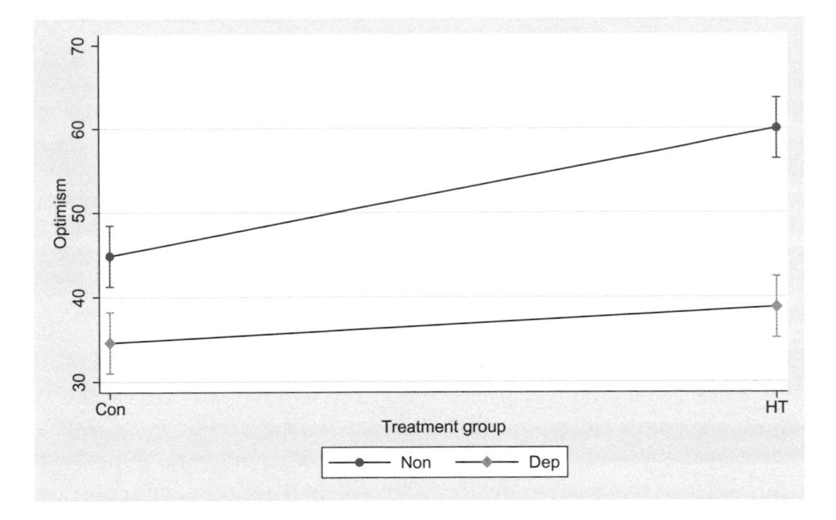

Figure 8.1. Graph of means

Each line in figure 8.1 reflects the effect of happiness therapy. The steepness of the line reflects the size of the effect of happiness therapy. The line for the nondepressed group looks steeper than the line for the depressed group, suggesting that happiness therapy might be more effective for those who are nondepressed. The test of the treatment by depression group interaction tests whether the difference in the steepness of the lines is significant.

Let's now run an analysis that predicts `opt` based on `treat`, `depstat`, and the interaction of these two variables. This analysis uses the `anova` command (instead of the `regress` command) because the `anova` command directly shows the significance tests for each of the main effects as well as the interaction.

3. For example, optimism might be more malleable for those who have not been diagnosed as clinically depressed than for those who have been diagnosed as clinically depressed. If so, then happiness therapy might be more effective for those who are not depressed than for those who are depressed.

```
. anova opt depstat##treat

                   Number of obs =        120    R-squared     =  0.4889
                   Root MSE      =    10.0136    Adj R-squared =  0.4757

          Source | Partial SS         df         MS        F    Prob>F
    -------------+----------------------------------------------------
           Model |      11126          3  3708.6667    36.99  0.0000
                 |
         depstat |  7426.1333          1  7426.1333    74.06  0.0000
           treat |  2803.3333          1  2803.3333    27.96  0.0000
   depstat#treat |  896.53333          1  896.53333     8.94  0.0034
                 |
        Residual |  11631.467        116  100.27126
    -------------+----------------------------------------------------
           Total |  22757.467        119  191.23922
```

Note! The anova and regress commands

You could use the `regress` command instead of the `anova` command to fit the previous model, as well as all models illustrated in this chapter. However, there are two caveats. First, the `regress` command will require an extra step using the `contrast` command to test the overall interaction (for example, `contrast depstat#treat`). Second, the tests of the main effects differ when using the `regress` command compared with the `anova` command, as explained in section 8.6.

The `depstat#treat` interaction is statistically significant ($F = 8.94$, $p = 0.0034$). To begin to understand the nature of this interaction, we can compute the means broken down by treatment group and depression status using the `margins` command below.[4] The average optimism for those in treatment 1 (control group) and depression status 1 (nondepressed) is 44.9. The average optimism for those in treatment 2 (happiness therapy) and depression status 1 (nondepressed) is 60.0. The effect of happiness therapy among those who are nondepressed is 15.1 (60.0 − 44.9). By contrast, for those who are depressed, the effect of happiness therapy is 4.2 (38.8 − 34.6). The significant interaction indicates that these effects are significantly different (that 15.1 significantly differs from 4.2). In other words, the effect of happiness therapy for those who are nondepressed is significantly different from the effect of happiness therapy for those who are depressed.

4. Note that the means are the same as those that we obtained from the `table` command earlier. This is because there are no additional predictors (covariates) in the model. Had there been additional predictors in the model, the means from the `margins` command would have been adjusted for those predictors and would have differed from the means produced by the `table` command.

```
. margins treat#depstat, nopvalues
Adjusted predictions                              Number of obs     =        120
Expression   : Linear prediction, predict()

                        Delta-method
                  Margin   Std. Err.     [95% Conf. Interval]

treat#depstat
     Con#Non  44.86667   1.828216     41.24565    48.48768
     Con#Dep      34.6   1.828216     30.97899    38.22101
      HT#Non        60   1.828216     56.37899    63.62101
      HT#Dep      38.8   1.828216     35.17899    42.42101
```

It can be easier to interpret the results by displaying them using a graph. The `marginsplot` command below creates a graph of the means computed by the `margins` command. This creates the graph that we saw in figure 8.1.

```
. marginsplot
  (output omitted)
```

The graph in figure 8.1 illustrated how the effect of happiness therapy is greater for those who are nondepressed than for those who are depressed. It also helps us see the effect of happiness therapy at each level of depression status. It appears that among those who are depressed, those who receive happiness therapy have similar optimism to those in the control group. We can assess the effect of treatment at each level of depression through the use of simple effects analysis, described below.

8.2.1 Simple effects

The significant interaction indicates that the effect of happiness therapy is different for those who are depressed versus nondepressed. Each of these effects is called a *simple effect*, because they reflect the effect of one variable while holding another variable constant. We can further probe the nature of this interaction by looking at the simple effect of happiness therapy separately for those who are depressed and nondepressed. Referring to figure 8.1, these simple effects correspond to the slopes of each of the lines. We can estimate and test these simple effects using the `contrast` command, as shown below. Note the use of the `@` symbol. This requests the simple effect of `treat` at each level of `depstat`.

```
. contrast treat@depstat
Contrasts of marginal linear predictions
Margins      : asbalanced

-----------------------------------------------------
                |         df           F        P>F
----------------+------------------------------------
treat@depstat   |
           Non  |          1       34.26     0.0000
           Dep  |          1        2.64     0.1070
         Joint  |          2       18.45     0.0000
                |
   Denominator  |        116
-----------------------------------------------------
```

The first test shows that the effect of happiness therapy is significant for those who are nondepressed ($F = 34.26$, $p = 0.0000$). The second test shows that the effect of happiness therapy is not significant for those who are depressed ($F = 2.64$, $p = 0.1070$).

Let's run this `contrast` command again, except this time we will add the `nowald` and `pveffects` options, as shown below.

```
. contrast treat@depstat, nowald pveffects
Contrasts of marginal linear predictions
Margins      : asbalanced
```

	Contrast	Std. Err.	t	P>|t|
treat@depstat				
(HT vs base) Non	15.13333	2.585489	5.85	0.000
(HT vs base) Dep	4.2	2.585489	1.62	0.107

By adding the **nowald** and **pveffects** options, the **contrast** command displays a table with the estimate of the simple effect, the standard error, and a significance test of the simple effect. This shows the effect of treatment among those who are nondepressed equals 15.1, and this effect is significant ($t = 5.85$, $p = 0.000$).[5] Among those who are depressed, the treatment effect is 4.2, and this difference is not significant ($t = 1.62$, $p = 0.107$).

Note! Contrast options

Combining the **nowald** and **pveffects** options provides a concise output that includes an estimate of the size of the contrast and a test of its significance. Many examples in this chapter will incorporate these options on the **contrast** command. See appendix D for more about the **nowald** and **pveffects** options for customizing the output from the **contrast** command.

8.2.2 Estimating the size of the interaction

As we saw in the analysis of the simple effects, the simple effect of treatment is 4.2 for those who are depressed and is 15.1 for those who are nondepressed. Taking the difference in these simple effects (4.2 – 15.1) gives us an estimate of the size of the interaction, which is −10.9. This is the same value that we obtain if we estimate the interaction using the **contrast** command below.

```
. contrast treat#depstat, nowald pveffects
Contrasts of marginal linear predictions
Margins      : asbalanced
```

	Contrast	Std. Err.	t	P>|t|
treat#depstat				
(HT vs base) (Dep vs base)	-10.93333	3.656433	-2.99	0.003

5. This significance test is reported as a t test, but it is equivalent to the F test we saw from the previous example.

This provides the estimate of the size of the interaction effect, the standard error, and a test of the significance of the interaction. Note that the p-value for this test matches the p-value of the `treat#depstat` interaction from the original `anova` command.

8.2.3 More about interaction

Before concluding this section on two by two interactions, let's further explore what we mean by an interaction by considering the hypothetical pattern of results shown in figure 8.2, where there is no interaction.

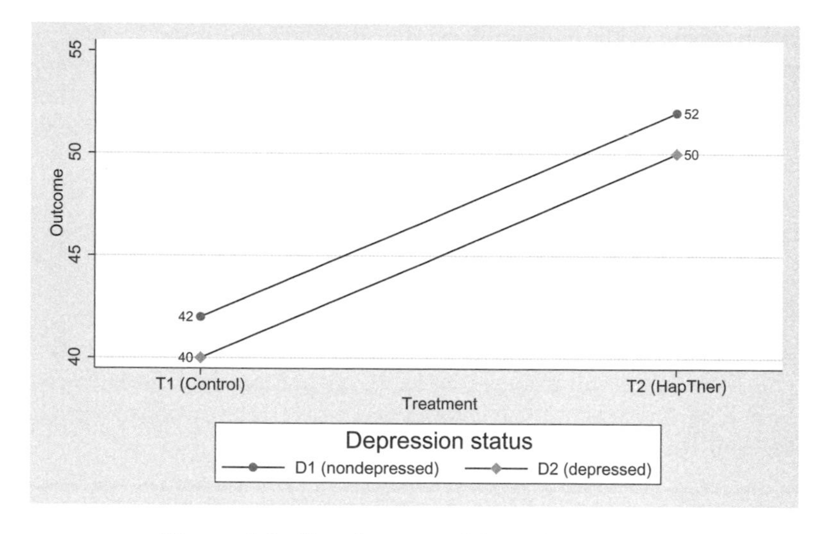

Figure 8.2. Two by two with no interaction

This is an example pattern in which there is no interaction between treatment and depression status. One way we can see the absence of an interaction is by seeing that the line for *D1* is parallel to the line for *D2*. What makes these lines parallel is the fact that the distance between *D1* and *D2* is the same at each level of treatment. At *T1*, the difference between *D1* and *D2* is 2 $(42-40)$. At level *T2*, the difference between *D1* and *D2* is also 2 $(52-50)$. The size of the interaction is the difference in these differences, $2-2$ or 0.

Another way to think about the absence of an interaction is to think in terms of comparing the simple effect of treatment for those who are nondepressed and depressed. The simple effect of treatment (*T2*− *T1*) for those who are nondepressed is 10 $(52-42)$. The simple effect of treatment for those who are depressed is also 10 $(50-40)$. The interaction is the difference in these simple effects, $10-10$ or 0.

8.2.4 Summary

This section has illustrated how to perform an analysis involving an interaction of two categorical variables that each have two levels. To help interpret the interaction, the `margins` command can be used to create a table of the outcome means broken down by the categorical variables, and the `marginsplot` command can be used to create a graph of the means. Furthermore, the `contrast` command can be used to test simple effects and can also be used to obtain an estimate of the size of the two by two interaction.

8.3 Two by three models

This section considers models where one of the categorical variables has two levels and the other categorical variable has three levels. Such models are often described as two by three models (also called three by two models). The examples from this section are an extension of the example from section 8.2.

8.3.1 Example 2

Referring to the example from the previous section, depression status had two levels, nondepressed and depressed. Suppose that we instead use three categories for depression status: nondepressed, mildly depressed, and severely depressed. The previous example found that the treatment effect was not significant among those who are depressed. Perhaps the effect of happiness therapy for those who are mildly depressed is significant, but it could not be detected in the previous study because they were pooled together with those who are severely depressed.

Let's begin by using the dataset for this example and showing a table of the mean of optimism broken down by treatment group and depression group.

```
. use opt-2by3-ex1, clear
. table depstat treat, contents(mean opt)
```

Depressio n group	Treatment group Con	 HT
Non	44.2	59.6
Mild	39.6	49.7
Sev	29.8	29.6

Let's look at the effect of happiness therapy (compared with the control group) for each depression group. For those who are nondepressed, the effect of happiness therapy is 59.6 compared with 44.2 for the control group ($59.6-44.2 = 15.4$). Compare that with the effect of happiness therapy for those who are mildly depressed ($49.7 - 39.6 = 10.1$) and for those who are severely depressed ($29.6 - 29.8 = -0.2$). This suggests that the effect of happiness therapy (compared with the control group) may be greater for those who are nondepressed and mildly depressed than for those who are severely depressed.

In other words, it appears that there may be a treatment group by depression group interaction.

Let's look at a graph of these means as shown in figure 8.3. This is another way to see that happiness therapy appears to be more effective for those who are nondepressed and mildly depressed than for those who are severely depressed.

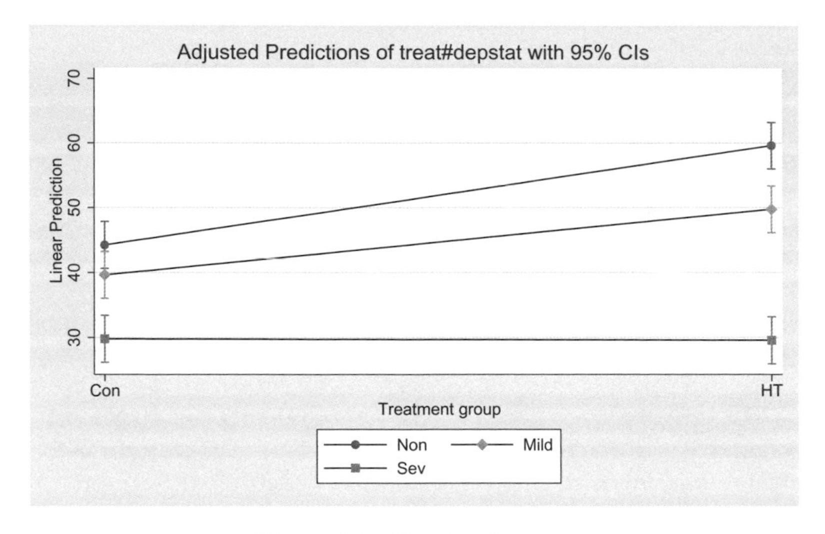

Figure 8.3. Graph of means

Let's now perform an analysis predicting optimism from treatment group, depression status, and the interaction of these two variables.

```
. anova opt depstat##treat

                    Number of obs =        180     R-squared     =  0.5402
                    Root MSE      =    10.0154     Adj R-squared =  0.5270

          Source  | Partial SS         df         MS           F    Prob>F
  ----------------+-------------------------------------------------------
           Model  |  20505.828          5    4101.1656      40.89   0.0000
                  |
         depstat  |  15432.844          2    7716.4222      76.93   0.0000
           treat  |  3183.6056          1    3183.6056      31.74   0.0000
   depstat#treat  |  1889.3778          2    944.68889       9.42   0.0001
                  |
        Residual  |  17453.567        174    100.30785
  ----------------+-------------------------------------------------------
           Total  |  37959.394        179    212.06366
```

As expected, the `depstat#treat` interaction is significant. We can compute the mean optimism as a function of depression status and treatment group by using the `margins` command below. Had there been additional predictors in the model, the `margins` command would have produced adjusted means, adjusting for the other predictors in the model.

```
. margins treat#depstat, nopvalues

Adjusted predictions                            Number of obs     =        180
Expression   : Linear prediction, predict()

------------------------------------------------------------------------
             |            Delta-method
             |     Margin   Std. Err.     [95% Conf. Interval]
-------------+----------------------------------------------------------
treat#depstat |
     Con#Non |   44.23333    1.82855      40.62434    47.84233
    Con#Mild |   39.63333    1.82855      36.02434    43.24233
     Con#Sev |       29.8    1.82855      26.19101    33.40899
      HT#Non |       59.6    1.82855      55.99101    63.20899
     HT#Mild |   49.73333    1.82855      46.12434    53.34233
      HT#Sev |   29.56667    1.82855      25.95767    33.17566
------------------------------------------------------------------------
```

The `marginsplot` command can be used to create a graph of the means created by the `margins` command. This produces the same graph that we saw in figure 8.3.

```
. marginsplot
  (output omitted)
```

Now that we know the interaction is significant, we can say that the effect of happiness therapy (compared with the control group) differs as a function of depression status. One way to further understand this interaction is through tests of simple effects.

Simple effects

We can ask whether the effect of happiness therapy is significant at each level of depression status. In other words, we can test the simple effect of `treat` at each level of `depstat`. We test this using the `contrast` command, as shown below.

```
. contrast treat@depstat, nowald pveffects

Contrasts of marginal linear predictions

Margins      : asbalanced

-------------------------------------------------------------------
                    |   Contrast   Std. Err.        t    P>|t|
--------------------+----------------------------------------------
      treat@depstat |
 (HT vs base) Non   |   15.36667    2.58596      5.94    0.000
 (HT vs base) Mild  |       10.1    2.58596      3.91    0.000
 (HT vs base) Sev   |  -.2333333    2.58596     -0.09    0.928
-------------------------------------------------------------------
```

For those who are nondepressed, the effect of happiness therapy versus the control group is 15.4, and that difference is significant. For those who are mildly depressed, the effect of happiness therapy is 10.1, and that difference is also significant. However, for those who are severely depressed, the effect of happiness therapy is −0.2, and that difference is not significant. This test of simple effects tells us that happiness therapy is significantly better than the control group for those who are nondepressed and for those who are mildly depressed. For those who are severely depressed, happiness therapy is not significantly different from being in the control group.

Partial interactions

Another way to dissect a two by three interaction is through the use of partial interactions. In this example, a partial interaction is constructed by applying a contrast operator to `depstat` and interacting that with `treat`. For example, applying the `a.` contrast operator to `depstat` yields two contrasts: group 1 versus 2 (`Non vs Mild` depression); and group 2 versus 3 (`Mild vs Sev` depression). Interacting `a.depstat` with `treat` forms two partial interactions. I used the means from the `margins` command to manually create graphs that illustrates these two partial interactions, shown in figure 8.4.

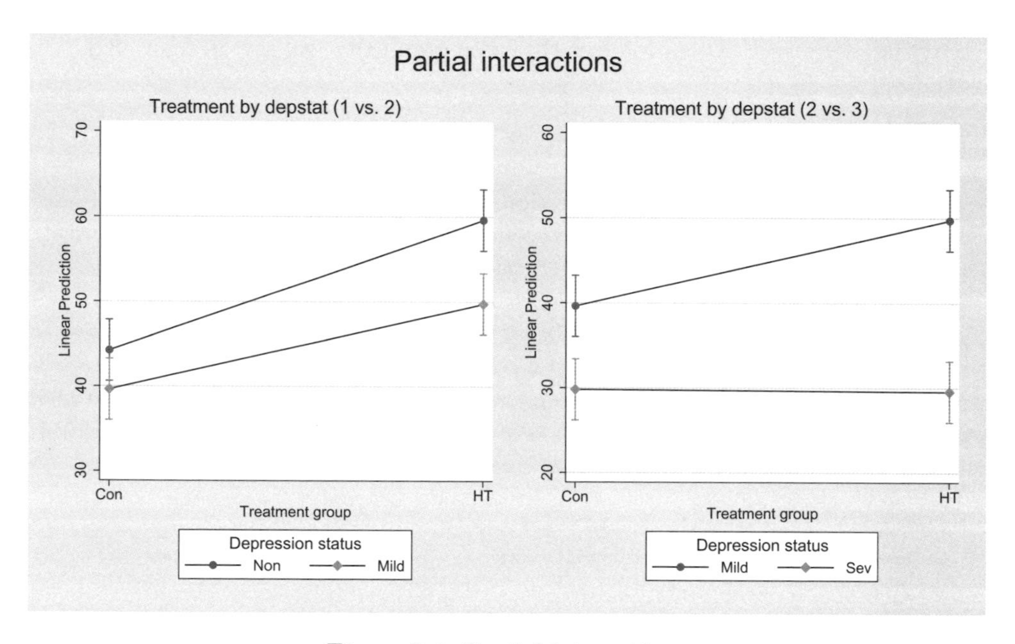

Figure 8.4. Partial interactions

The first partial interaction forms a two by two interaction of treatment by depression group 1 versus 2 (`Non vs Mild` depression), as pictured in the left panel of figure 8.4. This tests whether the treatment effect is the same for those who are nondepressed versus mildly depressed. The second partial interaction forms a two by two interaction of treatment group by depression group 2 versus 3 (`Mild vs Sev` depression), as pictured in the right panel of figure 8.4. This tests if the treatment effect is the same for those who are mildly depressed versus severely depressed. We can test these partial interactions using the `contrast` command below.

```
. contrast a.depstat#treat
Contrasts of marginal linear predictions
Margins      : asbalanced

                            |       df          F        P>F
----------------------------+--------------------------------
               depstat#treat|
(Non vs Mild) (joint)       |        1       2.07     0.1516
(Mild vs Sev) (joint)       |        1       7.98     0.0053
                       Joint|        2       9.42     0.0001
                            |
                 Denominator|      174
-------------------------------------------------------------
```

The first partial interaction is not significant ($F = 2.07$, $p = 0.1516$). The effect of happiness therapy (compared with the control group) is not significantly different for those who are nondepressed versus mildly depressed. Referring to the left panel of figure 8.4, we can see that the effect of happiness therapy (compared with the control group) is similar for those who are nondepressed and those who are mildly depressed.

The second partial interaction is significant ($F = 7.98$, $p = 0.0053$). The effect of happiness therapy (compared with the control group) is significantly different when comparing those who are mildly depressed versus severely depressed. Referring to the graph of the means in the right panel of figure 8.4, we can see that the effect of happiness therapy (compared with the control group) is greater for those who are mildly depressed than for those who are severely depressed.

We are not limited to using the `a.` contrast operator when forming such partial interactions. Any of the contrast operators illustrated in chapter 7 could have been used in place of the `a.` contrast operator. For example, the `j.` contrast operator (as described in section 7.8.1) is interacted with `treat` using the `contrast` command. The `j.` contrast operator compares each group with the mean of the previous groups. When applied to `depstat`, this yields a comparison of group 2 versus 1 (`Mild vs Non` depressed) and of group 3 versus groups 1 and 2 combined (`Sev vs <Sev`; severely depressed versus mildly and nondepressed combined).

```
. contrast j.depstat#treat
Contrasts of marginal linear predictions
Margins      : asbalanced

                            |       df          F        P>F
----------------------------+--------------------------------
               depstat#treat|
(Mild vs  Non) (joint)      |        1       2.07     0.1516
 (Sev vs <Sev) (joint)      |        1      16.76     0.0001
                       Joint|        2       9.42     0.0001
                            |
                 Denominator|      174
-------------------------------------------------------------
```

The first contrast shown by the `contrast` command is the same as the first test we saw using the `a.` contrast operator. However, the second contrast shows us the interaction of the comparison of those who are severely depressed versus those mildly depressed and nondepressed interacted with treatment. This partial interaction is significant. The effectiveness of happiness therapy (compared with the control group) is significantly different when comparing severely depressed people versus the combination of those who are mildly depressed or nondepressed.

8.3.2 Example 3

Let's consider another example of a two by three design. This study is an extension of the two by two study from section 8.2, but adds a third treatment group, traditional therapy. Each subject is assigned to one of the three treatments: a control group, traditional therapy, or happiness therapy. Furthermore, the participants are classified as nondepressed or depressed. Thus, the design includes depression status as a two-level factor and treatment group as a three-level factor, forming a two by three design. The dataset for this example is used below.

```
. use opt-2by3-ex2
```

The `table` command is used to show the mean optimism by treatment group and depression status.

```
. table depstat treat, contents(mean opt)

----------------------------------
Depressio |    Treatment group
n status  |    Con     TT     HT
----------+-----------------------
      Non |   44.6   54.7   59.3
      Dep |   34.8   44.3   39.2
----------------------------------
```

Let's analyze the data using the `anova` command predicting optimism from `depstat`, `treat`, and the interaction of these two variables. The analysis shows that the `depstat#treat` interaction is significant ($F = 5.00$, $p = 0.0077$).

```
. anova opt depstat##treat

                           Number of obs =        180    R-squared     =  0.4256
                           Root MSE      =    9.97535    Adj R-squared =  0.4091

                  Source | Partial SS         df       MS           F    Prob>F
          ---------------+----------------------------------------------------
                   Model |  12830.667          5   2566.1333       25.79  0.0000
                         |
                 depstat |  8107.0222          1   8107.0222       81.47  0.0000
                   treat |     3727.6          2      1863.8       18.73  0.0000
           depstat#treat |  996.04444          2   498.02222        5.00  0.0077
                         |
                Residual |  17314.333        174   99.507663
          ---------------+----------------------------------------------------
                   Total |      30145        179   168.40782
```

Let's use the **margins** command to show the mean of the outcome broken down by **treat** and **depstat**. This is followed by the **marginsplot** command to create a graph of these means, as shown in figure 8.5.

```
. margins treat#depstat, nopvalues
Adjusted predictions                          Number of obs     =        180
Expression   : Linear prediction, predict()

------------------------------------------------------------------------
             |            Delta-method
             |     Margin   Std. Err.      [95% Conf. Interval]
-------------+----------------------------------------------------------
treat#depstat|
     Con#Non |   44.63333   1.821242       41.03876     48.2279
     Con#Dep |   34.83333   1.821242       31.23876     38.4279
      TT#Non |   54.73333   1.821242       51.13876     58.3279
      TT#Dep |   44.33333   1.821242       40.73876     47.9279
      HT#Non |   59.26667   1.821242        55.6721    62.86124
      HT#Dep |       39.2   1.821242       35.60543    42.79457
------------------------------------------------------------------------

. marginsplot
  Variables that uniquely identify margins: treat depstat
```

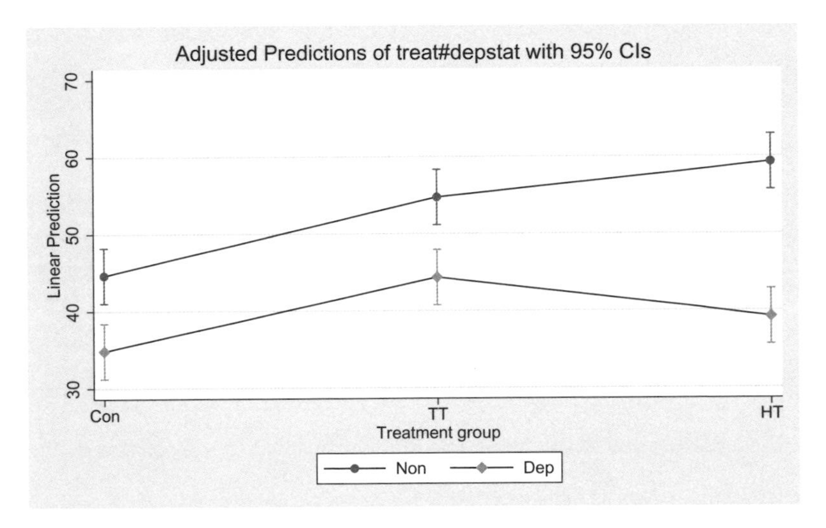

Figure 8.5. Graph of means

Let's inspect the graph of the means to begin to make sense of the interaction. If we focus on those who are nondepressed, it looks like both traditional therapy and happiness therapy are more effective than being in the control group. Among those who are depressed, it looks like traditional therapy yields larger optimism scores than the scores in the control group, but the optimism scores for happiness therapy and the control group seem similar.

Let's explore this further by looking at the effect of treatment group assignment separately for each level of depression.

Simple effects

To further understand this interaction, we can compute the simple effect of treatment at each level of depression status. This shows us whether the overall effect of treatment was significant at each level of depression status. The test of the simple effect of `treat` at each level of `depstat` is performed below using the `contrast` command.

```
. contrast treat@depstat
Contrasts of marginal linear predictions
Margins      : asbalanced

------------------------------------------------
              |         df           F        P>F
--------------+---------------------------------
treat@depstat |
         Non  |          2       16.92     0.0000
         Dep  |          2        6.82     0.0014
       Joint  |          4       11.87     0.0000
              |
  Denominator |        174
------------------------------------------------
```

The effect of treatment is significant for those who are depressed. The effect of treatment is also significant for those who are nondepressed. Let's further dissect these simple effects by applying contrasts to the `treat` factor variable, forming simple contrasts.

Simple contrasts

We can follow up on these simple effects by performing simple contrasts on treatment at each level of depression status. For example, let's apply the `r.` contrast operator to `treat` to compare each group with the reference group (that is, comparing traditional with the control group, and happiness therapy with the control group). These comparisons are performed separately for those who are nondepressed and for those who are depressed.

```
. contrast r.treat@depstat, nowald pveffects
Contrasts of marginal linear predictions
Margins      : asbalanced

---------------------------------------------------------------
                  |   Contrast   Std. Err.        t     P>|t|
------------------+--------------------------------------------
    treat@depstat |
(TT vs Con) Non   |       10.1    2.575625     3.92     0.000
(TT vs Con) Dep   |        9.5    2.575625     3.69     0.000
(HT vs Con) Non   |   14.63333    2.575625     5.68     0.000
(HT vs Con) Dep   |   4.366667    2.575625     1.70     0.092
---------------------------------------------------------------
```

This yields a total of four tests. The first test compares the traditional therapy group with the control group (group 2 versus 1) for those who are nondepressed. The optimism scores for those in traditional therapy are significantly larger than the control group (by 10.1 units) among those who are nondepressed. The second test forms this same comparison for those who are depressed. The mean optimism for those in traditional therapy is significantly larger (by 9.5 units) than the control group for those who are depressed.

The third and fourth tests compare the happiness therapy group with the control group (group 3 versus 1). The third test forms this comparison among those who are nondepressed, showing a significant difference (14.6) in favor of happiness therapy. The fourth test forms the same comparison for those who are depressed, finding a difference of 4.4, which is not significant.

Partial interaction

Another way to dissect a three by two interaction is through the use of a partial interaction. This applies a contrast to the three-level factor and interacts those contrasts with the two-level factor. Let's apply the `r.` contrast operator to the treatment factor variable. This yields two contrasts—group 2 versus 1 (traditional therapy versus control); and group 3 versus 1 (happiness therapy versus control). Interacting these contrasts with depression status forms two partial interactions. The first partial interaction forms a two by two interaction of treatment group 2 versus 1 with depression status (as pictured in the left panel of figure 8.6). The second partial interaction forms a two by two interaction of treatment group 3 versus 1 with depression status (as pictured in the right panel of figure 8.6).[6]

6. Note that I manually created the graph shown in figure 8.6.

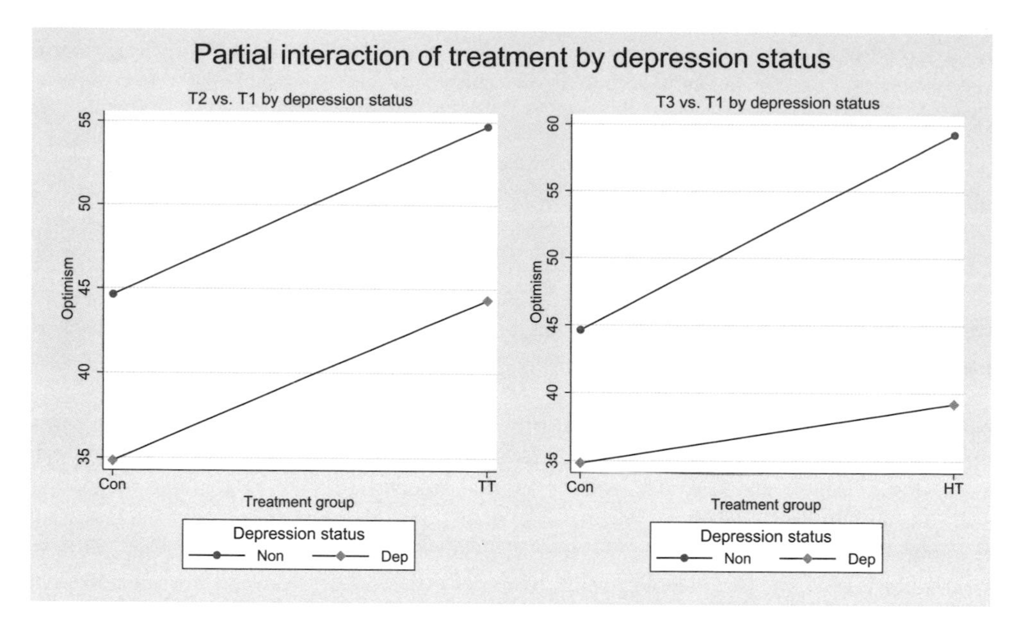

Figure 8.6. Partial interactions

These partial interactions can be estimated by applying the `r.` contrast operator to `treat` and interacting that with `depstat`, as shown below.

```
. contrast r.treat#depstat
Contrasts of marginal linear predictions
Margins      : asbalanced
```

	df	F	P>F
treat#depstat			
(TT vs Con) (joint)	1	0.03	0.8694
(HT vs Con) (joint)	1	7.94	0.0054
Joint	2	5.00	0.0077
Denominator	174		

The first partial interaction is not significant ($F = 0.03$, $p = 0.8694$). This shows that the effect of traditional therapy (when compared with the control group) is not significantly different for those who are depressed versus nondepressed. This is illustrated in the left panel of figure 8.6.

The second partial interaction is significant. This shows that the effect of happiness therapy (when compared with the control group) is significantly different for those who are nondepressed compared with those who are depressed. The results of the test are illustrated in the right panel of figure 8.6, showing that happiness therapy is significantly more effective for those who are nondepressed than for those who are depressed.

8.3.3 Summary

This section has used two examples to illustrate interactions of two categorical variables where one variable has two levels and the other has three levels. The margins command creates a table of means broken down by the categorical predictors and then the marginsplot command graphs the means. The interaction can be further explored through tests of simple effects that look at the effect of one factor at each of the levels of the other factor. We can further explore the simple effects by performing simple contrasts that apply contrasts to one factor at each of the levels of the other factor. Finally, we saw that we can perform partial interaction tests by applying contrasts to the three-level factor and interacting those contrasts with the two-level factor.

8.4 Three by three models: Example 4

Let's now consider an example that illustrates a three by three design. Like the previous examples, the two categorical variables are treatment group and depression group, and the outcome is optimism. In this example, there are three levels of treatment (control group, traditional therapy, and happiness therapy), and three depression groups (nondepressed, mildly depressed, and severely depressed).

The dataset for this analysis is used below, followed by a tabulation of the frequencies of participants by depstat and treat.

```
. use opt-3by3, clear
. tabulate depstat treat
```

Depression group	Treatment group Con	TT	HT	Total
Non	30	30	30	90
Mild	30	30	30	90
Sev	30	30	30	90
Total	90	90	90	270

There were a total of 270 participants in the study. Before the beginning of the study, 90 participants were nondepressed, 90 were mildly depressed, and 90 were severely depressed. Each of these 90 participants was randomly assigned (in equal numbers) to one of three treatments: a control group, traditional therapy, or happiness therapy. After treatment, optimism scores were measured for each participant. The table command is used below to compute the mean of the optimism scores by treatment group and level of depression.

```
. table depstat treat, contents(mean opt)

----------------------------
Depressio |  Treatment group
n group   |   Con     TT     HT
----------+-----------------
      Non |  44.2   54.5   59.3
     Mild |  39.7   49.5   49.9
      Sev |  29.9   39.8   30.1
----------------------------
```

I invite you to take a moment and informally compare the optimism scores among the three groups at each level of depression (nondepressed, mildly depressed, and severely depressed). Are you seeing any kind of pattern? Now, let's see what the statistical analysis test is. We will use the **anova** command to analyze optimism scores at the end of the study as a function of **treat**, **depstat**, and the interaction of these two variables.

```
. anova opt depstat##treat
                         Number of obs =        270    R-squared     =  0.4898
                         Root MSE      =     10.023    Adj R-squared =  0.4742

                  Source |  Partial SS         df         MS         F    Prob>F
          ---------------+----------------------------------------------------
                   Model |   25175.519          8   3146.9398     31.32  0.0000
                         |
                 depstat |   17664.096          2   8832.0481     87.92  0.0000
                   treat |    5250.363          2   2625.1815     26.13  0.0000
           depstat#treat |   2261.0593          4   565.26481      5.63  0.0002
                         |
                Residual |   26220.367        261   100.46117
          ---------------+----------------------------------------------------
                   Total |   51395.885        269   191.06277
```

The **depstat#treat** interaction is significant. Let's use the **margins** command to compute the mean optimism by **treat** and **depstat** and then use the **marginsplot** command to graph the means computed by the **margins** command (see figure 8.7). We can use this table and graph to help us interpret the **depstat#treat** interaction.

```
. margins treat#depstat, nopvalues
Adjusted predictions                            Number of obs     =        270
Expression   : Linear prediction, predict()

------------------------------------------------------------------------
              |            Delta-method
              |     Margin   Std. Err.     [95% Conf. Interval]
--------------+---------------------------------------------------------
treat#depstat |
      Con#Non |       44.2   1.829947      40.59666    47.80334
     Con#Mild |       39.7   1.829947      36.09666    43.30334
      Con#Sev |       29.9   1.829947      26.29666    33.50334
       TT#Non |   54.53333   1.829947      50.92999    58.13667
      TT#Mild |   49.53333   1.829947      45.92999    53.13667
       TT#Sev |       39.8   1.829947      36.19666    43.40334
       HT#Non |   59.33333   1.829947      55.72999    62.93667
      HT#Mild |   49.86667   1.829947      46.26333    53.47001
       HT#Sev |       30.1   1.829947      26.49666    33.70334
------------------------------------------------------------------------

. marginsplot
  Variables that uniquely identify margins: treat depstat
```

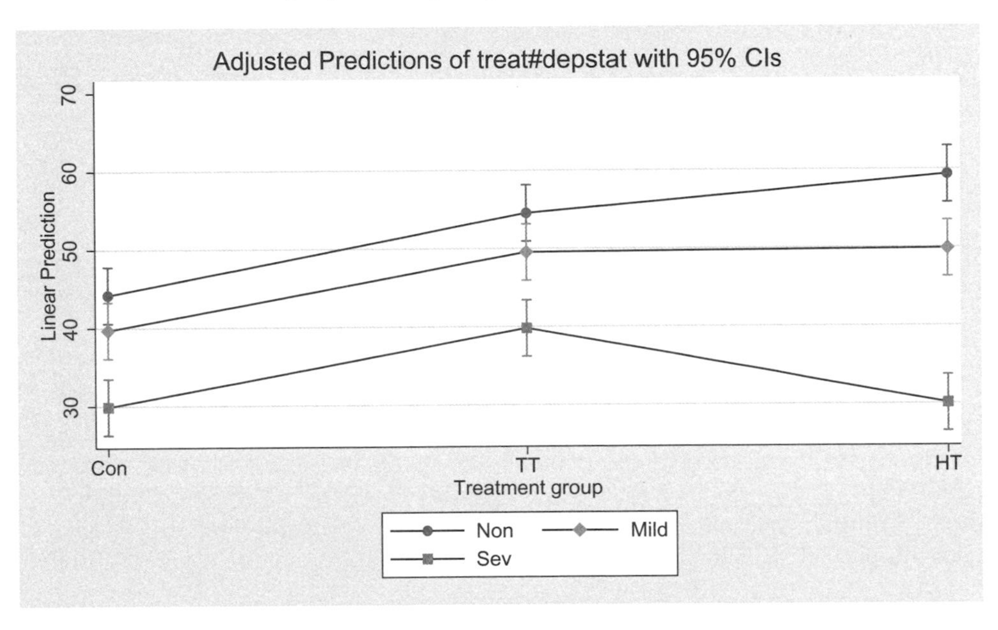

Figure 8.7. Mean optimism by treatment group and depression status

Let's focus on the two lines representing the nondepressed and mildly depressed groups in figure 8.7. The effect of treatment assignment appears similar for these two groups. Optimism scores look larger for traditional therapy than for the control group, and optimism scores for happiness therapy appear larger than the control group as well. Contrast this with the severely depressed group, where it appears that the optimism scores are larger for those in traditional therapy than for those in the control group but about the same for those in happiness therapy as for those in the control group.

We can statistically dissect this interaction in four different ways: using simple effects (see section 8.4.1), simple contrasts (see section 8.4.2), partial interactions (see section 8.4.3), or interaction contrasts (see section 8.4.4). Each of these techniques is illustrated below.

8.4.1 Simple effects

One way to dissect the interaction is by looking at the effect of treatment at each level of depression status. This is performed using the `contrast` command below.

```
. contrast treat@depstat
Contrasts of marginal linear predictions
Margins      : asbalanced

------------------------------------------------
             |         df           F        P>F
-------------+----------------------------------
treat@depstat|
         Non |          2       17.86     0.0000
        Mild |          2        9.96     0.0001
         Sev |          2        9.56     0.0001
       Joint |          6       12.46     0.0000
             |
 Denominator |        261
------------------------------------------------
```

Three tests are performed, showing the effect of `treat` at each of the three levels of `depstat`. These results show that the effect of `treat` is significant at each level of `depstat`. We can further dissect the simple effects by applying contrasts to `treat`, yielding simple contrasts.

8.4.2 Simple contrasts

Let's repeat the previous `contrast` command but apply the `r.` contrast operator to `treat`. This yields a comparison of each treatment group with the reference group (that is, group 1, the control group) at each level of depression status. To keep the output simple, let's begin by focusing on those who are nondepressed (by specifying `1.depstat`).

```
. contrast r.treat@1.depstat, nowald pveffects
Contrasts of marginal linear predictions
Margins      : asbalanced

--------------------------------------------------------------
                  |   Contrast   Std. Err.        t    P>|t|
------------------+-------------------------------------------
    treat@depstat |
(TT vs Con) Non   |   10.33333   2.587936      3.99    0.000
(HT vs Con) Non   |   15.13333   2.587936      5.85    0.000
--------------------------------------------------------------
```

For those who are nondepressed, the comparison of traditional therapy versus control group is significant ($t = 3.99$, $p < 0.0.001$). The mean optimism is 10.3 units larger for those receiving traditional therapy than those in the control group. For those who are nondepressed, the comparison of happiness therapy versus control group is also significant ($t = 5.85$, $p < 0.0.001$). The mean optimism scores are, on average, 15.1 units larger for those receiving happiness therapy than those in the control group.

Let's now perform these simple contrasts for those who are mildly depressed.

```
. contrast r.treat@2.depstat, nowald pveffects
Contrasts of marginal linear predictions
Margins      : asbalanced

─────────────────────────────────────────────────────────────
                     │   Contrast   Std. Err.        t    P>|t|
─────────────────────┼───────────────────────────────────────
       treat@depstat │
(TT vs Con) Mild     │   9.833333   2.587936      3.80    0.000
(HT vs Con) Mild     │   10.16667   2.587936      3.93    0.000
─────────────────────────────────────────────────────────────
```

Both of these contrasts are significant ($p < 0.001$). The mean optimism is 9.8 units larger for those in the traditional therapy group than those in the control group. Also, the happiness therapy group shows 10.2 units larger optimism than those in the control group.

Finally, let's perform these simple contrasts for those who are severely depressed.

```
. contrast r.treat@3.depstat, nowald pveffects
Contrasts of marginal linear predictions
Margins      : asbalanced

───────────────────────────────────────────────────────────
                   │   Contrast   Std. Err.        t    P>|t|
───────────────────┼───────────────────────────────────────
     treat@depstat │
(TT vs Con) Sev    │        9.9   2.587936      3.83    0.000
(HT vs Con) Sev    │         .2   2.587936      0.08    0.938
───────────────────────────────────────────────────────────
```

Among those who are severely depressed, there is a significant difference comparing traditional therapy with the control group ($t = 3.83$, $p < 0.0.001$) The average optimism for those in traditional therapy is 9.9 units larger than the control group. However, the difference between group happiness therapy and the control group is not significant ($t = 0.08$, $p < 0.0.938$).

If you prefer, you can obtain all six of these simple contrasts at once using the `contrast` command below (the output is omitted to save space).

```
. contrast r.treat@depstat, nowald pveffects
  (output omitted)
```

Let's now consider the use of partial interactions to further understand the `treat#depstat` interaction.

8.4.3 Partial interaction

The three by three interaction can also be further understood through the use of partial interactions. A contrast is applied to one of the factors, and those contrasts are interacted with the other three-level factor. For example, applying the `r.` contrast operator to the treatment factor variable yields two contrasts: group 2 versus 1 and group 3 versus 1. We can then interact these contrasts with depression status. This decomposes the overall three by three interaction into a pair of three by two interactions.

I used the means from the `margins` command to create a visual depiction of these two partial interactions, shown in figure 8.8. The left panel of figure 8.8 illustrates the comparison of treatment group 2 versus 1 (traditional therapy versus control) interacted with depression status. The right panel illustrates the comparison of treatment group 3 versus 1 (happiness therapy versus control) interacted with depression status.

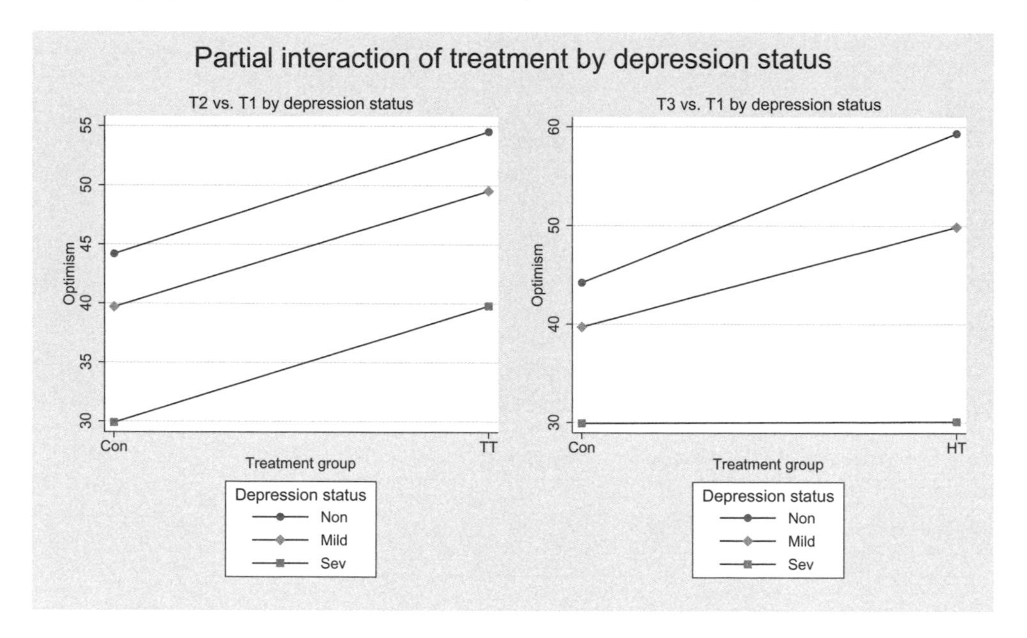

Figure 8.8. Partial interactions

The `contrast` command below tests the two partial interactions pictured in figure 8.8.

```
. contrast r.treat#depstat

Contrasts of marginal linear predictions

Margins      : asbalanced

------------------------------------------------
                   |         df           F        P>F
-------------------+----------------------------
     treat#depstat |
(TT vs Con) (joint)  |          2        0.01     0.9891
(HT vs Con) (joint)  |          2        8.64     0.0002
              Joint  |          4        5.63     0.0002
                   |
        Denominator  |        261
------------------------------------------------
```

The first partial interaction is not significant ($F = 0.01$, $p = 0.9891$). The difference in optimism between traditional therapy and the control group does not differ among the levels of depression status. We can see this in left panel of figure 8.8. The effect of traditional therapy (versus the control group) is similar for all three lines (representing the three levels of depression).

The second partial interaction is significant ($F = 8.64$, $p = 0.0.002$). The difference in optimism between happiness therapy and the control group depends on the level of depression. Looking at the right panel of figure 8.8, it appears that the effect of happiness therapy (compared with the control group) may be similar for those who are nondepressed and mildly depressed but different for those who are severely depressed. We can investigate these differences using interaction contrasts, illustrated in the next section.

8.4.4 Interaction contrasts

An interaction contrast is formed by applying contrasts to both of the factor variables and then interacting the resulting contrasts. Suppose we applied the `r.` contrast operator to treatment group and the `a.` contrast operator to depression status. This would create contrasts of each treatment group against the control group (that is, group 2 versus 1 and group 3 versus 1) interacted with contrasts of adjacent levels of depression groups (that is, group 1 versus 2 and group 2 versus 3). This yields a total of four interaction contrasts. I have created a visual representation of these four interaction contrasts based on the means from the `margins` command, pictured in figure 8.9.

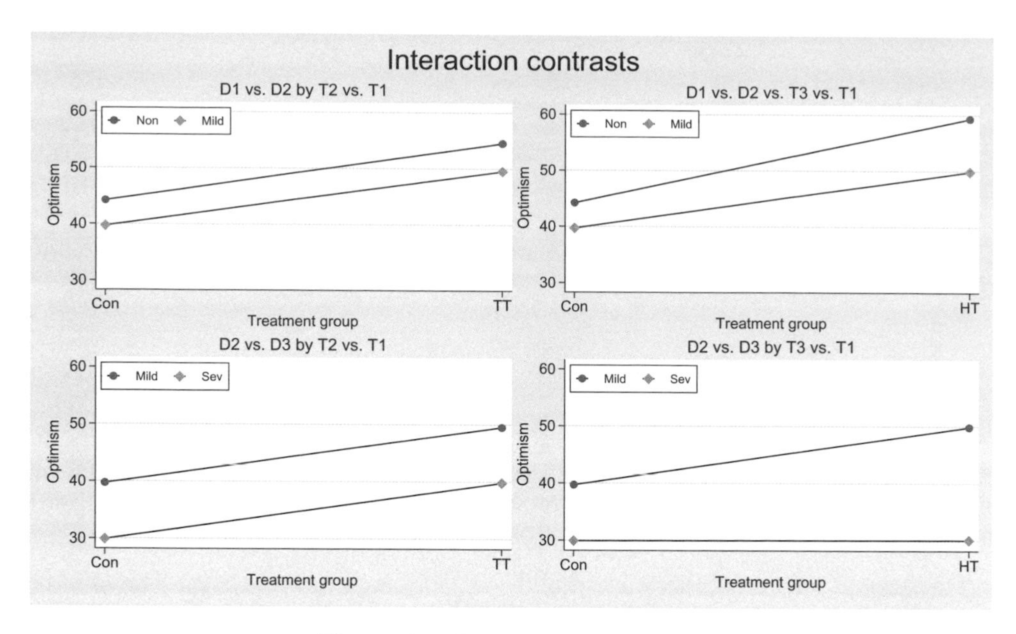

Figure 8.9. Interaction contrasts

Each panel of figure 8.9 forms a two by two interaction, selecting two levels of the treatment factor to be interacted with two levels of the depression factor. For example, the top left panel interacts the comparison of the traditional therapy group versus the control group by nondepressed versus mildly depressed. In the case of a three by three design, an interaction contrast dissects the interaction into four pieces. You can select the contrast operators for each factor to form comparisons that are of interest to you. In this example, we are applying the `r.` contrast operator to the treatment factor and the `a.` contrast operator to the depression status factor. The `contrast` command below tests each of these interaction contrasts.

```
. contrast a.depstat#r.treat, nowald pveffects
Contrasts of marginal linear predictions
Margins      : asbalanced
```

	Contrast	Std. Err.	t	P>|t|
depstat#treat				
(Non vs Mild) (TT vs Con)	.5	3.659894	0.14	0.891
(Non vs Mild) (HT vs Con)	4.966667	3.659894	1.36	0.176
(Mild vs Sev) (TT vs Con)	-.0666667	3.659894	-0.02	0.985
(Mild vs Sev) (HT vs Con)	9.966667	3.659894	2.72	0.007

The four tests created by the `contrast` command correspond to the top left, top right, bottom left, and bottom right panels of figure 8.9.

Let's begin by interpreting the fourth interaction contrast, the only one that was significant. This contrast, pictured in the bottom right panel of figure 8.9, is the same as a two by two interaction where treatment has two levels (control group versus happiness therapy) and depression status has two levels (mildly depressed versus severely depressed). The significance of this interaction contrast indicates that the effect of happiness therapy (compared with the control group) is different for those who are mildly depressed compared with those who are severely depressed. Based on the pattern of means we see in figure 8.9, we can say that happiness therapy is more effective for those who are mildly depressed than for those who are severely depressed.

Now, let's return to interpreting the first, second, and third contrasts. The first contrast shows that the effect of traditional therapy (versus the control group) is not significantly different for those who are nondepressed and mildly depressed. This is consistent with the pattern of means shown in the top left panel of figure 8.9.

The second contrast (which is depicted in the top right panel) focuses on the comparison of happiness therapy with the control group interacted with the comparison of nondepressed versus mildly depressed people. This nonsignificant interaction contrast indicates that the effect of happiness therapy does not differ (significantly) when comparing those who are nondepressed and mildly depressed.

The third contrast (depicted in the bottom left panel) shows that the effect of traditional therapy (versus the control group) is not significantly different when comparing those who are mildly and severely depressed.

8.4.5 Summary

This section has illustrated how you can interpret interactions that involve two categorical variables that both have three levels. Actually, the principles illustrated here can be applied to designs where each of the categorical variables has three or more levels. The interpretation of the interaction begins with using the `margins` command to compute the means as a function of the categorical variables followed by the `marginsplot` command to create a graph of the means. Then four different techniques can be used for dissecting the interaction: simple effects, simple contrasts, partial interactions, and interaction contrasts.

8.5 Unbalanced designs

The examples presented in this chapter have illustrated balanced designs—the number of observations was the same for all the cells. Even in the context of a randomized experiment, it is unusual to have the same number of observations in each cell. This section presents an example of an unbalanced design. This will allow us to consider two different strategies that can be used for estimating adjusted means, the as-observed strategy and the as-balanced strategy. As we will see, the `asbalanced` option can be used with the `margins` command to estimate margins as though the design were balanced, even if the actual design is not balanced.

For this example, let's use the GSS dataset with the variable **happy7** as the outcome. This Likert variable contains the respondent's rating of their happiness where 1 represents "completely unhappy" and 7 represents "completely happy". Let's predict happiness based on whether the respondent is married (**married**), whether the respondent is a college graduate (**cograd**), and the interaction of these two variables. Because **married** has two levels and **cograd** has two levels, this yields a two by two design. Let's assess whether this is a balanced design by displaying the cell frequencies of **married** by **cograd** when **happy7** is not missing.

```
. use gss_ivrm
. tabulate married cograd if !missing(happy7), row

+----------------+
| Key            |
|----------------|
|   frequency    |
| row percentage |
+----------------+

   marital:
married=1,
unmarried=     College graduate
         0       (1=yes, 0=no)
 (recoded)   Not CO Gr    CO Grad |     Total
-----------+----------------------+----------
 Unmarried         454        147 |       601
                 75.54      24.46 |    100.00
-----------+----------------------+----------
   Married         403        153 |       556
                 72.48      27.52 |    100.00
-----------+----------------------+----------
     Total         857        300 |     1,157
                 74.07      25.93 |    100.00
```

This tabulation clearly shows that this is not a balanced design. The number of observations is not equal in each of the cells. For example, there are 454 respondents who are unmarried and are not college graduates compared with 147 respondents who are unmarried and are college graduates. Note that this table also shows us that, overall, 25.93% of the respondents are college graduates and 74.07% are not college graduates. I will refer to these percentages when manually computing adjusted means later in this section.

Before performing the analysis, let's compute the mean of **happy7** by **married** and **cograd** using the **tabulate** command below.[7]

7. I use the **tabulate** command (instead of the **table** command) in this example because the variables **married** and **cograd** have missing values. The **table** command includes extra rows and columns for these missing values, while the **tabulate** command omits the missing values, producing a more readable table.

```
. tabulate married cograd, sum(happy7)
  Means, Standard Deviations and Frequencies of how happy R is (recoded)

  marital:   |
married=1,   |
unmarried=   |  College graduate
         0   |    (1=yes, 0=no)
 (recoded)   | Not CO Gr    CO Grad |     Total
-------------+----------------------+-----------
 Unmarried   | 5.3039648  5.5170068 | 5.3560732
             | 1.0590229  .99556373 | 1.0470596
             |       454        147 |       601
-------------+----------------------+-----------
   Married   | 5.7121588  5.6862745 |  5.705036
             |  .8986037  .79032462 | .86953016
             |       403        153 |       556
-------------+----------------------+-----------
     Total   |  5.495916  5.6033333 | 5.5237684
             | 1.0071217  .89926887 |  .9810472
             |       857        300 |      1157
```

This shows the mean of happy7 for each cell of the two by two design. It also shows the row mean of happy7 for each level of marital. I will refer to these means to illustrate how adjusted means are computed in an unbalanced design.

Let's now perform an analysis that predicts happy7 from married, cograd, and the interaction of these two variables. This is performed using the anova command below.

```
. anova happy7 married##cograd
                        Number of obs =      1,157    R-squared     =  0.0362
                        Root MSE      =    .964375    Adj R-squared =  0.0337

                Source | Partial SS         df         MS        F    Prob>F
        ---------------+----------------------------------------------------
                 Model |  40.284425          3   13.428142    14.44  0.0000
                       |
               married |  18.502336          1   18.502336    19.89  0.0000
                cograd |    1.94355          1     1.94355     2.09  0.1486
        married#cograd |  3.1674385          1   3.1674385     3.41  0.0652
                       |
              Residual |  1072.3119      1,153   .93001903
        ---------------+----------------------------------------------------
                 Total |  1112.5964      1,156   .96245361
```

We can see that the married#cograd interaction is not significant ($p = 0.0652$). Let's assume that we want to retain this interaction. (We may want to retain it based on theoretical considerations or because its p-value is close to 0.05.) Let's now turn our attention to married, which is significant ($p < 0.001$). To understand this significant result, let's use the margins command to compute the adjusted means by the levels of married.

```
. margins married, nopvalues
Predictive margins                              Number of obs     =      1,157
Expression   : Linear prediction, predict()

                           Delta-method
                    Margin   Std. Err.     [95% Conf. Interval]

     married
   Unmarried      5.359205   .0393607      5.281978    5.436431
     Married      5.705447   .0409245      5.625152    5.785742
```

Note how the adjusted means from the `margins` command are similar to, but not the same as, the row means from the `tabulate` command. Consider the adjusted mean for those who are not married, 5.359205. We can think of this adjusted mean as being computed by taking each cell mean of `happy7` among those who are not married and weighing it by the corresponding proportion of those are college graduates or noncollege graduates, as illustrated below.

```
. display 5.3039648*.7407 + 5.5170068*.2593
5.3592066
```

The cell mean for unmarried noncollege graduates (5.30) is multiplied by the overall proportion of respondents who are noncollege graduates (0.7407). The cell mean for unmarried college graduates (5.52) is multiplied by the overall proportion of college graduates (0.2593). When these weighted means are added together we obtain 5.3592, the adjusted mean for those who are not married.

We can likewise compute the adjusted mean for those who are married using the same strategy, as shown below.

```
. display 5.7121588*.7407 + 5.6862745*.2593
5.705447
```

The key point is that the adjusted means are computed by creating a weighted average of cell means that is weighted by the observed proportions of observations in the data (in this case, the observed proportions of `cograd`). Stata calls this the as-observed strategy. This is the default strategy, unless we specify otherwise. We can explicitly request this strategy by adding the `asobserved` option to the `margins` command, as shown below. This yields the same adjusted means we saw in the previous `margins` command.

```
. margins married, nopvalues asobserved
Predictive margins                          Number of obs     =      1,157
Expression   : Linear prediction, predict()

------------------------------------------------------------------
             |            Delta-method
             |     Margin   Std. Err.     [95% Conf. Interval]
-------------+----------------------------------------------------
     married |
   Unmarried |   5.359205   .0393607      5.281978    5.436431
     Married |   5.705447   .0409245      5.625152    5.785742
------------------------------------------------------------------
```

In some cases, we might want the adjusted means to be computed using an equal weighting of the cell means, as though the design had been balanced. In the context of this example, it would mean weighing college graduates and noncollege graduates equally.[8] This can be accomplished by using the **asbalanced** option. In the example below, the adjusted means are computed as though the design was balanced.

```
. margins married, nopvalues asbalanced
Adjusted predictions                        Number of obs     =      1,157
Expression   : Linear prediction, predict()
at           : married          (asbalanced)
               cograd           (asbalanced)

------------------------------------------------------------------
             |            Delta-method
             |     Margin   Std. Err.     [95% Conf. Interval]
-------------+----------------------------------------------------
     married |
   Unmarried |   5.410486    .045758      5.320708    5.500264
     Married |   5.699217   .0457884      5.609379    5.789054
------------------------------------------------------------------
```

These adjusted means reflect an equal weighting of college graduates and noncollege graduates. Let's illustrate this by manually computing the adjusted mean for those who are not married. The as-balanced adjusted mean for those who are unmarried can be computed by multiplying the cell mean for unmarried noncollege graduates by 0.5 and the cell mean of unmarried college graduates by 0.5. These equally weighted cell means are added together. This is illustrated using the **display** command below. This yields the as-balanced adjusted mean for those who are not married.

```
. display 5.3039648*.5 + 5.5170068*.5
5.4104858
```

8. Such a strategy may be desirable when analyzing the results from a designed experiment. For example, imagine that we have a variable that reflects experimental group assignment (for example, treatment versus control group). In computing adjusted means, the experimenter may want the treatment and control groups to be weighted equally, even if there were differing numbers of participants in each group.

Likewise, we can manually compute the as-balanced adjusted mean for those who are married, as shown below.

```
. display 5.7121588*.5 + 5.6862745*.5
5.6992167
```

This only scratches the surface regarding the ways in which you can use the options `asbalanced` and `asobserved`. For example, you can specify that certain variables be treated as-balanced while other variables be treated as-observed. For more details regarding these options, see [R] **margins**.

8.6 Main effects with interactions: anova versus regress

This section considers the meaning of main effects in the presence of an interaction when using the `regress` command compared with the `anova` command. The `regress` command uses dummy (0/1) coding compared with the `anova` command that uses effect (−1/1) coding (Pedhazur and Schmelkin 1991, 474). In the presence of interactions, this can lead to conflicting estimates of so-called main effects for the `regress` command versus the `anova` command. This is illustrated using the example from section 8.2 in which a two by two model was used predicting optimism from treatment (control group and happiness therapy) and depression status (nondepressed and depressed). Let's use the dataset for this example and show the mean optimism by treatment group and depression status.

```
. use opt-2by2
. tabulate depstat treat, summarize(opt)
          Means, Standard Deviations and Frequencies of Optimism

Depression |  Treatment group
    status |       Con         HT |     Total
-----------+----------------------+----------
       Non |      44.9       60.0 |      52.4
           |      10.1       10.0 |      12.5
           |        30         30 |        60
-----------+----------------------+----------
       Dep |      34.6       38.8 |      36.7
           |      10.0       10.1 |      10.2
           |        30         30 |        60
-----------+----------------------+----------
     Total |      39.7       49.4 |      44.6
           |      11.2       14.6 |      13.8
           |        60         60 |       120
```

Now, let's repeat the `anova` command that was used in section 8.2 predicting optimism from treatment, depression status, and the interaction of these two variables.

```
. anova opt treat##depstat

                         Number of obs =        120    R-squared     =  0.4889
                         Root MSE      =    10.0136    Adj R-squared =  0.4757

                  Source | Partial SS         df         MS        F    Prob>F
          ---------------+----------------------------------------------------
                   Model |      11126          3   3708.6667     36.99  0.0000
                         |
                   treat |  2803.3333          1   2803.3333     27.96  0.0000
                 depstat |  7426.1333          1   7426.1333     74.06  0.0000
           treat#depstat |  896.53333          1   896.53333      8.94  0.0034
                         |
                Residual |  11631.467        116   100.27126
          ---------------+----------------------------------------------------
                   Total |  22757.467        119   191.23922
```

Let's now perform this analysis but instead use the `regress` command.

```
. regress opt treat##depstat, vsquish

      Source |       SS           df       MS      Number of obs   =       120
-------------+----------------------------------   F(3, 116)       =     36.99
       Model |       11126         3  3708.66667   Prob > F        =    0.0000
    Residual |  11631.4667       116  100.271264   R-squared       =    0.4889
-------------+----------------------------------   Adj R-squared   =    0.4757
       Total |  22757.4667       119  191.239216   Root MSE        =    10.014

------------------------------------------------------------------------------
          opt |      Coef.   Std. Err.      t    P>|t|     [95% Conf. Interval]
--------------+----------------------------------------------------------------
        treat |
          Con |          0  (base)
           HT |   15.13333   2.585489     5.85   0.000     10.01245    20.25422
      depstat |
          Non |          0  (base)
          Dep |  -10.26667   2.585489    -3.97   0.000    -15.38755   -5.145781
treat#depstat |
       HT#Dep |  -10.93333   3.656433    -2.99   0.003    -18.17536   -3.691307
        _cons |   44.86667   1.828216    24.54   0.000     41.24565    48.48768
------------------------------------------------------------------------------
```

Let's now compare the results of the **anova** command with the results of the **regress** command, focusing on the significance tests. These comparisons are a bit tricky, because the **anova** command reports F statistics, whereas the **regress** command reports t statistics. But we can square the t value from the **regress** command to convert it into an equivalent of an F statistic.

Let's first compare the test of the **treat#depstat** interaction. The results of this test are the same for the **anova** and **regress** commands. If we square the t value of -2.99 from the **regress** command, we obtain the 8.94, the same value as the F statistic from the **anova** command.

Using the **contrast treat#depstat** command following the **regress** command also yields the same results as the **anova** command. The F value from the **contrast** command is the same as the F value from the **anova** command, 8.94.

```
. contrast treat#depstat
Contrasts of marginal linear predictions
Margins      : asbalanced
```

	df	F	P>F
treat#depstat	1	8.94	0.0034
Denominator	116		

Let's now compare the test of `treat` from the `anova` command with the `regress` command. We square of the t value for the `treat` effect from the `regress` command (5.85) and obtain 34.22. This is different from the F value for the `treat` effect from the `anova` command, 27.96.

Suppose we use the `contrast treat` command to test the `treat` effect. The F value from the `contrast` command is 27.96, which matches the F value as from the `anova` command.

```
. contrast treat
Contrasts of marginal linear predictions
Margins      : asbalanced
```

	df	F	P>F
treat	1	27.96	0.0000
Denominator	116		

This might seem perplexing, but there is a perfectly logical explanation for this. The reason for these discrepancies is because of differences in the coding used by the `anova` and `regress` commands. The `regress` command uses dummy (0/1) coding, whereas the `anova` command and the `contrast` command use effect ($-1/1$) coding. The interpretation of the interactions is the same whether you use effect coding or dummy coding, but the meaning of the main effects differ. When using dummy coding, the coefficient for `treat` represents the effect of `treat` when `depstat` is held constant at 0 (that is, for those who are nondepressed, the reference group). Referring back to the table of means, we see that the effect of `treat` for those who are nondepressed is $60.0 - 44.9 = 15.1$. This matches the coefficient for `treat` from the `regress` command. Likewise, the coefficient for `depstat` corresponds to the effect of `depstat` for the control group, which is $34.6 - 44.9 = -10.3$. This matches the coefficient of `depstat` from the `regress` command.

In the dummy-coded model used by the `regress` command, the tests of `treat` and `depstat` might be called main effects, but this is really a misnomer. These are really simple effects. The test of `treat` from the `regress` output is the effect of `treat` when `depstat` is held constant at the reference group, and the coefficient for `depstat` is the effect of `depstat` when `treat` is held constant at the reference group.

The anova and contrast commands instead use effect coding, and the terms associated with the main effects represent the classic analysis of variance main effects. Referring to the table of means, the main effect of treat concerns the differences in the column means (that is, 39.7 versus 49.4), and the main effect of depstat concerns the differences in the row means (that is, 52.4 versus 36.7).

This begs the question of whether we should even be interpreting main effects in the presence of an interaction. Whether the main effects are coded using dummy coding or using effect coding, main effects are not meaningful when they are part of a significant interaction term.

8.7 Interpreting confidence intervals

The marginsplot command displays margins and confidence intervals that were computed from the most recent margins command. The graphs produced by the command marginsplot, especially those created in the context of categorical by categorical interactions, can tempt you into falsely believing that the confidence intervals reflect comparisons between groups. Let's consider an example where this temptation can be very compelling. This example predicts happy7 from marital3, gender, and marital3#gender.[9]

. This model is run below, including c.health as a covariate.

```
. use gss_ivrm
. anova happy7 i.marital3##i.gender c.health
                         Number of obs =        783    R-squared     =  0.1246
                         Root MSE      =    .941504    Adj R-squared =  0.1178

                  Source | Partial SS         df         MS        F    Prob>F
         ----------------+----------------------------------------------------
                   Model |  97.864391          6   16.310732     18.40  0.0000
                         |
                marital3 |   26.17214          2    13.08607     14.76  0.0000
                  gender |  3.8126465          1   3.8126465      4.30  0.0384
         marital3#gender |  2.4708203          2   1.2354101      1.39  0.2488
                  health |  51.985684          1   51.985684     58.65  0.0000
                         |
                Residual |  687.86996        776   .88643037
         ----------------+----------------------------------------------------
                   Total |  785.73436        782   1.0047754
```

The interaction of marital3 by gender is not significant. Nevertheless, let's compute the adjusted means of happiness by marital3 by gender using the margins command and graph them using the marginsplot command (see figure 8.10).

9. In this chapter, I use the variable gender (coded: 1 = Male and 2 = Female). This yields output that clearly distinguishes the variable name (that is, gender) and its values (that is, Male, and Female).

```
. margins marital3#gender, nopvalues
Predictive margins                              Number of obs     =        783
Expression   : Linear prediction, predict()
```

	Margin	Delta-method Std. Err.	[95% Conf.	Interval]
marital3#gender				
Married#Male	5.681808	.0948106	5.495693	5.867924
Married#Female	5.687241	.0575204	5.574327	5.800154
Prevmarried#Male	5.12593	.1140485	4.902049	5.34981
Prevmarried#Female	5.293957	.082515	5.131978	5.455936
Never married#Male	5.267392	.0910344	5.088689	5.446095
Never married#Female	5.552688	.0912994	5.373465	5.731911

```
. marginsplot
  Variables that uniquely identify margins: marital3 gender
```

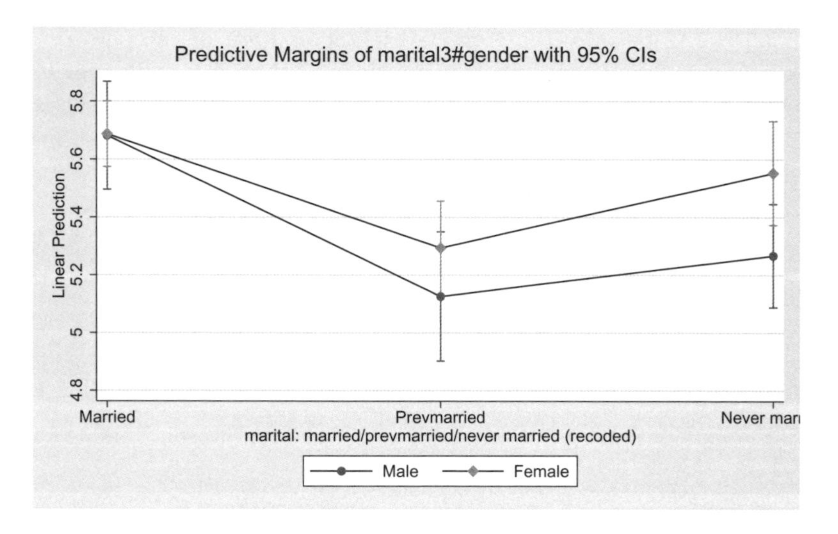

Figure 8.10. Adjusted means of happiness by marital status and gender

Looking at the confidence intervals in figure 8.10, your eye might be tempted to believe happiness is the same for males and females at each level of marital status because the confidence intervals for males and females highly overlap. However, we would need to test the effect of gender at each level of marital status using the `margins` command, as shown below.

```
. margins gender@marital3, contrast(nowald pveffects)
Contrasts of predictive margins                Number of obs     =        783
Expression   : Linear prediction, predict()

------------------------------------------------------------------------------
                                      |            Delta-method
                                      |   Contrast   Std. Err.      t    P>|t|
--------------------------------------+---------------------------------------
                      gender@marital3 |
       (Female vs base) Married       |   .0054323   .1111648    0.05   0.961
    (Female vs base) Prevmarried      |   .1680273    .140184    1.20   0.231
  (Female vs base) Never married      |   .2852962   .1288355    2.21   0.027
------------------------------------------------------------------------------
```

After adjusting for the covariates, these results actually show that the happiness of males and females significantly differs among those who were never married ($p = 0.027$).

In summary, the `marginsplot` command provides a graphical display of the results calculated by the `margins` command. Sometimes, the appearance of the confidence intervals of individual groups might tempt you to inappropriately make statistical inferences about the comparisons between the groups. To avoid this trap, you can directly form comparisons among the groups of interest to ascertain the significance of the group differences.

8.8 Summary

This chapter considered three types of designs involving two categorical variables: two by two, three by two, and three by three designs. Following each design, the `margins` command can be used to obtain the means as a function of the categorical variables, and the `marginsplot` command can be used to display a graph of the means. Following each of these designs, you can use a test of simple effects to assess the significance of one factor at each of the levels of the other factor.

In a two by three design, you can further perform simple contrasts by applying contrast coefficients to the three-level factor and examining the effect of the contrasts at each level of the two-level factor. You can also perform partial interaction tests by applying contrasts to the three-level factor, and interacting that with the two-level factor. These techniques can be applied in the same way for four by two designs, five by two designs, and so forth.

In a three by three design, you can use all techniques illustrated in the three by two design, namely the use of simple effects, simple contrasts, and partial interactions. In a three by three design, you can also use interaction contrasts by applying contrasts to each of the three-level factors and then interacting the contrasts.

I will reiterate a point I made at the beginning of the chapter. This chapter has illustrated a wide variety of methods for understanding and dissecting interactions. You do not need to apply every method illustrated in this chapter, and you do not need to apply them in the order they were illustrated in this chapter. Before looking at your data, I encourage you to describe the pattern of results that you expect and create an analysis plan (using the techniques described in this chapter) that will test for your predicted pattern of results.

For more information about the interaction of two categorical variables, I recommend Keppel and Wickens (2004), Maxwell et al. (2018), and Jaccard (1998). For a more detailed discussion of the use of partial interactions and interaction contrasts, see Abelson and Prentice (1997), Boik (1979), and Levin and Marascuilo (1972). For an example illustrating the application of an interaction contrast, see Mason, Prevost, and Sutton (2008).

9 Categorical by categorical by categorical interactions

9.1 Chapter overview

This chapter illustrates models involving interactions of three categorical variables, with an emphasis on how to interpret the interaction of the three categorical variables. The chapter begins with a two by two by two model (see section 9.2) followed by an example illustrating a two by two by three model (see section 9.3). The chapter then concludes with a discussion of models that have at least three levels for each factor, using a three by three by four model as an example (see section 9.4).

Like the previous chapter, this chapter illustrates the many ways in which you can dissect interactions. In the case of a three-way interaction, there are even more ways that such interactions can be dissected. The best practice would be to use your research questions to develop an analysis plan that describes how the interactions will be dissected. This plan would also consider the number of statistical tests that will be performed and consider whether a strategy is required to control the overall type I error rate. By contrast, an undesirable practice would be to include an interaction because it happened to be significant, dissect it every which way possible until an unpredicted significant test is obtained, and make no adjustment for the number of unplanned statistical tests that were performed.

The examples in the chapter illustrate the possible ways to dissect interactions, so you can understand how these techniques work and can choose among them for your analysis plan. This chapter focuses on teaching these techniques and, to that end, data are inspected prior to analyzing it, patterns of data are used to suggest interesting tests to perform, and every possible method is illustrated for dissecting interactions. I think this is a useful teaching strategy but not a research strategy to emulate.

9.2 Two by two by two models

This chapter begins with the simplest example of a model with an interaction of three categorical variables, a two by two by two model. Let's continue to use the same research example that we saw from chapter 8 that focused on the effect of treatment type and depression status on optimism. The same hypothetical optimism scale is used, in which scale scores could range from 0 to 100 and average about 50 for the general population.

Suppose that the researcher wants to further extend this work by considering the role of the time of year (that is, season), thinking that optimism may be less malleable during dark winter months than during cheerful summer months. To investigate this, the researcher conducts a new study using a two by two by two model in which treatment has two levels (control group versus happiness therapy), depression status has two levels (nondepressed versus mildly depressed), and season has two levels (winter and summer).

The results of this study are contained in the dataset named `opt-2by2by2.dta`. Let's use this dataset and show the mean of the outcome variable (optimism) by depression status, treatment, and season.

```
. use opt-2by2by2, clear
. table depstat treat season, contents(mean opt)

----------------------------------------
           |   Season and Treatment
           |          group
Depressio  |  - Winter -     - Summer -
n status   |   Con     HT     Con     HT
-----------+----------------------------
       Non |  44.4   59.9    51.7   64.6
      Mild |  39.2   50.9    45.0   64.8
----------------------------------------
```

The `anova` command is used below to fit a model that predicts optimism from treatment, depression status, season, all two-way interactions of these variables, and the three-way interaction of these variables.

```
. anova opt depstat##treat##season
                         Number of obs =        240    R-squared     =  0.5677
                         Root MSE      =    8.01794    Adj R-squared =  0.5546

                  Source | Partial SS         df         MS           F    Prob>F
             ------------+----------------------------------------------------
                   Model |   19584.517         7   2797.7881       43.52  0.0000
                         |
                 depstat |   1601.6667         1   1601.6667       24.91  0.0000
                   treat |   13470.017         1   13470.017      209.53  0.0000
           depstat#treat |   35.266667         1   35.266667        0.55  0.4596
                  season |   3713.0667         1   3713.0667       57.76  0.0000
          depstat#season |   220.41667         1   220.41667        3.43  0.0653
            treat#season |   112.06667         1   112.06667        1.74  0.1880
    depstat#treat#season |   432.01667         1   432.01667        6.72  0.0101
                         |
                Residual |   14914.667       232   64.287356
             ------------+----------------------------------------------------
                   Total |   34499.183       239   144.34805
```

Note! Three-way interaction shortcut

Specifying **depstat##treat##season** is a shortcut for specifying all main effects, two-way interactions, and the three-way interaction of **depstat**, **treat**, and **season**. This both saves time and helps ensure that you include all lower order effects. Even if not significant, these lower order effects should be included in the model.

The **depstat#treat#season** interaction is significant ($F = 6.72$, $p = 0.0101$). Let's use the **margins** command to show the mean of optimism broken down by these three factors.

```
. margins depstat#treat#season, nopvalues
Adjusted predictions                              Number of obs     =        240
Expression   : Linear prediction, predict()

------------------------------------------------------------------------
                     |            Delta-method
                     |     Margin   Std. Err.     [95% Conf. Interval]
---------------------+--------------------------------------------------
depstat#treat#season |
      Non#Con#Winter |       44.4    1.463869      41.51582    47.28418
      Non#Con#Summer |   51.66667    1.463869      48.78249    54.55084
       Non#HT#Winter |   59.93333    1.463869      57.04916    62.81751
       Non#HT#Summer |   64.56667    1.463869      61.68249    67.45084
     Mild#Con#Winter |   39.23333    1.463869      36.34916    42.11751
     Mild#Con#Summer |   44.96667    1.463869      42.08249    47.85084
      Mild#HT#Winter |   50.93333    1.463869      48.04916    53.81751
      Mild#HT#Summer |   64.76667    1.463869      61.88249    67.65084
------------------------------------------------------------------------
```

Let's then use the `marginsplot` command to make a graph of the means, showing `treat` on the x axis and the different seasons in separate panels (see figure 9.1).

```
. marginsplot, xdimension(treat) bydimension(season) noci
  Variables that uniquely identify margins: depstat treat season
```

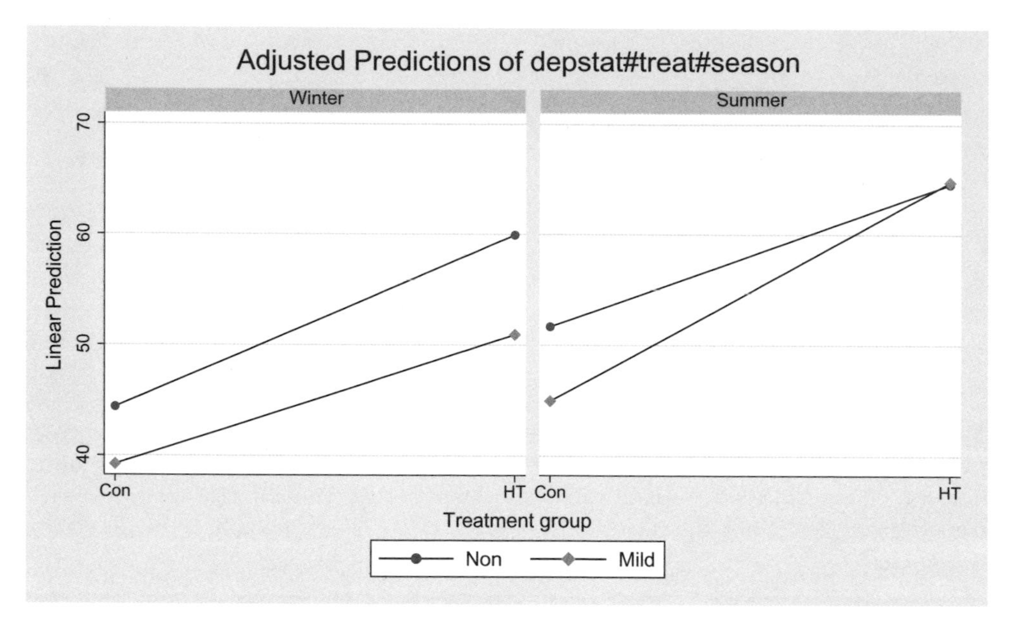

Figure 9.1. Optimism by treatment, depression status, and season

The graph in figure 9.1 illustrates the three-way interaction, showing that the size of the `treat#depstat` interaction differs for winter versus summer. It appears that there is no `treat#depstat` interaction during the winter but that there is such an interaction during the summer. In fact, we can perform such tests by looking at the simple interaction of `treat` by `depstat` at each level of `season`.

9.2.1 Simple interactions by season

One way that we can dissect the three-way interaction is by looking at the simple interactions of treatment by depression status at each season. Looking at figure 9.1, the left panel illustrates the simple interaction of treatment by depression status for the winter, and the right panel illustrates the simple interaction of treatment by depression status for the summer. It appears that the interaction is not significant during the winter and is significant during the summer. We test this using the `contrast` command, which tests the `treat#depstat` interaction at each level of `season`.

```
. contrast treat#depstat@season
Contrasts of marginal linear predictions
Margins      : asbalanced
```

	df	F	P>F
treat#depstat@season			
Winter	1	1.71	0.1917
Summer	1	5.55	0.0193
Joint	2	3.63	0.0279
Denominator	232		

Indeed, the interaction of treatment by depression status is not significant during the winter ($F = 1.71$, $p = 0.1917$) and is significant during the summer ($F = 5.55$, $p = 0.0193$). Looking at the left panel of figure 9.1, we see that in the winter, the effect of happiness therapy (compared with the control group) does not depend on depression status, hence the nonsignificant simple interaction. However, in the summer, the effect of happiness therapy does depend on depression status. Looking at the right panel of figure 9.1, we can see that in the summer happiness therapy is more effective for those who are mildly depressed than for those who are nondepressed.

9.2.2 Simple interactions by depression status

Another way to dissect this three-way interaction is by looking at the simple interaction of treatment by season at each level of depression status. To visualize this, let's rerun the `margins` command and then use the `marginsplot` command to graph the means showing `treat` on the x axis and separate panels for those who are nondepressed and mildly depressed. This is visualized in figure 9.2.

```
. margins depstat#treat#season
  (output omitted)
. marginsplot, xdimension(treat) bydimension(depstat) noci
  Variables that uniquely identify margins: depstat treat season
```

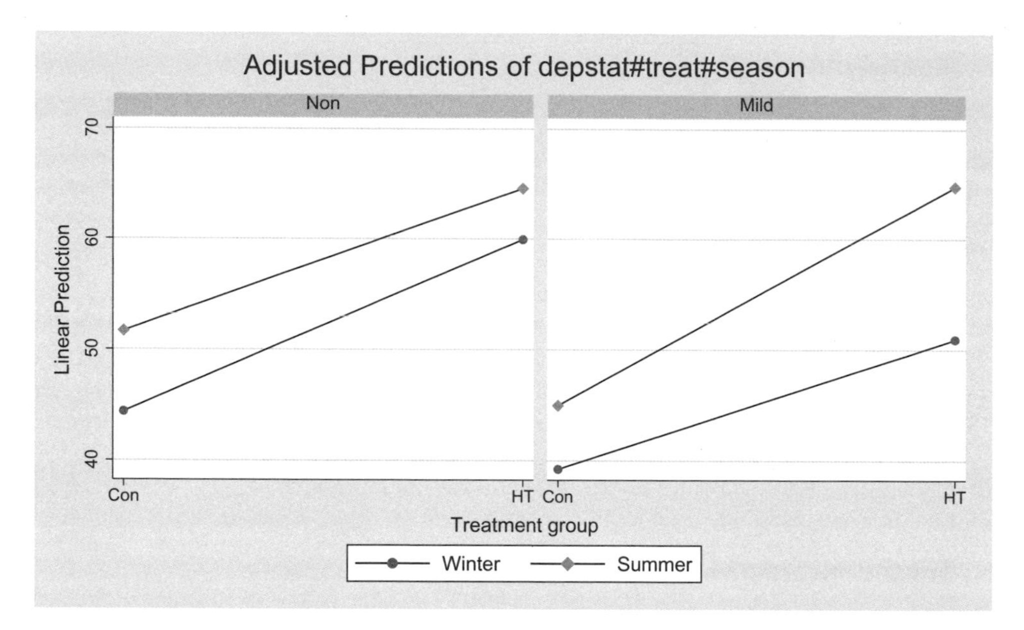

Figure 9.2. Optimism by treatment, season, and depression status

Among those who are nondepressed, it appears that happiness therapy is equally effective in the summer and winter (see the left panel of figure 9.2). By contrast, among those who are mildly depressed, it appears that the effectiveness of happiness therapy depends on the season (see the right panel of figure 9.2). In other words, it appears that there is an interaction of treatment by season for those who are mildly depressed, but no such interaction for those who are not depressed. We can test the interaction of treatment by season at each level of depression status using the following `contrast` command:

```
. contrast treat#season@depstat
Contrasts of marginal linear predictions
Margins      : asbalanced
```

	df	F	P>F
treat#season@depstat			
Non	1	0.81	0.3693
Mild	1	7.65	0.0061
Joint	2	4.23	0.0157
Denominator	232		

Indeed, the interaction of treatment by season is not significant for those who are not depressed ($F = 0.81$, $p = 0.3693$) but is significant for those who are mildly depressed ($F = 7.65$, $p = 0.0061$).

9.2.3 Simple effects

We have seen that the effect of happiness therapy depends on both season and depression status. We might want to know whether the effect of happiness therapy is significant for each combination of season and depression status. This simple effect can be tested using the `contrast` command shown below. It tests the effect of `treat` at each combination of `season` and `depstat`.

```
. contrast treat@season#depstat, nowald pveffects
Contrasts of marginal linear predictions
Margins      : asbalanced
```

	Contrast	Std. Err.	t	P>|t|
treat@season#depstat				
(HT vs base) Winter#Non	15.53333	2.070223	7.50	0.000
(HT vs base) Winter#Mild	11.7	2.070223	5.65	0.000
(HT vs base) Summer#Non	12.9	2.070223	6.23	0.000
(HT vs base) Summer#Mild	19.8	2.070223	9.56	0.000

The mean optimism for those in the happiness therapy group is always greater than the control group at each level of season and at each level of depression status. For example, among those who are nondepressed and treated during the winter, the mean optimism for those in happiness therapy is 15.5 points greater than the control group, a significant difference ($t = 7.50$, $p = 0.000$).

9.3 Two by two by three models

Let's now consider an example with three factors, two of which have two levels and one of which has three levels. This is an extension of the example shown in the previous section, involving treatment group, depression status, and season as factor variables. In this example, the treatment variable now has three levels: 1) control group, 2) traditional therapy, and 3) happiness therapy.

Let's use the dataset for this example and show the mean optimism by depression status, treatment group, and season.

```
. use opt-3by2by2, clear
. table depstat treat season, contents(mean opt)
```

Depression status	Winter (S1) Con	Winter (S1) TT	Winter (S1) HT	Summer (S2) Con	Summer (S2) TT	Summer (S2) HT
Non	44.7	54.3	59.8	49.7	59.4	64.6
Mild	39.6	49.2	49.3	44.2	54.7	64.2

Let's now use the **anova** command to predict **opt** from **depstat**, **treat**, **season**, all two-way interactions of these variables, and the three-way interaction.

```
. anova opt depstat##treat##season
```

Number of obs = 360 R-squared = 0.4912
Root MSE = 7.9879 Adj R-squared = 0.4751

Source	Partial SS	df	MS	F	Prob>F
Model	21435.233	11	1948.6576	30.54	0.0000
depstat	2423.2111	1	2423.2111	37.98	0.0000
treat	13811.217	2	6905.6083	108.23	0.0000
depstat#treat	3.9055556	2	1.9527778	0.03	0.9699
season	3960.1	1	3960.1	62.06	0.0000
depstat#season	253.34444	1	253.34444	3.97	0.0471
treat#season	469.11667	2	234.55833	3.68	0.0263
depstat#treat#season	514.33889	2	257.16944	4.03	0.0186
Residual	22204.667	348	63.806513		
Total	43639.9	359	121.55961		

The three-way interaction is significant ($F = 4.03$, $p = 0.0186$). Let's use the **margins** and **marginsplot** commands to display and graph the means by each of these categorical variables. In creating the graph of the means, let's display separate panels for each level of depression status (see figure 9.3). This allows us to focus on the way that the **treat#season** interaction varies by depression status.

```
. margins depstat#treat#season, nopvalues
Adjusted predictions                              Number of obs     =        360
Expression   : Linear prediction, predict()

------------------------------------------------------------------------------
                           |            Delta-method
                           |     Margin   Std. Err.     [95% Conf. Interval]
---------------------------+--------------------------------------------------
   depstat#treat#season    |
    Non#Con#Winter (S1)    |       44.7   1.458384      41.83164    47.56836
    Non#Con#Summer (S2)    |       49.7   1.458384      46.83164    52.56836
     Non#TT#Winter (S1)    |   54.33333   1.458384      51.46498    57.20169
     Non#TT#Summer (S2)    |       59.4   1.458384      56.53164    62.26836
     Non#HT#Winter (S1)    |   59.76667   1.458384      56.89831    62.63502
     Non#HT#Summer (S2)    |   64.56667   1.458384      61.69831    67.43502
   Mild#Con#Winter (S1)    |   39.63333   1.458384      36.76498    42.50169
   Mild#Con#Summer (S2)    |       44.2   1.458384      41.33164    47.06836
    Mild#TT#Winter (S1)    |   49.23333   1.458384      46.36498    52.10169
    Mild#TT#Summer (S2)    |       54.7   1.458384      51.83164    57.56836
    Mild#HT#Winter (S1)    |   49.33333   1.458384      46.46498    52.20169
    Mild#HT#Summer (S2)    |   64.23333   1.458384      61.36498    67.10169
------------------------------------------------------------------------------

. marginsplot, bydimension(depstat) noci
  Variables that uniquely identify margins: depstat treat season
```

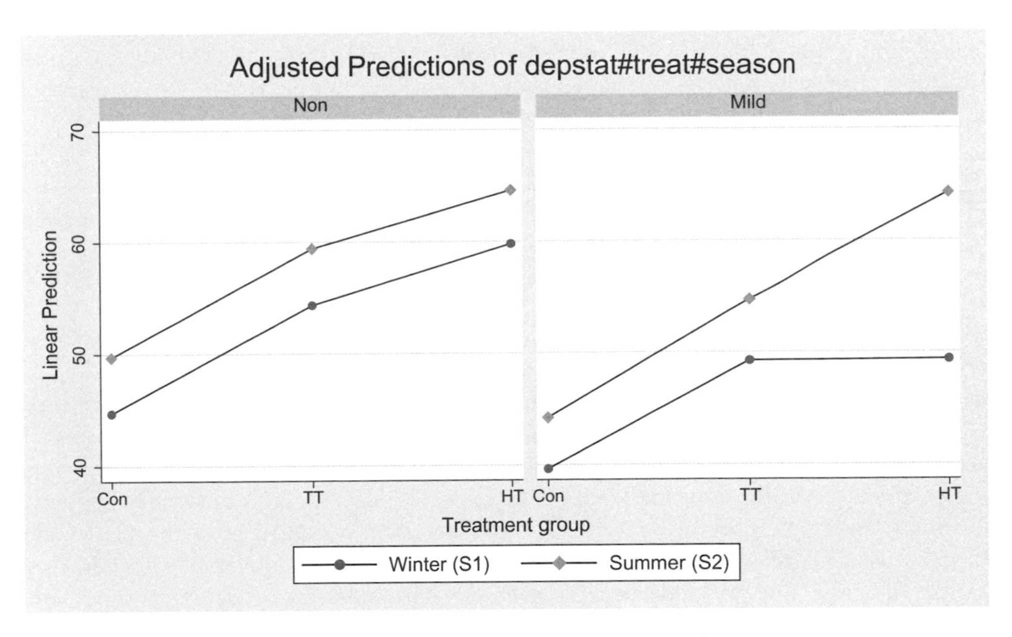

Figure 9.3. Optimism by treatment, season, and depression status

9.3.1 Simple interactions by depression status

It appears that the `treat#season` interaction might not be significant for those who are not depressed (see the left panel of figure 9.3) but might be significant for those who are mildly depressed (see the right panel of figure 9.3). We can explore this by assessing the `treat#season` interaction at each level of `depstat` using the `contrast` command below.

```
. contrast treat#season@depstat
Contrasts of marginal linear predictions
Margins      : asbalanced

                      |         df           F        P>F
----------------------+-----------------------------------
treat#season@depstat  |
                  Non |          2        0.00     0.9955
                 Mild |          2        7.70     0.0005
                Joint |          4        3.85     0.0044
                      |
          Denominator |        348
```

The treatment by season interaction is not significant among those who are nondepressed. The parallel lines in the left panel of figure 9.3 illustrate the absence of an interaction. By contrast, the treatment by season interaction is significant for those who are mildly depressed (see the right panel of figure 9.3).

Let's further dissect this simple interaction by applying contrasts to the treatment factor through the use of simple partial interactions.

9.3.2 Simple partial interaction by depression status

If we focus our attention on the right panel of figure 9.3, we can see that this simple interaction is really a two by three interaction, and we can dissect it using the tools from section 8.3 on two by three interactions. We can begin by applying contrasts to the treatment factor, forming a simple partial interaction. Say that we want to form two comparisons with respect to `treat`, comparing group 2 versus 1 (traditional therapy versus control group) and comparing group 3 versus 2 (happiness therapy versus traditional therapy). The interaction of treatment (traditional therapy versus control group) by season for those who are mildly depressed is visualized in the left panel of figure 9.4, and the interaction of treatment (happiness therapy versus traditional therapy) by season for those who are mildly depressed is visualized in the right panel of figure 9.4.

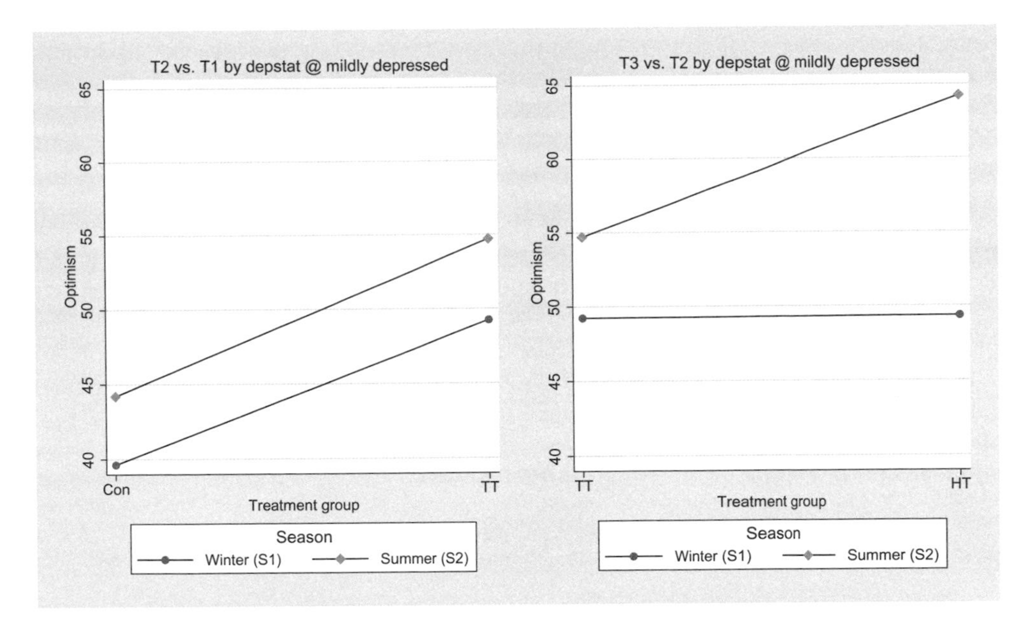

Figure 9.4. Simple partial interactions

The `contrast` command below tests the two partial interactions depicted in the left and right panels of figure 9.4 by specifying `ar.treat#season@2.depstat`. To understand this, let's break it into two parts. The first part, `ar.treat#season`, creates the interactions of treatment (traditional therapy versus control group) by season, and treatment (happiness therapy versus traditional therapy) by season. The second part, `@2.depstat` indicates the contrasts will be performed only for level 2 of `depstat` (the mildly depressed group).

```
. contrast ar.treat#season@2.depstat
Contrasts of marginal linear predictions
Margins      : asbalanced
```

	df	F	P>F
treat#season@depstat			
(TT vs Con) (joint) Mild	1	0.10	0.7578
(HT vs TT) (joint) Mild	1	10.46	0.0013
Joint	2	7.70	0.0005
Denominator	348		

The first test is not significant ($F = 0.10$, $p = 0.7578$). As depicted in the left panel of figure 9.4, the difference between traditional therapy and control group does not depend on season (focusing on those who are mildly depressed). The second test is significant ($F = 10.46$, $p = 0.0013$). As shown in the right panel of figure 9.4, the difference between happiness therapy and traditional therapy does depend on season.

Let's now further understand this simple partial interaction through the use of simple contrasts.

9.3.3 Simple contrasts

The significant simple partial interaction shown in the right panel of figure 9.4 is now a two by two analysis and can be dissected using simple contrasts. In particular, let's ask whether there is a difference between happiness therapy and traditional therapy separately for each season focusing only on those who are mildly depressed. We can perform this test by specifying `ar3.treat@season#2.depstat` on the `contrast` command. Let's break this into two parts. The first part, `ar3.treat`, requests the comparison of treatment group 3 versus 2 (happiness therapy versus traditional therapy). The second part, `@season#2.depstat`, requests that the contrasts be performed at each level of season and at level 2 of depression status (mildly depressed).

```
. contrast ar3.treat@season#2.depstat, nowald pveffects
Contrasts of marginal linear predictions
Margins      : asbalanced

------------------------------------------------------------------------------
                              |   Contrast   Std. Err.          t     P>|t|
------------------------------+-----------------------------------------------
         treat@season#depstat |
(HT vs TT) Winter (S1)#Mild   |         .1   2.062466        0.05    0.961
(HT vs TT) Summer (S2)#Mild   |   9.533333   2.062466        4.62    0.000
------------------------------------------------------------------------------
```

The results of these two tests are consistent with what we observe in the right panel of figure 9.4. The first test is not significant ($t = 0.05$, $p = 0.961$). There is no difference between happiness therapy and traditional therapy among those who are mildly depressed in the winter. The second test is significant ($t = 4.62$, $p = 0.000$). In the summer, there is a significant difference between happiness therapy and traditional therapy among those who are mildly depressed.

9.3.4 Partial interactions

Let's consider another strategy we could use for further understanding the three-way interaction. Let's look back to the original graph of the results of the three-way interaction shown in figure 9.3. Say that we wanted to further understand this interaction by applying adjacent group contrasts to the treatment factor. These contrasts compare group 2 versus 1 (traditional therapy versus control group) and group 3 versus 2 (happiness therapy versus traditional therapy). We could interact these contrasts with season and depression status. The contrast of `treat` (traditional therapy versus control group) by `depstat` by `season` is visualized in figure 9.5 and the contrast of `treat` (happiness therapy versus traditional therapy) by `depstat` by `season` is visualized in figure 9.6.

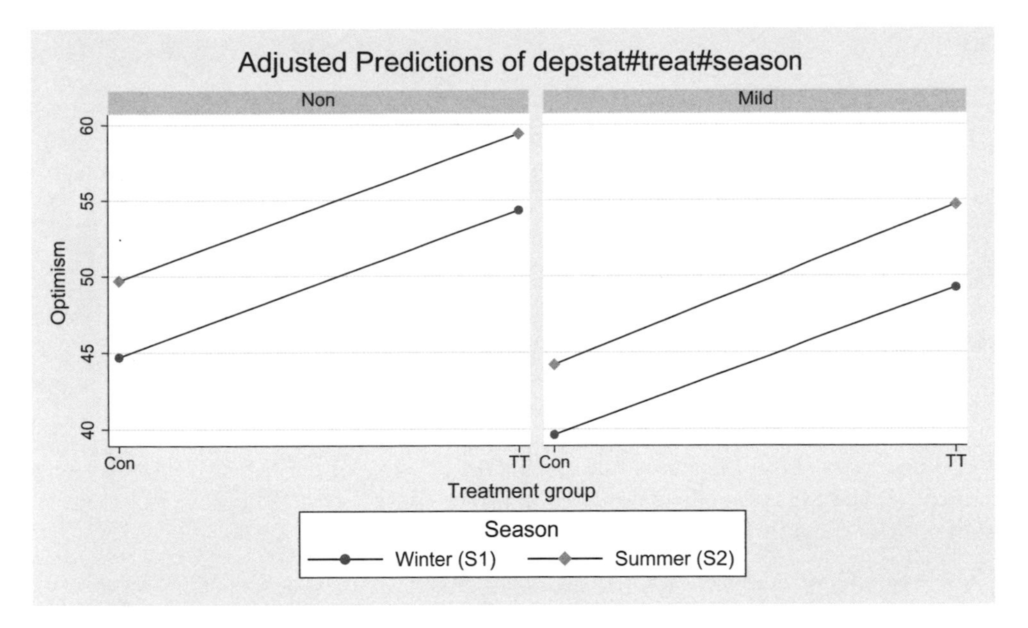

Figure 9.5. Simple partial interaction: T2 vs. T1 by `season` by `depstat`

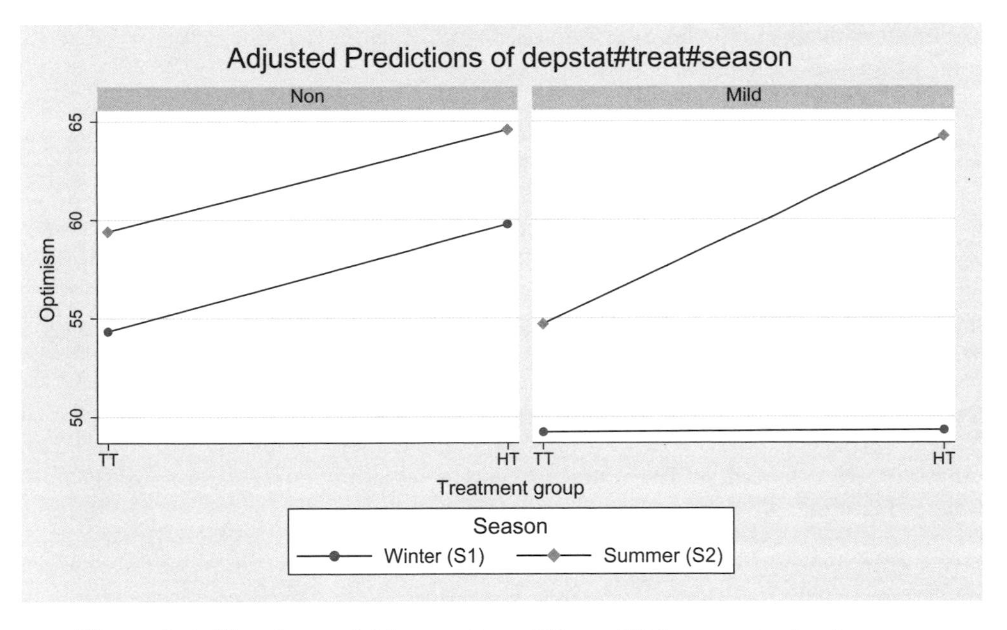

Figure 9.6. Simple partial interaction: T3 vs. T2 by `season` by `depstat`

We can test each of these partial interactions using the `contrast` command below.

```
. contrast ar.treat#season#depstat
Contrasts of marginal linear predictions
Margins      : asbalanced

─────────────────────────────────────────────────────────────
                              │        df           F        P>F
──────────────────────────────┼──────────────────────────────
        treat#season#depstat  │
(TT vs Con) (joint) (joint)   │         1        0.04     0.8400
 (HT vs TT) (joint) (joint)   │         1        5.53     0.0193
                      Joint   │         2        4.03     0.0186
                              │
                Denominator   │       348
─────────────────────────────────────────────────────────────
```

The first partial interaction is not significant ($F = 0.04$, $p = 0.8400$). As we can see in figure 9.5, the `treat` (traditional therapy versus control group) by `season` interaction is roughly the same for each level of depression status.

The second partial interaction is significant ($F = 5.53$, $p = 0.0193$). As shown in figure 9.6, the `treat` (happiness therapy versus traditional therapy) by `season` interaction differs by depression status. This partial interaction now resembles a two by two by two model, and you can further dissect it using the techniques illustrated in section 9.2.

Let's now move on to the most complex kind of three-way interactions, those that involve at least three levels for each factor.

9.4 Three by three by three models and beyond

Three-way interactions become increasingly complex as the number of levels of each factor increases. When a factor goes from having two levels to having three levels, it introduces the possibility of applying contrasts to the factor to further probe the interaction. For example, in the two by two by three model from section 9.3, we applied contrasts to the three-level `treat` factor. If we extended the `depstat` factor to also have three levels (for example, nondepressed, mildly depressed, and severely depressed), then it would be possible to apply contrasts to the `depstat` factor as well. If we take this one step further and include a third level of `season`, then the model would be a three by three by three model, and it would be possible to apply contrasts to any (or all) of the three factors.

Let's consider an extension of the example from 9.3 involving the factors `treat`, `depstat`, and `season`, except that these factors have three, three and four levels, respectively. The three levels of `treat` are 1) control group, 2) traditional therapy, and 3) happiness therapy. The three levels of `depstat` are 1) nondepressed, 2) mildly depressed, and 3) severely depressed. The four levels of `season` are 1) winter, 2) spring, 3) summer, and 4) fall.

Let's use the dataset for this hypothetical example and show the mean of the outcome by these three factors.

```
. use opt-3by3by4, clear
. table season treat depstat, contents(mean opt)

------------------------------------------------------------------------
         |          Depression status and Treatment group
         | ------- Non -------    ------- Mild -------    ----- Severe -----
  Season |  Con    TT    HT       Con    TT    HT       Con    TT    HT
---------+--------------------------------------------------------------
  Winter | 44.6  54.3  61.3      39.6  49.2  51.2      36.7  47.2  46.7
  Spring | 47.5  57.3  62.7      42.7  52.2  56.4      39.7  50.2  49.8
  Summer | 49.7  60.4  64.7      44.5  54.1  65.3      39.6  49.2  48.6
    Fall | 47.8  57.0  62.9      42.7  52.0  56.2      40.0  50.0  50.2
------------------------------------------------------------------------
```

Let's analyze the data for this example using the **anova** command below.

```
. anova opt depstat##treat##season

                         Number of obs =      1,080    R-squared     =  0.6172
                         Root MSE      =    6.00663    Adj R-squared =  0.6043

                  Source | Partial SS         df         MS        F    Prob>F
  -----------------------+----------------------------------------------------
                   Model |    60718.9         35   1734.8257    48.08  0.0000
                         |
                 depstat |  18782.039          2   9391.0194   260.29  0.0000
                   treat |  34636.039          2   17318.019   480.00  0.0000
           depstat#treat |  1702.0056          4   425.50139    11.79  0.0000
                  season |  3522.8333          3   1174.2778    32.55  0.0000
          depstat#season |  1004.2722          6    167.3787     4.64  0.0001
            treat#season |  185.11667          6   30.852778     0.86  0.5275
    depstat#treat#season |  886.59444         12    73.88287     2.05  0.0179
                         |
                Residual |  37667.067      1,044   36.079566
  -----------------------+----------------------------------------------------
                   Total |  98385.967      1,079   91.182546
```

The three-way interaction of **depstat#treat#season** is significant ($F = 2.05$, $p = 0.0179$). To begin to understand the nature of this interaction, let's first graph the interaction using the **margins** and **marginsplot** commands. The graph of the mean optimism by treatment group, depression status, and season is shown in figure 9.7.

```
. margins depstat#treat#season
  (output omitted)
. marginsplot, xdimension(treat) bydimension(season) noci legend(rows(1))
  Variables that uniquely identify margins: depstat treat season
```

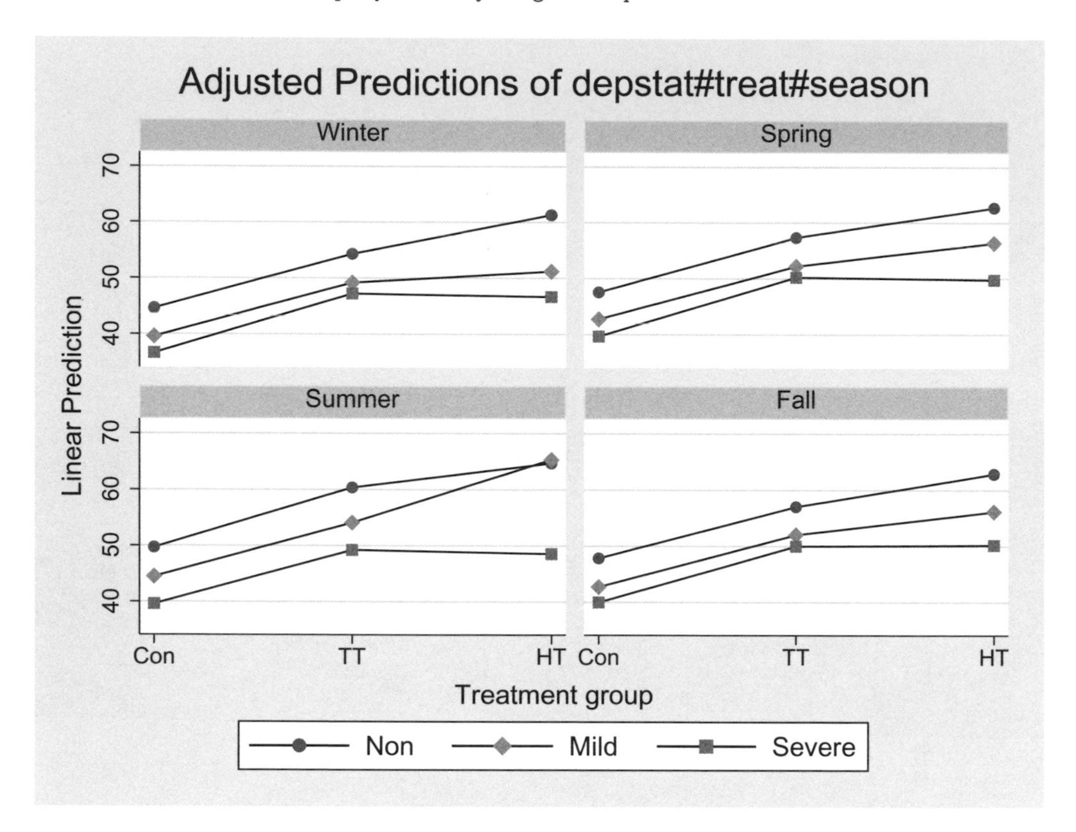

Figure 9.7. Optimism by treatment, season, and depression status

9.4.1 Partial interactions and interaction contrasts

Let's explore ways to dissect the three-way interaction of `depstat#treat#season` by applying contrasts to one or more of the factors. Let's begin by testing whether the interaction of `treat#depstat` is the same for each season compared with winter (season 1). This is performed by applying the `r.` contrast operator to season and interacting that with `treat` and `depstat`.

```
. contrast r.season#treat#depstat
Contrasts of marginal linear predictions
Margins      : asbalanced
```

	df	F	P>F
season#treat#depstat			
(Spring vs Winter) (joint) (joint)	4	0.46	0.7619
(Summer vs Winter) (joint) (joint)	4	5.36	0.0003
(Fall vs Winter) (joint) (joint)	4	0.40	0.8115
Joint	12	2.05	0.0179
Denominator	1044		

The comparison of spring versus winter (season 2 versus 1) with `treat` and `depstat` is not significant ($F = 0.46$, $p = 0.7619$). In figure 9.7, we can see how the two-way interaction of `treat` by `depstat` is similar for spring versus winter. Likewise, the comparison of fall versus winter (season 4 versus 1) with `treat` and `depstat` is not significant ($F = 0.40$, $p = 0.8115$). However, the comparison of summer versus winter (season 3 versus 1) with `treat` by `depstat` is significant ($F = 5.36$, $p = 0.0003$). Looking at the winter and summer panels from figure 9.7, we can see that the `treat#depstat` interaction differs for summer versus winter.

Note! Fall versus spring

Say that we wanted to test the interaction of `season` (spring versus fall) by `treat` by `depstat`. We could form that test using the following `contrast` command. A custom contrast is applied to `season` to obtain the comparison of fall versus spring (season 4 versus 2).

```
. contrast {season 0 -1 0 1}#treat#depstat
```

This test is not significant ($F = 0.01$, $p = 0.9996$). This is consistent with what we see in figure 9.7, in which the pattern of the `treat` by `depstat` interaction is rather similar for spring versus fall.

Let's further explore the comparison of summer versus winter (season 3 versus 1) with `treat` and `depstat`. Referring to the winter and summer panels from figure 9.7, let's focus on nondepressed and mildly depressed groups. Imagine testing the interaction formed by focusing on the interaction of summer versus winter (season 3 versus 1), by depression status (nondepressed versus mildly depressed) by treatment. I created a graph visualizing this partial interaction contrast, see figure 9.8. The `contrast` command below performs this comparison by applying the `r3.` contrast to `season` to compare summer with winter (season 3 versus 1) and the `r2.` contrast on `depstat` to compare those who are mildly depressed with those who are nondepressed (levels 2 versus 1).

These terms are all interacted (that is, `r3.season#r2.depstat#treat`) yielding a test of summer versus winter (season 3 versus 1) by mildly depressed versus nondepressed (levels 2 versus 1) by treatment.

```
. * Season(winter vs. summer) by depstat(nondepressed vs. mildly depressed) by
>   treat
. contrast r3.season#r2.depstat#treat
Contrasts of marginal linear predictions
Margins      : asbalanced
```

	df	F	P>F
season#depstat#treat	2	9.03	0.0001
Denominator	1044		

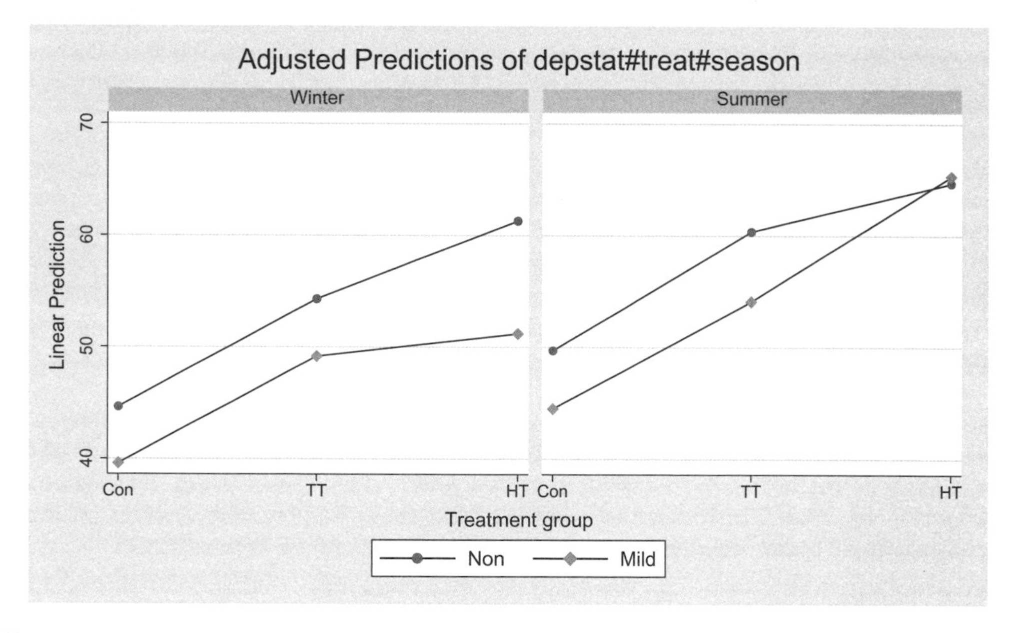

Figure 9.8. Interaction contrast of season (winter versus summer) by depression status (mildly depressed versus nondepressed) by treatment

This test is significant ($F = 9.03$, $p = 0.0001$). This is visualized by comparing the interaction in the left panel of figure 9.8 with the right panel of figure 9.8. This significant test indicates that the two-way interaction formed by interacting depression status (nondepressed versus mildly depressed) by treatment differs by season (winter versus summer).

Say that we wanted to take this test and focus on the contrast of happiness therapy versus traditional therapy (group 3 versus 2).[1] This yields an interaction of season (winter versus summer) by depression status (nondepressed versus mildly depressed) by treatment (happiness therapy versus traditional therapy). I created the graph shown in figure 9.9 that visualizes this interaction contrast. We can test this interaction contrast using the `contrast` command below.

```
. * season (winter vs. summer) by depstat (nondepressed vs. mildly depressed) by
> treat (HT vs. TT)
. contrast r3.season#r2.depstat#r3b2.treat, noeffects
Contrasts of marginal linear predictions
Margins      : asbalanced
```

	df	F	P>F
season#depstat#treat	1	14.64	0.0001
Denominator	1044		

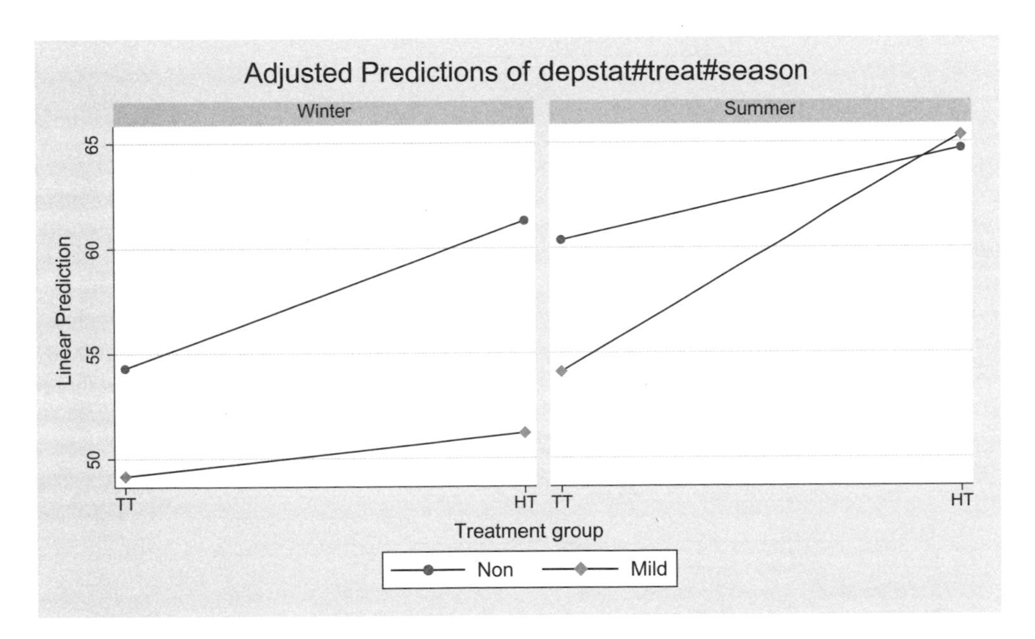

Figure 9.9. Interaction contrast of season (winter versus summer) by depression status (mildly depressed versus nondepressed) by treatment (HT versus TT)

This test is significant ($F = 14.64$, $p = 0.0001$). The two-way interaction of depression status (nondepressed versus mildly depressed) by treatment (happiness therapy versus traditional therapy) significantly differs between summer and winter.

1. This contrast is specified as `r3b2.treat`, which indicates to compare group 3 with group 2. This could have also been specified as a custom contrast, `{treat 0 -1 1}`.

9.4.2 Simple interactions

Let's now explore a different way to dissect the three-way interaction through the use of simple interaction tests. Consider the `treat#depstat` interaction at each of the four levels of season, as shown in the four panels of figure 9.7. We can assess the `treat#depstat` interaction for each season using the `contrast` command below.

```
. contrast depstat#treat@season
Contrasts of marginal linear predictions
Margins      : asbalanced

------------------------------------------------------
                     |         df           F        P>F
---------------------+--------------------------------
depstat#treat@season |
              Winter |          4        3.75     0.0049
              Spring |          4        2.29     0.0575
              Summer |          4        9.88     0.0000
                Fall |          4        2.01     0.0904
               Joint |         16        4.48     0.0000
                     |
         Denominator |       1044
------------------------------------------------------
```

The `treat#depstat` interaction is significant for winter (season 1) and summer (season 3). The `treat#depstat` interaction is not significant in the spring (season 2) or fall (season 4).

Let's perform this same `contrast` command, but apply the `r2.` contrast to `depstat` that compares those who are nondepressed versus mildly depressed. This yields four partial interactions of depression status (nondepressed versus mildly depressed) by treatment at each level of season. These tests are illustrated in figure 9.10. These tests are performed using the following `contrast` command.

```
. contrast r2.depstat#treat@season
Contrasts of marginal linear predictions
Margins      : asbalanced

---------------------------------------------------------------
                              |         df           F        P>F
------------------------------+--------------------------------
         depstat#treat@season |
(Mild vs Non) (joint) Winter  |          2        3.49     0.0309
(Mild vs Non) (joint) Spring  |          2        0.27     0.7631
(Mild vs Non) (joint) Summer  |          2        5.74     0.0033
  (Mild vs Non) (joint) Fall  |          2        0.38     0.6851
                        Joint |          8        2.47     0.0119
                              |
                  Denominator |       1044
---------------------------------------------------------------
```

The interaction of depression status (nondepressed versus mildly depressed) by treatment is significant in season 1, winter ($F = 3.49$, $p = 0.0309$). In the top left panel of figure 9.10, we can see how the lines are not parallel for during the winter. Jumping forward to summer (group 3), this test is also significant ($F = 5.74$, $p = 0.0033$). This is

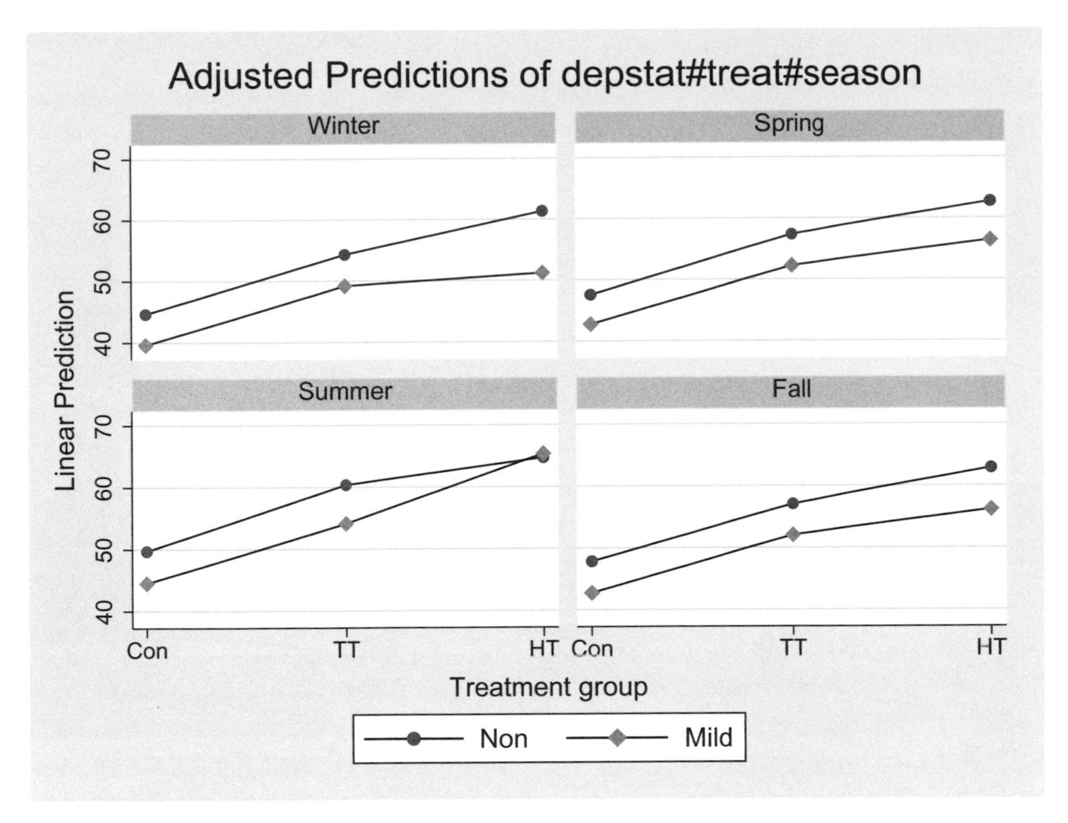

Figure 9.10. Optimism by treatment and season focusing on mildly depressed versus nondepressed

consistent with the nonparallel lines displayed in the bottom left panel in figure 9.10. By comparison, the lines are parallel for spring (group 2) and fall (group 4), reflected in the nonsignificant interaction of depression status (nondepressed versus mildly depressed) by treatment for those seasons.

Let's also apply the `r3b2.` contrast to `treat`, comparing group 3 with group 2 (happiness therapy to traditional therapy). This yields an interaction contrast of depression status (mildly depressed versus nondepressed) by treatment (happiness therapy versus traditional therapy) performed at each of the four seasons. These four simple interaction contrasts, visualized in figure 9.11, are tested using the `contrast` command below.

```
. contrast r2.depstat#r3b2.treat@season, noeffects
Contrasts of marginal linear predictions
Margins      : asbalanced
```

	df	F	P>F
depstat#treat@season			
(Mild vs Non) (HT vs TT) Winter	1	5.13	0.0238
(Mild vs Non) (HT vs TT) Spring	1	0.30	0.5844
(Mild vs Non) (HT vs TT) Summer	1	9.90	0.0017
(Mild vs Non) (HT vs TT) Fall	1	0.60	0.4385
Joint	4	3.98	0.0033
Denominator	1044		

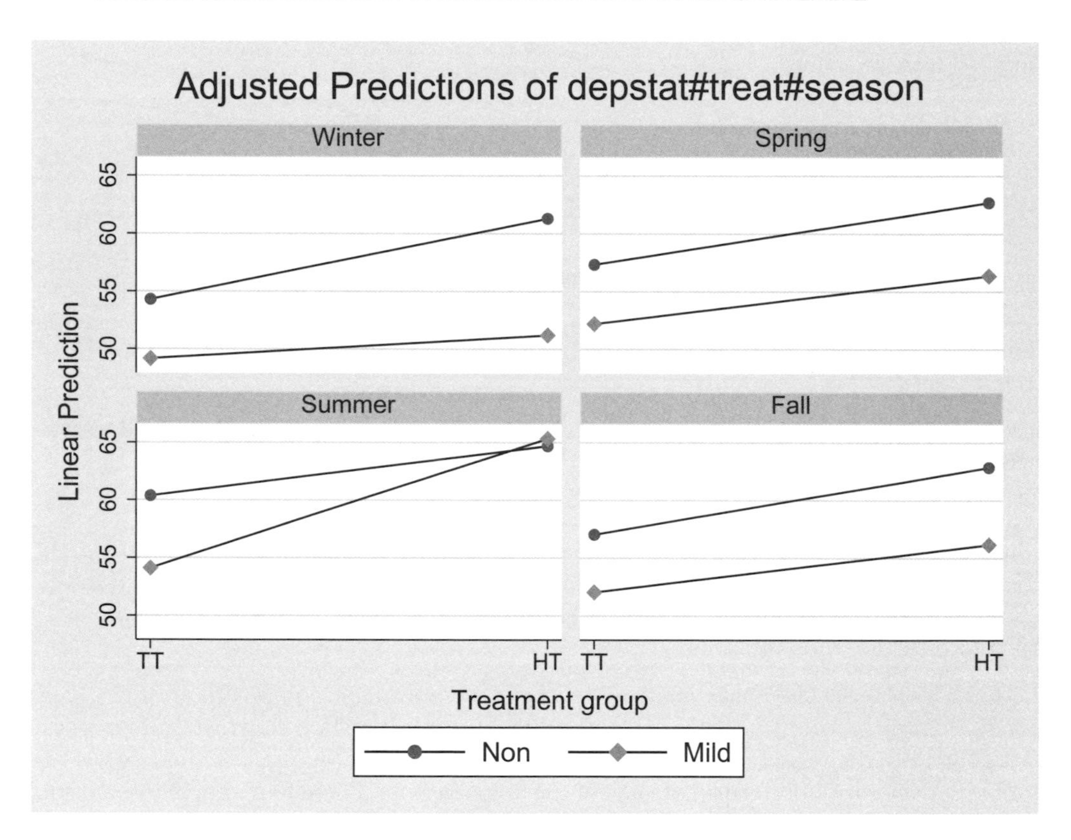

Figure 9.11. Simple interaction contrast of depression status by treatment at each season

The simple interaction contrast of depression status (nondepressed versus mildly depressed) by treatment (happiness therapy versus traditional therapy) is significant for season 1, winter ($F = 5.13$, $p = 0.0238$). The top left panel of figure 9.11 shows how the effect of treatment (happiness therapy versus traditional therapy) differs by depression status (nondepressed versus mildly depressed) during the winter. The simple interaction contrast is also significant for the season 3, summer ($F = 9.90$, $p = 0.0017$). As we see in the bottom left panel of figure 9.11, the effect of treatment (happiness therapy versus traditional therapy) differs by depression status (nondepressed versus mildly depressed) during the summer. The simple interaction contrasts for season 2 (spring) and season 4 (fall) are not significant, consistent with the parallel lines for these two seasons shown in figure 9.11.

9.4.3 Simple effects and simple comparisons

We might be interested in focusing on the simple effects of treatment across the levels of season and depression status. The following `contrast` command tests the effect of `treat` at each level of season and depression status. The output is lengthy, so it is omitted to save space.

```
. contrast treat@season#depstat
  (output omitted)
```

The results of this `contrast` command would show that the effect of `treat` is significant for every combination of season and depression status. We could further refine this test by focusing on the comparison of happiness therapy versus traditional therapy (group 3 versus 2) by applying the `r3b2.` contrast operator to `treat`, as shown below.

```
. contrast r3b2.treat@season#depstat, nowald pveffects
Contrasts of marginal linear predictions
Margins      : asbalanced
```

	Contrast	Std. Err.	t	P>|t|
treat@season#depstat				
(HT vs TT) Winter#Non	7	1.550904	4.51	0.000
(HT vs TT) Winter#Mild	2.033333	1.550904	1.31	0.190
(HT vs TT) Winter#Severe	-.5333333	1.550904	-0.34	0.731
(HT vs TT) Spring#Non	5.366667	1.550904	3.46	0.001
(HT vs TT) Spring#Mild	4.166667	1.550904	2.69	0.007
(HT vs TT) Spring#Severe	-.4666667	1.550904	-0.30	0.764
(HT vs TT) Summer#Non	4.333333	1.550904	2.79	0.005
(HT vs TT) Summer#Mild	11.23333	1.550904	7.24	0.000
(HT vs TT) Summer#Severe	-.6666667	1.550904	-0.43	0.667
(HT vs TT) Fall#Non	5.833333	1.550904	3.76	0.000
(HT vs TT) Fall#Mild	4.133333	1.550904	2.67	0.008
(HT vs TT) Fall#Severe	.2	1.550904	0.13	0.897

This shows that for some combinations of `season` and `depstat`, the difference between happiness and traditional therapy is significant. For example, in season 1 (winter) and depression status 1 (nondepressed), the difference between happiness and traditional therapy is significant, with happiness therapy yielding optimism scores that are 7 points greater than traditional therapy.

9.5 Summary

This chapter has illustrated many of the ways that you can dissect three-way interactions. It has illustrated common designs with three categorical variables, including two by two by two designs and two by two by three designs. The chapter also illustrated a three by three by four design, illustrating some of the other ways you can dissect an interaction that involves at least three levels for each factor.

For more information about models involving three-way interactions of categorical variables, I recommend Keppel and Wickens (2004).

Part III

Continuous and categorical predictors

This part of the book focuses on models that involve interactions of continuous and categorical variables. Such models blend (and build upon) the models illustrated in parts I and II.

Chapters 10 to 12 illustrate models with a continuous predictor interacted with a categorical variable.

Chapter 10 illustrates the interaction of a linear continuous variable with a categorical variable. Chapter 11 covers continuous variables fit using polynomial terms interacted with a categorical variable. Interactions of a continuous variable fit via a piecewise model interacted with a categorical variable are illustrated in chapter 12.

Chapters 13 and 14 cover models involving three-way interactions of continuous and categorical variables.

Chapter 13 illustrates models involving an interaction of two continuous predictors and a categorical variable. This includes linear by linear by categorical interactions and linear by quadratic by categorical interactions. Chapter 14 illustrates models involving an interaction of a linear continuous predictor and two categorical variables.

10 Linear by categorical interactions

10.1 Chapter overview

This chapter illustrates models that involve interactions of categorical and continuous variables. Section 10.2 begins with an analysis of a model including a two-level categorical predictor and a continuous predictor but no interaction. Next, section 10.3 illustrates the same model but includes an interaction of the continuous and categorical predictor. Finally, section 10.4 illustrates a model with an interaction of a continuous by categorical predictor where the categorical predictor has three levels.

In this chapter, the continuous variable is assumed to be linearly related to the outcome variable. Chapters 11 and 12 cover models with interactions of categorical and continuous variables where the continuous variable is nonlinearly related to the outcome variable. Chapter 11 covers models where the nonlinearity is in the form of a polynomial term (for example, quadratic or cubic), whereas chapter 12 illustrates models in which the nonlinearity is in the form of a piecewise model.

10.2 Linear and two-level categorical: No interaction

10.2.1 Overview

This section introduces the concepts involving models that combine continuous and categorical predictors with no interaction. Let's begin by considering a hypothetical simple regression model predicting income from age, focusing on people who are aged 22 to 55. Using the gss_ivrm.dta dataset, we can predict income (realrinc) from age (age). The fitted values from this hypothetical model are graphed in figure 10.1. In this model, the intercept is 9,000, and the slope is 400. The intercept reflects the predicted income for someone who is zero years old. Although the actual ages range from 22 to 55, I have extended the x axis to include zero to show the placement of the intercept. The equation for this regression is shown below.

$$\widehat{\texttt{realrinc}} = 9000 + 400\texttt{age}$$

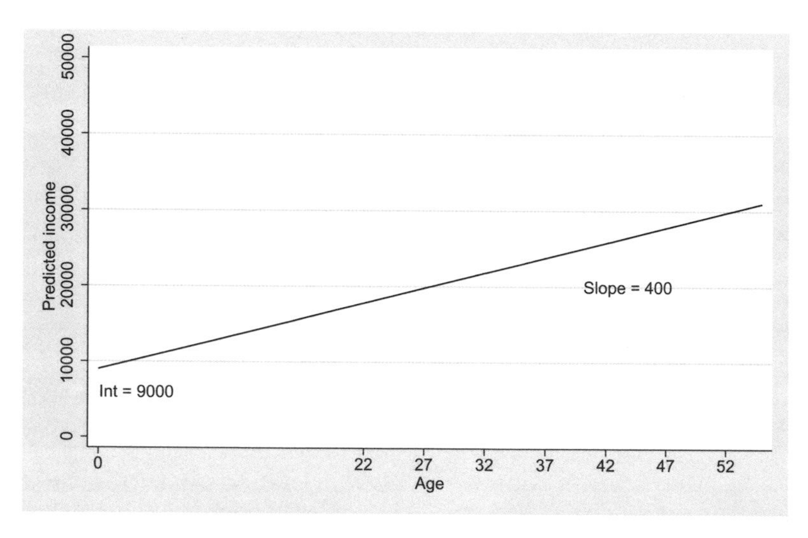

Figure 10.1. Simple linear regression predicting income from age

Let's now expand upon this model and introduce a categorical variable with two levels reflecting whether the respondent graduated college (named cograd in the dataset gss_ivrm.dta). It is coded 0 if the respondent did not graduate college and 1 if the respondent did graduate college. The hypothetical results of this fitted model are graphed in figure 10.2. The slope of the relationship between income and age is 400 for both groups. However, the intercept for those who did not graduate college is 4,000 and for those who did graduate college is 21,000. The difference in these intercepts is 17,000.

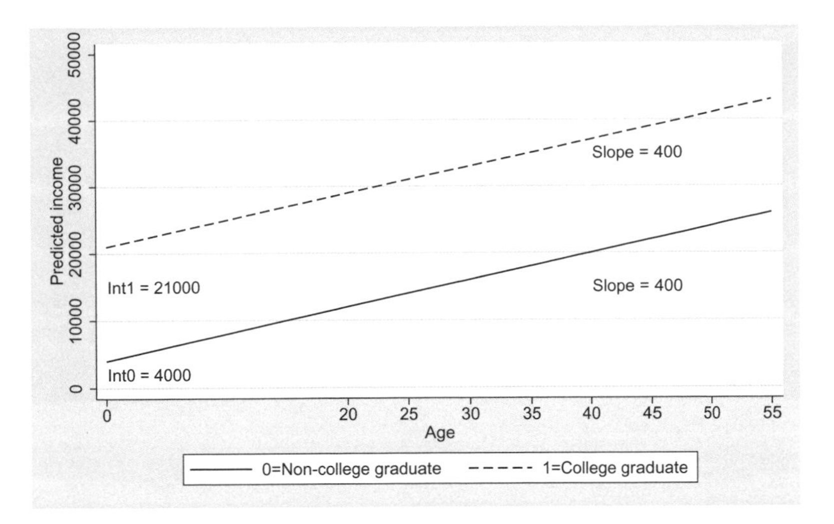

Figure 10.2. One continuous and one categorical predictor with labels for slopes and intercepts

This regression model can be expressed as separate equations for each group. One equation is given for noncollege graduates and another for college graduates, as shown below.

$$\text{Non-college graduates: } \widehat{\texttt{realrinc}} = 4000 + 400\texttt{age}$$
$$\text{College graduates: } \widehat{\texttt{realrinc}} = 21000 + 400\texttt{age}$$

This regression model can also be expressed as one equation, as shown below.

$$\widehat{\texttt{realrinc}} = 4000 + 17000\texttt{cograd} + 400\texttt{age}$$

Note how the intercept is expressed as the intercept for those who did not graduate college, and the main effect of `cograd` corresponds to the difference in the intercepts (of those who did versus those who did not graduate college). The `age` slope is identical for both groups so is expressed as one term in this combined model.

The predicted values at 30, 40, and 50 years of age have been computed both for those who did and for those who did not graduate college and are graphed in figure 10.3.

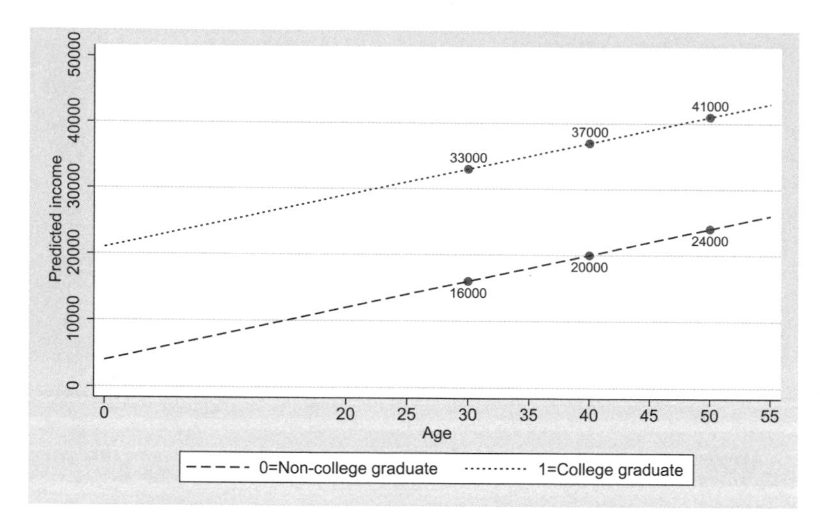

Figure 10.3. One continuous and one categorical predictor with labels for predicted values

Note how the predicted values for those who graduated college are always $17,000 more than the predicted values for their counterparts who did not graduate college. For example, a 40-year-old who graduated college has a predicted income of $37,000, compared with $20,000 for someone who did not graduate college. No matter what the age of the person is, graduating college has the same impact on income, leading to a predicted gain of $17,000. This is because this model has only main effects and does not include an interaction between age and college graduation status.

10.2.2 Examples using the GSS

This section applies the model predicting `realrinc` from `age` and `cograd` using the GSS dataset. Below, we use the `gss_ivrm.dta` dataset and use the `keep` command to keep only those aged 22 to 55. We fit a regression model predicting `realrinc` from `age` and `i.cograd`. The variable `female` is included as a covariate, entered as a dummy variable.[1]

1. From a substantive perspective, the variable `female` is included to account for gender in the prediction of income. For the purposes of the examples in this chapter, the variable `female` is included for the sake of having a covariate in the model so I can illustrate how to compute adjusted means after accounting for one or more covariates. For this reason, I will not focus on the results regarding `female`.

```
. use gss_ivrm
. keep if age>=22 & age<=55
(18,936 observations deleted)
. regress realrinc age i.cograd female, vce(robust)
Linear regression                               Number of obs     =     25,718
                                                F(3, 25714)       =     906.12
                                                Prob > F          =     0.0000
                                                R-squared         =     0.1569
                                                Root MSE          =      23938

------------------------------------------------------------------------------
             |               Robust
    realrinc |      Coef.   Std. Err.      t    P>|t|     [95% Conf. Interval]
-------------+----------------------------------------------------------------
         age |   539.9036   15.26786    35.36   0.000     509.9777    569.8295
             |
      cograd |
 Not CO Grad |          0  (base)
     CO Grad |   14176.56   425.1603    33.34   0.000     13343.22     15009.9
             |
      female |  -12119.58   295.2722   -41.05   0.000    -12698.33   -11540.83
       _cons |   4171.565   518.7608     8.04   0.000     3154.765    5188.366
------------------------------------------------------------------------------
```

Before interpreting the coefficients for this model, let's create a graph showing the adjusted means as a function of age and cograd.[2] First, we use the margins command to compute the adjusted means at ages 22 and 55 for each level of cograd.

```
. margins cograd, nopvalues at(age=(22 55)) vsquish
Predictive margins                              Number of obs     =     25,718
Model VCE    : Robust

Expression   : Linear prediction, predict()
1._at        : age             =          22
2._at        : age             =          55

--------------------------------------------------------------
               |            Delta-method
               |     Margin   Std. Err.     [95% Conf. Interval]
---------------+----------------------------------------------
     _at#cograd|
 1#Not CO Grad |   10048.09   216.1978       9624.33    10471.85
     1#CO Grad |   24224.65   398.8151      23442.95    25006.35
 2#Not CO Grad |   27864.91   344.2312       27190.2    28539.62
     2#CO Grad |   42041.46   553.3733      40956.82    43126.11
--------------------------------------------------------------
```

2. I am graphing the adjusted means as a function of age and cograd to illustrate the pattern of the adjusted means when there is no interaction between these variables. In section 10.3.2, the model will include an interaction of these variables and we will see how the graph of the adjusted means changes due to the age by cograd interaction.

The `marginsplot` command graphs these adjusted means, placing `age` on the x axis and using separate lines for those who have and have not graduated college. This graph is shown in figure 10.4.

```
. marginsplot, noci
Variables that uniquely identify margins: age cograd
```

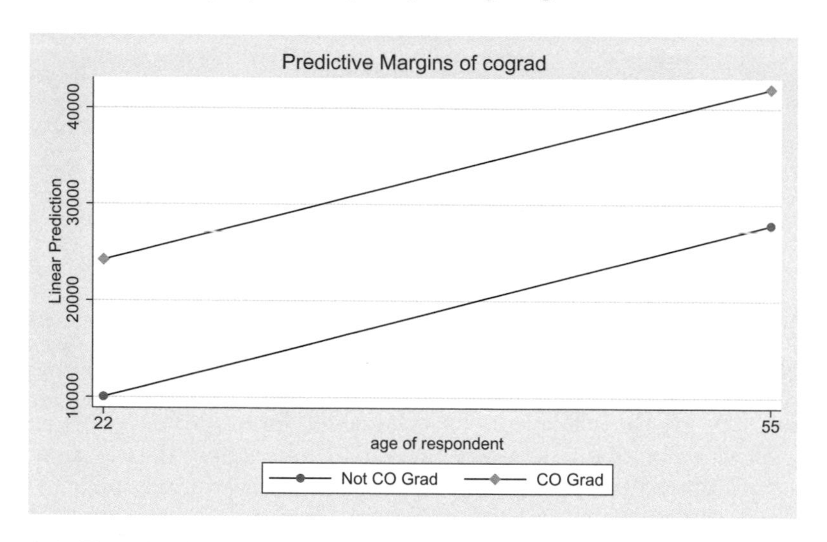

Figure 10.4. Fitted values of continuous and categorical model without interaction

Note how figure 10.4 shows two parallel lines, one for those who graduated college and one for those who did not graduate college. These parallel lines have the same slope, as represented by the coefficient for `age`. Regardless of whether you graduated college, the `age` slope is 539.90.

Although the lines for college graduates have the same slope as noncollege graduates, these lines do not have the same overall height. The fitted line for those who graduated college is higher than the fitted line for those who did not. In fact, the fitted line for those who graduated college is 14,176.56 higher, corresponding to the coefficient for `1.cograd`. This coefficient represents the difference in the fitted values for those who graduated college (group 1) compared with those who did not graduate college (group 0).

We can use the **margins** command to compute adjusted means based on this model. For example, the adjusted mean for someone who graduated college and was 40 years old, adjusting for gender, is 33,942.91.

```
. margins, nopvalues at(cograd=1 age=40)
Predictive margins                              Number of obs     =     25,718
Model VCE    : Robust
Expression   : Linear prediction, predict()
at           : age             =          40
               cograd          =           1

             |            Delta-method
             |     Margin   Std. Err.     [95% Conf. Interval]
-------------+------------------------------------------------
       _cons |   33942.91   419.9859      33119.71    34766.11
```

Let's repeat this command, but this time obtain the adjusted means separately for those who did and for those who did not graduate college.

```
. margins cograd, nopvalues at(age=40)
Predictive margins                              Number of obs     =     25,718
Model VCE    : Robust
Expression   : Linear prediction, predict()
at           : age             =          40

             |            Delta-method
             |     Margin   Std. Err.     [95% Conf. Interval]
-------------+------------------------------------------------
      cograd |
 Not CO Grad |   19766.35   151.4622      19469.48    20063.23
     CO Grad |   33942.91   419.9859      33119.71    34766.11
```

Note the difference between these two values corresponds to the main effect of `cograd`. Here $33942.91 - 19766.35 = 14176.56$. You could repeat the above command for any given level of `age` and the difference in the adjusted means would remain the same.

Adjusted means

In the context of this example, it would be common to express the difference between college graduates and noncollege graduates, adjusting for age and gender, using adjusted means. The adjusted means are computed by setting all the covariates (that is, `age` and `female`) to the average value of the entire sample. In this example, adding the `at((mean) age female)` option specifies that `age` and `female` should be held constant at their mean.

```
. margins cograd, nopvalues at((mean) age female)
Adjusted predictions                            Number of obs     =     25,718
Model VCE    : Robust

Expression   : Linear prediction, predict()
at           : age             =    37.28214 (mean)
               female          =    .4951785 (mean)

------------------------------------------------------------------
             |            Delta-method
             |     Margin   Std. Err.     [95% Conf. Interval]
-------------+----------------------------------------------------
      cograd |
Not CO Grad  |   18298.97     129.817      18044.52    18553.42
     CO Grad |   32475.53    405.0787      31681.55    33269.51
------------------------------------------------------------------
```

The adjusted mean for noncollege graduates is 18,298.97, compared with 32,475.53 for college graduates. The difference in these means is sometimes called the marginal effect at the mean, because the effect is computed at the mean of the covariates.

Based on this example, you might think that you would need to specify each of the covariates in the `at()` option to compute adjusted means. However, we can use the `margins` command below and obtain the same results.[3]

```
. margins cograd, nopvalues
Predictive margins                              Number of obs     =     25,718
Model VCE    : Robust

Expression   : Linear prediction, predict()

------------------------------------------------------------------
             |            Delta-method
             |     Margin   Std. Err.     [95% Conf. Interval]
-------------+----------------------------------------------------
      cograd |
Not CO Grad  |   18298.97     129.817      18044.52    18553.42
     CO Grad |   32475.53    405.0787      31681.55     33269.5
------------------------------------------------------------------
```

The difference in these means, although the same, is sometimes called the average marginal effect. In a linear model (for example, following the `regress` command), the marginal effect at the mean and average marginal effect methods provide the same results.

Unadjusted means

You might be interested in computing the unadjusted mean of income separately for college graduates and noncollege graduates. Such values can be computed using the `margins` command with the `over(cograd)` option, as shown below.

3. This is not the case for nonlinear models such as logistic regression (see chapter 18 for more details).

```
. margins, nopvalues over(cograd)

Predictive margins                              Number of obs     =     25,718
Model VCE    : Robust

Expression   : Linear prediction, predict()
over         : cograd

------------------------------------------------------------------------
             |            Delta-method
             |     Margin   Std. Err.      [95% Conf. Interval]
-------------+----------------------------------------------------------
      cograd |
 Not CO Grad |   18157.94    128.4178      17906.24    18409.65
     CO Grad |   32821.99    408.3032      32021.69    33622.28
------------------------------------------------------------------------
```

Among those included in the estimation sample, the average income for noncollege graduates is $18,157.94, compared with $32,821.99 for college graduates.

Summary

These examples have illustrated a model that predicts income from age and college graduation status, assuming that the `age` slope is the same for those who did and for those who did not graduate college. These examples illustrated how to use the `margins` command to compute adjusted means and unadjusted means.

Rather than assuming that the `age` slope is the same for those who did and for those who did not graduate college, we might hypothesize that the `age` slope might differ by college graduation status. Let's explore this in the following section.

10.3 Linear by two-level categorical interactions

10.3.1 Overview

Let's build upon the model that was illustrated in section 10.2.1 by including `age` and `cograd` as predictors, as well as the interaction between `age` and `cograd`. Such a model fits a separate slope for the relationship between income and age for those who graduated college versus those who did not graduate college. The hypothetical predicted values from such a model are shown in figure 10.5. This figure includes labels showing the intercept and slope for those who did not graduate college (labeled "Int0" and "Slope0") and the intercept and slope for those who did graduate college (labeled "Int1" and "Slope1").

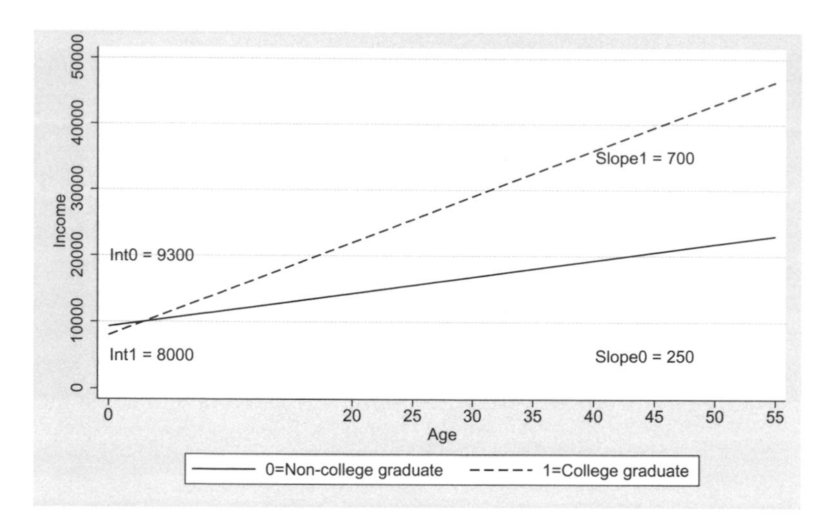

Figure 10.5. Linear by two-level categorical predictor with labels for intercepts and slopes

The regression equation for this hypothetical example can be written as shown below.

$$\widehat{\texttt{realrinc}} = 9300 + -1300\texttt{cograd} + 250\texttt{age} + 450\texttt{cograd*age}$$

The intercept in the regression equation corresponds to the intercept for those who did not graduate college (that is, 9,300). The coefficient for `cograd` is the difference in the intercepts (that is, $8000 - 9300 = -1300$). Note how this represents the difference between a college graduate versus a noncollege graduate when age is held constant at zero (which is implausible and completely absurd). This is often called the *main effect* of `cograd`, but that is misleading because in the presence of the interaction, this term represents the effect of `cograd` when age is held constant at zero.

The coefficient for `age` is 250, corresponding to the slope for those who did not graduate college. The interaction term (`cograd*age`) is the difference in the slopes comparing those who graduated college with those who did not graduate college (that is, $700 - 250 = 450$). This interaction term compares the slope of college graduates with the slope of noncollege graduates.

It can be easier to understand this model when it is written as two equations, one equation for those who did not graduate college (labeled as "Non-college graduate") and another for those who did graduate college (labeled as "College graduate").

$$\text{Non-college graduate: } \widehat{\texttt{realrinc}} = 9300 + 250\texttt{age}$$

$$\text{College graduate: } \widehat{\texttt{realrinc}} = 8000 + 700\texttt{age}$$

This allows us to think in terms of the slopes for each group. These equations make it clear that the slope is 250 for those who did not graduate college and is 700 for those who did graduate college. The meaning of the intercepts is also clear. For a noncollege graduate, the predicted income is \$9,300 for someone who is zero years old. Likewise, for a college graduate, the predicted income is \$8,000 for someone who is zero years old. Clearly, the intercepts are generally not meaningful in such a model.

A key question in this model is whether the `age` slope for college graduates is significantly different from the slope for noncollege graduates.[4] Furthermore, we might be interested in knowing whether each of the slopes is significantly different from 0—is the slope of 250 significantly different from 0, and is the slope of 700 significantly different from 0?

We might also be interested in the effect of graduating college on income. However, the size of this effect depends on a person's age. Referring to figure 10.5, the effect of `cograd` grows larger with increasing age. Because the lines for college graduates and noncollege graduates are not parallel, the difference in the predicted income for college graduates versus noncollege graduates varies as a function of age. In other words, there is no longer an overall estimate of the main effect of `cograd`. Instead, we can estimate the effect of `cograd` at a particular age (or ages). For example, we can estimate the effect of `cograd` for 30-year-olds and 50-year-olds (illustrated in figure 10.6). For 30-year-olds, college graduates are predicted to earn \$29,000 compared with \$16,800 for noncollege graduates, a difference of \$12,200. Compare that with 50-year-olds, where the college graduates are predicted to earn \$21,200 more than noncollege graduates (\$43,000 minus \$21,800).

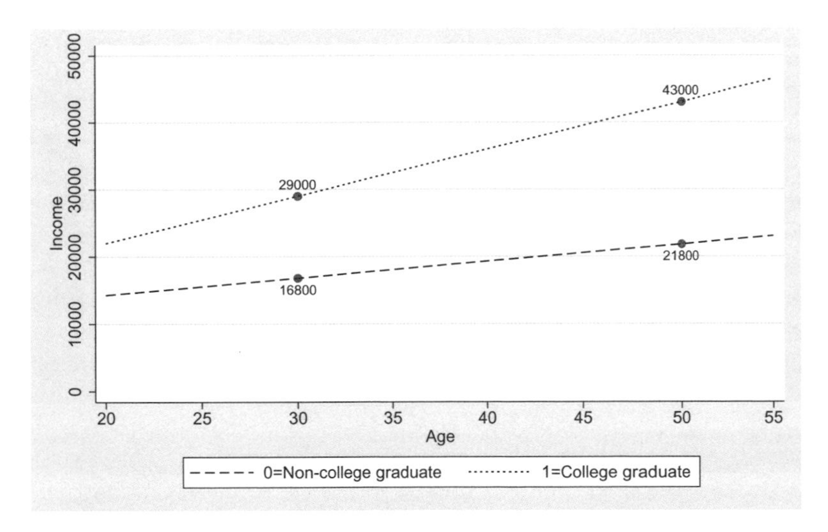

Figure 10.6. Linear by two-level categorical predictor with labels for fitted values

4. If these slopes are not significantly different, then you could estimate a common slope for both groups by removing the interaction term.

10.3.2 Examples using the GSS

Let's use the GSS dataset to explore and fit the model above. First, we use the gss_ivrm.dta dataset and keep only those who are aged 22 to 55.

```
. use gss_ivrm
. keep if age>=22 & age<=55
(18,936 observations deleted)
```

Now, let's fit a model that predicts realrinc from age, cograd, and the interaction of these two variables. Such a model is fit below. The interaction is included by specifying i.cograd#c.age. The variable female is included as a covariate.

```
. regress realrinc i.cograd age i.cograd#c.age female, vce(robust)
Linear regression                               Number of obs     =     25,718
                                                F(4, 25713)       =     710.76
                                                Prob > F          =     0.0000
                                                R-squared         =     0.1661
                                                Root MSE          =      23807

------------------------------------------------------------------------------
             |               Robust
    realrinc |      Coef.   Std. Err.      t    P>|t|     [95% Conf. Interval]
-------------+----------------------------------------------------------------
      cograd |
 Not CO Grad |          0  (base)
     CO Grad |  -8648.272   1426.594    -6.06   0.000    -11444.48   -5852.068
             |
         age |   375.1547   14.67547    25.56   0.000       346.39    403.9195
             |
cograd#c.age |
     CO Grad |   607.5224   42.23988    14.38   0.000     524.7299     690.315
             |
      female |  -12004.56   293.4602   -40.91   0.000    -12579.76   -11429.37
       _cons |   10224.97   480.3151    21.29   0.000     9283.523    11166.41
------------------------------------------------------------------------------
```

As a shorthand, we can specify i.cograd##c.age. The ## operator includes both the main effects and interactions of the variables specified.

```
. regress realrinc i.cograd##c.age female, vce(robust)
  (output omitted)
```

The coefficient estimates are expressed using the following regression equation:

$$\widehat{\texttt{realrinc}} = 10224.97 + -8648.27\texttt{cograd} + 375.15\texttt{age} + 607.5224\texttt{cograd*age} + -12004.56\texttt{female}$$

Let's create a graph to aid in the process of interpreting the results. The margins command is used to compute the adjusted means for ages 22 and 55 separately for each level of cograd.

```
. margins cograd, nopvalues at(age=(22 55))
Predictive margins                              Number of obs     =     25,718
Model VCE    : Robust
Expression   : Linear prediction, predict()
1._at        : age              =          22
2._at        : age              =          55

                    Delta-method
             Margin   Std. Err.     [95% Conf. Interval]

_at#cograd
1#Not CO Grad   12533.97   185.5671     12170.25    12897.69
     1#CO Grad   17251.19   549.5347     16174.07    18328.31
2#Not CO Grad   24914.08   349.7497     24228.55    25599.61
     2#CO Grad   49679.54   950.3805     47816.74    51542.34
```

Then, the **marginsplot** command is used to graph these adjusted means. The resulting graph is shown in figure 10.7. This shows the adjusted mean for income as a function of age (on the x axis) and with separate lines to represent whether one graduated college.

```
. marginsplot, noci
  Variables that uniquely identify margins: age cograd
```

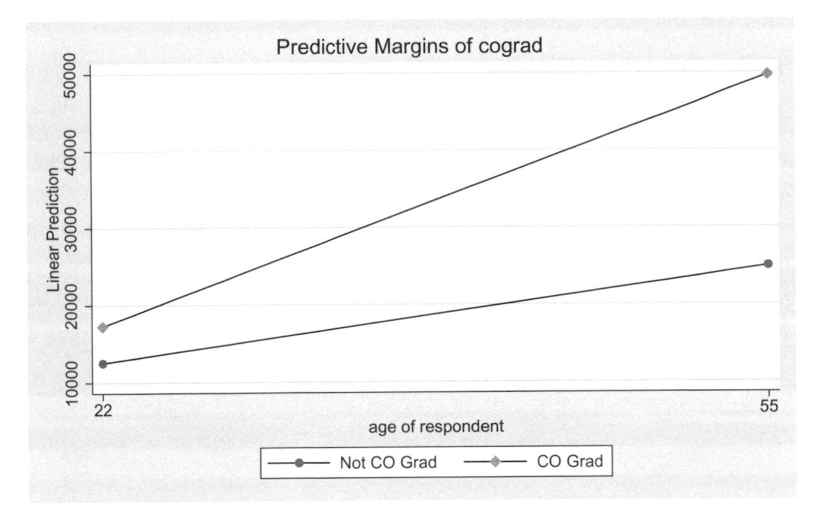

Figure 10.7. Fitted values for linear by two-level categorical predictor model

Figure 10.7 illustrates the slopes for the two groups. The interaction effect (`cograd#c.age`) from the original `regress` command tests the equality of the `age` coefficients for those who graduated college compared with those who did not graduate college. That test is significant ($t = 14.38$, $p = 0.000$), indicating that the slopes are significantly different.

Estimates of slopes

Figure 10.7 illustrates the age slope is larger for college graduates than noncollege graduates. Let's use the margins command to compute the age slope for college graduates and noncollege graduates. The dydx(age) option is used with the over(cograd) option to compute the age slope at each level of cograd. The slope for college graduates (that is, when cograd equals 1) is 982.68 and is statistically significant. The slope for noncollege graduates (that is, when cograd equals 0) is 375.15 and is also statistically significant.[5]

```
. margins, dydx(age) over(cograd)
Average marginal effects                        Number of obs     =     25,718
Model VCE    : Robust

Expression   : Linear prediction, predict()
dy/dx w.r.t. : age
over         : cograd

------------------------------------------------------------------------------
             |            Delta-method
             |      dy/dx   Std. Err.      t    P>|t|     [95% Conf. Interval]
-------------+----------------------------------------------------------------
age          |
      cograd |
 Not CO Grad |   375.1547   14.67547    25.56   0.000       346.39    403.9195
     CO Grad |   982.6772    39.5525    24.84   0.000      905.152    1060.202
------------------------------------------------------------------------------
```

Estimates and contrasts on means

Let's change our focus from the estimates and comparisons of the slopes to the estimates and comparisons of means. Referring to figure 10.7, you can see how the difference between these college graduates and noncollege graduates depends on age. As age increases, the differences in the adjusted mean for college graduates compared with noncollege graduates grows larger. Let's begin the investigation of the effect of cograd by computing the adjusted mean of income for each level of cograd, holding age constant at 30. The adjusted mean for 30-year-old college graduates is 25,112.61 compared with 15,535.21 for 30-year-old noncollege graduates.

5. Note how the slope for noncollege graduates corresponds to the coefficient for age from the regress command. The slope for college graduates corresponds to the age coefficient plus the cograd#c.age coefficient.

```
. margins cograd, nopvalues at(age=30)

Predictive margins                              Number of obs     =     25,718
Model VCE    : Robust

Expression   : Linear prediction, predict()
at           : age             =          30

------------------------------------------------------------
             |            Delta-method
             |     Margin   Std. Err.     [95% Conf. Interval]
-------------+----------------------------------------------
      cograd |
 Not CO Grad |   15535.21   112.5515      15314.6    15755.82
     CO Grad |   25112.61   367.2358     24392.81    25832.41
------------------------------------------------------------
```

Now, let's compare these two adjusted means by applying the `r.` contrast operator with **cograd**, which compares each group with the reference group. This compares group 1 (college graduates) with group 0 (noncollege graduates), yielding a difference of 9,577.40, which is significant. At age 30, college graduates earn $9,577.40 more than noncollege graduates.

```
. margins r.cograd, at(age=30) contrast(nowald pveffects)

Contrasts of predictive margins                 Number of obs     =     25,718
Model VCE    : Robust

Expression   : Linear prediction, predict()
at           : age             =          30

-----------------------------------------------------------------
                           |            Delta-method
                           |   Contrast   Std. Err.      t    P>|t|
---------------------------+-------------------------------------
                    cograd |
(CO Grad vs Not CO Grad)   |     9577.4   383.8656    24.95   0.000
-----------------------------------------------------------------
```

Let's repeat this **margins** command, but this time specify that the comparisons should be made at three ages—30, 40, and 50.

```
. margins r.cograd, at(age=(30 40 50)) contrast(nowald pveffects) vsquish

Contrasts of predictive margins                 Number of obs     =     25,718
Model VCE    : Robust

Expression   : Linear prediction, predict()
1._at        : age             =          30
2._at        : age             =          40
3._at        : age             =          50

-------------------------------------------------------------------
                             |            Delta-method
                             |   Contrast   Std. Err.      t    P>|t|
-----------------------------+-------------------------------------
                   cograd@_at |
(CO Grad vs Not CO Grad) 1   |     9577.4   383.8656    24.95   0.000
(CO Grad vs Not CO Grad) 2   |   15652.62   481.4179    32.51   0.000
(CO Grad vs Not CO Grad) 3   |   21727.85    820.375    26.49   0.000
-------------------------------------------------------------------
```

As we saw before, the difference in adjusted means for college graduates versus noncollege graduates at age 30 is 9,577.40, which is significant. At 40 years of age, this difference is 15,652.62, which is also significant ($z = 32.51$, $p = 0.000$). At 50 years of age, this difference is 21,727.85, which is also significant ($z = 26.49$, $p = 0.000$).

Let's now create a graph that visualizes the differences in the adjusted means for college graduates versus noncollege graduates for each year of age from 22 to 55. We first use the **margins** command to compute these differences.

```
. margins r.cograd, at(age=(22(1)55))
  (output omitted)
```

Then, we use the **marginsplot** command to graph these differences and the confidence interval for the difference, as shown in figure 10.8. If the confidence interval does not include zero, then the difference is significant. The income of college graduates is significantly higher than noncollege graduates across the entire spectrum of ages (ranging from 22 to 55).

```
. marginsplot
  Variables that uniquely identify margins: age
```

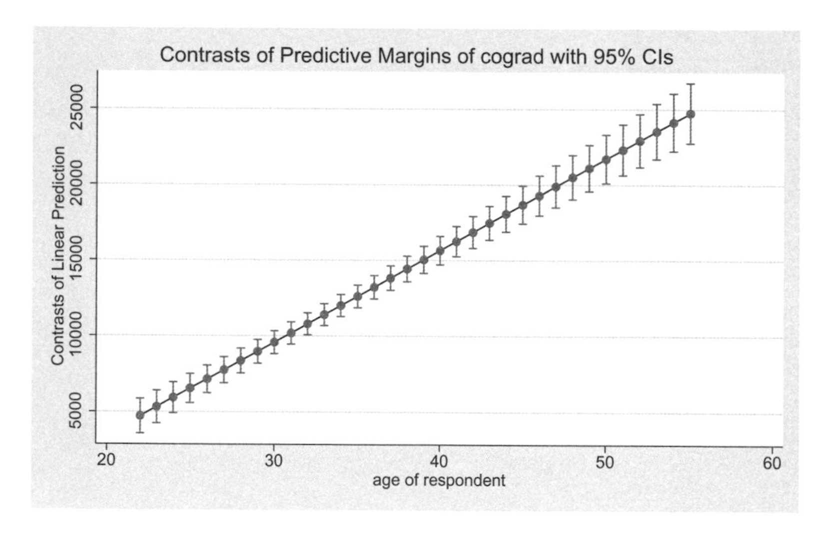

Figure 10.8. Contrasts of fitted values by age with confidence intervals

This section has illustrated an analysis that includes an interaction of a continuous and a two-level categorical predictor. We have seen that the interaction reflects the difference in the slopes between the two groups. Furthermore, the **margins** command can be used to estimate the slope for each group. The **margins** command also allows us to estimate and compare the adjusted means for each group, holding the continuous variable constant. The concepts and commands used for a two-level categorical variable apply to a three-level categorical variable, as illustrated in the next section.

10.4 Linear by three-level categorical interactions

10.4.1 Overview

Let's now explore a model in which a three-level categorical variable is interacted with a continuous variable. Let's extend the previous example by considering three educational groups: 1) non–high school graduates, 2) high school graduates, and 3) college graduates. The hypothetical predicted values from such a model are shown in figure 10.9. This figure includes labels showing the slope for each of the three educational groups. Although not shown in the graph, the intercept is 3,000 for group 1, 3,700 for group 2, and −5,000 for group 3.

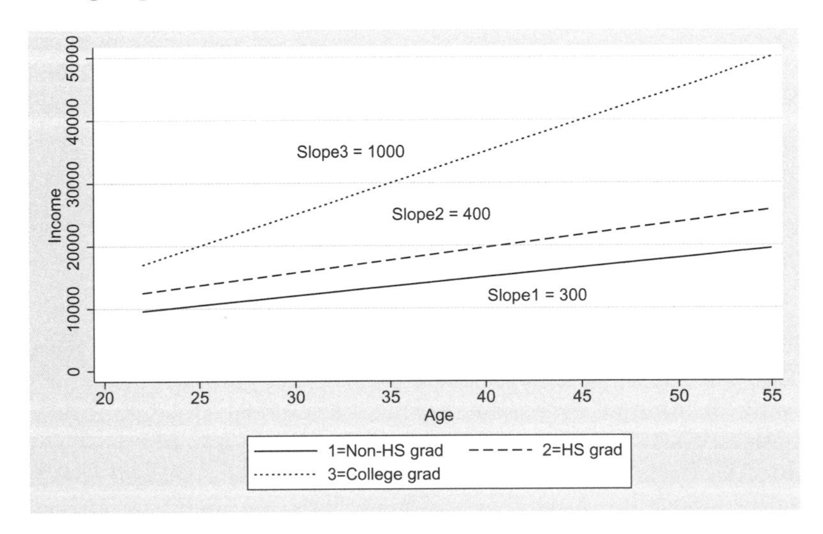

Figure 10.9. Linear by three-level categorical predictor with labels for slopes

The regression equation for this hypothetical example can be written as shown below. Note that non–high school graduates is the reference group, `hsgrad` represents the comparison of high school graduates with non–high school graduates, and `cograd` represents the comparison of college graduates with non–high school graduates.

$$\widehat{\texttt{realrinc}} = 3000 + 700\texttt{hsgrad} + -8000\texttt{cograd} \\ + 300\texttt{age} + 100\texttt{hsgrad*age} + 700\texttt{cograd*age}$$

First, let's consider the meaning of the intercept and the `hsgrad` and `cograd` coefficients. The value of 3,000 is the intercept for those who did not graduate college. The coefficient for `hsgrad` is the difference in the intercepts between high school graduates and non–high school graduates (that is, $3700 - 3000 = 700$). The coefficient for `cograd` is the difference in the intercepts between college graduates and non–high school graduates (that is, $-5000 - 3000 = -8000$). The terms for `hsgrad` and `cograd` represent differences among the education groups when age is held constant at zero. These are often called *main effects* with respect to education, but this is misleading. In the presence of the interaction with age, these terms represents comparisons among education groups when age is held constant at zero. The terms for the intercept, as well as `hsgrad` and `cograd`, are usually not interesting.

Now, let's consider the coefficients for `age` and the interactions of education with `age`. The coefficient for `age` is 300, corresponding to the slope for those who did not graduate high school. The `hsgrad*age` interaction is the difference in the `age` slopes comparing those who graduated high school with those who did not graduate high school (that is, $400 - 300 = 100$). The `cograd*age` interaction is the difference in the `age` slopes comparing those who graduated college with those who did not graduate high school (that is, $1000 - 300 = 700$).

It can be easier to understand this model when it is written as three equations: one equation for those who did not graduate high school, one for those who did graduate high school, and another for those who graduated college, as shown below.

$$\begin{aligned}
\text{Non-HS grad: } \widehat{\texttt{realrinc}} &= 3000 + 300\texttt{age} \\
\text{HS grad: } \widehat{\texttt{realrinc}} &= 3700 + 400\texttt{age} \\
\text{College grad: } \widehat{\texttt{realrinc}} &= -5000 + 1000\texttt{age}
\end{aligned}$$

This allows us to focus on the slopes for each group. Looking at these equations makes it clear that the slope is 300 for non–high school graduates, 400 for high school graduates, and 1,000 for college graduates. In such a model, it is interesting to test the null hypothesis that all of these slopes are equal. If we do not reject this null hypothesis, then the education by interaction terms may no longer be needed and could be omitted from the model. If we do reject this null hypothesis, we might be further interested in forming specific contrasts among the slopes, for example, comparing the slopes for high school graduates with non–high school graduates, and college graduates with high school graduates.

Because the lines for the three educational groups are not parallel (due to the inclusion of the age by education interaction), the difference in the predicted incomes among the education groups will vary as a function of age. We might be interested in comparing the different educational groups given different levels of `age`. As an illustration, the predicted mean of income for those aged 30, 40, and 50 for each educational group have been computed and are plotted in figure 10.10.

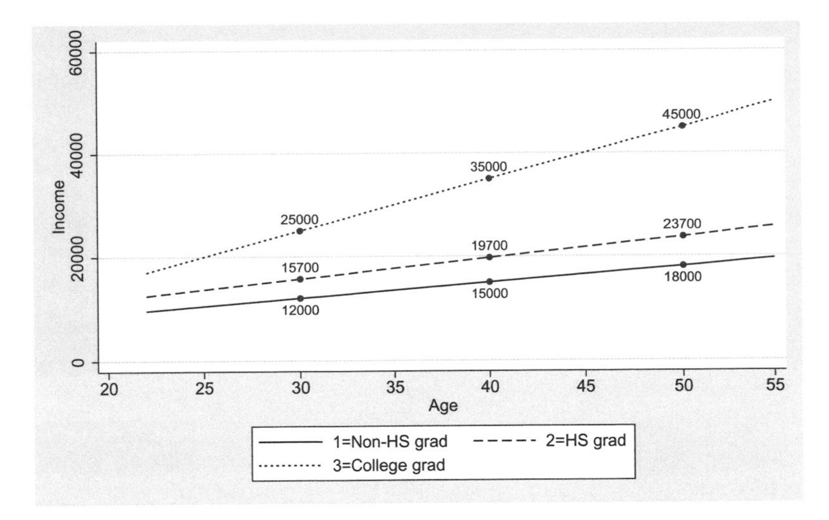

Figure 10.10. Linear by three-level categorical predictor with labels for fitted values

There are three means that can be compared at any specific level of age. For example, at age 30 the predicted mean income is $12,000 for non–high school graduates, $15,700 for high school graduates, and $25,000 for college graduates. There are several comparisons we could make among these three groups. We could compare the overall equality of the three means, we could compare each mean with a reference group (such as non–high school graduates), we could compare each educational level with the next highest educational level, and so forth.

10.4.2 Examples using the GSS

Let's now consider applying this kind of model using the GSS dataset. Let's continue to use `age` as the continuous predictor and `realrinc` as the outcome, but now we will use a three-category education variable, `educ3`. The variable `educ3` is coded: 1 = non–high school graduate, 2 = high school graduate, and 3 = college graduate. The `gss_ivrm.dta` dataset is used below and, as before, the analysis focuses on those aged 22 to 55.

```
. use gss_ivrm
. keep if age>=22 & age<=55
(18,936 observations deleted)
```

Let's begin by running an analysis that predicts realrinc from age, educ3, and the interaction of these two variables. The model also includes gender as a covariate.

```
. regress realrinc i.educ3##c.age female, vce(robust)
Linear regression                               Number of obs     =     25,718
                                                F(6, 25711)       =     563.58
                                                Prob > F          =     0.0000
                                                R-squared         =     0.1728
                                                Root MSE          =      23712

------------------------------------------------------------------------------
             |               Robust
    realrinc |      Coef.   Std. Err.      t    P>|t|     [95% Conf. Interval]
-------------+----------------------------------------------------------------
       educ3 |
     not hs  |          0  (base)
         HS  |   1906.981   1098.008     1.74   0.082    -245.1763    4059.137
       Coll  |  -6287.544   1643.872    -3.82   0.000    -9509.626   -3065.462
             |
         age |   299.1506   28.35131    10.55   0.000     243.5804    354.7208
             |
 educ3#c.age |
         HS  |   119.9961    32.9587     3.64   0.000     55.39518     184.597
       Coll  |   682.8823   48.70536    14.02   0.000     587.4171    778.3476
             |
      female |   -12258.8   293.4448   -41.78   0.000    -12833.96   -11683.63
       _cons |   8012.498    944.865     8.48   0.000     6160.509    9864.486
------------------------------------------------------------------------------
```

Before interpreting these results, let's make a graph showing the adjusted means by age and educ3. We first use the margins command to compute the adjusted means at ages 22 and 55 for each level of educ3.

```
. margins educ3, nopvalues at(age=(22 55))
Predictive margins                              Number of obs     =     25,718
Model VCE    : Robust

Expression   : Linear prediction, predict()
1._at        : age             =          22
2._at        : age             =          55

--------------------------------------------------------------
             |            Delta-method
             |     Margin   Std. Err.     [95% Conf. Interval]
-------------+------------------------------------------------
    _at#educ3 |
   1#not hs  |   8523.519   361.1772      7815.591    9231.447
       1#HS  |   13070.41   208.6737       12661.4    13479.43
     1#Coll  |   17259.39   549.7561      16181.83    18336.94
   2#not hs  |   18395.49   655.3127      17111.04    19679.94
       2#HS  |   26902.25   405.9435      26106.58    27697.93
     2#Coll  |   49666.47   950.0856      47804.25    51528.69
--------------------------------------------------------------
```

Then, the marginsplot command is used to graph the adjusted mean on the y axis, graph age on the x axis, and indicate educ3 using separate lines (see figure 10.11).

```
. marginsplot, noci
  Variables that uniquely identify margins: age educ3
```

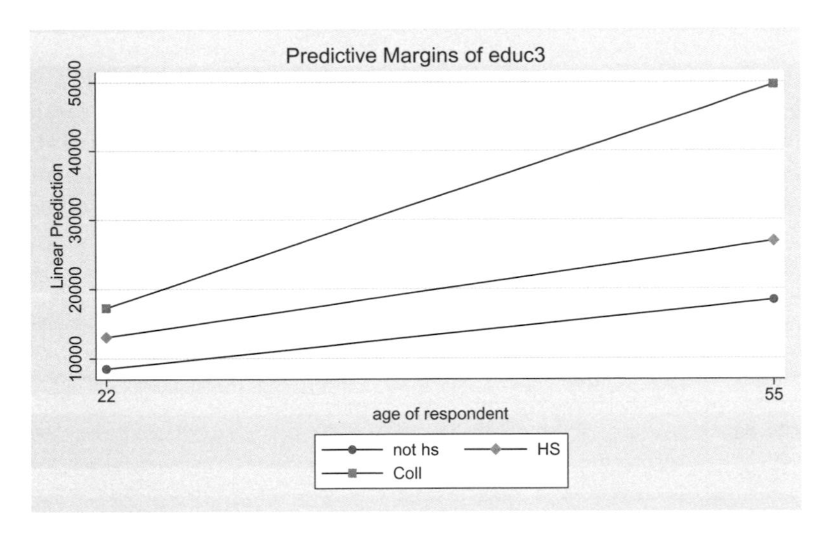

Figure 10.11. Fitted values for linear by three-level categorical predictor model

Estimates and contrasts on slopes

The graph in figure 10.11 illustrates the slopes for each group. The **margins** command below computes the **age** slope at each level of **educ3**. The **dydx(age)** is combined with the **over(educ3)** option to compute the **age** slope separately for each level of **educ3**.

```
. margins, dydx(age) over(educ3)
Average marginal effects                        Number of obs     =     25,718
Model VCE    : Robust

Expression   : Linear prediction, predict()
dy/dx w.r.t. : age
over         : educ3
```

	dy/dx	Delta-method Std. Err.	t	P>|t|	[95% Conf.	Interval]
age						
educ3						
not hs	299.1506	28.35131	10.55	0.000	243.5804	354.7208
HS	419.1467	16.85162	24.87	0.000	386.1166	452.1768
Coll	982.0329	39.54918	24.83	0.000	904.5143	1059.552

The output not only shows the **age** slope at each level of **educ3** but also includes the standard error, confidence interval, and a test of whether the slope is significantly different from 0. The **age** slope is significantly different from 0 at each level of **educ3**.

Let's test whether these slopes are equal to each other. In other words, let's test the null hypothesis,

$$H_0\colon\ \beta_1 = \beta_2 = \beta_3$$

where β_1 is the slope for non–high school graduates, β_2 is the slope for high school graduates, and β_3 is the slope for college graduates. The following `contrast` command tests this null hypothesis by testing the interaction of `educ3` and `age`. This interaction is significant, so we can reject the null hypothesis that these slopes are all equal.

```
. contrast educ3#c.age
Contrasts of marginal linear predictions
Margins      : asbalanced

             |         df           F        P>F
-------------+----------------------------------
 educ3#c.age |          2      105.89     0.0000
             |
 Denominator |      25711
```

Suppose we wanted to further dissect this interaction by comparing the slope for each level of `educ3` with the third group (that is, college graduates). Expressed in terms of null hypotheses, say, we want to test the following two null hypotheses:

$$H_0\#1\colon\ \beta_1 = \beta_3$$
$$H_0\#2\colon\ \beta_2 = \beta_3$$

The following `contrast` command tests these two null hypotheses. Specifying `rb3.educ3` uses reference group comparisons with group 3 as the baseline (comparison) group. We can reject both null hypotheses. The difference in the `age` slopes for group 1 versus 3 (non–high school graduates to college graduates) is significant. Likewise, the difference in the `age` slopes for group 2 versus 3 (high school graduates versus college graduates) is also significant.

```
. contrast rb3.educ3#c.age, nowald pveffects
Contrasts of marginal linear predictions
Margins      : asbalanced

                   |   Contrast   Std. Err.       t    P>|t|
-------------------+----------------------------------------
       educ3#c.age |
(not hs vs Coll)   |  -682.8823   48.70536   -14.02   0.000
    (HS vs Coll)   |  -562.8863   43.03201   -13.08   0.000
```

We can specify any of the contrast operators covered in chapter 7 in place of the **rb3.** operator to form contrasts among the **age** slopes. Let's try using the **j.** contrast operator that compares each group with the mean of the previous levels, as shown below.

```
. contrast j.educ3#c.age, nowald pveffects
Contrasts of marginal linear predictions
Margins      : asbalanced

                   |  Contrast   Std. Err.        t    P>|t|
-------------------+----------------------------------------
         educ3#c.age |
(HS vs   not hs)   |  119.9961     32.9587     3.64    0.000
(Coll vs <Coll)    |  622.8843    42.90004    14.52    0.000
```

This shows that the **age** slope for group 2 is significantly different from group 1 (high school graduates versus non–high school graduates). Also, the **age** slope for group 3 (college graduates) is significantly different from the average of groups 1 and 2.

Estimates and contrasts on means

The previous section focused on comparisons of the slopes as a means of understanding and interpreting the interaction of **age** and **educ3**. Another way we can understand this interaction is to examine the differences in the adjusted means among the levels of **educ3** for a specified level (or levels) of **age**.

Referring back to figure 10.11, imagine estimating the adjusted means at specific values of **age** for the different levels of **educ3**. Let's begin by computing the adjusted means when **age** equals 30 for the different levels of **educ3** using the **margins** command (below).

```
. margins educ3, nopvalues at(age=30)
Predictive margins                              Number of obs     =     25,718
Model VCE    : Robust

Expression   : Linear prediction, predict()
at           : age             =          30

             |            Delta-method
             |     Margin   Std. Err.     [95% Conf. Interval]
-------------+------------------------------------------------
       educ3 |
      not hs |   10916.72   205.3558      10514.21    11319.23
          HS |   16423.59   127.0515      16174.56    16672.61
        Coll |   25115.65   367.3734      24395.58    25835.72
```

Let's test the equality of these means by adding the **contrast(overall)** option. This tests the equality of the adjusted means across the levels of **educ3** when **age** is held constant at 30. This test is significant ($p = 0.0000$).

```
. margins educ3, at(age=(30)) contrast(overall)
Contrasts of predictive margins                 Number of obs     =     25,718
Model VCE    : Robust

Expression   : Linear prediction, predict()
at           : age             =          30
```

	df	F	P>F
educ3	2	593.45	0.0000
Denominator	25711		

We can specify any of the contrast operators described in chapter 7 to form contrasts on the categorical variable (that is, `educ3`). For example, let's use the `ar.` contrast operator to perform adjacent group contrasts on `educ3` while holding age constant at 30. The output shows that, at 30 years of age, there is a significant difference in the adjusted mean of income comparing group 2 versus 1 (high school graduates compared with non–high school graduates, $t = 22.67$, $p < 0.001$). Likewise, there is a significant difference in the comparison of group 3 versus 2 (college graduates compared with high school graduates, $t = 22.45$, $p < 0.001$).

```
. margins ar.educ3, at(age=(30)) contrast(nowald pveffects)
Contrasts of predictive margins                 Number of obs     =     25,718
Model VCE    : Robust

Expression   : Linear prediction, predict()
at           : age             =          30
```

	Contrast	Delta-method Std. Err.	t	P>|t|
educ3				
(HS vs not hs)	5506.863	242.8756	22.67	0.000
(Coll vs HS)	8692.063	387.2089	22.45	0.000

Let's repeat these contrasts, but do so instead at ages 30, 40, and 50. These contrasts are performed using the margins command below.

```
. margins ar.educ3, at(age=(30 40 50)) contrast(nowald pveffects)
Contrasts of predictive margins                Number of obs     =     25,718
Model VCE    : Robust

Expression   : Linear prediction, predict()
1._at        : age             =          30
2._at        : age             =          40
3._at        : age             =          50

------------------------------------------------------------------------
                     |             Delta-method
                     |   Contrast   Std. Err.          t    P>|t|
---------------------+--------------------------------------------------
           educ3@_at |
   (HS vs not hs) 1  |   5506.863   242.8756      22.67    0.000
   (HS vs not hs) 2  |   6706.824   333.5987      20.10    0.000
   (HS vs not hs) 3  |   7906.785   617.1246      12.81    0.000
     (Coll vs HS) 1  |   8692.063   387.2089      22.45    0.000
     (Coll vs HS) 2  |   14320.93   488.5892      29.31    0.000
     (Coll vs HS) 3  |   19949.79   835.3795      23.88    0.000
------------------------------------------------------------------------
```

This provides a total of six tests. There are two tests of the comparisons among the levels of educ3 (high school graduates versus non–high school graduates; and college graduates compared versus high school graduates) at the three different levels of age (30, 40, and 50). All of these comparisons are significant.

Say that we wanted to form the same comparisons on education (that is, specifying ar.educ3) but spanning the ages ranging from 22 to 55. We can create a graph that illustrates such comparisons using the margins and marginsplot commands. First, the margins command is used, applying the ar. contrast operator to educ combined with the at() option. This performs the contrasts of education groups 2 versus 1 (high school graduates versus non–high school graduates) and education group 3 versus 2 (college graduates versus high school graduates) holding age constant at every value ranging from 22 to 55 years of age.

```
. margins ar.educ3, at(age=(22(1)55))
  (output omitted)
```

Then, the marginsplot command graphs the contrasts computed by the margins command (see figure 10.12). The bydimension(educ3) option produces two panels, the left panel that compares high school graduates versus non–high school graduates and the right panel that compares college graduates versus high school graduates. When the confidence interval for a contrast excludes zero, the difference is significant at the 5% level.

```
. marginsplot, recast(line) recastci(rarea) bydimension(educ3)
  Variables that uniquely identify margins: age educ3
```

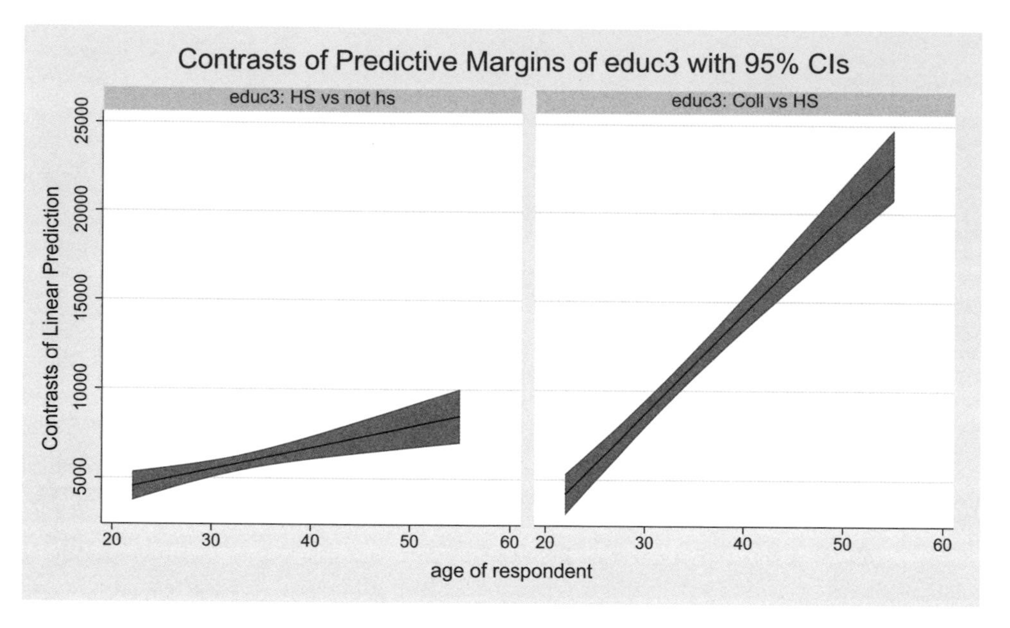

Figure 10.12. Adjacent contrasts on education by age with confidence intervals, as two graph panels

Across the entire spectrum of age, these graphs show that the difference in the adjusted means comparing group 2 versus 1 are significant (see left panel of figure 10.12). Likewise, the difference in the adjusted means comparing group 3 versus 2 is also all significant (see right panel of figure 10.12).

10.5 Summary

This chapter has illustrated the interpretation of interactions of continuous and categorical variables using `age` as a continuous variable and `educ` as a categorical variable. In the presence of an `age` by `educ` interaction, the `age` slope differs as a function of the categorical variable, `educ`. When `educ` was treated as a two-level categorical variable, a significant interaction implied that the `age` slope differed for the two levels of education. When `educ` was treated as a three-level categorical variable, the `age` by `educ` interaction was more complex. The interaction signifies that the slopes of the three groups are not equal. The `contrast` command can be used to form contrasts among the slopes for the three different groups. The `margins` command can also be used to form contrasts among the levels of education, holding `age` constant at one or more levels. This chapter illustrated the use of the `marginsplot` command to visualize the pattern of results.

For more information about the interaction of continuous and categorical variables, consult your favorite regression book because this is a topic is covered in most regression books. In particular, I recommend Fox (2016) and Cohen et al. (2003) for their treatments of this topic.

11 Polynomial by categorical interactions

11.1 Chapter overview

This chapter illustrates models that involve a continuous variable modeled using a polynomial term interacted with a categorical variable. This chapter covers two types of polynomial terms: quadratic and cubic. The first section involves interactions of quadratic terms with categorical variables (see section 11.2). This includes models involving the interaction of a quadratic term with a two-level categorical variable (see section 11.2.2) and a quadratic term interacted with a three-level categorical variable (see section 11.2.3). Section 11.3 describes models involving the interaction of a cubic term with a categorical variable, illustrating a cubic term interacted with a two-level categorical variable.

11.2 Quadratic by categorical interactions

This section covers models that include a two-level categorical variable interacted with a continuous variable modeled with a quadratic term. The examples presented are an extension of the examples shown in section 3.2.2, which showed a quadratic relationship between age and income (`age` and `realrinc`). That section illustrated a quadratic model in which income rises with increasing age until around age 50 and then income declines thereafter. This section will illustrate how the degree of curvature in the relationship between `age` and `realrinc` depends on one's educational status (for example, whether one is a college graduate). Let's first consider an overview of such models before considering specific examples.

11.2.1 Overview

Models that involve interactions of a categorical variable with a continuous variable modeled using a quadratic term blend two modeling techniques that we have previously seen.

The first modeling technique is the use of a quadratic term to account for nonlinearity in the relationship between the predictor and outcome. We have seen the modeling of quadratic terms in section 3.2. A hypothetical example of a quadratic model predicting income from age is shown in figure 11.1. The quadratic term for `age` is −25, hence the inverted U-shape in the relationship between age and income.

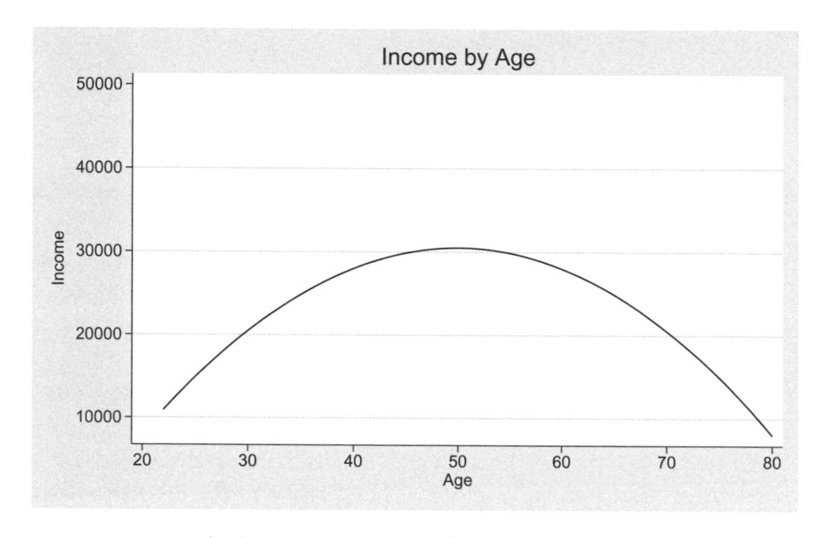

Figure 11.1. Predicted values from quadratic regression

The second modeling technique involves the interactions of categorical and continuous variables, as illustrated in chapter 10. A hypothetical example of a model with a categorical by continuous interaction is shown in figure 11.2. A person's income is predicted by age and whether one is a college graduate, as well as the interaction of these two variables. The interaction yields different slopes for the relationship between income and age depending on whether one is a college graduate. The slope is 250 for noncollege graduates and is 800 for college graduates.

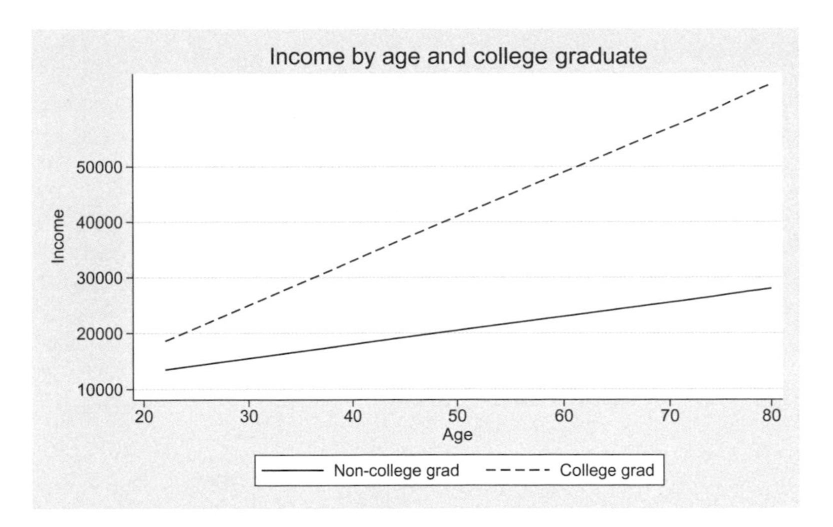

Figure 11.2. Predicted values from linear by two-level categorical variable model

These two modeling techniques can be blended together, yielding an interaction of a categorical variable with a continuous variable modeled using a quadratic term. The fitted values from such a hypothetical model are graphed in figure 11.3. The model predicts income from age, age squared, and college graduation status (0 = noncollege graduate and 1 = college graduate). The model also includes the interaction of age with college graduation status, and age squared with college graduation status. The interaction of age squared with college graduation status permits the degree of curvature in age to differ for college graduates versus noncollege graduates. In this example, the quadratic term is −40 for college graduates and is −15 for noncollege graduates. This is reflected in the greater degree of curvature in the relationship between age and income for the college graduates. Section 11.2.2 provides examples of how to fit these kinds of models.

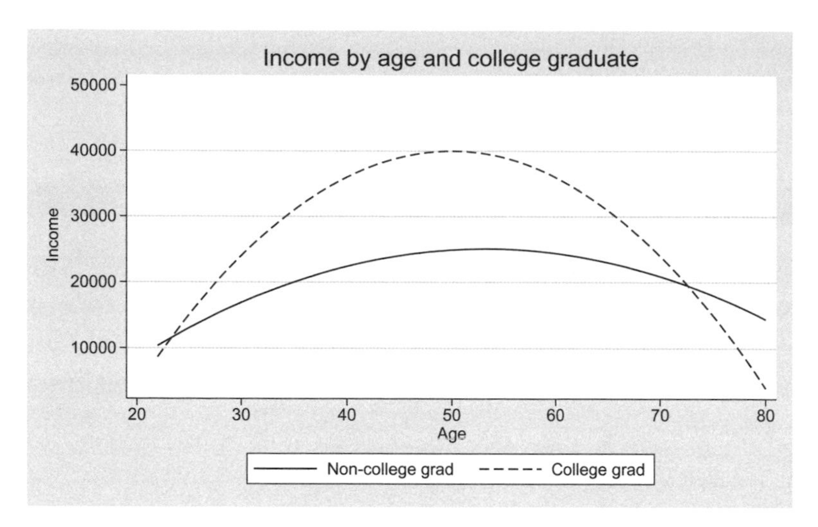

Figure 11.3. Predicted values from quadratic by two-level categorical variable model

It is not difficult to extend the model illustrated in figure 11.3 to include three or more groups. The hypothetical fitted values of such a model with three groups is illustrated in figure 11.4.

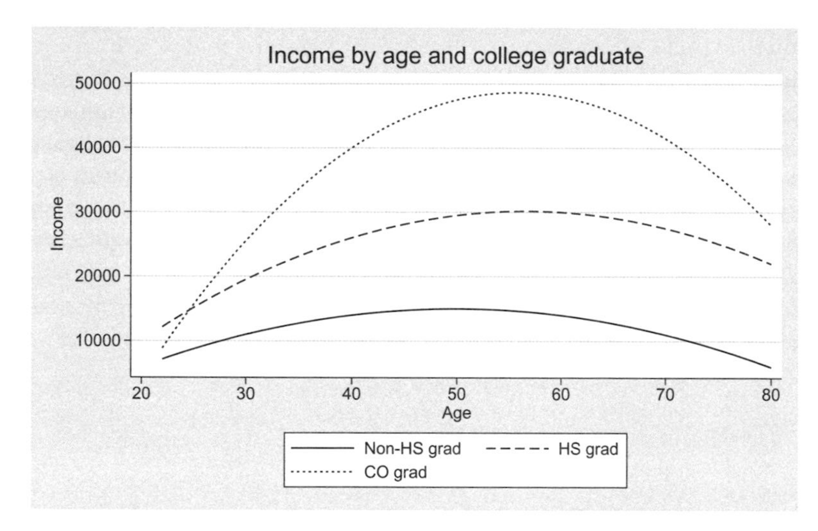

Figure 11.4. Predicted values from quadratic by three-level categorical variable model

Education is broken up into three groups based on whether one is a non–high school graduate, a high school graduate, or a college graduate. In this example, the non–high school graduates and high school graduates do not differ much with respect to the quadratic term. The quadratic term for age is -10 for non–high school graduates, is -15 for high school graduates, and is -35 for college graduates. This is reflected in the greater curvature in the relationship between age and income for college graduates. Section 11.2.3 of this chapter provides examples of how to fit models in which a three-level categorical variable is interacted with a continuous variable involving a quadratic term.

11.2.2 Quadratic by two-level categorical

Let's now use the GSS dataset to fit a model predicting income from age (modeled using a quadratic term), whether one is a college graduate, as well as the interaction of these variables. The examples in this section use the `gss_ivrm.dta` dataset focusing only on those who are aged 22 to 80.

```
. use gss_ivrm
. keep if age>=22 & age<=80
(4,541 observations deleted)
```

Let's begin by looking at the relationship between age and income using a lowess smoother. The `lowess` command is used to create the variable `yhat_lowess` that contains the lowess smoothed values of `realrinc`. The `graph` command creates a graph showing the lowess smoothed values across `age`. The graph (see figure 11.5) shows that incomes rise with increasing age until around age 50, where incomes peak. As age increases beyond 50, incomes decline.

```
. lowess realrinc age, generate(yhat_lowess) nograph
. graph twoway line yhat_lowess age, sort
```

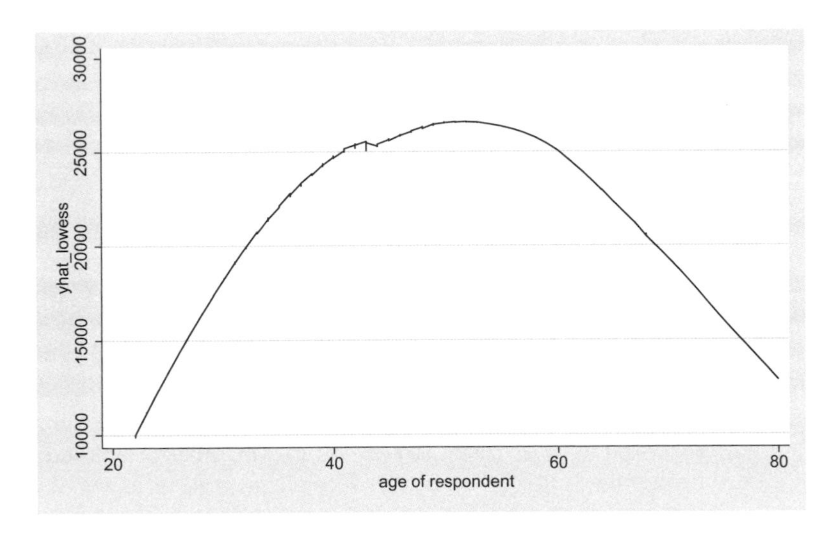

Figure 11.5. Lowess smoothed values of income by age

Let's create this same kind of graph but separating people based on the variable `cograd`, which is coded: 0 = noncollege graduate and 1 = college graduate. The lowess command is issued twice, each with an if specification. The first lowess command is restricted to people who did not graduate college and generates the variable yhat_lowess0, representing the lowess smoothed value of income for those who have not graduated college. The second lowess command creates yhat_lowess1, the lowess smoothed value of income for those who have graduated college.

```
. lowess realrinc age if cograd == 0, generate(yhat_lowess0) nograph
. lowess realrinc age if cograd == 1, generate(yhat_lowess1) nograph
```

The graph command is then used to create a graph of the lowess smoothed values of income as a function of age, separately for those who did and for those who did not graduate college. This graph, shown in figure 11.6, suggests that the degree of curvature in the relationship between age and income differs based on college graduation status. The lowess smoothed values for those who graduated college appear to show a greater degree of curvature than those who did not graduate college.

```
. graph twoway line yhat_lowess0 yhat_lowess1 age, sort
```

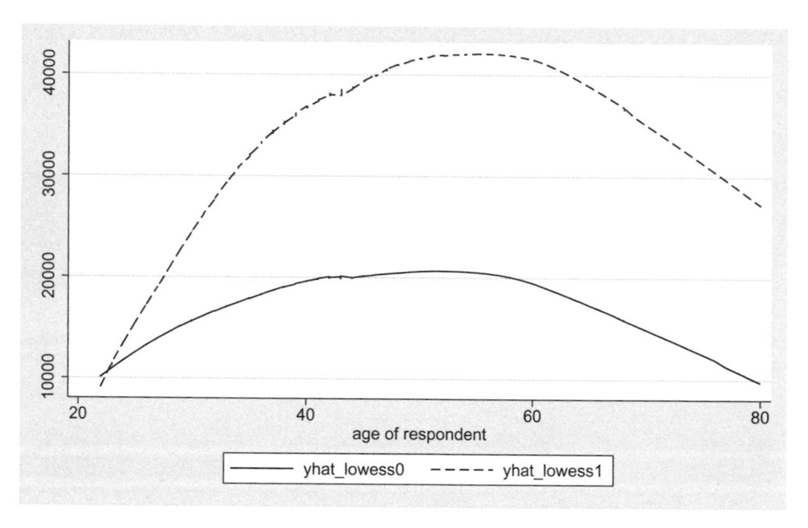

Figure 11.6. Lowess smoothed values of income predicted from age by college graduation status

Based on this visual inspection of the data, a regression model predicting realrinc from age and cograd would not only need to account for the quadratic trend in age, but also the difference in the quadratic trend in age for college graduates versus noncollege graduates.

Let's fit a model that includes an intercept, age, and age squared for noncollege graduates, and a separate intercept, age, and age squared for college graduates. This model is specified using the regress command below. Specifying ibn.cograd with the noconstant option yields separate intercept estimates by college graduation status.

Specifying `ibn.cograd#c.age` yields separate **age** estimates by college graduation status, and `ibn.cograd#c.age#c.age` yields separate **age#age** estimates by college graduation status.[1] (The variable **female** is included as a covariate, entered as a dummy variable.)

```
. regress realrinc ibn.cograd ibn.cograd#c.age ibn.cograd#c.age#c.age
> female, noconstant vce(robust)

Linear regression                               Number of obs     =     30,576
                                                F(7, 30569)       =    5127.56
                                                Prob > F          =     0.0000
                                                R-squared         =     0.5088
                                                Root MSE          =      24865

------------------------------------------------------------------------------
             |               Robust
    realrinc |      Coef.   Std. Err.      t    P>|t|     [95% Conf. Interval]
-------------+----------------------------------------------------------------
      cograd |
 Not CO Grad |  -11096.99   1107.783   -10.02   0.000    -13268.29   -8925.686
     CO Grad |  -52730.93   3688.047   -14.30   0.000    -59959.66   -45502.21
             |
 cograd#c.age|
 Not CO Grad |   1594.873   57.25945    27.85   0.000     1482.642    1707.104
     CO Grad |   3941.463   194.3691    20.28   0.000     3560.491    4322.434
             |
 cograd#c.age#|
       c.age |
 Not CO Grad |  -16.19906   .6617172   -24.48   0.000    -17.49606   -14.90207
     CO Grad |  -37.80665   2.297263   -16.46   0.000    -42.30938   -33.30392
             |
      female |  -12457.11   278.5685   -44.72   0.000    -13003.12   -11911.11
------------------------------------------------------------------------------
```

Note! Model shortcut

The previous model can also be specified as shown below.

```
. regress realrinc ibn.cograd ibn.cograd#(c.age c.age#c.age) female,
> noconstant vce(robust)
```

Stata expands the expression `ibn.cograd#(c.age c.age#c.age)` to become `ibn.cograd#c.age ibn.cograd#c.age#c.age`, yielding the same model shown previously.

1. I specified the **regress** command in this way for pedagogical reasons. This coding method makes it easy to write separate regression equations for college graduates and noncollege graduates.

We can express these results by writing two regression equations: one for those who did not graduate college and one for those who did graduate college.

$$\begin{aligned}\text{Non-college graduate: } \widehat{\texttt{realrinc}} &= -11096.99 + 1594.87\texttt{age} + -16.20\texttt{age}^2 \\ &\quad + -12457.11\texttt{female} \\ \text{College graduate: } \widehat{\texttt{realrinc}} &= -52730.93 + 3941.46\texttt{age} + -37.81\texttt{age}^2 \\ &\quad + -12457.11\texttt{female}\end{aligned}$$

Note the differences in the quadratic coefficient for age. This coefficient is −37.81 for those who graduated college and is −16.20 for those who did not graduate college. Let's visualize the impact of the differences in these quadratic coefficients by graphing the adjusted means as a function of age and college graduation status (that is, cograd), adjusting for gender. First, we use the margins command to compute the adjusted means as a function of age and cograd. The margins command below computes the adjusted means for ages ranging from 22 to 80 (in one-unit increments), separately for each level of cograd.

```
. margins cograd, nopvalues at(age=(22(1)80))
  (output omitted)
```

The marginsplot command[2] graphs the adjusted means from the margins command; see figure 11.7. The graph shows that the college graduates have a greater inverted U-shape than the noncollege graduates. This arises because the quadratic coefficient for age is more negative for college graduates than for noncollege graduates.

2. This command incorporates the recast(line) option to display each line without marker symbols and the scheme(s1mono) option to display the graph using the s1mono scheme, which displays each line using a different pattern.

```
. marginsplot, noci recast(line) scheme(s1mono)
Variables that uniquely identify margins: age cograd
```

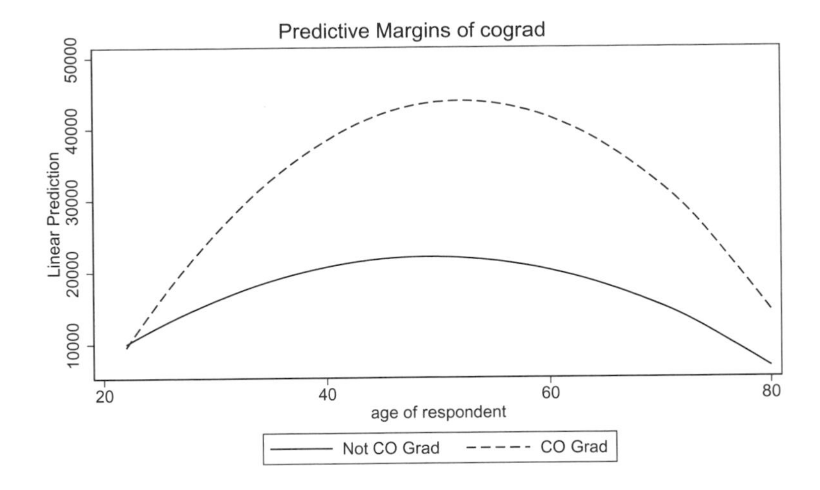

Figure 11.7. Fitted values from quadratic by two-level categorical model

We can ask whether the degree of curvature between college graduates and noncollege graduates is significantly different. The `contrast` command below tests whether the quadratic term for noncollege graduates is equal to the quadratic term for college graduates. The output shows the two quadratic coefficients are significantly different from each other ($F = 81.75$, $p = 0.000$). College graduates show a significantly greater level of curvature in the relationship between age and income than noncollege graduates.

```
. contrast cograd#c.age#c.age
Contrasts of marginal linear predictions
Margins      : asbalanced
```

	df	F	P>F
cograd#c.age#c.age	1	81.75	0.0000
Denominator	30569		

Let's now see how the `margins` command can be used to estimate adjusted means from this model. The `margins` command below computes the adjusted mean of income for a 30-year-old who did not graduate college, adjusting for gender. The adjusted mean for such a person is 16,044.57.

```
. margins, nopvalues at(cograd=0 age=30)

Predictive margins                              Number of obs     =     30,576
Model VCE    : Robust

Expression   : Linear prediction, predict()
at           : cograd          =           0
               age             =          30

------------------------------------------------------------------------
             |            Delta-method
             |     Margin   Std. Err.     [95% Conf. Interval]
-------------+----------------------------------------------------------
       _cons |   16044.57   112.3878      15824.29    16264.86
------------------------------------------------------------------------
```

The **margins** command below estimates the adjusted mean for 40-year-olds, separately by the levels of cograd. A college graduate who is 40 years old has a predicted income of $38,311.44, and a noncollege graduate who is 40 years old has a predicted income of $20,653.96.

```
. margins cograd, nopvalues at(age=40)

Predictive margins                              Number of obs     =     30,576
Model VCE    : Robust

Expression   : Linear prediction, predict()
at           : age             =          40

------------------------------------------------------------------------
             |            Delta-method
             |     Margin   Std. Err.     [95% Conf. Interval]
-------------+----------------------------------------------------------
      cograd |
 Not CO Grad |   20653.96   175.1034      20310.75    20997.17
     CO Grad |   38311.44   568.5894      37196.98     39425.9
------------------------------------------------------------------------
```

The at() option allows us to specify multiple ages at once, as shown in the example below. This example computes the adjusted means for ages 30 to 70 (in 10-year increments) separately for each level of cograd.

```
. margins cograd, nopvalues at(age=(30(10)70)) vsquish
Predictive margins                              Number of obs     =     30,576
Model VCE    : Robust
Expression   : Linear prediction, predict()
1._at        : age             =          30
2._at        : age             =          40
3._at        : age             =          50
4._at        : age             =          60
5._at        : age             =          70

------------------------------------------------------------------
                |            Delta-method
                |     Margin   Std. Err.     [95% Conf. Interval]
----------------+-------------------------------------------------
     _at#cograd |
1#Not CO Grad   |   16044.57   112.3878      15824.29    16264.86
    1#CO Grad   |   25361.47   340.9874      24693.12    26029.82
2#Not CO Grad   |   20653.96   175.1034      20310.75    20997.17
    2#CO Grad   |   38311.44   568.5894      37196.98     39425.9
3#Not CO Grad   |   22023.54   206.4344      21618.92    22428.16
    3#CO Grad   |   43700.08    638.665      42448.27    44951.89
4#Not CO Grad   |    20153.3    269.911      19624.26    20682.33
    4#CO Grad   |   41527.39   877.7023      39807.05    43247.72
5#Not CO Grad   |   15043.25   500.8186      14061.62    16024.87
    5#CO Grad   |   31793.36   1754.301      28354.86    35231.86
------------------------------------------------------------------
```

The `margins` command can be used to compare the adjusted means of those who did and those who did not graduate college at specified ages. For example, we can compare the means for college graduates versus noncollege graduates for 30-year-olds using the `margins` command below. Specifying `r.cograd` yields reference group comparisons, comparing college graduates with noncollege graduates.

```
. margins r.cograd, at(age=30) contrast(nowald pveffects)

Contrasts of predictive margins                 Number of obs     =     30,576
Model VCE    : Robust

Expression   : Linear prediction, predict()
at           : age             =          30

------------------------------------------------------------------------------
                        |            Delta-method
                        |   Contrast   Std. Err.          t    P>|t|
------------------------+-----------------------------------------------------
                 cograd |
(CO Grad vs Not CO Grad)|     9316.9   358.4768       25.99    0.000
------------------------------------------------------------------------------
```

The difference in the adjusted means of college graduates and noncollege graduates who are 30 years old is 9,316.9, and that difference is statistically significant. Note how this corresponds to the difference in the adjusted means at 30 years of age computed by the previous `margins` command (that is, $25361.47 - 16044.57 = 9316.9$).

Let's repeat the last `margins` command, but this time specifying multiple ages in the `at()` option. The example below compares college graduates with noncollege graduates at 30 to 70 years of age in 10-year increments. The difference is significant at each of the specified ages.

```
. margins r.cograd, at(age=(30(10)70)) contrast(nowald pveffects) vsquish

Contrasts of predictive margins                 Number of obs     =     30,576
Model VCE    : Robust

Expression   : Linear prediction, predict()
1._at        : age             =          30
2._at        : age             =          40
3._at        : age             =          50
4._at        : age             =          60
5._at        : age             =          70

------------------------------------------------------------------------------
                          |            Delta-method
                          |   Contrast   Std. Err.          t    P>|t|
--------------------------+---------------------------------------------------
                cograd@_at|
(CO Grad vs Not CO Grad) 1|     9316.9   358.4768       25.99    0.000
(CO Grad vs Not CO Grad) 2|   17657.48   594.3632       29.71    0.000
(CO Grad vs Not CO Grad) 3|   21676.54   671.4589       32.28    0.000
(CO Grad vs Not CO Grad) 4|   21374.09   919.4648       23.25    0.000
(CO Grad vs Not CO Grad) 5|   16750.11   1825.292        9.18    0.000
------------------------------------------------------------------------------
```

Let's use the `margins` command to compute the difference between college graduates and high school graduates for 22- to 80-year-olds in one-year increments so we can then graph the differences using the `marginsplot` command. The graph of the differences and a shaded confidence interval for the differences is shown in figure 11.8.

```
. margins r.cograd, at(age=(22(1)80)) contrast(nowald effects)
  (output omitted)
. marginsplot, yline(0) recast(line) recastci(rarea)
  Variables that uniquely identify margins: age
```

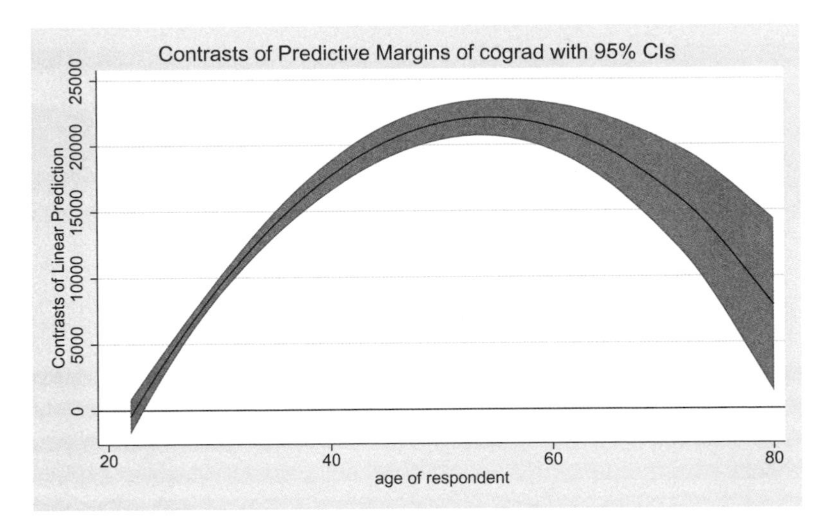

Figure 11.8. Contrasts on college graduation status by age

This graph shows the difference in the adjusted means between college graduates and noncollege graduates. Where the confidence interval excludes zero, the difference is significant at the 5% level. The confidence interval includes zero only at the lowest ages; thus, the difference is significant for those aged 24 to 80. You might notice that the width of the confidence intervals grows with increasing age. This is because of both smaller sample sizes and increased variability in incomes for those who are older.[3]

11.2.3 Quadratic by three-level categorical

Let's now consider a quadratic effect that interacts with a three-level categorical variable. This builds upon the analysis from the previous section by replacing the variable `cograd` with `educ3`. The variable `educ3` is a three-level variable that is coded: 1 = non–high school graduate, 2 = high school graduate, and 3 = college graduate.

The examples in this section will use the `gss_ivrm.dta` dataset focusing on those who are aged 22 to 80.

```
. use gss_ivrm
. keep if age>=22 & age<=80
(4,541 observations deleted)
```

3. Recall that the `vce(robust)` option was included when fitting the regression model, which accommodates heterogeneity of variance (that is, heteroskedasticity).

Let's begin our exploration by using the lowess command to generate variables that contain the smoothed relationship between realrinc and age, separately for each of the three levels of educ3. The graph command is then used to graph the lowess smoothed values of income as a function of age, separately for the three levels of educ3. The resulting graph is shown in figure 11.9.

```
. lowess realrinc age if educ3 == 1, generate(yhat_lowess1) nograph
. lowess realrinc age if educ3 == 2, generate(yhat_lowess2) nograph
. lowess realrinc age if educ3 == 3, generate(yhat_lowess3) nograph
. graph twoway line yhat_lowess1 yhat_lowess2 yhat_lowess3 age, sort
> legend(order(1 "Non-HS grad" 2 "HS grad" 3 "CO grad"))
```

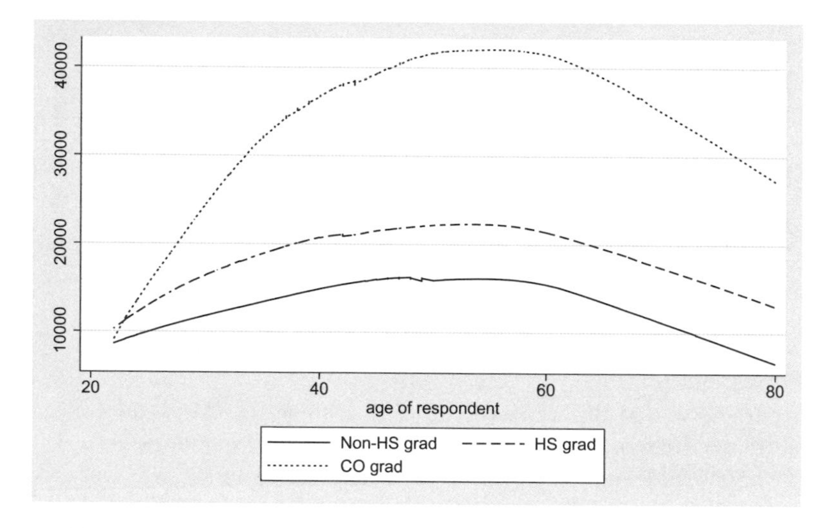

Figure 11.9. Lowess smoothed values of income by age, separated by three levels of education

Figure 11.9 shows that the relationship between age and realrinc is curved for all three levels of educ3. The degree of curvature seems to be strongest for those who graduated college. The degrees of curvature seem smaller, and similar, for those who graduated high school and those who did not graduate high school.

Let's now fit a model that includes a quadratic term for age as well as an interaction of educ3 and the quadratic term for age. The model is constructed in the same manner as the model from section 11.2.2, except that cograd is replaced with educ3.

```
. regress realrinc ibn.educ3
> ibn.educ3#(c.age c.age#c.age) female, noconstant vce(robust)
Linear regression                          Number of obs     =     30,576
                                           F(10, 30566)      =    3691.50
                                           Prob > F          =     0.0000
                                           R-squared         =     0.5129
                                           Root MSE          =      24761

------------------------------------------------------------------------------
             |               Robust
    realrinc |      Coef.   Std. Err.      t    P>|t|     [95% Conf. Interval]
-------------+----------------------------------------------------------------
       educ3 |
      not hs |  -8609.968   1907.078    -4.51   0.000    -12347.92   -4872.015
          HS |   -11115.2   1375.926    -8.08   0.000    -13812.07   -8418.325
        Coll |  -52556.01   3687.815   -14.25   0.000    -59784.28   -45327.74
             |
 educ3#c.age |
      not hs |   1245.109   93.00407    13.39   0.000     1062.818    1427.401
          HS |   1625.585   72.44602    22.44   0.000     1483.587    1767.582
        Coll |   3940.661   194.3347    20.28   0.000     3559.757    4321.565
             |
educ3#c.age# |
       c.age |
      not hs |  -12.45935   1.004618   -12.40   0.000    -14.42845   -10.49026
          HS |  -16.04013   .8650308   -18.54   0.000    -17.73562   -14.34463
        Coll |  -37.80793    2.29672   -16.46   0.000    -42.30959   -33.30626
             |
      female |  -12750.67   278.8573   -45.72   0.000    -13297.24    -12204.1
------------------------------------------------------------------------------
```

We can express the regression equation for this model as though there are three equations corresponding to the three different levels of educ3. The regression equation is written in this fashion as shown below.

$$\text{Non-HS graduate}: \widehat{\texttt{realrinc}} = -8609.97 + 1245.11\text{age} + -12.46\text{age}^2 + -12750.67\texttt{female}$$

$$\text{HS graduate}: \widehat{\texttt{realrinc}} = -11115.2 + 1625.59\text{age} + -16.04\text{age}^2 + -12750.67\texttt{female}$$

$$\text{College graduate}: \widehat{\texttt{realrinc}} = -52556.01 + 3940.66\text{age} + -37.81\text{age}^2 + -12750.67\texttt{female}$$

To help interpret the regression coefficients, let's use the margins and marginsplot commands to visualize the adjusted means as a function of age and educ3. The margins command is used to estimate the adjusted means for ages 22 to 80 (in one-unit increments), separately by the levels of educ3. The marginsplot command is used to graph these adjusted means, creating the graph shown in figure 11.10.

```
. margins educ3, at(age=(22(1)80))
  (output omitted)
. marginsplot, noci recast(line) scheme(s1mono)
  Variables that uniquely identify margins: age educ3
```

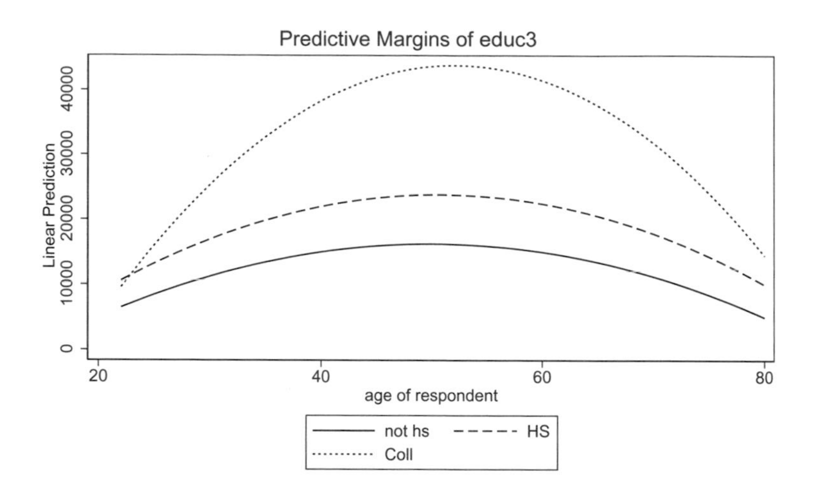

Figure 11.10. Fitted values from age (quadratic) by education level

Let's relate the quadratic coefficients for age to the degree of the curvature in the relationship between age and the adjusted means of income shown in figure 11.10. The coefficient for the quadratic term is −12.46 for those who did not graduate high school, −16.04 for those who graduated high school, and −37.81 for those who graduated college. The college graduates have the most negative coefficient and that group shows the greatest degree of curvature (that is, the strongest inverted U-shape). The quadratic coefficients for the non–high school graduates and high school graduates are similar in magnitude, and they show a similar level of curvature.

The `contrast` command can be used to compare the quadratic coefficients among the educational groups. The `contrast` command below tests the equality of the quadratic coefficient for all three levels of `educ3`.

```
. contrast educ3#c.age#c.age
Contrasts of marginal linear predictions
Margins      : asbalanced
```

	df	F	P>F
educ3#c.age#c.age	2	51.24	0.0000
Denominator	30566		

This test is significant, indicating that the quadratic term for age is not equal across all three groups. Contrast operators can be applied to educ3 in the contrast command to perform specific comparisons among the groups. For example, the a. contrast operator is used below to compare the quadratic terms for adjacent education groups.

```
. contrast a.educ3#c.age#c.age, nowald pveffects
Contrasts of marginal linear predictions
Margins      : asbalanced

------------------------------------------------------------
                    |   Contrast   Std. Err.      t    P>|t|
--------------------+---------------------------------------
educ3#c.age#c.age   |
   (not hs vs HS)   |   3.580775   1.325447    2.70   0.007
     (HS vs Coll)   |    21.7678   2.454231    8.87   0.000
------------------------------------------------------------
```

The first contrast compares the quadratic coefficient for group 1 (non–high school graduates) versus group 2 (high school graduates), and this contrast is significant ($p = 0.007$). Although the degree of curvature seems similar for high school graduates (group 2) and non–high school graduates (group 1), the quadratic term for these two groups are significantly different. The second contrast compares the quadratic coefficient for group 2 versus 3 (high school graduates versus college graduates). The quadratic coefficient for high school graduates is significantly different from the quadratic coefficient for high school graduates ($p = 0.000$).

Let's now see how to use the margins command to compute adjusted means. The margins command below computes the adjusted mean for a college graduate who is 25 years old, adjusting for gender.

```
. margins, nopvalues at(educ3=3 age=25)
Predictive margins                              Number of obs     =     30,576
Model VCE    : Robust

Expression   : Linear prediction, predict()
at           : educ3           =           3
               age             =          25

-------------------------------------------------------------------
             |            Delta-method
             |     Margin   Std. Err.     [95% Conf. Interval]
-------------+-----------------------------------------------------
       _cons |   16060.73   434.2163      15209.65    16911.81
-------------------------------------------------------------------
```

The margins command below computes the adjusted mean holding age constant at 30, separately for each education group.

```
. margins educ3, nopvalues at(age=30)

Predictive margins                              Number of obs     =     30,576
Model VCE    : Robust

Expression   : Linear prediction, predict()
at           : age             =          30

------------------------------------------------------------------
             |            Delta-method
             |     Margin   Std. Err.     [95% Conf. Interval]
-------------+----------------------------------------------------
       educ3 |
      not hs |   11260.07   201.4479      10865.22    11654.91
          HS |    16946.4   129.0068      16693.54    17199.26
        Coll |   25366.86   341.2143      24698.06    26035.65
------------------------------------------------------------------
```

The `margins` command can be used to make comparisons among the levels of `educ3` at different ages. Let's compare adjacent levels of `educ3` when age is held constant at 30. The `ar.` contrast operator yields comparisons of adjacent levels of education in reverse order (that is, group 2 versus 1 and group 3 versus 2).[4]

```
. margins ar.educ3, at(age=30) contrast(nowald pveffects)

Contrasts of predictive margins                 Number of obs     =     30,576
Model VCE    : Robust

Expression   : Linear prediction, predict()
at           : age             =          30

------------------------------------------------------------
                 |            Delta-method
                 |   Contrast   Std. Err.      t    P>|t|
-----------------+------------------------------------------
           educ3 |
(HS vs not hs)   |   5686.332   240.6051   23.63   0.000
  (Coll vs HS)   |   8420.458   363.2567   23.18   0.000
------------------------------------------------------------
```

At 30 years of age, the difference in adjusted means comparing high school graduates with non–high school graduates (group 2 versus 1) is 5,686.33, and this difference is significant ($z = 23.63$, $p = 0.000$). The comparison of college graduates with high school graduates (group 3 versus 2) is also significant ($z = 23.18$, $p = 0.000$). At 30 years of age, the adjusted mean for college graduates is 8,420.46 more than the adjusted mean for high school graduates.

Let's repeat the above contrasts at each age level ranging from 22 to 80 using the `margins` command. We can then use the `marginsplot` command to graph each of the contrasts, as shown below. Two graphs are created: one for the contrast of high school graduates with non–high school graduates (group 2 versus 1) and the other for the contrast of college graduates with high school graduates (group 3 versus 2); see figure 11.11. Each graph includes a confidence interval for the contrast—when the confidence interval excludes zero, the contrast is significant at the 5% level.

4. We can select from any of the contrast operators illustrated in chapter 7.

```
. margins ar.educ3, at(age=(22(1)80))
  (output omitted)
. marginsplot, bydimension(educ3) yline(0) recast(line) recastci(rarea)
  Variables that uniquely identify margins: age educ3
```

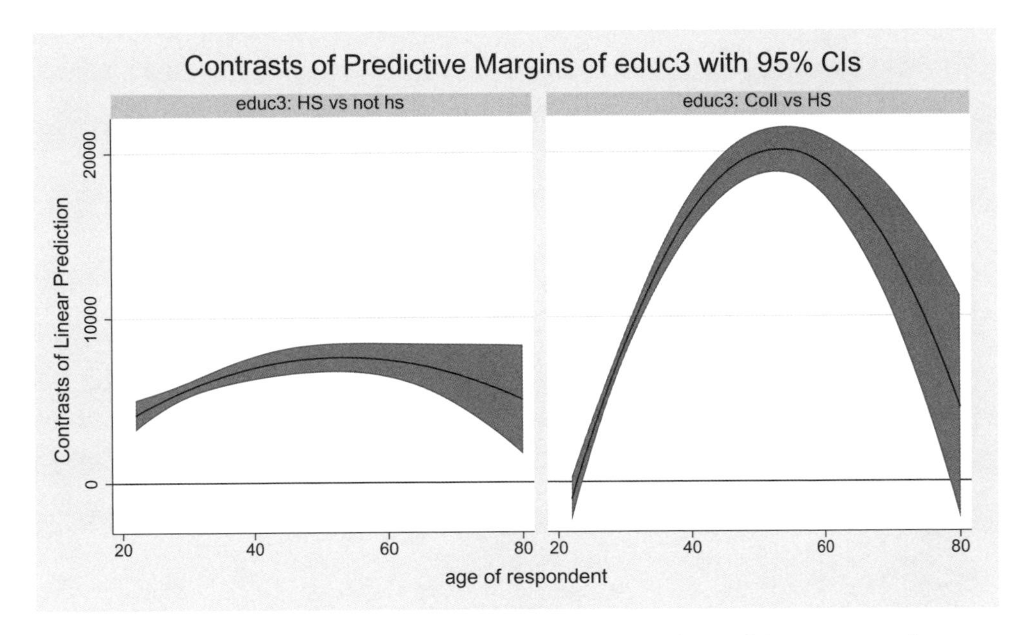

Figure 11.11. Contrasts on education by age, with confidence intervals

The contrast of high school graduates versus non–high school graduates shows that high school graduates have significantly higher incomes across all ages. The contrast of college graduates to high school graduates is significant for all ages except the youngest and oldest.[5]

11.3 Cubic by categorical interactions

This section describes models that involve interactions of a categorical variable with a continuous variable where the continuous variable is fit using a cubic polynomial term. As we saw in section 3.3, cubic terms accounts for two bends in the relationship between the predictor and outcome. A cubic by categorical interaction allows the groups formed by the categorical variable to differ in cubic trend.

Let's build upon the example we saw in section 3.3, where we looked at the year of a woman's birth as a predictor of her number of children. In that example, we saw a cubic relationship between year of birth and number of children. Suppose that we divided women into two groups: those who graduated college and those who did not

5. Inspecting the output of the `margins` command (not shown) reveals nonsignificant differences at ages 22, 23, 79, and 80.

graduate college. Those who graduated college might show a different kind of cubic trend across years of birth compared with noncollege graduates. Figure 11.12 shows a graph of hypothetical fitted values for such a model. The women who did not graduate college show a greater cubic trend (a greater rise and fall) over the years of birth than the women who did graduate college.

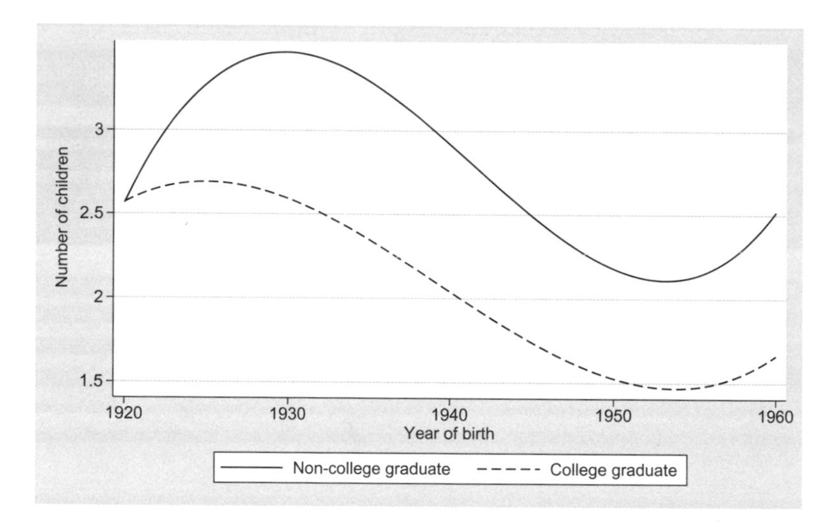

Figure 11.12. Predicted values from cubic by two-level categorical variable model

Let's illustrate a cubic by categorical interaction using the `gss_ivrm.dta` dataset. This example will focus on women aged 45 to 55 born between 1920 and 1960.

```
. use gss_ivrm
. keep if (age>=45 & age<=55) & (yrborn>=1920 & yrborn<=1960) & female==1
(50,021 observations deleted)
```

Let's begin by visualizing the nature of the relationship between year of birth and number of children separately for college graduates and noncollege graduates. We can do this using a lowess smoothed regression relating year of birth to number of children. The `lowess` command is used twice: once for those who have not graduated college and again for those who have graduated college. The `generate()` option is used to create the variables `yhatlowess0` and `yhatlowess1`, which contain the smoothed value of `children` for noncollege graduates and college graduates, respectively.

```
. lowess children yrborn if cograd==0, generate(yhatlowess0) nograph
. lowess children yrborn if cograd==1, generate(yhatlowess1) nograph
```

The `graph` command is then used to graph the lowess smoothed values by year of birth, showing a separate line for noncollege graduates and college graduates. This is shown in figure 11.13.

```
. graph twoway line yhatlowess0 yhatlowess1 yrborn, sort
```

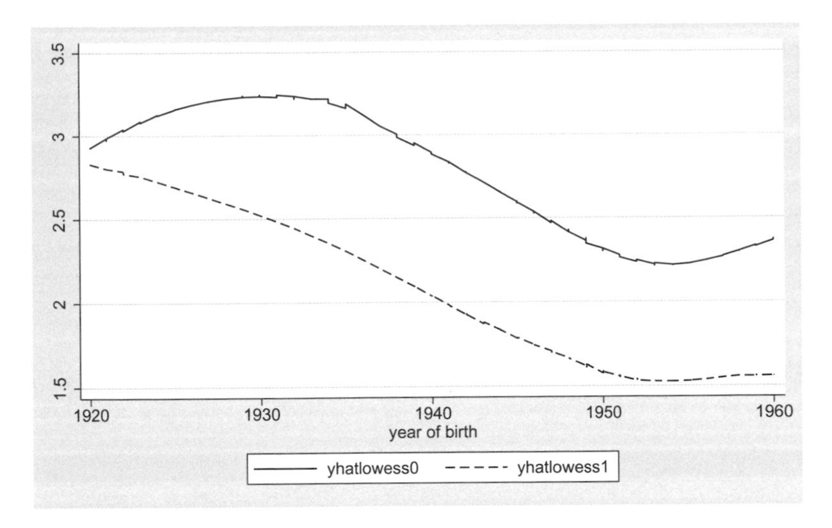

Figure 11.13. Lowess smoothed values of number of children by year of birth, separated by college graduation status

Focus on those who have not graduated college. Note how the relationship between year of birth and number of children has two bends. The number of children increases until a peak at around 1930, then decreases until around 1952 and then starts to increase again. This curve seems to exhibit a cubic trend. By contrast, those who have graduated college show a decline in the number of children from 1920 to 1950, at which point the number of children reaches a low. For this group, there seems to be much less of a tendency for a cubic trend relating year of birth to number of children.

The graph in figure 11.13 suggests that the relationship between year of birth and number of children may show a cubic trend. Furthermore, the cubic trend may differ based on whether the woman graduated college, suggesting an interaction of the cubic trend for year of birth and whether the woman graduated college. Let's fit a model using a separate slope and separate intercept strategy that allows us to compare the cubic trend in year of birth for college graduates with noncollege graduates. Such a model is fit using the `regress` command below. The first term in the model is `ibn.cograd`. When specified in combination with `noconstant` option, the model fits separate intercepts by college graduation status. The model also includes `ibn.cograd` interacted with the linear term for year of birth, the quadratic term for year of birth, and the cubic term for year of birth.[6] This yields separate estimates of these terms by college graduation status.[7]

6. Note that the centered variable, `yrborn40`, is used to represent year of birth to reduce collinearity. See section 3.3 for more details.
7. The `noci` option is included to make the output more readable for this example. See the callout titled *Using the* `noci` *option for clearer output* in section 2.5.1 for more details.

```
. regress children ibn.cograd
> ibn.cograd#(c.yrborn40 c.yrborn40#c.yrborn40
>             c.yrborn40#c.yrborn40#c.yrborn40) i.race, noconstant noci

      Source |       SS           df       MS      Number of obs   =     5,037
-------------+----------------------------------   F(10, 5027)     =   1223.60
       Model |  34370.3789        10  3437.03789   Prob > F        =    0.0000
    Residual |  14120.6211     5,027  2.80895585   R-squared       =    0.7088
-------------+----------------------------------   Adj R-squared   =    0.7082
       Total |       48491     5,037  9.62696049   Root MSE        =     1.676

------------------------------------------------------------------
                    children |      Coef.   Std. Err.      t    P>|t|
-----------------------------+------------------------------------
                      cograd |
                 Not CO Grad |   2.801449   .0426759    65.64   0.000
                     CO Grad |   1.961068   .0874233    22.43   0.000
                             |
             cograd#c.yrborn40 |
                 Not CO Grad |  -.0883989   .0058118   -15.21   0.000
                     CO Grad |  -.0626583   .0117141    -5.35   0.000
                             |
  cograd#c.yrborn40#c.yrborn40 |
                 Not CO Grad |  -.0009296   .0002285    -4.07   0.000
                     CO Grad |   .0002384   .0005872     0.41   0.685
                             |
 cograd#c.yrborn40#c.yrborn40#|
                  c.yrborn40 |
                 Not CO Grad |   .0002103   .0000223     9.42   0.000
                     CO Grad |   .0000919   .0000489     1.88   0.060
                             |
                        race |
                       white |          0  (base)
                       black |   .5677129   .0668376     8.49   0.000
                       other |   .6662013   .1265935     5.26   0.000
------------------------------------------------------------------
```

We can write this regression model as though there are two regression equations: one for those who did not graduate college and one for those who did graduate college.

$$
\begin{aligned}
\text{Non-college graduate: } \widehat{\texttt{children}} &= 2.80 + -0.088\texttt{yrborn40} \\
&\quad + -0.00093\texttt{yrborn40}^2 + 0.00021\texttt{yrborn40}^3 \\
&\quad + 0.57\texttt{black} + 0.67\texttt{other} \\
\text{College graduate: } \widehat{\texttt{children}} &= 1.96 + -0.063\texttt{yrborn40} \\
&\quad + 0.00024\texttt{yrborn40}^2 + 0.00009\texttt{yrborn40}^3 \\
&\quad + 0.57\texttt{black} + 0.67\texttt{other}
\end{aligned}
$$

Note the differences in the cubic coefficient for `yrborn`. This coefficient is 0.00021 for those who did not graduate college and is 0.00009 for those who graduated college. To help interpret these results, let's create a graph of the fitted values using the `margins` and `marginsplot` commands. The `margins` command is used to calculate the adjusted means for years of birth from 1920 to 1960 in one-year increments, separately for college graduates and noncollege graduates. Then, the **marginsplot** command is used to graph these adjusted means, as shown in figure 11.14.

```
. margins cograd, at(yrborn40=(-20(1)20))
  (output omitted)
. marginsplot, noci recast(line) scheme(s1mono)
  Variables that uniquely identify margins: yrborn40 cograd
```

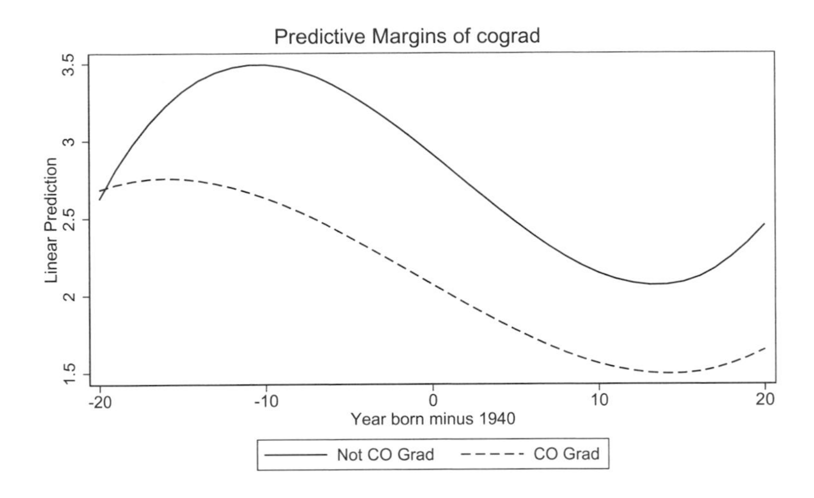

Figure 11.14. Fitted values of cubic by three-level categorical model

We can see that the noncollege graduates show a greater degree of rising and falling in the adjusted means across years than the college graduates. The larger cubic term for the noncollege graduates accounts for this greater degree of rising and falling. We can test the cubic trend for the college graduates compared with the noncollege graduates. The **contrast** command below compares the cubic trend for the college graduates with the noncollege graduates. The cubic trend is significantly different comparing noncollege graduates with college graduates ($F = 4.86$, $p = 0.0276$).

```
. contrast cograd#c.yrborn40#c.yrborn40#c.yrborn40
Contrasts of marginal linear predictions
Margins      : asbalanced
```

	df	F	P>F
cograd#c.yrborn40#c.yrborn40#c.yrborn40	1	4.86	0.0276
Denominator	5027		

The process of computing and comparing adjusted means is the same for the cubic model as it is for the quadratic model (see section 11.2.2). I refer you to that section for more information on computing and comparing adjusted means.

Note! Cubic by three-level interaction

The example from this section can be easily generalized to an interaction involving a three-level categorical predictor. The example below illustrates such a model, specifying educ3 as a categorical variable with three levels.

```
. regress children ibn.educ3
> ibn.educ3#(c.yrborn40 c.yrborn40#c.yrborn40
> c.yrborn40#c.yrborn40#c.yrborn40) i.race, noconstant
```

You can graph and dissect this model using the same kinds of techniques illustrated in section 11.2.3, which illustrated quadratic by three-level models.

11.4 Summary

This chapter has illustrated categorical by polynomial interactions. We saw examples of categorical by quadratic interactions, in which the curvature of the relationship between the predictor and outcome varied as a function of the categorical predictor. We also saw an example of a categorical by cubic interaction, in which one group showed a greater tendency to have two bends in the relationship between the predictor and outcome.

For more information about polynomial by categorical interactions, I recommend Cohen et al. (2003), West, Aiken, and Krull (1996), and Keppel and Wickens (2004) for a discussion of such interactions in the context of a designed experiment.

12 Piecewise by categorical interactions

12.1 Chapter overview

This chapter illustrates how to fit models in which a continuous variable, fit in a piecewise manner, is interacted with a categorical variable. This blends piecewise models (which were illustrated in chapter 4) with models that involve interactions of categorical and continuous variables (which were illustrated in chapter 10).

In chapter 4, we saw how piecewise models use two or more line segments, each connected by a knot, to model the relationship between the predictor and outcome.

The knot signifies a change in slope and can also signify a change in intercept (jump). Chapter 4 illustrated such models predicting income from education, modeling education in a piecewise manner. This chapter extends these models by introducing a categorical variable, gender, that is interacted with education.

Figure 12.1 shows a graph of the predicted values from a hypothetical piecewise model predicting income from education. The `educ` slope is modeled in two pieces—one slope for those with fewer than 12 years of education, and another slope for those with 12 or more years of education. The `educ` slope for those with fewer than 12 years of education is 500. For those with 12 or more years of education, the `educ` slope is 3,000. This model has one knot at 12 years of education that signifies a change in slope.

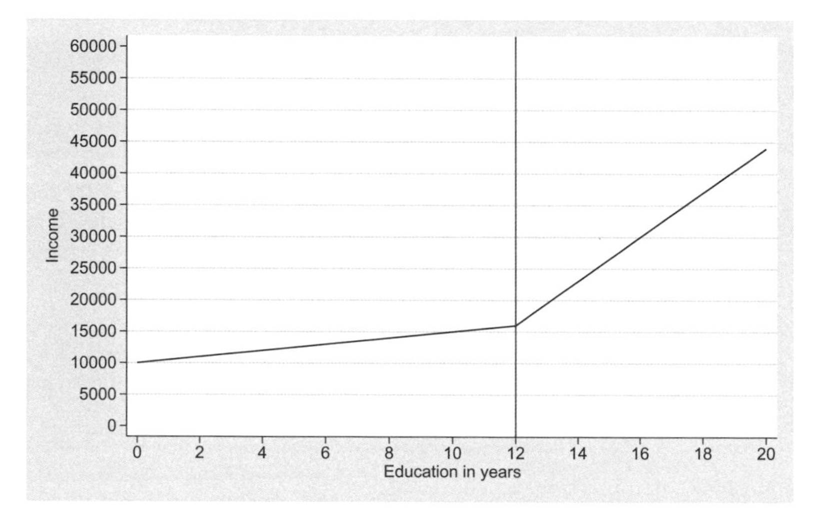

Figure 12.1. Piecewise regression with one knot at 12 years of education

Let's now consider a model that includes gender and interacts gender with the education terms (both before and after the knot). This permits the `educ` slope to differ by gender, modeling a different `educ` slope for men and women before the knot, and a different `educ` slope for men and women after the knot. A graph of the predicted values of such a hypothetical model is shown in the left panel of figure 12.2. The graph shows one line for men and another line for women. For men, the `educ` slope is 600 prior to graduating high school and 4,000 after graduating high school. Among women, the `educ` slope is 400 prior to graduating high school, and 2,000 after graduating high school.

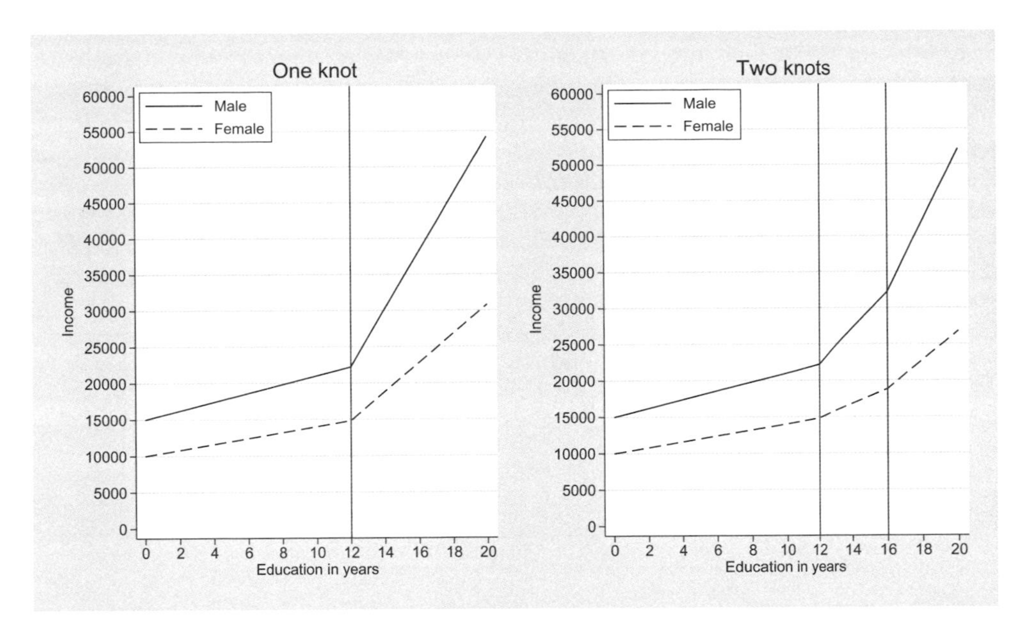

Figure 12.2. Piecewise model with one knot (left) and two knots (right), each by a categorical predictor

Suppose that we included two knots with respect to education: one at 12 years of education (corresponding to graduating high school) and a second at 16 years of education (corresponding to graduating college). The hypothetical results from such a model are depicted in the right panel of figure 12.2. There are two lines: one for men and one for women, with a change of slope at 12 and 16 years of education. Among non–high school graduates, the slope is 600 for men and is 400 for women. Among those with 12–15 years of education (who graduated high school, but not college), the slope is 2,500 for men and is 1,000 for women. Finally, for college graduates, the slope is 5,000 for men and is 2,000 for women.

A knot can signify a change of intercept as well as a change in slope. A change in intercept indicates a sudden jump (or drop) in the outcome at the knot. For example, the left panel of figure 12.3 shows a model where achieving 12 years of education results in not only a change of slope but also a sudden jump in income. The jump in income upon graduating high school is 4,500 for men and is 1,500 for women. The `educ` slope for men is 600 prior to graduating high school and is 2,500 after graduating high school. For women, the `educ` slope is 400 for non–high school graduates and is 2,000 for high school graduates.

The right panel of figure 12.3 shows a model with two knots at 12 and 16 years of education. Both knots result in a change of slope and change of intercept, each of which are estimated separately for men and women. The `educ` slope is estimated separately for non–high school graduates, high school graduates, and college graduates. For men, these three slopes are (respectively) 600, 2,200, and 4,000. For women, these slopes are

400, 1,000, and 2,000. The jump in income due to graduating high school and graduating college are also estimated separately by gender. For men, the jump in income due to graduating high school is \$3,000, and the jump in income due to graduating college is \$6,000. For women, the jump in income due to graduating high school is \$1,300, and the jump in income due to graduating college is \$2,400.

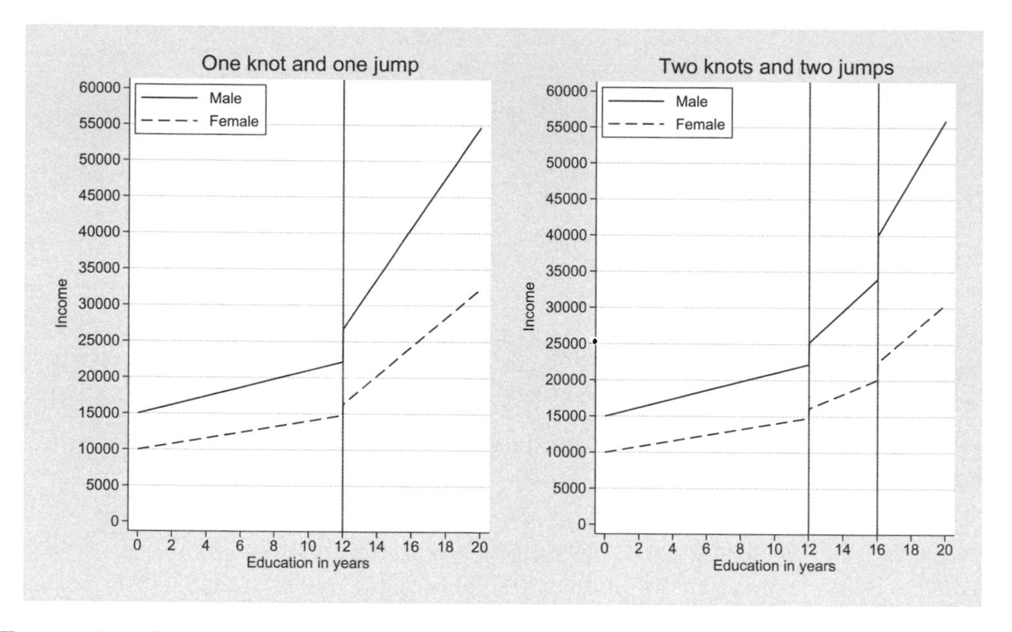

Figure 12.3. Piecewise model with one knot and one jump (left) and two knots and two jumps (right), each by a categorical predictor

This chapter illustrates how to fit the models depicted in the left and right panels of figure 12.3. Each of these models includes gender (a categorical predictor) interacted with education, where education is fit in a piecewise manner and includes a jump at each knot. Note that I will not be covering the models illustrated in the left and right panels of figure 12.2 (that is, models that do not include a jump). Once you see how to fit models that include a jump, you will easily understand how to fit a model that excludes a jump.

12.2 One knot and one jump

Consider the example shown in the left panel of figure 12.3. In that example, education was modeled in a piecewise manner including one knot at 12 years of education. This knot signifies both a change of slope and change in intercept (jump). The model also includes gender as a categorical variable and estimates separate slopes and intercepts for each level of gender (that is, for men and for women). I fit this kind of model using the GSS dataset and the results are depicted in figure 12.4 (we will see the analysis shortly).

Figure 12.4 shows the adjusted means of income as a function of education with labels for each of the slopes. For men, the slope for non–high school graduates is labeled β_{M1} and the slope for high school graduates is labeled β_{M2}. For women, the slope for non–high school graduates is labeled β_{F1} and the slope for high school graduates is labeled β_{F2}. The income jump at 12 years of education for men is labeled as α_{M1}. The corresponding jump at 12 years of education for women is labeled as α_{F1}. Note how an arrow head points to a sudden jump in income at 12 years of education.

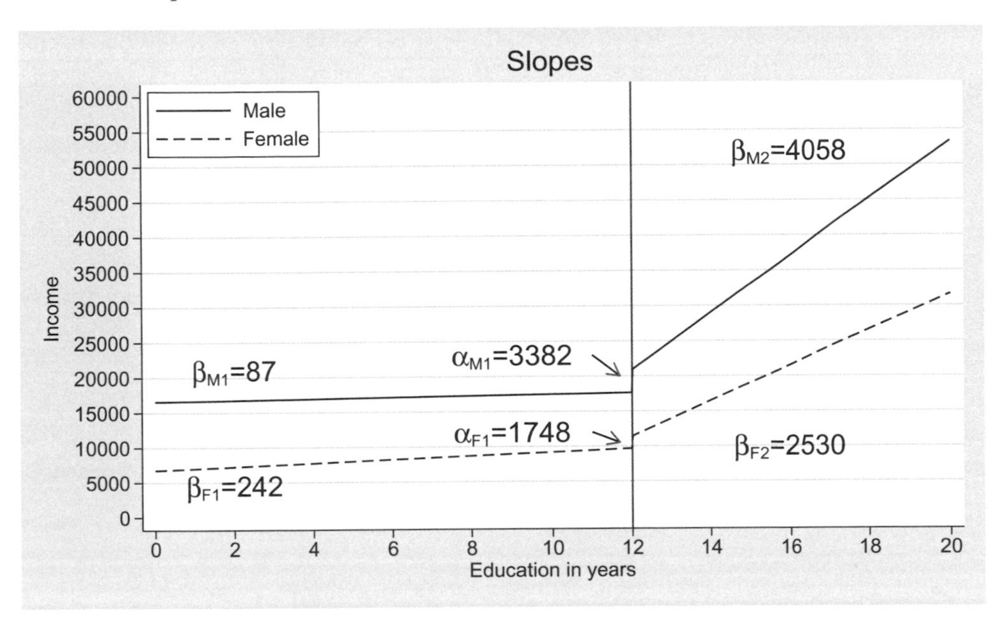

Figure 12.4. Piecewise regression with one knot and one jump, labeled with estimated slopes

To fit this model, we will use *separate slope and separate intercept* coding with respect to the gender groups. This means that we will estimate separate intercept terms for men and women (this includes separate jumps, α_{M1} and α_{F1}). It will also estimate separate slope terms for men (that is, β_{M1} and β_{M2}) and separate slope terms for women (that is, β_{F1} and β_{F2}).

We first use the `mkspline` command to create the variables `ed1` and `ed2` with a knot at 12 years of education. The `marginal` option is omitted, so the variables `ed1` and `ed2` will reflect the slopes of the individual line segments before and after the knot. (For more details, see chapter 4, especially the sections on *individual slope* coding.)

```
. use gss_ivrm
. mkspline ed1 12 ed2 = educ
```

We also need a variable to indicate the jump due to graduating high school. The dataset already includes the variable `hsgrad` that is coded 0 if one has not graduated high school and 1 if one has graduated high school. We will use that variable to signify the jump in income due to graduating high school. Additionally, I will use the variable **gender** (coded: 1 = Male and 2 = Female).[1]

We are now ready to run the piecewise regression model, shown in the **regress** command below. The **regress** command includes **ibn.gender** used in combination with the **noconstant** option to yield separate estimates of the intercept for each gender.[2] To estimate separate jumps for men and women, we include **ibn.gender#i.hsgrad**. The model also includes **ibn.gender#c.ed1** to fit separate **educ** slopes for the line segment before the knot and **ibn.gender#c.ed2** to fit separate **educ** slopes for the line segment after the knot. Finally, the model includes **race** as a covariate.

```
. regress realrinc ibn.gender ibn.gender#i.hsgrad
> ibn.gender#c.ed1 ibn.gender#c.ed2 i.race,
> vce(robust) noconstant noci
Linear regression                               Number of obs     =     32,183
                                                F(10, 32173)      =    3425.59
                                                Prob > F          =     0.0000
                                                R-squared         =     0.4864
                                                Root MSE          =      24948

                                 Robust
       realrinc       Coef.   Std. Err.       t    P>|t|

         gender
           Male    17144.66    1389.515    12.34   0.000
         Female    7325.482    922.5224     7.94   0.000

  gender#hsgrad
    Male#HS Grad   3382.005     661.547     5.11   0.000
  Female#HS Grad   1748.096    389.3635     4.49   0.000

    gender#c.ed1
           Male    86.70762     151.379     0.57   0.567
         Female    241.6081     98.9163     2.44   0.015

    gender#c.ed2
           Male    4057.639    150.3836    26.98   0.000
         Female    2529.547    110.4318    22.91   0.000

           race
          white           0  (base)
          black   -3521.336    246.6606   -14.28   0.000
          other   -1946.159    839.7251    -2.32   0.020
```

1. The variable **gender** will be used as a factor variable in this chapter. I chose to use **gender** (instead of **female**) because it leads to output that clearly distinguishes the variable name (that is, **gender**) and its values (that is, **Male** and **Female**). By contrast, the dummy variable **female** has values labeled **Male** and **Female**, thus creating confusion between the variable **female** and the value **Female**.
2. The **noci** option is included to make the output more readable for this example. See the callout titled *Using the* **noci** *option for clearer output* in section 2.5.1 for more details.

Note! Model shortcut

The previous model can also be specified, as shown below.

```
. regress realrinc ibn.gender ibn.gender#(i.hsgrad c.ed1 c.ed2) i.race,
> vce(robust) noconstant noci
```

Stata expands the expression `ibn.gender#(i.hsgrad c.ed1 c.ed2)` to become `ibn.gender#i.hsgrad ibn.gender#c.ed1 ibn.gender#c.ed2`, yielding the same model we saw before.

Key coefficients from the regression output are shown and annotated in table 12.1. The first two columns of the table repeat the name and estimate of the coefficient from the `regress` output. The third column shows the symbol used to represent the coefficient in figure 12.4, providing a cross reference between the output of the regression model and figure 12.4. The last column shows the symbolic name of the regression coefficient that we can use with the `lincom` command for making comparisons among the coefficients.[3] Let's refer to figure 12.4 and table 12.1 to help interpret the output of this model.

Table 12.1. Summary of piecewise regression results with one knot

	Coefficient	Symbol	Symbolic name
`gender#hsgrad`			
`Male#HS Grad`	3382.01	α_{M1}	`1.gender#1.hsgrad`
`Female#HS Grad`	1748.10	α_{F1}	`2.gender#1.hsgrad`
`gender#c.ed1`			
`Male`	86.71	β_{M1}	`1.gender#ed1`
`Female`	241.61	β_{F1}	`2.gender#ed1`
`gender#c.ed2`			
`Male`	4057.64	β_{M2}	`1.gender#ed2`
`Female`	2529.55	β_{F2}	`2.gender#ed2`

Let's begin by interpreting the change in intercept (jump) terms. The jump in the adjusted mean of income for men at 12 years of education is 3,382.01, and the corresponding jump for women is 1,748.10. Each of these jumps is statistically significant.

Let's now interpret the slopes. For men, the `educ` slope is 86.71 for non–high school graduates and is 4,057.64 for high school graduates. For women, the slope is 241.61 for non–high school graduates and is 2,529.55 for high school graduates. Aside from the `educ` slope for male non–high school graduates (86.71), all of these slopes are significantly different from 0.

3. You can add the `coeflegend` option to the `regress` command to include a coefficient legend in the output. The resulting regression output would look similar to table 12.1.

The following four sections will illustrate how to make a number of comparisons, showing how to

- compare slopes between women and men (see section 12.2.1),
- compare slopes across the levels of education, comparing high school graduates with non–high school graduates (see section 12.2.2),
- compare the changes in slope between men and women due to graduating high school (see section 12.2.3), and
- compare the jump in income due to graduating high school between men and women (see section 12.2.4).

12.2.1 Comparing slopes across gender

Let's begin by testing the equality of the educ slopes of women and men who have not graduated high school. In other words, is $\beta_{F1} = \beta_{M1}$? To make that comparison, we can use the contrast command, as shown below. The difference in the slope between women and men before graduating high school is 154.90 but is not statistically significant ($t = 0.86$, $p = 0.388$).

```
. contrast gender#c.ed1, pveffects nowald
Contrasts of marginal linear predictions
Margins      : asbalanced
```

	Contrast	Std. Err.	t	P>|t|
gender#c.ed1				
(Female vs base)	154.9005	179.3207	0.86	0.388

Now, let's compare the educ slopes of men and women for those who graduated high school by testing whether $\beta_{F2} = \beta_{M2}$. The contrast command below performs this test.

```
. contrast gender#c.ed2, pveffects nowald
Contrasts of marginal linear predictions
Margins      : asbalanced
```

	Contrast	Std. Err.	t	P>|t|
gender#c.ed2				
(Female vs base)	-1528.092	187.0884	-8.17	0.000

For those who graduated high school, the educ slope is significantly lower for women than men. The difference in the slopes (women versus men) is −1,528.09. For every year of education beyond the 12th year, the income for men increases by $1,528.09 more than for women.

12.2.2 Comparing slopes across education

This section focuses on comparing the educ slopes before and after graduating high school. First, let's examine the change in slope between high school graduates and non–high school graduates for men. Put another way, is $\beta_{M2} = \beta_{M1}$? As shown in table 12.1, the symbolic names for these coefficients are 1.gender#ed2 and 1.gender#ed1. We can compare these coefficients using the lincom command, as shown below.

```
. lincom 1.gender#c.ed2 - 1.gender#c.ed1
 ( 1)  - 1bn.gender#c.ed1 + 1bn.gender#c.ed2 = 0
```

realrinc	Coef.	Std. Err.	t	P>|t|	[95% Conf.	Interval]
(1)	3970.931	212.6975	18.67	0.000	3554.036	4387.826

Men show a significantly higher slope after graduating high school than men who have not graduated high school. The difference (comparing high school graduates with non–high school graduates) is 3,970.93. For men, each additional year of education is worth $3,970.93 more for high school graduates than for non–high school graduates.

We can formulate the same kind of test for women, comparing the educ slope for high school graduates with that for non–high school graduates. Referring to table 12.1, we can ask if $\beta_{F2} = \beta_{F1}$ using the lincom command below.

```
. lincom 2.gender#c.ed2 - 2.gender#c.ed1
 ( 1)  - 2.gender#c.ed1 + 2.gender#c.ed2 = 0
```

realrinc	Coef.	Std. Err.	t	P>|t|	[95% Conf.	Interval]
(1)	2287.939	147.2609	15.54	0.000	1999.302	2576.576

The result of this test is significant. For women who have completed 12 or more years of education, income rises by $2,287.94 per additional year of education compared with women who have not completed 12 years of education.

12.2.3 Difference in differences of slopes

The previous section showed that, for men, the educ slope after graduating high school minus the slope before graduating high school equals 3,970.93. Let's call this the *gain in slope* due to graduating high school. For women, the gain in slope due to graduating high school is 2,287.94. We might ask if the gain in slope due to graduating high school differs by gender. The lincom command below tests the gain in slope for men compared with the gain in slope for women.

```
. lincom (1.gender#c.ed2 - 1.gender#c.ed1) - (2.gender#c.ed2 - 2.gender#c.ed1)
 ( 1)  - 1bn.gender#c.ed1 + 2.gender#c.ed1 + 1bn.gender#c.ed2 - 2.gender#c.ed2 = 0
```

realrinc	Coef.	Std. Err.	t	P>|t|	[95% Conf. Interval]	
(1)	1682.992	259.2855	6.49	0.000	1174.783	2191.201

The lincom command shows that the gain in slope for men, compared with women, is 1,682.99, and this is statistically significant. The gain in slope due to graduating high school for men is significantly higher than the gain in slope for women.

12.2.4 Comparing changes in intercepts

Let's now ask whether the jump in income at 12 years of education is equal for men and women. As shown in figure 12.4, the income for men jumps by $3,382.01 at 12 years of education, whereas the corresponding jump for women is $1,748.10. Are these jumps equal? In other words, is $\alpha_{F2} = \alpha_{M2}$? This is tested using the contrast command below.

```
. contrast gender#hsgrad, pveffects nowald
Contrasts of marginal linear predictions
Margins      : asbalanced
```

	Contrast	Std. Err.	t	P>|t|
gender#hsgrad (Female vs base) (HS Grad vs base)	-1633.908	766.9848	-2.13	0.033

This jump in income due to graduating high school for women versus men is −$1,633.91, and this difference is statistically significant. In other words, the jump in income for women due to graduating high school is $1,633.91 less than the jump for men due to graduating high school.

Some people might find that this test is more intuitive if performed using the following lincom command below. This yields the same results as the contrast command.

```
. lincom 2.gender#1.hsgrad - 1.gender#1.hsgrad
  (output omitted)
```

12.2.5 Computing and comparing adjusted means

Let's now turn our attention to how we can compute adjusted means for this regression model. Before we can compute adjusted means with respect to education (that is, educ), we need to know how to express the level of education in terms of hsgrad, ed1, and ed2. The showcoding command (which you can download, see section 1.1) shows the correspondence between educ and the values of hsgrad, ed1, and ed2.

```
. showcoding educ hsgrad ed1 ed2
```

educ	hsgrad	ed1	ed2
0	0	0	0
1	0	1	0
2	0	2	0
3	0	3	0
4	0	4	0
5	0	5	0
6	0	6	0
7	0	7	0
8	0	8	0
9	0	9	0
10	0	10	0
11	0	11	0
12	1	12	0
13	1	12	1
14	1	12	2
15	1	12	3
16	1	12	4
17	1	12	5
18	1	12	6
19	1	12	7
20	1	12	8

The variable **hsgrad** contains 0 for those with fewer than 12 years of education, and 1 for those with 12 or more years of education. The variable **ed1** contains the number of years of education for those with 12 or fewer years of education, and contains 12 for those with more than 12 years of education. The variable **ed2** contains 0 for those with 12 or fewer years of education, and contains **educ** minus 12 for those with more than 12 years of education.

Let's use this information to estimate the adjusted mean for a male who has 10 years of education. Note that we code **gender** as 1 to indicate a male. We indicate 10 years of education by specifying that **hsgrad** equals 0, **ed1** equals 10, and **ed2** equals 0. This adjusted mean is 17,443.44.

```
. margins, nopvalues at(gender=1 hsgrad=0 ed1=10 ed2=0)
Predictive margins                              Number of obs     =     32,183
Model VCE    : Robust

Expression   : Linear prediction, predict()
at           : gender          =           1
               hsgrad          =           0
               ed1             =          10
               ed2             =           0

------------------------------------------------------------------
             |            Delta-method
             |     Margin   Std. Err.     [95% Conf. Interval]
-------------+----------------------------------------------------
       _cons |   17443.44   389.9541      16679.11    18207.76
------------------------------------------------------------------
```

Let's now estimate the adjusted mean for a female (that is, **gender**=2) with 15 years of education, as shown below. To indicate 15 years of education, we specify that **hsgrad** equals 1, **ed1** equals 12, and **ed2** equals 3.

```
. margins, nopvalues at(gender=2 hsgrad=1 ed1=12 ed2=3)
Predictive margins                              Number of obs     =     32,183
Model VCE    : Robust

Expression   : Linear prediction, predict()
at           : gender          =           2
               hsgrad          =           1
               ed1             =          12
               ed2             =           3

------------------------------------------------------------
             |            Delta-method
             |     Margin   Std. Err.     [95% Conf. Interval]
-------------+----------------------------------------------
       _cons |   18993.22   234.1696      18534.24     19452.2
------------------------------------------------------------
```

We can estimate the adjusted mean of income for men and women with 15 years of education using the **margins** command below.

```
. margins gender, nopvalues at(hsgrad=1 ed1=12 ed2=3)
Predictive margins                              Number of obs     =     32,183
Model VCE    : Robust

Expression   : Linear prediction, predict()
at           : hsgrad          =           1
               ed1             =          12
               ed2             =           3

------------------------------------------------------------
             |            Delta-method
             |     Margin   Std. Err.     [95% Conf. Interval]
-------------+----------------------------------------------
      gender |
        Male |   33171.77    351.092      32483.62    33859.93
      Female |   18993.22   234.1696      18534.24     19452.2
------------------------------------------------------------
```

Let's compare the adjusted means between men and women from the previous **margins** command. Specifying the **r.** contrast operator indicates we want to use reference group comparisons, comparing the adjusted mean of income for women versus men.

```
. margins r.gender, at(hsgrad=1 ed1=12 ed2=3) contrast(nowald pveffects)
Contrasts of predictive margins                 Number of obs     =     32,183
Model VCE    : Robust

Expression   : Linear prediction, predict()
at           : hsgrad          =           1
               ed1             =          12
               ed2             =           3

------------------------------------------------------------
                  |            Delta-method
                  |   Contrast   Std. Err.      t    P>|t|
------------------+-----------------------------------------
           gender |
(Female vs Male)  |  -14178.55   422.0554   -33.59   0.000
------------------------------------------------------------
```

At 15 years of education, the difference in the adjusted mean of income for women and that for men is $-14,178.55$, and this difference is significant. Note how this difference corresponds to the difference in the adjusted means for men and women we previously computed ($18993.22 - 33171.77 = -14178.55$). This same technique can be used to compare the adjusted means between men and women at any level of education.

12.2.6 Graphing adjusted means

Let's graph the adjusted means as a function of education and gender. To make this graph, we need to compute the adjusted means separately for men and women with 0 years of education, 12 years of education (assuming the absence and presence of a high school diploma), and 20 years of education.[4] The `margins` command for computing these adjusted means is shown below.

4. Obtaining the adjusted mean assuming the absence and presence of a high school diploma at 12 years of education illustrates the jump in income due to graduating high school.

```
. margins gender, at(hsgrad=0 ed1=0  ed2=0)
>                 at(hsgrad=0 ed1=12 ed2=0)
>                 at(hsgrad=1 ed1=12 ed2=0)
>                 at(hsgrad=1 ed1=12 ed2=8) nopvalues
Predictive margins                             Number of obs     =     32,183
Model VCE    : Robust

Expression   : Linear prediction, predict()

1._at        : hsgrad          =           0
               ed1             =           0
               ed2             =           0

2._at        : hsgrad          =           0
               ed1             =          12
               ed2             =           0

3._at        : hsgrad          =           1
               ed1             =          12
               ed2             =           0

4._at        : hsgrad          =           1
               ed1             =          12
               ed2             =           8

------------------------------------------------------------------
             |            Delta-method
             |     Margin   Std. Err.     [95% Conf. Interval]
-------------+----------------------------------------------------
  _at#gender |
      1#Male |   16576.36   1384.618      13862.46    19290.26
    1#Female |   6757.187   910.6287      4972.321    8542.054
      2#Male |   17616.85   590.7778      16458.91     18774.8
    2#Female |   9656.485   345.8282      8978.648    10334.32
      3#Male |   20998.86   297.4975      20415.75    21581.96
    3#Female |   11404.58   179.7705      11052.22    11756.94
      4#Male |   53459.97   1042.022      51417.56    55502.37
    4#Female |   31640.96   761.6903      30148.02     33133.9
------------------------------------------------------------------
```

We can then manually input the adjusted means into a dataset. The `graph` command is used to graph the adjusted means, as shown below.[5] The resulting graph is shown in figure 12.5.

```
. preserve

. clear

. input educ yhatm yhatf

          educ      yhatm      yhatf
  1. 0  16576.36 6757.187
  2. 12 17616.85 9656.485
  3. 12 20998.86 11404.58
  4. 20 53459.97 31640.96
  5. end

. graph twoway line yhatm yhatf educ,
> xline(12) legend(label(1 "Men") label(2 "Women"))
> xtitle(Education) ytitle(Adjusted mean)

. restore
```

5. The graphs created by the `marginsplot` command are not very useful when graphing adjusted means from piecewise models. This is illustrated in section 4.9.

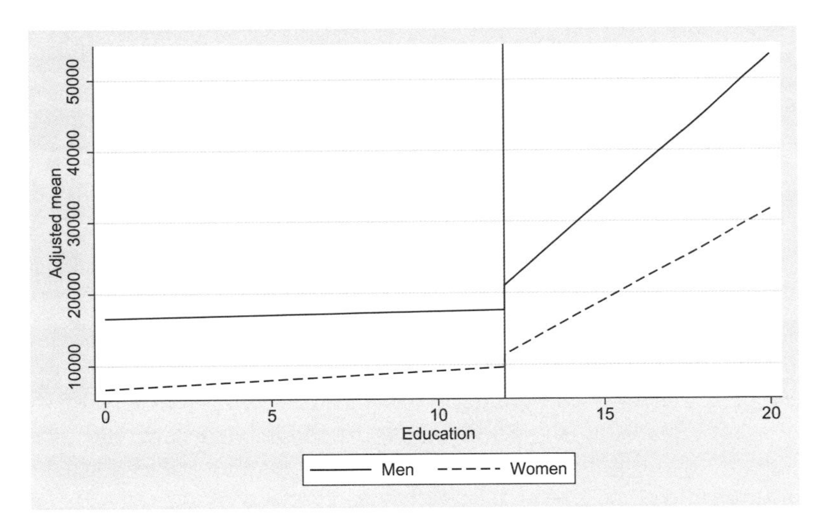

Figure 12.5. Fitted values from piecewise model with one knot and one jump at `educ` = 12

We can automate the creation of this graph by extending the strategy we saw in section 4.10. First, we use the **matrix** command to create a matrix named **yhat** that contains the adjusted means computed by the **margins** command.

```
. * First, rerun the -margins- command from above
. quietly margins gender, at(hsgrad=0 ed1=0  ed2=0)
>                        at(hsgrad=0 ed1=12 ed2=0)
>                        at(hsgrad=1 ed1=12 ed2=0)
>                        at(hsgrad=1 ed1=12 ed2=8)
. * store the adjusted means in a matrix named -yhat-
. matrix yhat = r(b)' // NOTE: We must use transpose here
```

Now, we encounter a twist because the adjusted means are computed as a function two variables (education and gender), whereas the example in section 4.10 computed the adjusted means as a function of one variable (`educ`). Looking at the output of the **margins** command, let's focus on the order in which the adjusted means are displayed with respect to education and gender. Eight adjusted means are shown, corresponding to the following levels of `educ`, `hsgrad`, and `gender`:

- `educ=0, hsgrad=0, gender=1`
- `educ=0, hsgrad=0, gender=2`
- `educ=12, hsgrad=0, gender=1`
- `educ=12, hsgrad=0, gender=2`
- `educ=12, hsgrad=1, gender=1`
- `educ=12, hsgrad=1, gender=2`
- `educ=20, hsgrad=1, gender=1`
- `educ=20, hsgrad=1, gender=2`

The **matrix** command is used to create a matrix named **educ** that reflects the levels of education shown in the bulleted list above. The **matrix** command is used again, this time to create a matrix called **gender** that contains the levels of gender shown in the bulleted list above.

```
. * store levels of education in a matrix named -educ-
. matrix educ = (0 \ 0 \ 12 \ 12 \ 12 \ 12 \ 20 \ 20)
. * store levels of gender in a matrix a named -gender-
. matrix gender = (1 \ 2 \ 1 \ 2 \ 1 \ 2 \ 1 \ 2)
```

The **svmat** command is then used three times, to save the matrices named **yhat**, **educ**, and **gender** to the current dataset. The **list** command is then used to show the variables **yhat1**, **educ1**, and **gender1** for the first 10 observations of the dataset. The variable **yhat1** contains the adjusted means, **educ1** contains the corresponding values for education, and **gender1** contains the corresponding values for gender. These variables contain valid data for the first eight observations of the dataset, and the rest of the observations are missing.

```
. svmat yhat     // save the matrix -yhat- to the current dataset
. svmat educ     // save the matrix -educ- to the current dataset
. svmat gender     // save the matrix -gender- to the current dataset
. list  yhat1 educ1 gender1 in 1/10, sep(2)
```

	yhat1	educ1	gender1
1.	16576.36	0	1
2.	6757.188	0	2
3.	17616.85	12	1
4.	9656.484	12	2
5.	20998.86	12	1
6.	11404.58	12	2
7.	53459.96	20	1
8.	31640.96	20	2
9.	.	.	.
10.	.	.	.

We are now ready to use the **graph** command to graph the adjusted means as a function of education and gender. The **graph** command below produces the same graph as the one displayed in figure 12.5.

```
. graph twoway (line yhat1 educ1 if gender1==1)
>              (line yhat1 educ1 if gender1==2),
>               xline(12) legend(label(1 "Men") label(2 "Women"))
>               xtitle(Education) ytitle(Adjusted mean)
```

Note! Omitting the jump

Say that you wanted to fit this model but exclude the jump in income at 12 years of education. You can simply omit `hsgrad` from the model. The way that you interpret the model remains the same except for the omission in the jump in income at the knot.

This section has illustrated a model with one knot where there was both a change in slope and a change in intercept (jump) at the knot. The next section illustrates a model with two knots with a change in slope and intercept at each knot.

12.3 Two knots and two jumps

This section covers piecewise regression models with two knots and two jumps interacted with a categorical variable. An example of such a model was shown in the right panel of figure 12.3 in which income was predicted from education modeled in a piecewise fashion with two knots and two jumps at 12 and 16 years of education. Furthermore, these piecewise terms were interacted with gender. I ran an analysis for this kind of model using the GSS dataset and created a graph of the adjusted means as a function of education and gender (see figure 12.6). The graph also shows the slope for each piece of the model, labeling the slope for male non–high school graduates as β_{M1}, male high school graduates as β_{M2}, and male college graduates as β_{M3}. The corresponding slopes for women for these three educational groups are labeled β_{F1}, β_{F2}, and β_{F3}.

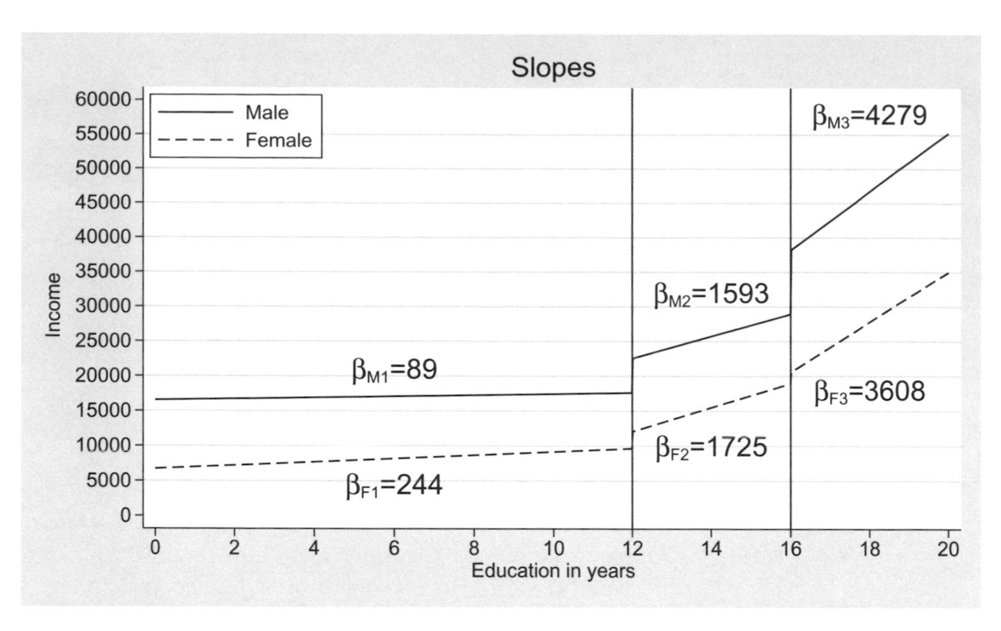

Figure 12.6. Piecewise regression with two knots and two jumps, labeled with estimated slopes

> **Note! High school and college graduates**
>
> In this section, I use the term *high school graduate* to refer to someone with 12 to 15 years of education, *college graduate* to refer to someone with 16 to 20 years of education, *non–high school graduate* to refer to someone with 11 or fewer years of education.

Figure 12.7 also shows the adjusted means but includes labels showing the jump (change in intercept) at each knot for men and women. The jump at 12 years of education for men is labeled as α_{M1}. The jump at 16 years of education for men is labeled as α_{M2}. These corresponding jumps for women are labeled α_{F1} and α_{F2}.

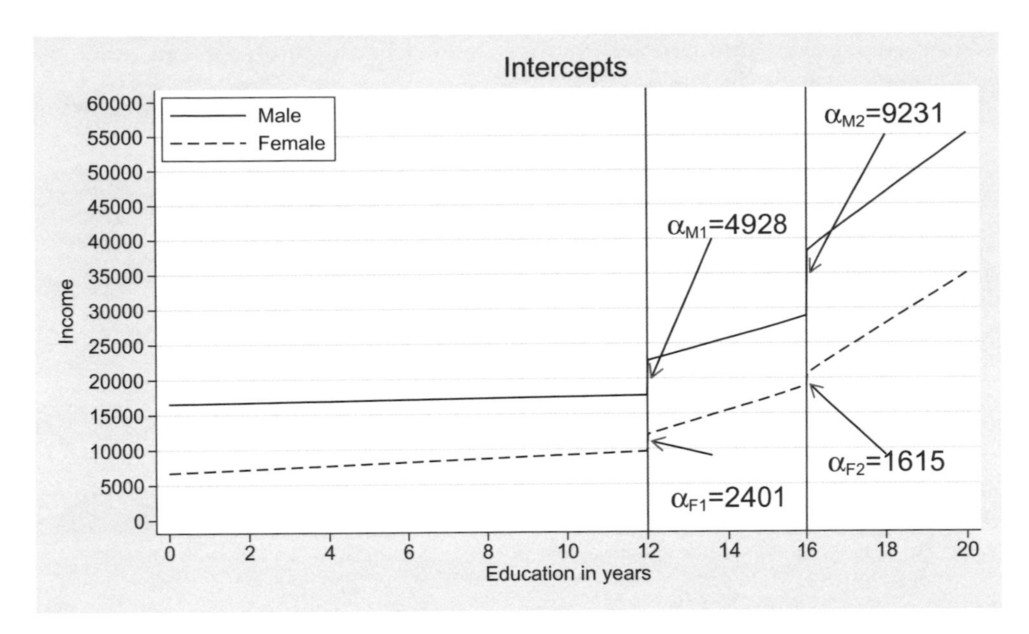

Figure 12.7. Piecewise regression with two knots and two jumps, labeled with estimated intercepts

Let's now illustrate how to perform this analysis. First, the `mkspline` command is used to create the variables `ed1`, `ed2`, and `ed3` based on the knots that are specified at 12 and 16 years of education. Like the analysis from the previous section, the `marginal` option is omitted.

```
. use gss_ivrm
. mkspline ed1 12 ed2 16 ed3 = educ
```

To account for the change in intercept (that is, jump in income) at 12 and 16 years of education, we need two dummy variables—one that indicates graduating high school, and one that indicates graduating college. The variables `hsgrad` and `cograd` already exist in the dataset and can be used to model the jump in income due to graduating high school and the jump in income due to graduating college.

The variables are now ready to run the piecewise regression model. The `regress` command, shown below, includes `ibn.gender` used in combination with the `noconstant` option. This models separate intercepts for men and women. The model also includes the interaction of `ibn.gender` with `i.hsgrad` and `i.cograd`. This models the jump in income due to graduating high school and graduating college separately for men and women. Finally, the model includes the interaction of `ibn.gender` with `c.ed1`, `c.ed2`, and `c.ed3`. This models the `educ` slope for non–high school graduates, high school graduates, and college graduates, estimating these slopes separately for men and women. The variable `i.race` is included as a covariate.[6]

```
. regress realrinc ibn.gender
> ibn.gender#(hsgrad cograd c.ed1 c.ed2 c.ed3) i.race,
> vce(robust) noconstant noci

Linear regression                               Number of obs     =     32,183
                                                F(14, 32169)      =    2602.08
                                                Prob > F          =     0.0000
                                                R-squared         =     0.4885
                                                Root MSE          =      24897

------------------------------------------------------------
                 |               Robust
        realrinc |      Coef.   Std. Err.      t    P>|t|
-----------------+------------------------------------------
          gender |
            Male |   17084.37   1389.704    12.29   0.000
          Female |   7262.505   921.9029     7.88   0.000
                 |
   gender#hsgrad |
    Male#HS Grad |   4927.532   655.3309     7.52   0.000
  Female#HS Grad |   2400.894   372.2095     6.45   0.000
                 |
   gender#cograd |
    Male#CO Grad |   9230.717   1224.498     7.54   0.000
  Female#CO Grad |   1615.195     789.62     2.05   0.041
                 |
    gender#c.ed1 |
            Male |   89.39533   151.3973     0.59   0.555
          Female |   243.6431   98.84487     2.46   0.014
                 |
    gender#c.ed2 |
            Male |   1592.636     255.81     6.23   0.000
          Female |   1724.742   182.7012     9.44   0.000
                 |
    gender#c.ed3 |
            Male |   4279.358   521.8313     8.20   0.000
          Female |    3608.45   496.9856     7.26   0.000
                 |
            race |
           white |          0  (base)
           black |  -3361.238   246.8406   -13.62   0.000
           other |  -1813.494   840.4705    -2.16   0.031
------------------------------------------------------------
```

The key coefficients from the regression output are shown and annotated in table 12.2. The first two columns of this table repeat the description and value of the regression coefficient from the `regress` command. The third column shows the symbol used to represent the coefficient in figures 12.6 and 12.7 (to help show the correspondence between the figures and the output of the regression model). The last column shows the symbolic name of the regression coefficient that we will use later with the `lincom` command for making comparisons among the coefficients.

6. Like an example earlier in this chapter, the `noci` option is included to make the output more readable. See the callout titled *Using the* `noci` *option for clearer output* in section 2.5.1 for more details.

Table 12.2. Summary of piecewise regression results with two knots

	Coef.	Symbol	Symbolic name
gender#hsgrad			
Male#HS Grad	4927.53	α_{M1}	1.gender#1.hsgrad
Female#HS Grad	2400.89	α_{F1}	2.gender#1.hsgrad
gender#cograd			
Male#CO Grad	9230.72	α_{M2}	1.gender#1.cograd
Female#CO Grad	1615.20	α_{F2}	2.gender#1.cograd
gender#c.ed1			
Male	89.40	β_{M1}	1.gender#ed1
Female	243.64	β_{F1}	2.gender#ed1
gender#c.ed2			
Male	1592.64	β_{M2}	1.gender#ed2
Female	1724.74	β_{F2}	2.gender#ed2
gender#c.ed3			
Male	4279.36	β_{M3}	1.gender#ed3
Female	3608.45	β_{F3}	2.gender#ed3

Let's refer to figures 12.6 and 12.7 and table 12.2 to help interpret the output of this model, beginning with the coefficients associated with the (jump) in income at 12 and 16 years of education (see figure 12.7). The jump in income for men at 12 years of education is 4,927.53, and the jump in income for women at 12 years of education is 2,400.89. The jump in income for men at 16 years of education is 9,230.72, and the jump in income for women at 16 years of education is 1,615.20. Each of these jumps is significantly different from 0. For example, the jump in income due to graduating high school for a female is 2,400.89 and is significant ($t = 6.45$, $p = 0.000$).

Let's interpret the slope coefficients by referring to figure 12.6. The `educ` slope is 89.40 for men who have not graduated high school and is 243.64 for such women. Note that the slope for men is not significantly different from 0 ($t = 0.59$, $p = 0.555$), but the slope is significantly different from zero for women ($t = 2.46$, $p = 0.014$). For high school graduates, the `educ` slope for men is 1,592.64 and for women is 1,724.74. Both of these coefficients are significantly different from 0. For college graduates, the `educ` slope for men is 4,279.36 and for women is 3,608.45. Both of these slopes are significantly different from 0. Focusing on the last result, each additional year of education (beyond the 16th year) is worth an additional $4,279.36 of income for men and is worth an additional $3,608.45 of income for women.

Let's now see how to form comparisons among these coefficients. Specifically, let's learn how to

- compare slopes between men and women (see section 12.3.1),
- compare slopes across the levels of education, comparing college graduates, high school graduates, and non–high school graduates (see section 12.3.2),
- compare changes in slope between men and women due to graduating high school and due to graduating college (see section 12.3.3),
- compare changes in intercept (the jumps in income due to graduating high school and college) by gender (see section 12.3.4), and
- compare changes in intercept (the jumps in income due to graduating high school and college) across levels of education (see section 12.3.5).

12.3.1 Comparing slopes across gender

Let's begin by testing the equality of the slopes between women and men who have not graduated high school—testing whether $\beta_{F1} = \beta_{M1}$. We can use the following **contrast** command to perform this test. The difference in the **educ** slope between women and men before graduating high school is 154.25. However, this difference is not statistically significant. In other words, prior to graduating high school, the **educ** slope is not significantly different for men and women.

```
. contrast gender#c.ed1, pveffects nowald
Contrasts of marginal linear predictions
Margins      : asbalanced

                   |  Contrast   Std. Err.        t    P>|t|
-------------------+-----------------------------------------
     gender#c.ed1  |
 (Female vs base)  |  154.2477   179.2953      0.86    0.390
```

Now, let's compare the slopes of women and men for those who have graduated high school, testing whether $\beta_{F2} = \beta_{M2}$. Using the **contrast** command below, we see that the difference in these slopes for women versus men is 132.11, but this difference is not statistically significant.

```
. contrast gender#c.ed2, pveffects nowald
Contrasts of marginal linear predictions
Margins      : asbalanced

                   |  Contrast   Std. Err.        t    P>|t|
-------------------+-----------------------------------------
     gender#c.ed2  |
 (Female vs base)  |  132.1066   314.0413      0.42    0.674
```

Finally, let's compare the slopes of women versus men for those who have graduated college. This asks the question: is $\beta_{F3} = \beta_{M3}$? The `contrast` command below shows difference in the slopes (women versus men) is −670.91, but this difference is not statistically significant. In other words, among college graduates, the `educ` slope is not significantly different for women compared with men.

```
. contrast gender#c.ed3, pveffects nowald
Contrasts of marginal linear predictions
Margins      : asbalanced
```

	Contrast	Std. Err.	t	P>|t|
gender#c.ed3 (Female vs base)	-670.9077	721.07	-0.93	0.352

The `lincom` command also works for performing these tests. For example, results of the previous `contrast` command could have been performed using the following `lincom` command.

```
. lincom 2.gender#c.ed3 - 1.gender#c.ed3
  (output omitted)
```

12.3.2 Comparing slopes across education

This section focuses on comparisons of the `educ` slopes, within gender, before and after each knot in education. Let's begin by comparing the slopes for high school graduates with non–high school graduates, starting with men. Expressed in terms of the slopes from figure 12.7, we are testing whether $\beta_{M2} = \beta_{M1}$. This comparison is shown below.

```
. lincom 1.gender#c.ed2 - 1.gender#c.ed1
 (1)  - 1bn.gender#c.ed1 + 1bn.gender#c.ed2 = 0
```

realrinc	Coef.	Std. Err.	t	P>|t|	[95% Conf.	Interval]
(1)	1503.24	297.1315	5.06	0.000	920.8512	2085.629

After graduating high school, each additional year of education is associated with a greater increase in income compared with those who have not graduated high school. The difference (comparing high school graduates with non–high school graduates) is 1,503.24 and this is significant.

We can formulate the same kind of test for women—testing whether $\beta_{F2} = \beta_{F1}$. This comparison is shown below.

```
. lincom 2.gender#c.ed2 - 2.gender#c.ed1
 ( 1)  - 2.gender#c.ed1 + 2.gender#c.ed2 = 0

    realrinc       Coef.   Std. Err.      t    P>|t|     [95% Conf. Interval]

         (1)    1481.099   208.8135     7.09   0.000     1071.817    1890.381
```

This test is statistically significant. For female high school graduates, income rises by $1,481.10 more per additional year of education compared with female non–high school graduates.

Let's now compare the educ slope for college graduates with high school graduates. These comparisons are shown below for men and for women.

```
. * men: cograd vs. hsgrad
. lincom 1.gender#c.ed3 - 1.gender#c.ed2
 ( 1)  - 1bn.gender#c.ed2 + 1bn.gender#c.ed3 = 0

    realrinc       Coef.   Std. Err.      t    P>|t|     [95% Conf. Interval]

         (1)    2686.722   581.2696     4.62   0.000     1547.412    3826.033

. * women: cograd vs. hsgrad
. lincom 2.gender#c.ed3 - 2.gender#c.ed2
 ( 1)  - 2.gender#c.ed2 + 2.gender#c.ed3 = 0

    realrinc       Coef.   Std. Err.      t    P>|t|     [95% Conf. Interval]

         (1)    1883.708   530.2042     3.55   0.000     844.4879    2922.928
```

Both of these tests are significant. Among male college graduates, an additional year of education is worth $2,686.72 more than an additional year of education for a high school graduate. For females, an additional year of education is worth $1,883.71 more than an additional year of education for college graduates than high school graduates.

12.3.3 Difference in differences of slopes

In the previous section, we compared the educ slope for male high school graduates with male non–high school graduates. To form this comparison, we estimated the difference between β_{M2} and β_{M1}, which equaled 1,503.24. Let's call this the *gain in slope* due to graduating high school for males. We also estimated the gain in slope due to graduating high school for females (that is, β_{F2} versus β_{F1}), which equaled $1,481.10. Let's now test the gain in slope due to graduating high school for men compared with the gain in slope due to graduating high school for women, which is $(\beta_{M2} - \beta_{M1}) - (\beta_{F2} - \beta_{F1})$. This is computed using the `lincom` command below. (Note the importance of the parentheses.) This shows that the gain in the slope for men due to graduating high school is 22.14 units more than the gain in slope for women due to graduating high school. However, this is not statistically significant.

```
. lincom (1.gender#c.ed2 - 1.gender#c.ed1) - (2.gender#c.ed2 - 2.gender#c.ed1)
 (1)  - 1bn.gender#c.ed1 + 2.gender#c.ed1 + 1bn.gender#c.ed2 - 2.gender#c.ed2 = 0
```

realrinc	Coef.	Std. Err.	t	P>|t|	[95% Conf.	Interval]
(1)	22.14114	361.966	0.06	0.951	-687.3259	731.6082

We can also compute the gain in the slope due to graduating college (compared with high school graduates) for men versus women. This difference is 803.01 but is not statistically significant.

```
. lincom (1.gender#c.ed3 - 1.gender#c.ed2) - (2.gender#c.ed3 - 2.gender#c.ed2)
 (1)  - 1bn.gender#c.ed2 + 2.gender#c.ed2 + 1bn.gender#c.ed3 - 2.gender#c.ed3 = 0
```

realrinc	Coef.	Std. Err.	t	P>|t|	[95% Conf.	Interval]
(1)	803.0143	786.6449	1.02	0.307	-738.8393	2344.868

2.3.4 Comparing changes in intercepts by gender

Let's now ask whether the jump in income at 12 years of education is the same for men and women. As shown in figure 12.7, the income for women jumps by $2,400.89 at 12 years of education, whereas the corresponding jump for men is $4,927.53. Let's test whether these jumps are equal (whether $\alpha_{F1} = \alpha_{M1}$). This is tested using the `contrast` command below. The jump in income that women receive due to graduating high school is $2,526.64 less than the jump that men receive, and this is significant ($t = -3.36$, $p = 0.001$).

```
. contrast gender#hsgrad, pveffects nowald
Contrasts of marginal linear predictions
Margins      : asbalanced
```

	Contrast	Std. Err.	t	P>|t|
gender#hsgrad (Female vs base) (HS Grad vs base)	-2526.638	752.5331	-3.36	0.001

Let's now compare the jump in income due to graduating college for women versus men. As shown in figure 12.7, the income for women jumps by $1,615.20 at 16 years of education, whereas the corresponding jump for men is $9,230.72. Are these jumps equal (that is, is $\alpha_{F2} = \alpha_{M2}$)? Using the `contrast` command below, we see that the difference in the jump in income due to graduating high school for women versus men is $-\$7,615.52$ and this difference is statistically significant ($t = -5.23$, $p = 0.000$).

```
. contrast gender#cograd, pveffects nowald
Contrasts of marginal linear predictions
Margins      : asbalanced
```

	Contrast	Std. Err.	t	P>|t|
gender#cograd				
(Female vs base) (CO Grad vs base)	-7615.523	1455.047	-5.23	0.000

The lincom command could also have been used for these tests. For example, the previous test could have been performed using the following lincom command.

```
. lincom 2.gender#1.cograd - 1.gender#1.cograd
  (output omitted)
```

12.3.5 Comparing changes in intercepts by education

We have seen that, for men, the jump in income at 12 years of education is $4,927.53, and at 16 years of education the jump is $9,230.72. We might ask whether the jump due to graduating college ($9,230.72) is equal to the jump due to graduating high school ($4,927.53). In terms of figure 12.7, this asks whether $\alpha_{M2} = \alpha_{M1}$. Using the lincom command (below), we see that the jump in income men receive due to graduating college is $4,303.19 greater than the jump they receive due to graduating high school. This difference is statistically significant.

```
. lincom 1.gender#1.cograd - 1.gender#1.hsgrad
 (1)  - 1bn.gender#1.hsgrad + 1bn.gender#1.cograd = 0
```

realrinc	Coef.	Std. Err.	t	P>|t|	[95% Conf.	Interval]
(1)	4303.185	1327.654	3.24	0.001	1700.934	6905.436

Let's now make the same comparison for women. Looking at figure 12.7, we are asking whether $\alpha_{F2} = \alpha_{F1}$. As shown in the lincom command below, the jump in income women receive due to graduating college is $785.70 less than the jump they receive for graduating high school. This difference, however, is not statistically significant.

```
. lincom 2.gender#1.cograd - 2.gender#1.hsgrad
 (1)  - 2.gender#1.hsgrad + 2.gender#1.cograd = 0
```

realrinc	Coef.	Std. Err.	t	P>|t|	[95% Conf.	Interval]
(1)	-785.6994	831.2172	-0.95	0.345	-2414.917	843.5177

12.3.6 Computing and comparing adjusted means

Let's now see how to compute adjusted means in the context of this model. To compute adjusted means with respect to education, we need to express education in terms of `hsgrad`, `cograd`, `ed1`, `ed2`, and `ed3`. Below, we can see the correspondence between `educ` and the values of `hsgrad`, `cograd`, `ed1`, `ed2`, and `ed3`.

```
. showcoding educ hsgrad cograd ed1 ed2 ed3
```

educ	hsgrad	cograd	ed1	ed2	ed3
0	0	0	0	0	0
1	0	0	1	0	0
2	0	0	2	0	0
3	0	0	3	0	0
4	0	0	4	0	0
5	0	0	5	0	0
6	0	0	6	0	0
7	0	0	7	0	0
8	0	0	8	0	0
9	0	0	9	0	0
10	0	0	10	0	0
11	0	0	11	0	0
12	1	0	12	0	0
13	1	0	12	1	0
14	1	0	12	2	0
15	1	0	12	3	0
16	1	1	12	4	0
17	1	1	12	4	1
18	1	1	12	4	2
19	1	1	12	4	3
20	1	1	12	4	4

We can use the output of the `showcoding` command to compute the adjusted mean for any given level of education. For example, let's compute the adjusted mean for a male who has 10 years of education. Note we code `gender` as 1 to indicate a male, and the rest of the values with respect to education are drawn from the `showcoding` command corresponding to `educ=10`. The `margins` command below shows that the adjusted mean for a man with 10 years of education is $17,438.20.

```
. margins, nopvalues at(gender=1 hsgrad=0 cograd=0 ed1=10 ed2=0 ed3=0)
Predictive margins                              Number of obs     =     32,183
Model VCE    : Robust

Expression   : Linear prediction, predict()
at           : gender          =           1
               hsgrad          =           0
               cograd          =           0
               ed1             =          10
               ed2             =           0
               ed3             =           0

------------------------------------------------------------------
             |            Delta-method
             |     Margin   Std. Err.     [95% Conf. Interval]
-------------+----------------------------------------------------
       _cons |    17438.2   390.0421       16673.7    18202.69
------------------------------------------------------------------
```

Using the **margins** command below, we estimate the adjusted mean for a female with 15 years of education as $17,221.21.

```
. margins, nopvalues at(gender=2 hsgrad=1 cograd=0 ed1=12 ed2=3 ed3=0)
Predictive margins                              Number of obs     =     32,183
Model VCE    : Robust

Expression   : Linear prediction, predict()
at           : gender          =           2
               hsgrad          =           1
               cograd          =           0
               ed1             =          12
               ed2             =           3
               ed3             =           0

------------------------------------------------------------------
             |            Delta-method
             |     Margin   Std. Err.     [95% Conf. Interval]
-------------+----------------------------------------------------
       _cons |   17221.21   488.5276      16263.68    18178.75
------------------------------------------------------------------
```

We can estimate the adjusted mean for men and women (separately) in one **margins** command. The example below computes the adjusted mean for men and women, holding education constant at 15 years of education.

```
. margins gender, nopvalues at(hsgrad=1 cograd=0 ed1=12 ed2=3 ed3=0)
Predictive margins                              Number of obs     =     32,183
Model VCE    : Robust

Expression   : Linear prediction, predict()
at           : hsgrad          =           1
               cograd          =           0
               ed1             =          12
               ed2             =           3
               ed3             =           0

------------------------------------------------------------------
             |            Delta-method
             |     Margin   Std. Err.     [95% Conf. Interval]
-------------+----------------------------------------------------
      gender |
        Male |   27322.43   652.7205      26043.07    28601.78
      Female |   17221.21   488.5276      16263.68    18178.75
------------------------------------------------------------------
```

Let's now use the **margins** command to compare the adjusted mean of income for women versus men among those with 15 years of education. Specifying the **r.** contrast operator indicates we want to use reference group comparisons. The difference in this adjusted mean for women and men with 15 years of education is $-10,101.21$, and this difference is significant. Thus, at 15 years of education, the adjusted mean of income for men is $10,101.21 higher than for women.

```
. margins r.gender, at(hsgrad=1 cograd=0 ed1=12 ed2=3 ed3=0)
> contrast(nowald pveffects)
Contrasts of predictive margins                 Number of obs     =     32,183
Model VCE    : Robust

Expression   : Linear prediction, predict()
at           : hsgrad          =           1
               cograd          =           0
               ed1             =          12
               ed2             =           3
               ed3             =           0

--------------------------------------------------------------
                 |            Delta-method
                 |   Contrast   Std. Err.      t    P>|t|
-----------------+--------------------------------------------
          gender |
(Female vs Male) |  -10101.21   813.7234   -12.41   0.000
--------------------------------------------------------------
```

12.3.7 Graphing adjusted means

In figure 12.6, we showed a graph of the adjusted means for this model as a function of education and gender. Let's show how to make such a graph. We first need to compute the adjusted means separately for men and women for the following levels of education—0, 12 (assuming the absence and presence of a high school diploma), 16 (assuming the absence and presence of a college degree), and 20.[7] These adjusted means are computed using the `margins` command, as shown below. The `noatlegend` option is used to save space.

```
. margins gender, at(hsgrad=0 cograd=0 ed1=0  ed2=0 ed3=0)
>                 at(hsgrad=0 cograd=0 ed1=12 ed2=0 ed3=0)
>                 at(hsgrad=1 cograd=0 ed1=12 ed2=0 ed3=0)
>                 at(hsgrad=1 cograd=0 ed1=12 ed2=4 ed3=0)
>                 at(hsgrad=1 cograd=1 ed1=12 ed2=4 ed3=0)
>                 at(hsgrad=1 cograd=1 ed1=12 ed2=4 ed3=4)
>                 vsquish noatlegend nopvalues
Predictive margins                              Number of obs     =     32,183
Model VCE    : Robust

Expression   : Linear prediction, predict()
```

	Margin	Delta-method Std. Err.	[95% Conf.	Interval]
_at#gender				
1#Male	16544.24	1384.772	13830.04	19258.45
1#Female	6722.377	909.9192	4938.901	8505.853
2#Male	17616.99	590.8879	16458.82	18775.15
2#Female	9646.094	345.6314	8968.643	10323.54
3#Male	22544.52	282.6604	21990.49	23098.54
3#Female	12046.99	142.2014	11768.27	12325.71
4#Male	28915.06	896.2039	27158.47	30671.66
4#Female	18945.96	667.0764	17638.46	20253.45
5#Male	38145.78	830.9111	36517.16	39774.4
5#Female	20561.15	422.3495	19733.33	21388.97
6#Male	55263.21	1736.329	51859.94	58666.48
6#Female	34994.95	1760.662	31543.99	38445.92

We can then manually input these adjusted means into a dataset and graph them, as shown below. The resulting graph is shown in figure 12.8.

7. Obtaining the adjusted mean for 12 years of education assuming the absence and presence of a high school diploma illustrates the jump in income due to graduating high school. Similarly, computing the adjusted mean at 16 years of education assuming the absence and presence of a college degree illustrates the income jump due to graduating college.

```
. preserve
. clear
. input educ yhatm yhatf
          educ       yhatm        yhatf
  1. 0  16544.24 6722.377
  2. 12 17616.99 9646.094
  3. 12 22544.52 12046.99
  4. 16 28915.06 18945.96
  5. 16 38145.78 20561.15
  6. 20 55263.21 34994.95
  7. end
. graph twoway line yhatm yhatf educ, xlabel(0(4)20) xline(12 16)
. restore
```

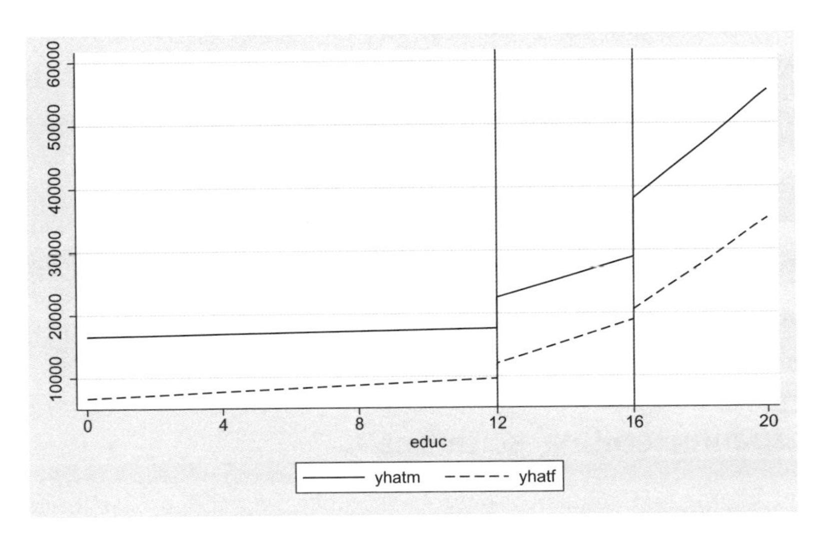

Figure 12.8. Fitted values from piecewise model with two knots and two jumps

With some extra effort, we can automate the creation of the graph shown in figure 12.8. Using the strategy from section 12.2.6, the `matrix` and `svmat` commands are used, followed by the `graph` command to graph the adjusted means as a function of `educ` and `gender`. These commands create the same graph that is shown in figure 12.8.

```
. * First, rerun the -margins- command from above
. quietly margins gender, at(hsgrad=0 cograd=0 ed1=0  ed2=0 ed3=0)
>                         at(hsgrad=0 cograd=0 ed1=12 ed2=0 ed3=0)
>                         at(hsgrad=1 cograd=0 ed1=12 ed2=0 ed3=0)
>                         at(hsgrad=1 cograd=0 ed1=12 ed2=4 ed3=0)
>                         at(hsgrad=1 cograd=1 ed1=12 ed2=4 ed3=0)
>                         at(hsgrad=1 cograd=1 ed1=12 ed2=4 ed3=4)
. * Store the adjusted means (from the -margins- command) in a matrix named -yhat-
. matrix yhat = r(b)' // NOTE:  We must transpose r(b) here!
. * store levels of education in a matrix named -educ-
. matrix educ = (0 \ 0 \ 12 \ 12 \ 12 \ 12 \ 16 \ 16 \ 16 \ 16 \ 20 \ 20)
. * store levels of gender in a matrix named -gender-
. matrix gender = (1 \ 2 \ 1 \ 2 \ 1 \ 2 \ 1 \ 2 \ 1 \ 2 \ 1 \ 2 )
. svmat yhat    // save the matrix -yhat- to the current dataset
. svmat educ    // save the matrix -educ- to the current dataset
. svmat gender // save the matrix -gender- to the current dataset
. graph twoway (line yhat1 educ1 if gender1==1)
>              (line yhat1 educ1 if gender1==2),
>               xline(12 16) legend(label(1 "Men") label(2 "Women"))
>               xtitle(Education) ytitle(Adjusted mean)
```

This section has illustrated a piecewise model with two knots, with a change in slope and change in intercept at each knot. This example (as well as all previous examples in this chapter) have used a separate intercept and separate slope coding scheme. The next section illustrates other possible coding schemes that can be used for these models.

12.4 Comparing coding schemes

This chapter has focused on one coding scheme for fitting models that interact a categorical variable and a continuous variable fit in a piecewise manner. I have focused on a coding scheme that I believe is the easiest to use, but there are other coding schemes that you could use. Depending on your research question, another coding scheme might be more useful. This section illustrates four coding schemes that can be used to fit the model illustrated in section 12.2. The coding scheme used in section 12.2 is repeated and will be called *coding scheme #1*. Then, three additional coding schemes will be illustrated.

12.4.1 Coding scheme #1

Let's begin by repeating the analysis from section 12.2. This coding scheme will be called coding scheme #1. The `noheader` option is used in this and subsequent examples to save space.[8]

```
. * Coding scheme #1.
. use gss_ivrm
. mkspline ed1 12 ed2 = educ
. regress realrinc ibn.gender ibn.gender#(i.hsgrad c.ed1 c.ed2) i.race,
> vce(robust) noconstant noheader noci
```

realrinc	Coef.	Robust Std. Err.	t	P>|t|
gender				
Male	17144.66	1389.515	12.34	0.000
Female	7325.482	922.5224	7.94	0.000
gender#hsgrad				
Male#HS Grad	3382.005	661.547	5.11	0.000
Female#HS Grad	1748.096	389.3635	4.49	0.000
gender#c.ed1				
Male	86.70762	151.379	0.57	0.567
Female	241.6081	98.9163	2.44	0.015
gender#c.ed2				
Male	4057.639	150.3836	26.98	0.000
Female	2529.547	110.4318	22.91	0.000
race				
white	0	(base)		
black	-3521.336	246.6606	-14.28	0.000
other	-1946.159	839.7251	-2.32	0.020

Figure 12.9 shows a graph of the adjusted means as a function of education and gender. Labels are included showing the key coefficients from the regression output. For example, the intercept for males is 17,145 and is labeled α_{M1}. The jump in income due to graduating high school for males is 3,382 and is labeled α_{M2}. For males, the `educ` slope before graduating high school is labeled β_{M1}, and the `educ` slope after graduating high school is labeled β_{M2}. The figure also includes coefficients associated with females, labeled α_{F1}, α_{F2}, β_{F1}, and β_{F2}.

8. As with earlier examples, the `noci` option is included in this example to make the output more readable. It will be used in additional examples in this chapter as well.

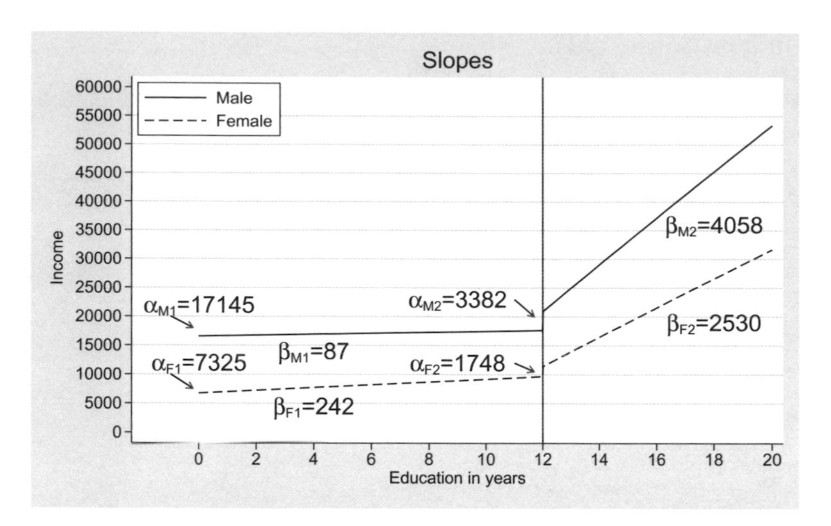

Figure 12.9. Intercept and slope coefficients from piecewise regression fit using coding scheme #1

12.4.2 Coding scheme #2

Let's now fit a model using what I call coding scheme #2. This coding scheme is the same as coding scheme #1, except that the `marginal` option is used on the `mkspline` command.

```
. * Coding scheme #2
. use gss_ivrm
. mkspline ed1m 12 ed2m = educ, marginal
. regress realrinc ibn.gender ibn.gender#(i.hsgrad c.ed1m c.ed2m) i.race,
> vce(robust) noconstant noheader noci

------------------------------------------------------------
                 |               Robust
        realrinc |      Coef.   Std. Err.      t    P>|t|
-----------------+------------------------------------------
          gender |
            Male |   17144.66   1389.515    12.34   0.000
          Female |   7325.482   922.5224     7.94   0.000
                 |
   gender#hsgrad |
    Male#HS Grad |   3382.005    661.547     5.11   0.000
  Female#HS Grad |   1748.096   389.3635     4.49   0.000
                 |
   gender#c.ed1m |
            Male |   86.70762    151.379     0.57   0.567
          Female |   241.6081    98.9163     2.44   0.015
                 |
   gender#c.ed2m |
            Male |   3970.931   212.6975    18.67   0.000
          Female |   2287.939   147.2609    15.54   0.000
                 |
            race |
           white |          0  (base)
           black |  -3521.336   246.6606   -14.28   0.000
           other |  -1946.159   839.7251    -2.32   0.020
------------------------------------------------------------
```

Table 12.3 summarizes the key results by using coding scheme #1 and coding scheme #2. It shows the name of the coefficient, the value of the coefficient, and the meaning of the coefficient, expressed in terms of the labels shown in figure 12.9. This illustrates the difference in the meaning of the coefficients when using coding scheme #1 versus coding scheme #2. You can see that the only difference is in the final row of the table. Coding scheme #1 estimates β_{M2} and β_{F2}, whereas coding scheme #2 estimates $\beta_{M2} - \beta_{M1}$ and $\beta_{F2} - \beta_{F1}$.

Table 12.3. Summary of regression results and meaning of coefficients for coding schemes #1 and #2

Coding scheme #1			Coding scheme #2		
Coef. name	Value	Meaning	Coef. name	Value	Meaning
gender			gender		
Male	17144.66	α_{M1}	Male	17144.66	α_{M1}
Female	7325.48	α_{F1}	Female	7325.48	α_{F1}
gender#hsgrad			gender#hsgrad		
Male#HS Grad	3382.01	α_{M2}	Male#HS Grad	3382.01	α_{M2}
Female#HS Grad	1748.10	α_{F2}	Female#HS Grad	1748.10	α_{F2}
gender#c.ed1			gender#c.ed1m		
Male	86.71	β_{M1}	Male	86.71	β_{M1}
Female	241.61	β_{F1}	Female	241.61	β_{F1}
gender#c.ed2			gender#c.ed2m		
Male	4057.64	β_{M2}	Male	3970.93	$\beta_{M2} - \beta_{M1}$
Female	2529.55	β_{F2}	Female	2287.94	$\beta_{F2} - \beta_{F1}$

12.4.3 Coding scheme #3

Let's now consider a third coding scheme. This coding scheme is like coding scheme #1, in that the `marginal` option is omitted from the `mkspline` command. Unlike coding scheme #1, coding scheme #3 specifies `i.gender` (instead of `ibn.gender`) and omits the `noconstant` option.

```
. * Coding scheme #3
. use gss_ivrm

. mkspline ed1 12 ed2 = educ

. regress realrinc i.gender##(i.hsgrad c.ed1 c.ed2) i.race,
> vce(robust) noheader noci

------------------------------------------------------------------
                 |               Robust
        realrinc |      Coef.   Std. Err.        t    P>|t|
-----------------+------------------------------------------------
          gender |
            Male |          0  (base)
          Female |  -9819.174   1638.556     -5.99    0.000
                 |
          hsgrad |
     Not HS Grad |          0  (base)
         HS Grad |   3382.005    661.547      5.11    0.000
                 |
             ed1 |   86.70762    151.379      0.57    0.567
             ed2 |   4057.639   150.3836     26.98    0.000
                 |
   gender#hsgrad |
  Female#HS Grad |  -1633.908   766.9848     -2.13    0.033
                 |
    gender#c.ed1 |
          Female |   154.9005   179.3207      0.86    0.388
                 |
    gender#c.ed2 |
          Female |  -1528.092   187.0884     -8.17    0.000
                 |
            race |
           white |          0  (base)
           black |  -3521.336   246.6606    -14.28    0.000
           other |  -1946.159   839.7251     -2.32    0.020
                 |
           _cons |   17144.66   1389.515     12.34    0.000
------------------------------------------------------------------
```

The key coefficients of coding scheme #3 are summarized in table 12.4. This illustrates the interpretation of the coefficients with respect to the symbols illustrated in figure 12.9. Let's now consider a fourth coding scheme, and then we can compare coding schemes #3 and #4.

12.4.4 Coding scheme #4

Finally, let's consider a fourth coding scheme. This coding scheme is like coding scheme #3, except that the `marginal` option is included on the `mkspline` command.

```
. * Coding scheme #4
. use gss_ivrm
. mkspline ed1m 12 ed2m = educ, marginal
. regress realrinc i.gender##(i.hsgrad c.ed1m c.ed2m) i.race,
> vce(robust) noheader noci
```

realrinc	Coef.	Robust Std. Err.	t	P>|t|
gender				
Male	0	(base)		
Female	-9819.174	1638.556	-5.99	0.000
hsgrad				
Not HS Grad	0	(base)		
HS Grad	3382.005	661.547	5.11	0.000
ed1m	86.70762	151.379	0.57	0.567
ed2m	3970.931	212.6975	18.67	0.000
gender#hsgrad				
Female#HS Grad	-1633.908	766.9848	-2.13	0.033
gender#c.ed1m				
Female	154.9005	179.3207	0.86	0.388
gender#c.ed2m				
Female	-1682.992	259.2855	-6.49	0.000
race				
white	0	(base)		
black	-3521.336	246.6606	-14.28	0.000
other	-1946.159	839.7251	-2.32	0.020
_cons	17144.66	1389.515	12.34	0.000

The key coefficients of coding schemes #3 and #4 are summarized in table 12.4. This shows the name of the coefficient for these two coding schemes, the value of the coefficient, and the meaning of the coefficient expressed in terms of the labels shown in figure 12.9. The coefficients are ordered differently than the output of the `regress` command to group related coefficients and to facilitate comparisons between tables 12.3 and 12.4.

Table 12.4. Summary of regression results and meaning of coefficients for coding schemes #3 and #4

Coding scheme #3			Coding scheme #4		
Coef. name	Value	Meaning	Coef. name	Value	Meaning
`_cons`	17144.66	α_{M1}	`_cons`	17144.66	α_{M1}
`gender`			`gender`		
`Female`	-9819.17	$\alpha_{F1} - \alpha_{M1}$	`Female`	-9819.17	$\alpha_{F1} - \alpha_{M1}$
`hsgrad`			`hsgrad`		
`HSgrad`	3382.01	α_{M2}	`HSgrad`	3382.01	α_{M2}
`gender#hsgrad`			`gender#hsgrad`		
`Female#HS Grad`	-1633.91	$\alpha_{F2} - \alpha_{M2}$	`Female#HS grad`	-1633.91	$\alpha_{F2} - \alpha_{M2}$
`ed1`	86.71	β_{M1}	`ed1m`	86.71	β_{M1}
`gender#c.ed1`			`gender#c.ed1m`		
`Female`	154.90	$\beta_{F1} - \beta_{M1}$	`Female`	154.91	$\beta_{F1} - \beta_{M1}$
`ed2`	4057.64	β_{M2}	`ed2m`	3970.93	$\beta_{M2} - \beta_{M1}$
`gender#c.ed2`			`gender#c.ed2m`		
`Female`	-1528.09	$\beta_{F2} - \beta_{M2}$	`Female`	-1682.99	$(\beta_{F2} - \beta_{F1}) - (\beta_{M2} - \beta_{M1})$

In comparing coding schemes #3 and #4, the only difference is in the final row of the table. Coding scheme #3 estimates β_{M2} and $\beta_{F2} - \beta_{M2}$. By comparison, coding scheme #4 estimates $\beta_{M2} - \beta_{M1}$ and $(\beta_{F2} - \beta_{F1}) - (\beta_{M2} - \beta_{M1})$.

2.4.5 Choosing coding schemes

Now that we understand how the results produced by these four coding schemes differ, we can deliberately choose the coding scheme that might make the most sense given our research question.

Say that the emphasis of your research study was to test gender differences in the `educ` slope among high school graduates (that is, $\beta_{F2} - \beta_{M2}$). In that case, coding scheme #3 might be the most useful, because the coefficient associated with `gender#ed2` directly estimates this difference ($\beta_{F2} - \beta_{M2}$). Had you chosen coding scheme #1, you could still estimate this difference, but would need to also use the `contrast gender#c.ed1` command to compute this difference.

Instead, imagine that your research question focused on gender differences in the `educ` slope for high school graduates versus non–high school graduates. In that case, coding system #4 might be the most useful because the coefficient associated with `gender#ed2m` directly estimates this difference $(\beta_{F2} - \beta_{F1}) - (\beta_{M2} - \beta_{M1})$.

12.5 Summary

This chapter has illustrated models that involve the interaction of a categorical variable with a continuous variable modeled in a piecewise manner. This included a piecewise model with one knot and one jump (in section 12.2) and two knots and two jumps (in section 12.3). For simplicity, the examples showed a categorical variable that had two levels. However, these examples can be generalized to cases where the categorical variable has three or more levels.

For more information, I highly recommend Singer and Willett (2003), who provide examples and additional details about interactions of a categorical variable with a continuous piecewise variable.

3 Continuous by continuous by categorical interactions

3.1 Chapter overview

This chapter explores models that include a categorical variable interacted with two continuous variables. This can be viewed as an extension of chapter 5 which illustrated interactions of two continuous variables. In fact, the examples shown in this chapter mirror those shown in chapter 5 but add a categorical variable to the interaction.

The examples shown in section 13.2 illustrate a continuous linear by continuous linear by categorical interaction. The example is an extension of the example from section 5.2, which showed an interaction of `age` and `educ` in the prediction of `realrinc`. The example in section 13.2 will show how the size of the interaction of these two continuous variables differs by a categorical variable, namely, `gender`.[1]

The examples illustrated in section 13.3 show a continuous linear by continuous quadratic by categorical interaction. This builds upon the example from section 5.3, which showed an interaction of `educ` and `age` in the prediction of `realrinc`, where `educ`

1. I use the variable `gender` (coded: 1 = Male and 2 = Female), because it leads to output that clearly distinguishes the variable name (that is, `gender`) and its values (that is, `Male`, and `Female`).

was modeled linearly and age was modeled using a quadratic term. The example in section 13.3 will examine whether the size of this interaction differs by gender.

13.2 Linear by linear by categorical interactions

In section 5.2, we saw an interaction of two continuous variables, age and educ, in the prediction of realrinc. In this section, we will explore whether the size of the c.age##c.educ interaction depends on gender. As with the analysis in section 5.2, this example will use the gss_ivrm.dta dataset, focusing on those who are aged 22 to 55 and who have at least 12 years of education.

```
. use gss_ivrm
. keep if (age>=22 & age<=55) & (educ>=12)
(25,024 observations deleted)
```

13.2.1 Fitting separate models for males and females

To get a general sense of the size of the c.age#c.educ interaction separately for males and females, let's fit separate models for males and females. The following regress command predicts realrinc from age, educ, and the interaction of these two variables, doing so separately for each level of gender by including the by gender, sort: prefix. (The vsquish and noheader options are used to save space.) The first set of results is restricted to analyzing only males, and the second set is restricted to analyzing only females. We can use this as a warm-up analysis to see whether it looks like the size of the c.age#c.educ interaction differs for males and females.

```
. by gender, sort: regress realrinc c.age##c.educ, vce(robust) vsquish noheader

-------------------------------------------------------------------------------
-> gender = Male

------------------------------------------------------------------------------
             |               Robust
    realrinc |      Coef.   Std. Err.      t    P>|t|     [95% Conf. Interval]
-------------+----------------------------------------------------------------
         age |  -1092.287   195.7135    -5.58   0.000    -1475.921   -708.6537
        educ |  -1831.065   526.0086    -3.48   0.001    -2862.136   -799.9943
c.age#c.educ |   134.4611   14.35327     9.37   0.000     106.3261     162.596
       _cons |   25275.91   7128.772     3.55   0.000     11302.24    39249.58
------------------------------------------------------------------------------

-------------------------------------------------------------------------------
-> gender = Female

------------------------------------------------------------------------------
             |               Robust
    realrinc |      Coef.   Std. Err.      t    P>|t|     [95% Conf. Interval]
-------------+----------------------------------------------------------------
         age |  -876.5621   158.2391    -5.54   0.000    -1186.738    -566.386
        educ |  -966.1149   426.7839    -2.26   0.024    -1802.686   -129.5442
c.age#c.educ |   87.94522   11.97356     7.34   0.000     64.47496    111.4155
       _cons |    17020.7   5661.946     3.01   0.003     5922.301     28119.1
------------------------------------------------------------------------------
```

The c.age#c.educ interaction is 134.46 for males and is 87.95 for females. Let's visualize the difference in the size of the c.age#c.educ interaction for males and females. We can visualize this in two different ways, by focusing on the slope in the direction of age or by focusing on the slope in the direction of educ. Let's begin by making a graph that focuses on the slope in the direction of age by placing age on the x axis, with separate lines for educ. This is visualized in figure 13.1, showing males in the left panel and females in the right panel. (I will show how to make this kind of graph later in this chapter.)

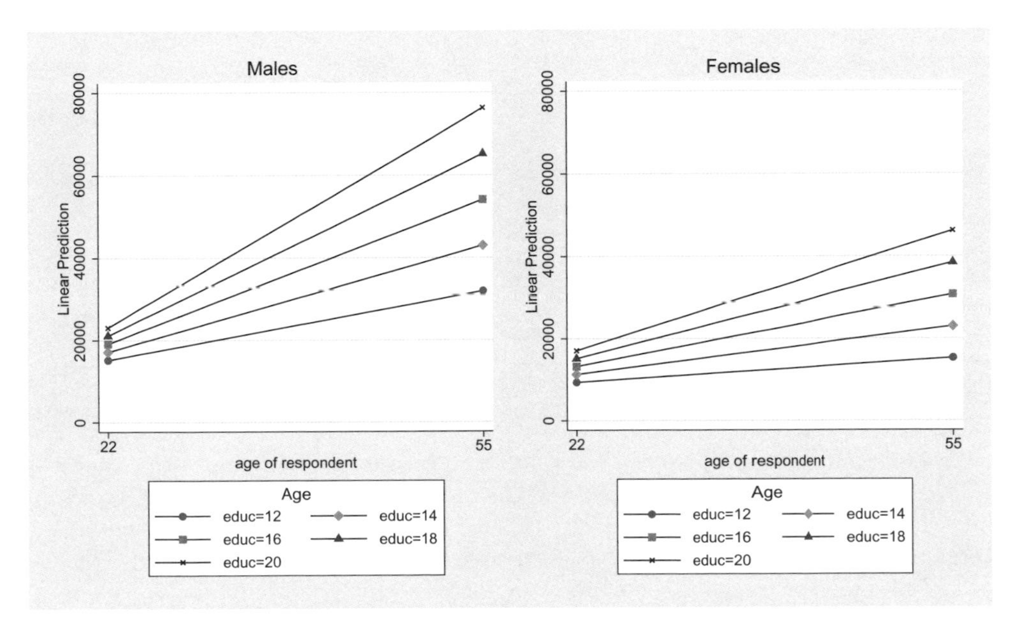

Figure 13.1. Fitted values for age by education interaction for males (left) and females (right)

Looking at figure 13.1, we can see that, for both males and females, the age slope grows as educ increases. We can also see that the degree of this growth is greater for males than for females. As we learned in section 5.2, the c.age#c.educ coefficient is the degree to which the age slope increases for every one-unit increase in educ. This quantifies the increase in the age slope for every one-year increase in educ for males and females. Among males, the age slope increases by 134.46 units for every one-year increase in educ. Among females, the age slope increases by 87.95 units for every one-year increase in educ.

We can visualize the c.age#c.educ interaction another way, focusing on the educ slope by placing educ on the x axis, with separate lines for age. This is visualized in figure 13.2, showing males in the left panel and females in the right panel. (I will show how to make this kind of graph later in this chapter.)

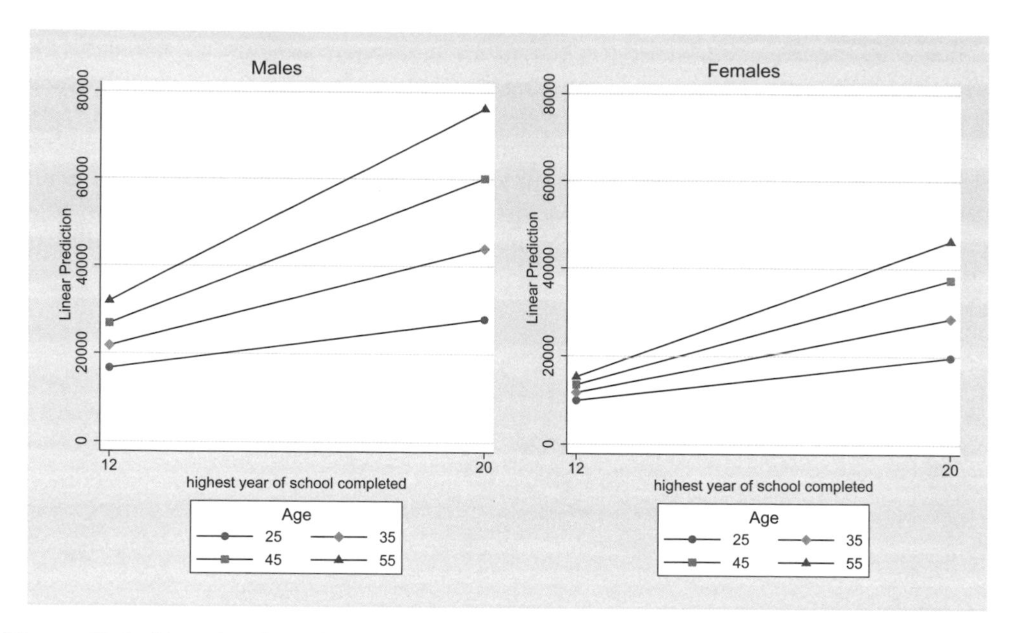

Figure 13.2. Fitted values for age by education interaction for males (left) and females (right) with education on the x axis

Looking at figure 13.2, we can see that for both males and females the `educ` slope grows as `age` increases. We can also see that the degree of this growth is greater for males than for females and can quantify this using the coefficient for the `c.age#c.educ` interaction. For males, the `educ` slope increases by 134.46 units for every one-year increase in `age`. For females, the `educ` slope increases by 87.95 units for every one-year increase in `age`.

To summarize, the coefficient for `c.age#c.educ` is 134.46 for males and is 87.95 for females. This suggests that the size of the `c.age#c.educ` coefficient may be significantly greater for males than for females. Let's test this by analyzing males and females together in a single model.

13.2.2 Fitting a combined model for males and females

We now create a regression model that includes both males and females and predicts `realrinc` from `age`, `educ`, and the interaction of these two continuous variables. The `regress` command (shown below) is constructed to provide separate estimates for the intercepts for males and females, as well as separate estimates for `age`, `educ`, and `age#educ` by gender. By specifying `ibn.gender` in conjunction with the `noconstant` option, the model fits separate intercepts by gender. By specifying the interaction of `ibn.gender` with `c.age`, with `c.educ`, and with `c.age#c.educ`, the model fits separate estimates of `age`, `educ`, and `age#educ` by gender. The model also includes `i.race`, which is treated as a covariate.[2]

```
. regress realrinc ibn.gender ibn.gender#(c.age c.educ c.age#c.educ)
> i.race, vce(robust) noconstant noci

Linear regression                               Number of obs     =     22,367
                                                F(10, 22357)      =    3126.19
                                                Prob > F          =     0.0000
                                                R-squared         =     0.5339
                                                Root MSE          =      24543

------------------------------------------------------------------
                     |               Robust
            realrinc |      Coef.   Std. Err.      t    P>|t|
---------------------+--------------------------------------------
              gender |
                Male |   27044.09     7125.42     3.80   0.000
              Female |   19020.31    5695.437     3.34   0.001
                     |
        gender#c.age |
                Male |  -1116.239    195.4295    -5.71   0.000
              Female |  -903.4975    158.7596    -5.69   0.000
                     |
       gender#c.educ |
                Male |   -1929.04    526.0811    -3.67   0.000
              Female |  -1063.549    428.2508    -2.48   0.013
                     |
gender#c.age#c.educ  |
                Male |   136.0001    14.34342     9.48   0.000
              Female |   89.51634    12.00574     7.46   0.000
                     |
                race |
               white |          0  (base)
               black |  -3067.109    305.0699   -10.05   0.000
               other |   493.0117    1192.162     0.41   0.679
------------------------------------------------------------------
```

We can express these results as two though there are two regression equations, one for males and one for females.

$$\text{Males: } \widehat{\texttt{realrinc}} = 27044.09 + -1116.24\texttt{age} + -1929.04\texttt{educ} + 136.00\texttt{age*educ} + -3067.11\texttt{black} + 493.01\texttt{other}$$

$$\text{Females: } \widehat{\texttt{realrinc}} = 19020.31 + -903.50\texttt{age} + -1063.55\texttt{educ} + 89.52\texttt{age*educ} + -3067.11\texttt{black} + 493.01\texttt{other}$$

The estimate of the `c.age#c.educ` interaction is 136.00 for males and is 89.52 for females. Let's now test the difference of the `c.age#c.educ` interaction for females versus males using the `contrast` command below.

2. The `noci` option is included to make the output more readable for this example. See the callout titled *Using the* `noci` *option for clearer output* in section 2.5.1 for more details. Note that I will use this option in later examples in this chapter where it makes the output easier to read.

```
. contrast gender#c.age#c.educ, nowald pveffects
Contrasts of marginal linear predictions
Margins      : asbalanced

------------------------------------------------------------------
                     |   Contrast   Std. Err.        t    P>|t|
---------------------+--------------------------------------------
gender#c.age#c.educ  |
   (Female vs base)  |  -46.48377     18.706    -2.48    0.013
------------------------------------------------------------------
```

This test is significant. It compared the size of the `c.age#c.educ` interaction for group 1 with the base group (that is, comparing females with males). This means that the `c.age#c.educ` interaction is significantly lower for females than for males. Let's explore how to interpret this interaction.

Note! What about the lower order effects?

We have been focusing on the `gender#c.age#c.educ` interaction, but you might wonder about the lower order effects, such as `gender#c.age` or `gender#c.educ`. It is important to include these effects in the model to preserve the interpretation of the `gender#c.age#c.educ` interaction. However, there is little to gain by trying to interpret these effects.

13.2.3 Interpreting the interaction focusing in the age slope

To help interpret this interaction, let's visualize it by making a graph that focuses on the `age` slope. We do this using the `margins` and `marginsplot` commands, as shown below. The resulting graph is shown in figure 13.3.

```
. margins gender, at(age=(22 55) educ=(12(2)20))
  (output omitted)
. marginsplot, bydimension(gender) plotdimension(educ, allsimple)
> legend(subtitle(Education) rows(2)) noci
  Variables that uniquely identify margins: age educ gender
```

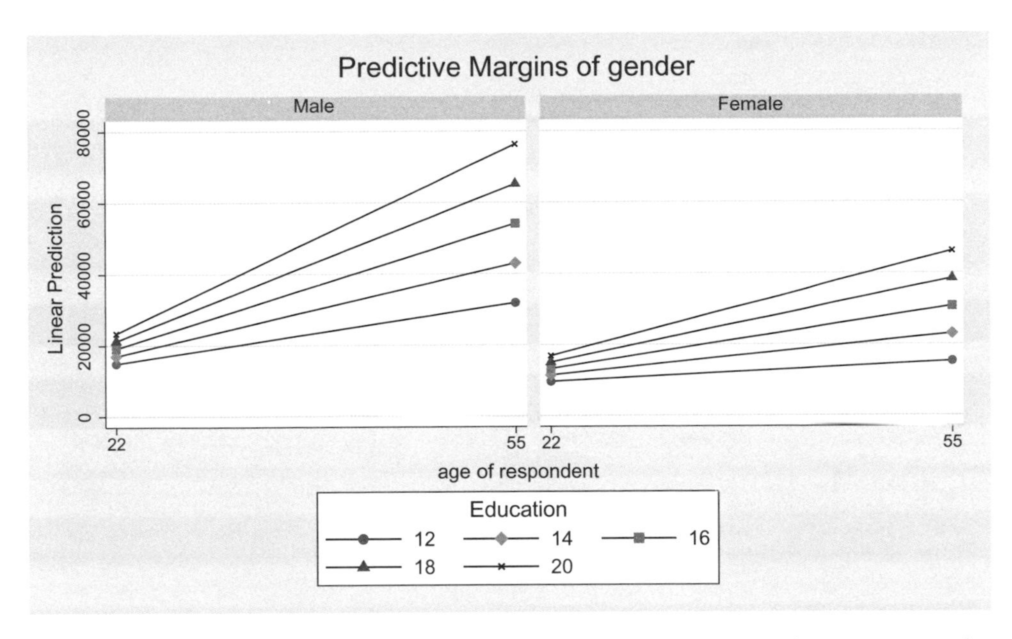

Figure 13.3. Fitted values by age (x axis), education (separate lines), and gender (separate panels)

Figure 13.3 resembles the graph that was created in figure 13.1, which resulted from the models which analyzed males and females separately. We see that the `age` slope increases more rapidly with increases in education for males than for females. For males, the `age` slope increases by 136.00 units for every one-unit increase in education. For females, the `age` slope increases by 89.52 units for every one-unit increase in education. The test of the `gender#c.age#c.educ` effect represents the difference in these interaction terms and this graph shows one way to visualize this.

The `margins` command can be used to show the size of the `age` slope by specifying the `dydx(age)` option. Let's use the `margins` command to estimate the `age` slope for each of the levels of education expressed as a separate line in figure 13.3. In other words, let's estimate the `age` slope for those with 12, 14, 16, 18, and 20 years of education, separately for males and females.

```
. margins gender, at(educ=(12(2)20)) dydx(age)
Average marginal effects                        Number of obs     =     22,367
Model VCE    : Robust

Expression   : Linear prediction, predict()
dy/dx w.r.t. : age

1._at        : educ            =          12

2._at        : educ            =          14

3._at        : educ            =          16

4._at        : educ            =          18

5._at        : educ            =          20

------------------------------------------------------------------------------
             |            Delta-method
             |      dy/dx   Std. Err.      t    P>|t|     [95% Conf. Interval]
-------------+----------------------------------------------------------------
age          |
  _at#gender |
     1#Male  |   515.7626   34.60947    14.90   0.000     447.9256    583.5995
   1#Female  |   170.6985    19.5576     8.73   0.000     132.3643    209.0328
     2#Male  |   787.7628   28.15217    27.98   0.000     732.5826     842.943
   2#Female  |   349.7312   16.77319    20.85   0.000     316.8546    382.6078
     3#Male  |   1059.763   45.09039    23.50   0.000     971.3827    1148.143
   3#Female  |   528.7639   36.51414    14.48   0.000     457.1936    600.3342
     4#Male  |   1331.763   70.13993    18.99   0.000     1194.284    1469.242
   4#Female  |   707.7966   59.48384    11.90   0.000     591.2041    824.3891
     5#Male  |   1603.763   97.22112    16.50   0.000     1413.203    1794.324
   5#Female  |   886.8292   83.04501    10.68   0.000     724.0552    1049.603
------------------------------------------------------------------------------
```

The age slope for a male with 12 years of education is 515.76. For a female with 12 years of education, the age slope is 170.70. At 20 years of education, the age slope is 1,603.76 for males and is 886.83 for females.

13.2.4 Interpreting the interaction focusing on the educ slope

Now, let's visualize this interaction by making a graph that focuses on the educ slope. We do this using the margins and marginsplot commands, as shown below. The resulting graph is shown in figure 13.4.

```
. margins gender, at(age=(25(10)55) educ=(12 20))
  (output omitted)
. marginsplot, bydimension(gender) xdimension(educ) noci legend(rows(2)
> subtitle(Age))
  Variables that uniquely identify margins: age educ gender
```

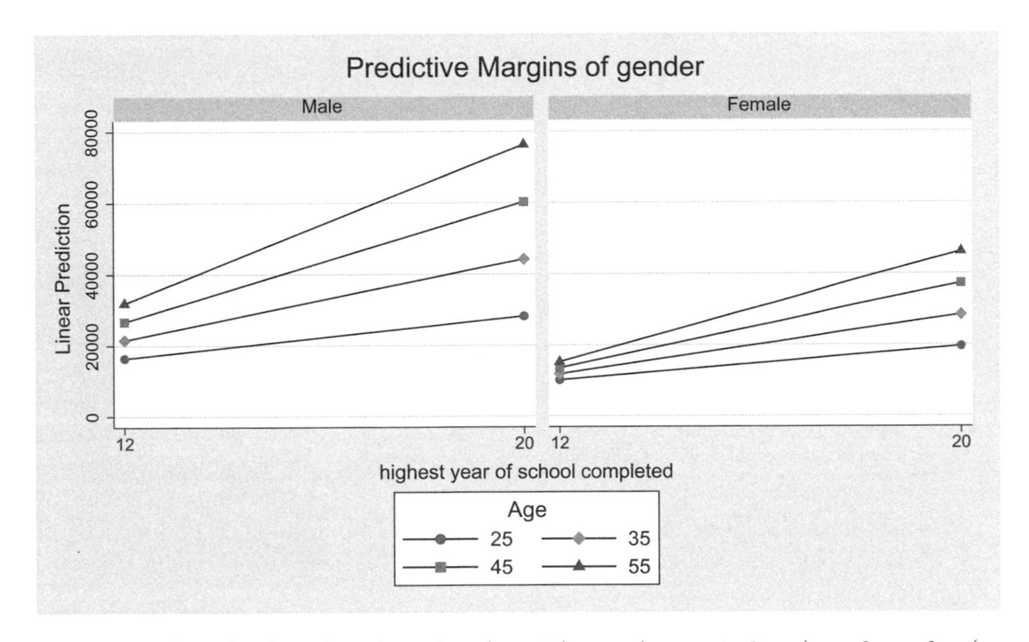

Figure 13.4. Fitted values by education (x axis), age (separate lines), and gender (separate panels)

Figure 13.4 resembles the graph that was created in figure 13.2, which resulted from the models that analyzed males and females separately. We see that the `educ` slope increases more rapidly with increases in age for males than for females. For males, the `educ` slope increases by 136.00 units for every one-unit increase in `age`. For females, the `educ` slope increases by 89.52 units for every one-unit increase in `age`. This graph illustrates the `gender#c.age#c.educ` interaction by showing how the `educ` slope increases more rapidly as a function of `age` for males than for females.

The `dydx(educ)` option can be used with the `margins` command to compute the `educ` slope for each of the lines displayed in figure 13.4, as shown below.

```
. margins gender, at(age=(25(10)55)) dydx(educ)

Average marginal effects                        Number of obs     =     22,367
Model VCE    : Robust

Expression   : Linear prediction, predict()
dy/dx w.r.t. : educ

1._at        : age             =          25

2._at        : age             =          35

3._at        : age             =          45

4._at        : age             =          55

------------------------------------------------------------------------------
             |            Delta-method
             |      dy/dx   Std. Err.      t    P>|t|     [95% Conf. Interval]
-------------+----------------------------------------------------------------
educ         |
  _at#gender |
     1#Male  |   1470.963   206.1855     7.13   0.000     1066.825    1875.101
   1#Female  |    1174.36   155.3249     7.56   0.000     869.9118    1478.807
     2#Male  |   2830.964   144.2903    19.62   0.000     2548.145    3113.783
   2#Female  |   2069.523   104.2348    19.85   0.000     1865.215    2273.831
     3#Male  |   4190.965    200.683    20.88   0.000     3797.613    4584.318
   3#Female  |   2964.686   162.5777    18.24   0.000     2646.023     3283.35
     4#Male  |   5550.966   317.6072    17.48   0.000     4928.434    6173.499
   4#Female  |    3859.85    266.131    14.50   0.000     3338.214    4381.485
------------------------------------------------------------------------------
```

The educ slope is 1,470.96 for a male who is 25 years old and is 1,174.36 for a female who is 25 years old. The educ slope is 5,550.97 for a male who is 55 years old and is 3,859.85 for a female who is 55 years old.

13.2.5 Estimating and comparing adjusted means by gender

The difference in the adjusted means by gender varies as a function of age and educ due to the interaction of gender with c.age#c.educ. Thus, any comparisons by gender should be performed by specifying a particular value of age and educ. For instance, let's begin by estimating the adjusted means for males and females who have 16 years of education and are 30 years old.

```
. margins gender, at(educ=16 age=30)

Predictive margins                              Number of obs     =     22,367
Model VCE    : Robust

Expression   : Linear prediction, predict()
at           : age             =          30
               educ            =          16

------------------------------------------------------------------------------
             |            Delta-method
             |     Margin   Std. Err.      t    P>|t|     [95% Conf. Interval]
-------------+----------------------------------------------------------------
      gender |
       Male  |   27599.71   480.2542    57.47   0.000     26658.37    28541.04
     Female  |    17493.8   332.3472    52.64   0.000     16842.37    18145.22
------------------------------------------------------------------------------
```

After adjusting for race, males with 16 years of education and 30 years of age are estimated to have an income of $27,599.71, compared with $17,493.80 for such females. By adding the `r.` contrast operator to `gender`, we see an estimate of the difference between these adjusted means (females versus males). For this combination of age and education, the difference between the adjusted means for females and for males is −10,105.91 and is significant.

```
. margins r.gender, at(educ=16 age=30) contrast(nowald pveffects)
Contrasts of predictive margins                 Number of obs     =     22,367
Model VCE    : Robust

Expression   : Linear prediction, predict()
at           : age             =          30
               educ            =          16

------------------------------------------------------------
                   |            Delta-method
                   |   Contrast   Std. Err.       t    P>|t|
-------------------+----------------------------------------
            gender |
  (Female vs Male) |  -10105.91   586.8291   -17.22   0.000
------------------------------------------------------------
```

Note! More than two levels of the categorical variable?

Had there been more than two levels of the categorical variable, you could specify any of the contrast operators illustrated in chapter 7. For example, if the categorical variable had four levels, the `a.` contrast operator would provide comparisons of each level with the next level (that is, comparing groups 1 versus 2, 2 versus 3, and 3 versus 4)

The size of the difference (as well as the significance of the difference) between males and females depends on both `educ` and `age`. You could repeat the `margins` commands above to obtain comparisons between males and females for a variety of values of `age` and `educ`. Or you can specify multiple values at once within the `margins` command. For example, the `margins` command below estimates the adjusted mean for males and females for the combinations of 12, 15, and 20 years of education and 30, 40, and 50 years of age (the output is omitted to save space).

```
. margins gender, at(educ=(12 16 20) age=(30 40 50))
  (output omitted)
```

This `margins` command adds the `r.` contrast operator, comparing females with males for these nine combinations of `educ` and `age`. This output is also omitted.

```
. margins r.gender, at(educ=(12 16 20) age=(30 40 50))
  (output omitted)
```

Now that we have explored models with categorical by linear by linear interactions, let's turn to a model that involves a categorical by linear by quadratic interaction.

13.3 Linear by quadratic by categorical interactions

In section 5.3, we included an interaction of educ with the quadratic term for age in the prediction of realrinc. In this section, we will explore whether the size of the c.educ#c.age#c.age interaction depends on gender. As with the analysis in section 5.3, this example will use the gss_ivrm.dta dataset, focusing on those who are aged 22 to 80 and who have at least 12 years of education.

```
. use gss_ivrm
. keep if (age>=22 & age<=80) & (educ>=12)
(15,934 observations deleted)
```

13.3.1 Fitting separate models for males and females

Let's begin by fitting a model that estimates the c.educ#c.age#c.age interaction in two separate models: one fit for males and another fit for females. This is performed by using the by gender, sort: prefix before the regress command that predicts realrinc from c.educ#c.age#c.age. This shortcut expands to include the interaction of all of these variables, as well as all two-way interactions and main effects. (The vsquish and noheader options are included to save space.)

```
. by gender, sort: regress realrinc c.educ##c.age##c.age, vce(robust)
> vsquish noheader

-> gender = Male

------------------------------------------------------------------------------
             |               Robust
    realrinc |      Coef.   Std. Err.      t    P>|t|     [95% Conf. Interval]
-------------+----------------------------------------------------------------
        educ |  -12420.54   1305.492    -9.51   0.000     -14979.5   -9861.579
         age |  -6160.363   876.5618    -7.03   0.000    -7878.554   -4442.172
 c.educ#c.age|    661.988   65.00427    10.18   0.000       534.57    789.4059
  c.age#c.age|   56.95601   9.962637     5.72   0.000     37.42776    76.48425
c.educ#c.age#|
       c.age |  -6.211378   .7370351    -8.43   0.000    -7.656076    -4.76668
       _cons |   131388.6   17570.69     7.48   0.000     96947.42    165829.8
------------------------------------------------------------------------------

-> gender = Female

------------------------------------------------------------------------------
             |               Robust
    realrinc |      Coef.   Std. Err.      t    P>|t|     [95% Conf. Interval]
-------------+----------------------------------------------------------------
        educ |    -6103.4   867.4264    -7.04   0.000    -7803.682   -4403.118
         age |  -3703.102   599.4866    -6.18   0.000    -4878.183   -2528.021
 c.educ#c.age|   366.2879   45.81099     8.00   0.000     276.4917    456.0841
  c.age#c.age|   35.82633   7.231316     4.95   0.000      21.6519    50.00076
c.educ#c.age#|
       c.age |  -3.573265   .5555728    -6.43   0.000    -4.662268   -2.484261
       _cons |   69959.52   11406.57     6.13   0.000     47600.98    92318.06
------------------------------------------------------------------------------
```

Let's focus on the c.educ#c.age#c.age interaction. For males, this coefficient is −6.21 and is significant. For females, this coefficient is −3.57 and is also statistically significant. This suggests that the size of this interaction might be more negative for males than for females. Before pursuing whether this difference is significant, let's first visualize the c.educ#c.age#c.age interaction by gender to gain a further understanding of what this interaction means. Figure 13.5 shows the adjusted means graphed as a function of age and educ, with age on the x axis and using separate lines for educ. The left panel shows the results for males, and the right panel shows the results for females.

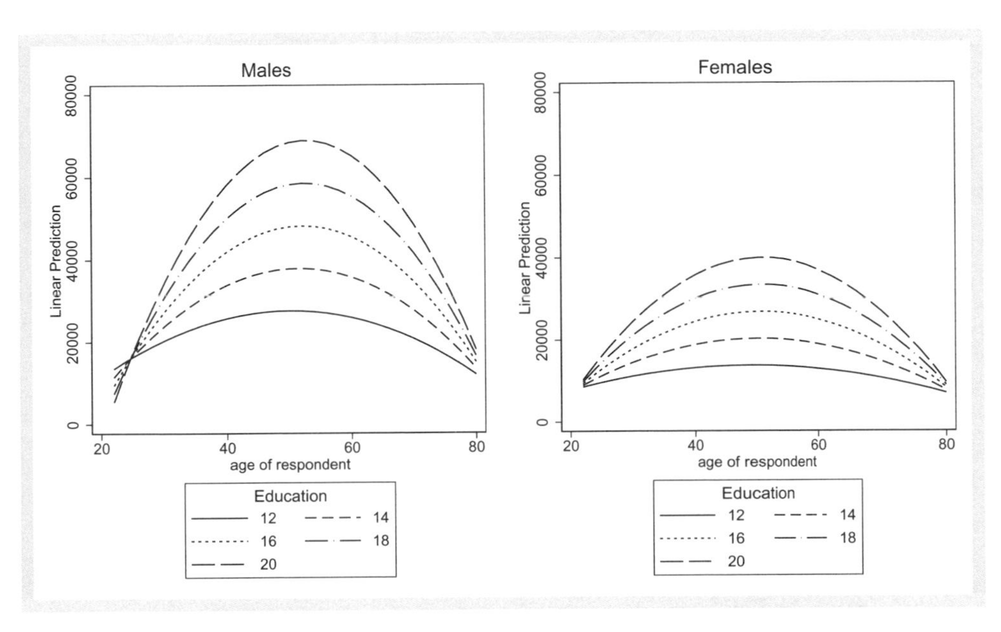

Figure 13.5. Fitted values for education by age-squared interaction for males (left) and females (right)

The coefficient for the c.educ#c.age#c.age interaction term is −6.21 for males and −3.57 for females. As we saw in section 5.3, a more negative coefficient for this interaction term reflects a greater curvature in the relationship between age and realrinc with increasing levels of educ. Because males have a more negative coefficient (−6.21), they exhibit a greater increase in the curvature in the relationship between age and income as a function of education than do females.

Let's test whether the c.educ#c.age#c.age interaction is significantly different for males versus females. To do this, we fit a combined model that includes both males and females to permit a statistical test of c.educ#c.age#c.age by gender.

13.3.2 Fitting a common model for males and females

Let's fit one model for males and females together using a separate intercept and separate slopes coding system that provides separate intercept and slope estimates for males and females. Separate intercepts by gender are obtained by specifying `ibn.gender` in conjunction with the `noconstant` option. Then, the model uses the shortcut notation to interact `ibn.gender` with each term created by `c.educ#c.age#c.age`. This model is shown below with `i.race` included as a covariate.

```
. regress realrinc ibn.gender
>                 ibn.gender#(c.educ##c.age##c.age) i.race,
>                 vce(robust) noconstant noci

Linear regression                               Number of obs     =     25,964
                                                F(14, 25950)      =    2544.35
                                                Prob > F          =     0.0000
                                                R-squared         =     0.5263
                                                Root MSE          =      25748

------------------------------------------------------------------------------
                        |               Robust
               realrinc |      Coef.   Std. Err.      t    P>|t|
------------------------+-----------------------------------------------------
                 gender |
                   Male |     133605   17548.76     7.61   0.000
                 Female |   73711.44    11412.1     6.46   0.000
                        |
          gender#c.educ |
                   Male |  -12550.64   1305.364    -9.61   0.000
                 Female |  -6329.009    867.974    -7.29   0.000
                        |
           gender#c.age |
                   Male |  -6202.103   875.0618    -7.09   0.000
                 Female |  -3809.816   598.9912    -6.36   0.000
                        |
    gender#c.educ#c.age |
                   Male |   665.2086   64.93261    10.24   0.000
                 Female |   374.2358   45.78292     8.17   0.000
                        |
     gender#c.age#c.age |
                   Male |   57.18022   9.946567     5.75   0.000
                 Female |   36.75597   7.223002     5.09   0.000
                        |
gender#c.educ#c.age#c.age |
                   Male |  -6.233645   .7361518    -8.47   0.000
                 Female |  -3.649852   .5549963    -6.58   0.000
                        |
                   race |
                  white |          0  (base)
                  black |  -3474.026   291.3494   -11.92   0.000
                  other |   -299.115   1110.196    -0.27   0.788
------------------------------------------------------------------------------
```

We can express results as separate regression equations for males and females, as shown below.

$$\begin{aligned}
\text{Male: } \widehat{\texttt{realrinc}} &= 133605 + -12550.64\texttt{educ} + -6202.10\texttt{age} \\
&+ 665.21\texttt{educ*age} + 57.18\texttt{age}^2 + -6.23\texttt{educ*age}^2 \\
&+ -3474.03\texttt{black} + -299.12\texttt{other} \\
\text{Female: } \widehat{\texttt{realrinc}} &= 73711.44 + -6329.01\texttt{educ} + -3809.82\texttt{age} \\
&+ 374.24\texttt{educ*age} + 36.76\texttt{age}^2 + -3.65\texttt{educ*age}^2 \\
&+ -3474.03\texttt{black} + -299.12\texttt{other}
\end{aligned}$$

The coefficient for the c.educ#c.age#c.age interaction is −6.23 for males and is −3.65 for females. We can compare these coefficients using the contrast command, as shown below.

```
. contrast gender#c.educ#c.age#c.age, nowald pveffects
Contrasts of marginal linear predictions
Margins      : asbalanced
```

	Contrast	Std. Err.	t	P>|t|
gender#c.educ#c.age#c.age (Female vs base)	2.583793	.9233154	2.80	0.005

This tests the interaction of gender#c.educ#c.age#c.age and is statistically significant ($p = 0.005$). Let's interpret the meaning of this interaction.

Note! Including all lower order effects

In this example, we have focused on the gender#c.educ#c.age#c.age interaction. We have included several lower order effects in the model as well, such as gender#c.educ or gender#c.age. It is important to include all of these lower order effects, significant or not, to preserve the interpretation of the gender#c.educ#c.age#c.age interaction.

13.3.3 Interpreting the interaction

To help us understand this interaction, let's visualize it by making a graph showing the c.educ#c.age#c.age interaction separately for males and females. First, the margins command is used to compute adjusted means for ages 22 to 80 (in one-year increments) and for 12 to 20 years of education (in two-year increments), separately for males and females.

```
. margins gender, at(age=(22(1)80) educ=(12(2)20))
  (output omitted)
```

Then, the `marginsplot` command is used to graph these adjusted means. The `bydimension(gender)` option is used to create the graphs separately for males and females. The `xdimension(age)` option is used to ensure that `age` is graphed on the x axis. The resulting graph is shown in figure 13.6.

```
. marginsplot, bydimension(gender) xdimension(age) plotdimension(educ, allsimp)
> noci recast(line) scheme(s1mono) legend(subtitle(Education) rows(1))
  Variables that uniquely identify margins: age educ gender
```

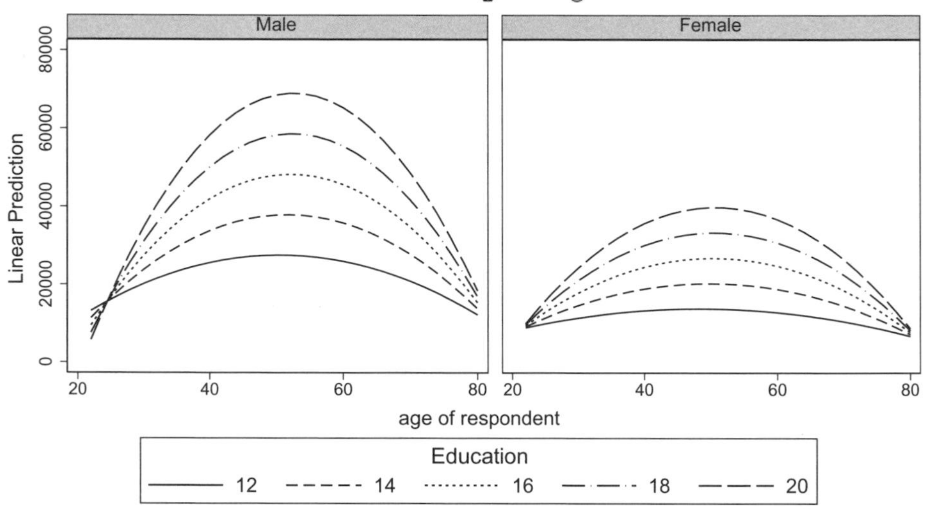

Figure 13.6. Fitted values by age (x axis), education (separate lines), and gender (separate panels)

Figure 13.6 illustrates the `gender#c.educ#c.age#c.age` interaction term. We see a similar pattern for both males and females; as `educ` increases, the inverted U-shape for the relationship between income and age becomes more pronounced. However, this effect appears stronger for males than for females. This is confirmed by the significant `gender#c.educ#c.age#c.age` interaction. In other words, this interaction shows that the degree to which the quadratic effect of `age` changes as a function of `educ` is stronger for males than it is for females.

13.3.4 Estimating and comparing adjusted means by gender

The difference in income for males and females depends on both `age` and `gender`. The `margins` command can be used to compute estimates of, and differences in, the adjusted means by gender. First, let's use the `margins` command to estimate the adjusted mean of income for those with 16 years of education and 30 years of age, separately for males and females.

```
. margins gender, at(educ=16 age=30)
Predictive margins                              Number of obs     =     25,964
Model VCE    : Robust

Expression   : Linear prediction, predict()
at           : educ            =          16
               age             =          30

------------------------------------------------------------------------------
             |            Delta-method
             |     Margin   Std. Err.      t    P>|t|     [95% Conf. Interval]
-------------+----------------------------------------------------------------
      gender |
        Male |   27288.24   430.8355    63.34   0.000     26443.78    28132.71
      Female |    17867.2   303.5242    58.87   0.000     17272.28    18462.12
------------------------------------------------------------------------------
```

After adjusting for race, males who are 30 years old with 16 years of education are estimated to have an income of $27,288.24, compared with $17,867.20 for such females. By adding the `r.` contrast operator to `gender`, we estimate the difference in these adjusted means, comparing females with males. This difference is significant.

```
. margins r.gender, at(educ=16 age=30) contrast(nowald pveffects)
Contrasts of predictive margins                 Number of obs     =     25,964
Model VCE    : Robust

Expression   : Linear prediction, predict()
at           : educ            =          16
               age             =          30

--------------------------------------------------------------
                   |            Delta-method
                   |   Contrast   Std. Err.      t    P>|t|
-------------------+------------------------------------------
            gender |
  (Female vs Male) |  -9421.044   529.9201   -17.78   0.000
--------------------------------------------------------------
```

The size and significance of the difference in the means between males and females depends on both `educ` and `age`. We can specify multiple values in the `at()` option to estimate the mean for males and females for various combinations of `age` and `educ`. The `margins` command below estimates the mean for males and females for the combinations of 12, 16, and 20 years of education and 30, 40, and 50 years of age. This yields adjusted means at nine combinations of `age` and `educ` for males and females. (The output is omitted to save space.)

```
. margins gender, at(educ=(12 16 20) age=(30 40 50))
  (output omitted)
```

The following `margins` command adds the `r.` contrast operator, comparing females with males for these nine combinations of `educ` and `age`.

```
. margins r.gender, at(educ=(12 16 20) age=(30 40 50))
  (output omitted)
```

Had there been more than two levels of the categorical variable, you could specify any of the contrast operators illustrated in chapter 7 to perform the comparisons of your choosing.

13.4 Summary

This chapter has illustrated models that involve interactions of a categorical variable with two continuous variables. Section 13.2 illustrated a categorical by linear by linear interaction using gender as the categorical variable and age and education as the linear variables. This analysis showed that males exhibit a stronger age by education interaction than do females. This was interpreted two ways, by focusing on the `age` slope and by focusing on the `educ` slope. Focusing on the `age` slope, males showed a greater increase in the `age` slope with increasing levels of education than females. Focusing on the `educ` slope, males showed a greater increase in the `educ` slope with increasing levels of age than females.

In section 13.3, we saw an example of a categorical by linear by quadratic interaction. This involved the variables gender (categorical) by education (linear) by age (quadratic). The three-way interaction of these three variables was statistically significant. Visualizing the three-way interaction showed that the quadratic effect of age increased more as a function of education for males than for females.

For more information, I recommend West, Aiken, and Krull (1996), who provide an excellent introduction to models involving the interaction of categorical and continuous variables and include an example of a continuous by continuous by categorical interaction.

14 Continuous by categorical by categorical interactions

14.1 Chapter overview

This chapter considers models that involve the interaction of two categorical predictors with a linear continuous predictor. Such models blend ideas from chapter 10 on categorical by continuous interactions and ideas from chapter 8 on categorical by categorical interactions. As we saw in chapter 10, interactions of categorical and continuous predictors describe how the slope of the continuous variable differs as a function of the categorical variable. In chapter 8, we saw models that involve the interaction of two categorical variables. This chapter blends these two modeling techniques by exploring how the slope of the continuous variable varies as a function of the interaction of the two categorical variables.

Let's consider a hypothetical example of a model with income as the outcome variable. The predictors include gender (a two-level categorical variable), education (treated as a three-level categorical variable), and age (a continuous variable). Income can be modeled as a function of each of the predictors, as well as the interactions of all the predictors. A three-way interaction of age by gender by education would imply that the effect of age interacts with gender by education. One way to visualize such an interaction would be to graph age on the x axis, with separate lines for the levels of education and separate graphs for gender. Figure 14.1 shows such an example, illustrating how the slope of the relationship between income and age varies as a function of education and gender.

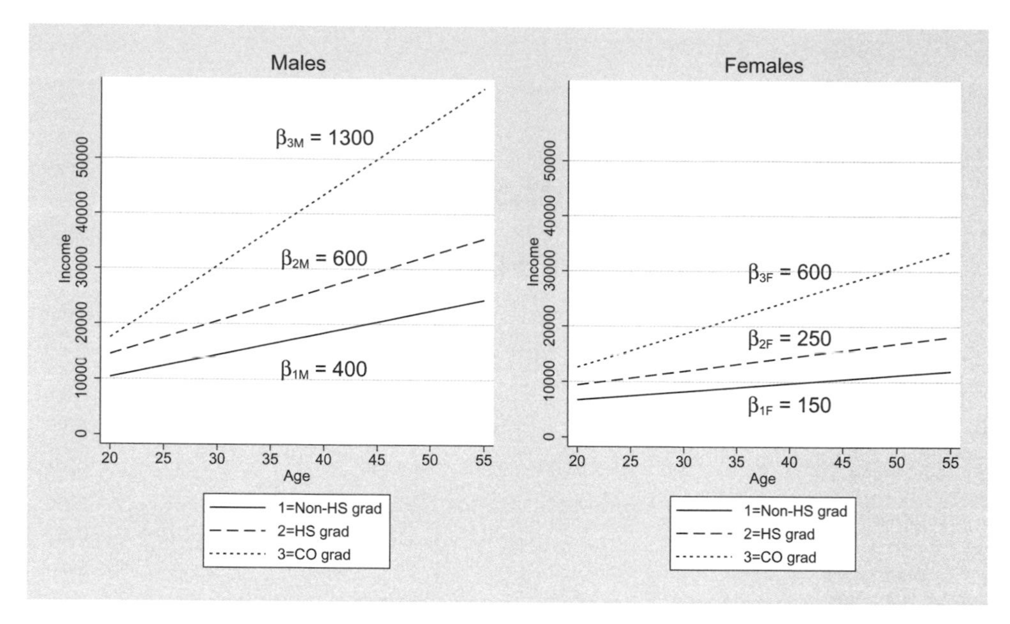

Figure 14.1. Fitted values of income as a function of age, education, and gender

The graph can be augmented by a table that shows the age slope broken down by education and gender. Such a table is shown in 14.1. The age slope shown in each cell of table 14.1 reflects the slope of the relationship between income and age for each of the lines illustrated in figure 14.1. For example, β_{3M} represents the age slope for male college graduates, and this slope is 1,300.

Table 14.1. The age slope by level of education and gender

	Non-HS grad	HS grad	CO grad
Male	$\beta_{1M} = 400$	$\beta_{2M} = 600$	$\beta_{3M} = 1{,}300$
Female	$\beta_{1F} = 150$	$\beta_{2F} = 250$	$\beta_{3F} = 600$

The age by education by gender interaction described in table 14.1 can be understood and dissected like the two by three interactions illustrated in chapter 8. The key difference is that table 14.1 is displaying the slope of the relationship between income and age, and the three-way interaction refers to the way that the slope varies as a function of education and gender.[1]

If there were no three-way interaction of age by gender by education, we would expect (for example) that the gender difference in the age slope would be approximately the

1. More precisely, how the slope varies as a function of the interaction of age and gender.

same at each level of education. But consider the differences in the age slopes between females and males at each level of education. This difference is −250 (150 − 400) for non–high school graduates, whereas this difference is −350 (250 − 600) for high school graduates, and the difference is −700 (600−1300) for college graduates. The difference in the age slopes between females and males seems to be much larger for college graduates than for high school graduates and non–high school graduates. This pattern of results appears consistent with a three-way interaction of age by education by gender.

Let's explore this in more detail with an example using the GSS dataset. To focus on the linear effect of age, we will keep those who are 22 to 55 years old.

```
. use gss_ivrm
. keep if age>=22 & age<=55
(18,936 observations deleted)
```

In this example, let's predict income as a function of gender[2] (gender), a three-level version of education (educ3), and age. The regress command below predicts realrinc from i.gender, i.educ3, and c.age (as well as all interactions of the predictors). The variable i.race is also included as a covariate.[3]

2. I use the variable gender (coded: 1 = Male and 2 = Female), because it leads to output that clearly distinguishes the variable name (that is, gender) and its values (that is, Male, and Female).
3. The noci option is included to make the output more readable for this example. See the callout titled *Using the noci option for clearer output* in section 2.5.1 for more details.

```
. regress realrinc i.gender##i.educ3##c.age i.race, vce(robust) noci
Linear regression                               Number of obs     =     25,718
                                                F(13, 25704)      =     411.30
                                                Prob > F          =     0.0000
                                                R-squared         =     0.1839
                                                Root MSE          =      23556

------------------------------------------------------------------
                     |               Robust
            realrinc |      Coef.   Std. Err.      t    P>|t|
---------------------+--------------------------------------------
              gender |
                Male |          0  (base)
              Female |   1337.125   1693.694     0.79   0.430
                     |
               educ3 |
              not hs |          0  (base)
                  HS |    550.476   1782.192     0.31   0.757
                Coll |   -11156.1   2618.976    -4.26   0.000
                     |
        gender#educ3 |
           Female#HS |   783.0991   2021.654     0.39   0.698
         Female#Coll |   7657.907   3164.299     2.42   0.016
                     |
                 age |   413.8695   45.62015     9.07   0.000
                     |
        gender#c.age |
              Female |  -264.9842   50.65695    -5.23   0.000
                     |
         educ3#c.age |
                  HS |   175.8497    54.7504     3.21   0.001
                Coll |   897.3326   77.47101    11.58   0.000
                     |
  gender#educ3#c.age |
           Female#HS |  -80.30545   60.94575    -1.32   0.188
         Female#Coll |  -414.6562   93.26714    -4.45   0.000
                     |
                race |
               white |          0  (base)
               black |  -2935.138   273.3294   -10.74   0.000
               other |   185.3956    956.338     0.19   0.846
                     |
               _cons |    2691.23   1495.778     1.80   0.072
------------------------------------------------------------------
```

Let's test the interaction of gender, education, and age using the `contrast` command below. The three-way interaction is significant.

```
. contrast i.gender#i.educ3#c.age
Contrasts of marginal linear predictions
Margins      : asbalanced

------------------------------------------------
                    |         df           F        P>F
--------------------+---------------------------
gender#educ3#c.age  |          2       10.17     0.0000
                    |
        Denominator |      25704
------------------------------------------------
```

To begin the process of interpreting the three-way interaction, let's create a graph of the adjusted means as a function of age, education, and gender. First, the `margins` command below is used to compute the adjusted means by gender and education for ages 22 and 55 (the output is omitted to save space). Then, the `marginsplot` command is used to graph the adjusted means, as shown in figure 14.2.

```
. margins gender#educ3, at(age=(22 55))
  (output omitted)
. marginsplot, bydimension(gender) noci
  Variables that uniquely identify margins: age gender educ3
```

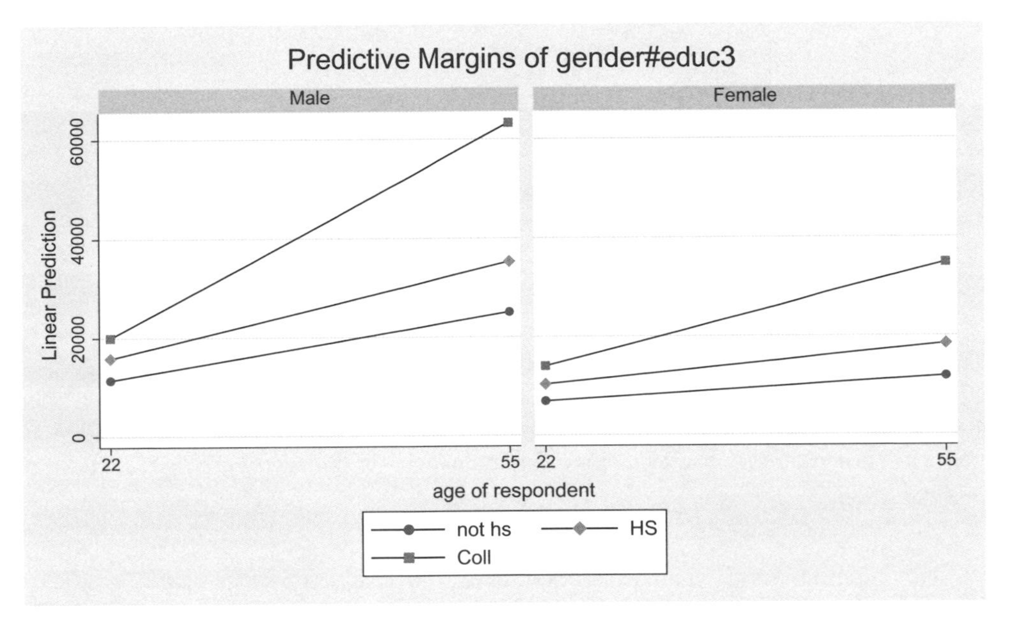

Figure 14.2. Fitted values of income as a function of age, education, and gender

The graph in figure 14.2 illustrates how the age slope varies as a function of gender and education. Let's compute the age slope for each of the lines shown in this graph. The margins command is used with the dydx(age) and over() options to compute the age slopes separately for each combination of gender and education.

```
. margins, noci dydx(age) over(gender educ3)
Average marginal effects                        Number of obs     =     25,718
Model VCE      : Robust

Expression     : Linear prediction, predict()
dy/dx w.r.t. : age
over           : gender educ3

------------------------------------------------------------------------
                |            Delta-method
                |      dy/dx   Std. Err.          t    P>|t|
----------------+-------------------------------------------------------
age             |
  gender#educ3  |
   Male#not hs  |   413.8695   45.62015        9.07    0.000
       Male#HS  |   589.7192   30.37993       19.41    0.000
     Male#Coll  |   1311.202   62.88374       20.85    0.000
 Female#not hs  |   148.8854   22.09037        6.74    0.000
     Female#HS  |   244.4296   15.25412       16.02    0.000
   Female#Coll  |   631.5618   46.90854       13.46    0.000
------------------------------------------------------------------------
```

Let's reformat the output of the margins command to emphasize how the age slope varies as a function of the interaction of gender and education (see table 14.2). Each cell of table 14.2 shows the age slope for the particular combination of gender and education. For example, the age slope for males with a college degree is 1,311.20 and is labeled as β_{3M}.

Table 14.2. The age slope by level of education and gender

	Non-HS grad	HS grad	CO grad
Male	$\beta_{1M} = 413.87$	$\beta_{2M} = 589.72$	$\beta_{3M} = 1{,}311.20$
Female	$\beta_{1F} = 148.89$	$\beta_{2F} = 244.43$	$\beta_{3F} = 631.56$

We can dissect the three-way interaction illustrated in table 14.2 using the techniques from section 8.3 on two by three models. Specifically, we can use simple effects analysis, simple contrasts, and partial interactions.

14.2 Simple effects of gender on the age slope

We can use the contrast command to test the simple effect of gender on the age slope. This is illustrated below.

```
. contrast gender#c.age@educ3, nowald pveffects
Contrasts of marginal linear predictions
Margins      : asbalanced

------------------------------------------------------------------
                         |  Contrast   Std. Err.        t    P>|t|
-------------------------+----------------------------------------
    gender@educ3#c.age   |
(Female vs base) not hs  | -264.9842   50.65695     -5.23   0.000
    (Female vs base) HS  | -345.2896   33.98931    -10.16   0.000
  (Female vs base) Coll  | -679.6404    78.4498     -8.66   0.000
------------------------------------------------------------------
```

Each of these tests represents the comparison of females versus males in terms of the age slope. The first test compares the age slope for females versus males among non–high school graduates. Referring to table 14.2, this test compares β_{1F} with β_{1M}. The difference in these age slopes is -264.98 $(148.89 - 413.87)$, and this difference is significant. The age slope for females who did not graduate high school is 264.98 units smaller than the age slope for males who did not graduate high school. The second test is similar to the first, except the comparison is made among high school graduates, comparing β_{2F} with β_{2M} from table 14.2. This test is also significant. The third test compares the age slope between females and males among college graduates (that is, comparing β_{3F} with β_{3M}). This test is also significant. In summary, the comparison of the age slope for females versus males is significant at each level of education.

14.3 Simple effects of education on the age slope

We can also look at the simple effects of education on the age slope at each level of gender. This test is performed using the contrast command below.

```
. contrast educ3#c.age@gender
Contrasts of marginal linear predictions
Margins      : asbalanced

------------------------------------------------
                    |         df           F        P>F
--------------------+---------------------------
educ3@gender#c.age  |
              Male  |          2       70.96     0.0000
            Female  |          2       43.37     0.0000
             Joint  |          4       57.21     0.0000
                    |
       Denominator  |      25704
------------------------------------------------
```

The first test compares the age slope among the three levels of education for males. Referring to table 14.2, this tests the following null hypothesis.

$$H_0\colon\ \beta_{1M} = \beta_{2M} = \beta_{3M}$$

This test is significant. The age slope significantly differs as a function of education among males.

The second test is like the first test, except that the comparisons are made for females. This tests the following null hypothesis.

$$H_0: \beta_{1F} = \beta_{2F} = \beta_{3F}$$

This test is also significant. Among females, the age slope significantly differs among the three levels of education.

14.4 Simple contrasts on education for the age slope

We can further dissect the simple effects tested above by applying contrast coefficients to the education factor. For example, say that we used the `ar.` contrast operator to form reverse adjacent group comparisons. This would yield comparisons of group 2 versus 1 (high school graduates with non–high school graduates) and group 3 versus 2 (college graduates with high school graduates). Applying this contrast operator yields simple contrasts on education at each level of gender, as shown below.

```
. contrast ar.educ3#c.age@gender, nowald pveffects
Contrasts of marginal linear predictions
Margins      : asbalanced

------------------------------------------------------------------
                        |  Contrast   Std. Err.        t    P>|t|
------------------------+-----------------------------------------
     educ3@gender#c.age |
     (HS vs not hs) Male |  175.8497    54.7504      3.21   0.001
   (HS vs not hs) Female |  95.54426   26.83611      3.56   0.000
        (Coll vs HS) Male |  721.4829   69.74939     10.34   0.000
      (Coll vs HS) Female |  387.1322   49.38976      7.84   0.000
------------------------------------------------------------------
```

The first test compares the age slope for male high school graduates with the age slope for males who did not graduate high school. In terms of table 14.2, this is the comparison of β_{2M} with β_{1M}. The difference in these age slopes is 175.85 and is significant. The second test is the same as the first test, except the comparison is made for females, comparing β_{2F} with β_{1F}. The difference is 95.54 and is significant. The third and fourth tests compare college graduates with high school graduates. The third test forms this comparison among males and is significant, and the fourth test forms this comparison among females and is also significant.

14.5 Partial interaction on education for the age slope

The three-way interaction can be dissected by forming contrasts on the three-level categorical variable. Say that we use reverse adjacent group comparisons on education, which compares high school graduates with non–high school graduates and college graduates with high school graduates. We can interact that contrast with gender and age, as shown in the `contrast` command below.

```
. contrast ar.educ3#r.gender#c.age, nowald pveffects
Contrasts of marginal linear predictions
Margins      : asbalanced
```

	Contrast	Std. Err.	t	P>|t|
educ3#gender#c.age				
(HS vs not hs) (Female vs Male)	-80.30545	60.94575	-1.32	0.188
(Coll vs HS) (Female vs Male)	-334.3507	85.48932	-3.91	0.000

The first comparison tests the interaction of the contrast of high school graduates versus non–high school graduates by gender by `age`. The difference in the `age` slope between high school graduates and non–high school graduates for females is 244.43 minus 148.89 (95.54). For males, this difference is 589.72 minus 413.87 (175.85). The difference in these differences is −80.31, which is not significant (see the first comparison from the `margins` command). The difference in the `age` slope comparing high school graduates with non–high school graduates is not significantly different for males and females.

The second test forms the same kind of comparison, but compares college graduates with high school graduates. The difference in the `age` slope comparing female college graduates with female high school graduates is 631.56 minus 244.43 (387.13). This difference for males is 1,311.20 minus 589.72 (721.48). The difference of these differences is −334.35 and is statistically significant (see the second comparison from the `margins` output). The increase in the `age` slope comparing college graduates with high school graduates is greater for males than it is for females.

14.6 Summary

This chapter has illustrated models that involve the interaction of a continuous variable with two categorical variables. When such an interaction is significant, the slope associated with the continuous variable varies as a function of the interaction of the two categorical variables. As we saw, the interaction can be dissected and understood using the methods covered in chapter 8 on categorical by categorical interactions, forming contrasts with respect to the slope term.

I am not aware of any books that directly cover this kind of interaction, so I recommend the references provided in chapter 10 (linear by categorical interactions) and chapter 8 (categorical by categorical interactions).

Part IV

Beyond ordinary linear regression

This part covers models that go beyond ordinary linear regression. These include multilevel models, longitudinal models, nonlinear models, and the analysis of complex survey data.

Chapter 15 covers multilevel models (also known as hierarchical linear models), such as models where students are nested within classrooms.

Chapters 16 and 17 cover longitudinal models that involve multiple observations over time. Chapter 16 focuses on models in which time is treated as a continuous predictor, and chapter 17 covers models where time is treated as a categorical predictor.

Chapter 18 covers nonlinear models. This includes examples illustrating binary logistic regression, multinomial logistic regression, ordinal logistic regression, and Poisson models.

Chapter 19 illustrates how to use the commands `margins`, `marginsplot`, `contrast`, and `pwcompare` with complex survey data.

15 Multilevel models

15.1 Chapter overview

Multilevel models are described using a variety of names, including hierarchical linear models, nested models, mixed models, and random-coefficient models. One of the key features of such models is the nesting of observations, for example, the nesting of students within classrooms. In such an example, variables that describe students are called level-1 variables, and variables that describe classrooms are level-2 variables.

One of the unique features of multilevel models is the ability to study cross-level interactions—the interactions of a level-1 variable with a level-2 variable. Such interactions allow you to explore the extent to which the effect of a level-1 variable is moderated by a level-2 variable. But we should not let the fact that these effects are called cross-level interactions distract us from the fact that they are interactions and can be interpreted and visualized like any interaction. In fact, my goal in writing this chapter is to show that cross-level interactions from multilevel models can be interpreted and visualized using the same techniques illustrated in chapters 5, 8, and 10. With this limited focus, the chapter is written assuming you are familiar with multilevel models and are also familiar with how to fit such models using Stata. References are provided at the end of this chapter describe resources for learning multilevel modeling as well as how to fit multilevel models using Stata.

This chapter contains four examples, all illustrating multilevel models where students are nested within schools. Each example focuses on the cross-level interaction of a student-level (level-1) variable with a school-level (level-2) variable. These four examples provide the opportunity to explore four kinds of cross-level interactions: continuous by continuous (example 1), continuous by categorical (example 2), categorical by continuous (example 3), and categorical by categorical (example 4). All of these examples are completely hypothetical and have been constructed to simplify the interpretation and visualization of the results.

15.2 Example 1: Continuous by continuous interaction

Consider a two-level multilevel model where students are nested within schools. One hundred schools were randomly sampled from a population of schools, and students were randomly sampled from each of the schools. Two student-level variables were measured: socioeconomic status (`ses`) and a standardized writing test score (`write`). Furthermore, a school-level variable was measured: the number students per computer within the school, `stucomp`.

To familiarize ourselves with this dataset, summary statistics and a listing of the first five observations are shown below.

```
. use school_write
. summarize
    Variable |        Obs        Mean    Std. Dev.       Min        Max
-------------+---------------------------------------------------------
    schoolid |      3,026    50.71481    29.01184          1        100
       stuid |      3,026    17.08824    10.81501          1         52
       write |      3,026    542.1325    191.3663          0       1200
         ses |      3,026    49.78352    10.18169    14.1897   85.06909
     stucomp |      3,026    5.857066    3.533295   1.149443   16.50701
. list in 1/5, abbreviate(30)

     +---------------------------------------------------+
     | schoolid   stuid   write        ses     stucomp   |
     |---------------------------------------------------|
  1. |        1       1     553   55.38129    3.350059   |
  2. |        1       2     530   61.13125    3.350059   |
  3. |        1       3     604   47.61407    3.350059   |
  4. |        1       4     433   48.26278    3.350059   |
  5. |        1       5     370    47.9762    3.350059   |
     +---------------------------------------------------+
```

The variable `schoolid` uniquely identifies each school, and the variable `stuid` uniquely identifies each student within each school. The variable `write` contains the score on the standardized writing test. The writing scores range from 0 to 1200. The variable `ses` is a continuous measure of socioeconomic status of the student, which ranges from 14.19 to 85.07. Finally, the variable `stucomp` is the ratio of the number of students to computers (measured at the school level) and ranges from 1.15 to 16.51.

The aim of this hypothetical study is to determine if the greater availability of computers at a school reduces the strength of the relationship between socioeconomic status and writing test scores. In other words, the goal is to determine if there is a cross-level interaction of `ses` and `stucomp` in the prediction of `write`.

The `mixed` command predicts `write` from `c.ses`, `c.stucomp`, and the interaction `c.stucomp#c.ses`. The random-effects portion of the model indicates that `ses` is a random effect across levels of `schoolid`. The `covariance(un)` (`un` is short for `unstructured`) option permits the random intercept and `ses` slope to be correlated. (In this and all subsequent examples in this chapter, the `nolog` and `noheader` options are used to save space. These options suppress the iteration log and the header information.)

```
. mixed write c.stucomp##c.ses || schoolid: ses, covariance(un) nolog noheader
```

write	Coef.	Std. Err.	z	P>|z|	[95% Conf.	Interval]
stucomp	-26.74503	2.687083	-9.95	0.000	-32.01162	-21.47845
ses	.3583319	.6772738	0.53	0.597	-.9691003	1.685764
c.stucomp# c.ses	.2357494	.1036246	2.28	0.023	.0326489	.4388499
_cons	602.2904	18.18306	33.12	0.000	566.6523	637.9286

Random-effects Parameters	Estimate	Std. Err.	[95% Conf.	Interval]
schoolid: Unstructured				
var(ses)	8.916766	1.734181	6.090615	13.0543
var(_cons)	372.3454	1229.73	.5750648	241087.8
cov(ses,_cons)	19.19714	35.82069	-51.01013	89.40441
var(Residual)	9820.887	261.1257	9322.197	10346.25

```
LR test vs. linear model: chi2(3) = 3253.82                Prob > chi2 = 0.0000
Note: LR test is conservative and provided only for reference.
```

Note! Fixed and random effects

When using multilevel modeling commands like `mixed` or other mixed-effects commands (see `help me`), the command has two portions, the fixed-effects and random-effects portions. (Note that in the context of multilevel modeling, fixed effects has a different meaning than in econometrics.) The fixed effects are specified after the dependent variable and before the `||`. The random effects are specified after the `||`. For the fixed-effects portion of the model, you can generally use what we learned in previous chapters for specifying the predictors and interpreting the output. The key difference for multilevel models is the random-effects portion. In these examples, I will not interpret or discuss the random effects.

As expected, the `c.stucomp#c.ses` interaction is significant. The prediction was that increasing availability of computers (that is, lower values of `stucomp`) would be associated with a diminished relationship between `ses` and `write`. We can use the `margins` command below with the `dydx(ses)` option to compute the `ses` slope for schools that have between one and eight students per computer.

```
. margins, dydx(ses) at(stucomp=(1(1)8)) vsquish
Average marginal effects                        Number of obs     =      3,026

Expression     : Linear prediction, fixed portion, predict()
dy/dx w.r.t. : ses
1._at          : stucomp         =           1
2._at          : stucomp         =           2
3._at          : stucomp         =           3
4._at          : stucomp         =           4
5._at          : stucomp         =           5
6._at          : stucomp         =           6
7._at          : stucomp         =           7
8._at          : stucomp         =           8

------------------------------------------------------------------------------
             |            Delta-method
             |      dy/dx   Std. Err.      z    P>|z|     [95% Conf. Interval]
-------------+----------------------------------------------------------------
ses          |
         _at |
          1  |   .5940814   .5908737     1.01   0.315    -.5640098    1.752173
          2  |   .8298308   .5109206     1.62   0.104    -.1715551    1.831217
          3  |    1.06558   .4409354     2.42   0.016     .2013628    1.929798
          4  |    1.30133   .3863733     3.37   0.001      .544052    2.058607
          5  |   1.537079   .3544302     4.34   0.000     .8424087     2.23175
          6  |   1.772829   .3513308     5.05   0.000     1.084233    2.461424
          7  |   2.008578   .3777857     5.32   0.000     1.268132    2.749024
          8  |   2.244328   .4283536     5.24   0.000      1.40477    3.083885
------------------------------------------------------------------------------
```

We can see that as the number of students per computer increases (that is, as the availability of computers decreases), the `ses` slope increases. In fact, for schools that have two (or fewer) students per computer, the `ses` slope is not significantly different from zero.

We can graphically see the `c.stucomp#c.ses` interaction using the `margins` and `marginsplot` commands below. The graph created by the `marginsplot` command shows the relationship between socioeconomic status and standardized writing test scores, with separate lines corresponding to the different levels of students per computer (see figure 15.1). This graph illustrates how the `ses` slope increases as the number of students per computer increases.

The output of the previous `margins` command shows the slope of each of the lines displayed in figure 15.1. For example, when `stucomp` is one, the `ses` slope is 0.59, and when `stucomp` is eight, the `ses` slope is 2.24.

```
. margins, at(stucomp=(1(1)8) ses=(20(5)80))
  (output omitted)
. marginsplot, noci xdimension(ses) plotdimension(stucomp, allsimple)
> legend(subtitle(Students per computer) rows(2)) recast(line) scheme(s1mono)
  Variables that uniquely identify margins: stucomp ses
```

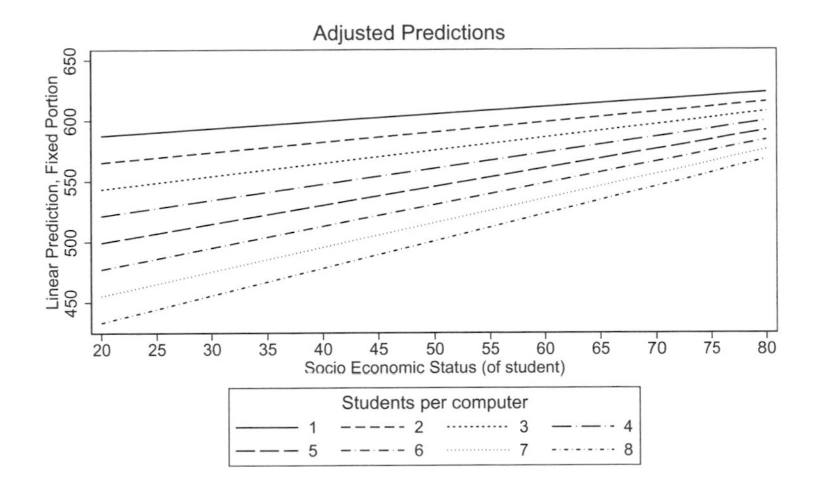

Figure 15.1. Writing score by socioeconomic status and students per computer

You can further explore the `stucomp` by `ses` interaction using the techniques illustrated in chapter 5 about continuous by continuous interactions.

15.3 Example 2: Continuous by categorical interaction

Like the first example, this example also involves students nested within schools (although not the same students or schools). One hundred schools were randomly sampled, and then students were randomly sampled within the schools. In this study, standardized reading scores were measured as well as the socioeconomic status of the student. Of the 100 schools, 35 were private (non-Catholic), 35 were public, and 30 were Catholic schools.

The summary statistics for this dataset are shown below along with a listing of the first five observations. The variable `schoolid` is the school identifier and `stuid` is the student identifier (within each school). The standardized reading scores range from 0 to 1485, with a mean of 705.33. The mean of `ses` is 49.77, and it ranges from 13.87 to 85.81. The variable `schtype` is the type of school, coded: 1 = Private, 2 = Public, and 3 = Catholic.

```
. use school_read
. summarize
```

Variable	Obs	Mean	Std. Dev.	Min	Max
schoolid	2,973	51.43592	28.77028	1	100
stuid	2,973	17.25193	11.33727	1	61
read	2,973	705.3266	214.8079	0	1485
ses	2,973	49.7656	9.845242	13.86614	85.81158
schtype	2,973	1.964346	.8021297	1	3

```
. list in 1/5, abb(30)
```

	schoolid	stuid	read	ses	schtype
1.	1	1	728	51.10166	Private
2.	1	2	696	54.53889	Private
3.	1	3	676	57.86998	Private
4.	1	4	722	46.70633	Private
5.	1	5	766	51.34241	Private

The goal of this hypothetical study is to examine the relationship between socioeconomic status (`ses`) and reading scores (`read`), and to determine if the strength of that relationship varies as a function of the type of school (private, public, or Catholic). This involves examining the cross-level interaction of `ses` and `schtype`.

The `mixed` command for performing this analysis is shown below. The variable `read` is predicted from `ses`, `schtype`, and the interaction of these two variables. The variable `ses` is specified as a random coefficient that varies across schools.

```
. mixed read i.schtype##c.ses || schoolid: ses, covariance(un) nolog noheader

        read      Coef.   Std. Err.      z    P>|z|     [95% Conf. Interval]

     schtype
     Private          0  (base)
      Public   41.96651   23.26979    1.80   0.071    -3.641439    87.57447
    Catholic   152.4775   23.78712    6.41   0.000     105.8556    199.0994

         ses   4.042064   .6734925    6.00   0.000     2.722043    5.362085

schtype#c.ses
      Public  -1.240951   .9512141   -1.30   0.192    -5.105296    .6233947
    Catholic    -3.3688   .9865231   -3.41   0.001     -5.30235   -1.435251

       _cons   519.2829   16.57457   31.33   0.000     486.7974    551.7685


  Random-effects Parameters    Estimate   Std. Err.     [95% Conf. Interval]

schoolid: Unstructured
                   var(ses)    12.13332   2.128861      8.602635    17.11307
                 var(_cons)    4.916466   73.91741      7.84e-13    3.08e+13
             cov(ses,_cons)    6.756672   31.29828     -54.58682    68.10016

              var(Residual)    10015.67   264.2554      9510.902    10547.23

LR test vs. linear model: chi2(3) = 4052.81               Prob > chi2 = 0.0000
Note: LR test is conservative and provided only for reference.
```

The `contrast` command is used below to test the overall `schtype#c.ses` interaction. This tests the following null hypothesis:

$$H_0\colon \beta_1 = \beta_2 = \beta_3$$

β_1 is the average `ses` slope for private schools, β_2 is the average `ses` slope for public schools, and β_3 is the average `ses` slope for Catholic schools. This test is significant, indicating that the `ses` slopes differ by `schtype`.

```
. contrast schtype#c.ses
Contrasts of marginal linear predictions
Margins      : asbalanced

                      df        chi2     P>chi2

read
schtype#c.ses          2       11.82     0.0027
```

Let's create a graph that illustrates the `ses` slopes by `schtype`. We do this using the `margins` command to compute the adjusted means of reading scores as a function of `ses` and `schtype`, and then graphing these adjusted means using the `marginsplot` command (see figure 15.2). This graph shows that the `ses` slope is steepest for private schools and is weakest for Catholic schools.

```
. margins schtype, at(ses=(20(5)80))
  (output omitted)
. marginsplot, noci
  Variables that uniquely identify margins: ses schtype
```

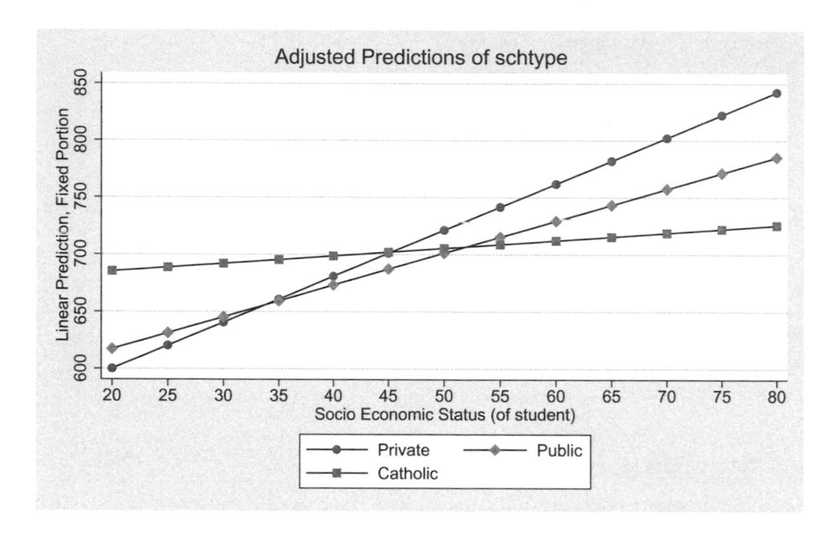

Figure 15.2. Reading score by socioeconomic status and school type

Let's now use the **margins** command combined with the **dydx(ses)** option to estimate the **ses** slope for each of the three different types of schools. The **ses** slope is 4.04 for private schools, and this slope is significantly different from 0. The **ses** slope is 2.80 for public schools, and this slope is also significantly different from 0. The **ses** slope is 0.67 for Catholic schools, but this slope is not significantly different from 0.

```
. margins, dydx(ses) over(schtype) vsquish
Average marginal effects                        Number of obs     =      2,973
Expression   : Linear prediction, fixed portion, predict()
dy/dx w.r.t. : ses
over         : schtype

------------------------------------------------------------------------------
             |            Delta-method
             |      dy/dx   Std. Err.      z    P>|z|     [95% Conf. Interval]
-------------+----------------------------------------------------------------
ses          |
     schtype |
    Private  |   4.042064   .6734925     6.00   0.000     2.722043    5.362085
     Public  |   2.801114   .6717262     4.17   0.000     1.484555    4.117673
   Catholic  |   .6732639   .7208576     0.93   0.350    -.7395911    2.086119
------------------------------------------------------------------------------
```

Let's now use the **contrast** command to form contrasts among the **ses** slopes for the three different school types. Let's compare the **ses** slope for each group with group 2 (public schools)—that is, comparing private schools with public schools; and comparing Catholic schools with public schools. These tests are computed using the `contrast` command below. Note that the `rb2.` contrast operator compares each group with group 2 (that is, public schools).

```
. contrast rb2.schtype#c.ses, pveffects nowald
Contrasts of marginal linear predictions
Margins      : asbalanced

                       |   Contrast   Std. Err.          z    P>|z|
-----------------------+-------------------------------------------
read                   |
         schtype#c.ses |
  (Private vs Public)  |   1.240951    .9512141       1.30   0.192
  (Catholic vs Public) |   -2.12785    .9853181      -2.16   0.031
```

The comparison of the `ses` slope for private versus public schools is not significant ($p = 0.192$). However, the test comparing the `ses` slope of Catholic versus public schools is significant ($p = 0.031$).

See chapter 10 for further details about interpreting categorical by continuous interactions.

15.4 Example 3: Categorical by continuous interaction

This example also involves a study of 100 randomly chosen schools, from which students were randomly sampled. A standardized math score was measured for each student (`math`). The aim of this study is to look at gender differences in math performance and to determine if smaller class sizes are associated with smaller gender differences in math scores. In other words, the aim is to determine if there is a cross-level interaction between gender (a level-1 predictor) and class size (a level-2 predictor).

The dataset for this study is named `school_math.dta`. The summary statistics for this dataset are shown below along with a listing of the first five observations.

```
. use school_math

. summarize

    Variable |        Obs        Mean    Std. Dev.       Min        Max
-------------+---------------------------------------------------------
    schoolid |      2,926    51.70813    28.54301          1        100
       stuid |      2,926    16.65584    10.73641          1         52
        math |      2,926    412.7519    104.0544         24        788
      gender |      2,926    .5075188    .5000289          0          1
      clsize |      2,926    24.41871     8.53494   9.853713   48.90636

. list in 1/5

     +----------------------------------------------+
     | schoolid   stuid   math   gender     clsize |
     |----------------------------------------------|
  1. |        1       1    417   Female   24.18711 |
  2. |        1       2    350   Female   24.18711 |
  3. |        1       3    425     Male   24.18711 |
  4. |        1       4    448   Female   24.18711 |
  5. |        1       5    375     Male   24.18711 |
     +----------------------------------------------+
```

The variable **schoolid** is the identifier for each school, and **stuid** is the identifier for each student within a school. The standardized math scores have a mean of 412.75 and range from 24 to 788. The variable **gender** is coded: 0 = Male and 1 = Female. The variable **clsize** is the average class size at the school level and ranges from 9.85 to 48.91.

The **mixed** command is used to predict **math** from **gender**, **clsize**, and the interaction of these variables. The variable **gender** is specified as a random coefficient at the school level.

```
. mixed math i.gender##c.clsize || schoolid: gender, covariance(un) nolog
> noheader
```

math	Coef.	Std. Err.	z	P>|z|	[95% Conf.	Interval]
gender						
Male	0	(base)				
Female	13.83252	12.4097	1.11	0.265	-10.49003	38.15508
clsize	-1.338499	.3223871	-4.15	0.000	-1.970366	-.7066318
gender# c.clsize						
Female	-1.10358	.4793638	-2.30	0.021	-2.043115	-.164044
_cons	451.9378	8.342056	54.18	0.000	435.5877	468.288

Random-effects Parameters	Estimate	Std. Err.	[95% Conf.	Interval]
schoolid: Unstructured				
var(gender)	244.7894	231.4543	38.36731	1561.795
var(_cons)	35.46479	96.32662	.1729147	7273.825
cov(gender,_cons)	78.9754	114.9984	-146.4174	304.3682
var(Residual)	10273.41	277.0736	9744.46	10831.07

```
LR test vs. linear model: chi2(3) = 15.08                 Prob > chi2 = 0.0018
Note: LR test is conservative and provided only for reference.
```

The `gender#c.clsize` interaction is significant ($p = 0.021$). To help interpret this effect, we can graph the results using the `margins` and `marginsplot` commands below (see figure 15.3).

```
. margins gender, at(clsize=(15(5)40))
  (output omitted)
. marginsplot, noci
  Variables that uniquely identify margins: clsize gender
```

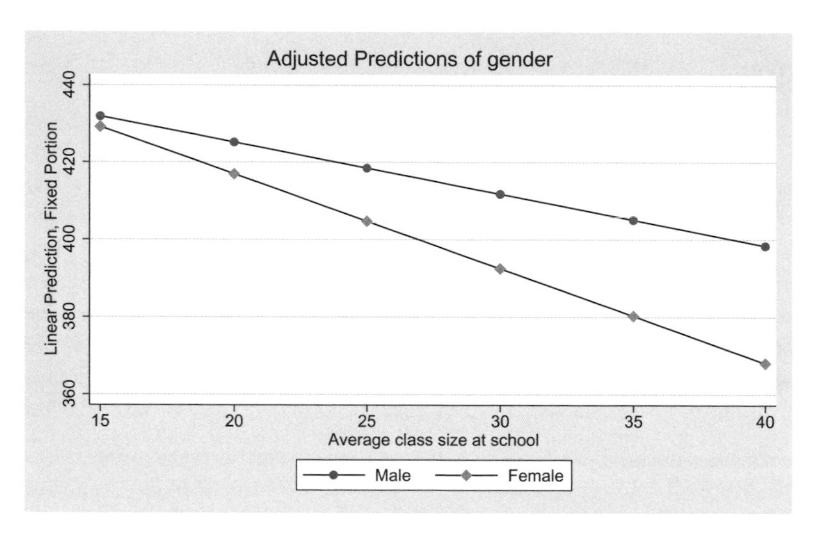

Figure 15.3. Math scores by gender and average class size

Focusing on the gender differences, we see that the gender differences increase as a function of class size. For a class size of 15, the math scores for females are close to the math scores for males. For a class size of 40, the difference is much greater. In fact, let's assess the significance of the gender difference for class sizes ranging from 15 to 40 in five-student increments using the margins command below.

```
. margins r.gender, at(clsize=(15(5)40)) contrast(pveffects nowald) vsquish
Contrasts of adjusted predictions                Number of obs     =      2,926
Expression    : Linear prediction, fixed portion, predict()
1._at         : clsize          =          15
2._at         : clsize          =          20
3._at         : clsize          =          25
4._at         : clsize          =          30
5._at         : clsize          =          35
6._at         : clsize          =          40

------------------------------------------------------------
                    |            Delta-method
                    |   Contrast   Std. Err.        z    P>|z|
--------------------+---------------------------------------
         gender@_at |
(Female vs Male) 1  |  -2.721172   6.101257    -0.45   0.656
(Female vs Male) 2  |   -8.23907   4.612941    -1.79   0.074
(Female vs Male) 3  |  -13.75697   4.101536    -3.35   0.001
(Female vs Male) 4  |  -19.27487   4.884204    -3.95   0.000
(Female vs Male) 5  |  -24.79277   6.509822    -3.81   0.000
(Female vs Male) 6  |  -30.31066   8.508207    -3.56   0.000
------------------------------------------------------------
```

The gender difference in **math** scores is not significant for schools with an average of 15 students per class ($p = 0.656$), and the gender difference remains nonsignificant for schools with an average of 20 students per class ($p = 0.074$). However, for schools with an average of 25 students (or more) per class, female math scores are significantly lower than male math scores.

The **margins** command is repeated below for 15 to 40 students per class in one-student increments (the output is omitted to save space). The **marginsplot** command is then used to visualize the gender differences (with a confidence interval) across the entire spectrum of class sizes (see figure 15.4). Where the confidence interval includes 0, the gender differences are not significant at the 5% level.

```
. margins r.gender, at(clsize=(15(1)40)) contrast(effects)
  (output omitted)
. marginsplot, recastci(rarea) ciopts(fcolor(%20)) yline(0)
  Variables that uniquely identify margins: clsize
```

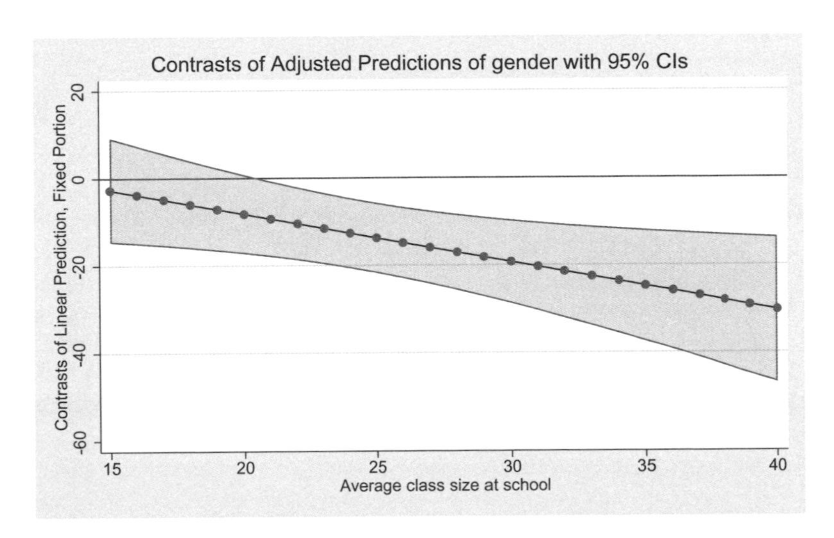

Figure 15.4. Gender difference in reading score by average class size

This example has illustrated how you can interpret a cross-level interaction that involves a categorical level-1 variable interacted with a continuous level-2 variable. Chapter 10 provides more details about the interpretation of such interactions, as well as examples that involve a three-level categorical variable interacted with a continuous variable.

15.5 Example 4: Categorical by categorical interaction

This last example also involves 100 randomly sampled schools, with students randomly sampled within the schools. This study focuses on gender differences in standardized science scores, and whether such differences vary by school size. In this study, each school is classified into one of three sizes: small, medium, or large. Thus, the focus of this study is on the cross-level interaction of gender by school size, where both gender and school size are categorical variables.

The dataset for this example is used below. The summary statistics are shown for the variables in the dataset, along with a listing of the first five observations.

```
. use school_science

. summarize

    Variable |        Obs        Mean    Std. Dev.       Min        Max
-------------+---------------------------------------------------------
    schoolid |      2,764    45.80246     26.6763          1         90
       stuid |      2,764    17.19247     10.8986          1         54
     science |      2,764    409.1914    102.4961         88        779
      gender |      2,764    .4916787    .5000212          0          1
     schsize |      2,764    2.017004    .8279852          1          3

. list in 1/5

     +-----------------------------------------------------+
     | schoolid   stuid   science   gender   schsize |
     |-----------------------------------------------------|
  1. |        1       1       312     Male     Small |
  2. |        1       2       556   Female     Small |
  3. |        1       3       372     Male     Small |
  4. |        1       4       333     Male     Small |
  5. |        1       5       408     Male     Small |
     +-----------------------------------------------------+
```

As with the previous examples, the variable `schid` identifies schools and the variable `stuid` identifies students within the schools. The standardized science score is called `science`, which has a mean of 409.19 and ranges from 88 to 779. The variable `gender` is coded: 0 = Male and 1 = Female. The variable `schsize` is a three-level categorical variable describing the size of the school and is coded: 1 = Small, 2 = Medium, and 3 = Large.

The `mixed` command below is used to predict `science` from `gender`, `schsize`, and `gender#schsize`. This last term represents the cross-level interaction of gender by school size. The variable `gender` is included as a random effect.

```
. mixed science i.gender##i.schsize || schoolid: gender, covariance(un)
> nolog noheader
```

science	Coef.	Std. Err.	z	P>|z|	[95% Conf.	Interval]
gender						
Male	0	(base)				
Female	-19.64777	6.64678	-2.96	0.003	-32.67522	-6.62032
schsize						
Small	0	(base)				
Medium	18.56062	7.360262	2.52	0.012	4.134774	32.98647
Large	40.16295	7.196425	5.58	0.000	26.05821	54.26768
gender# schsize						
Female # Medium	1.469545	9.540647	0.15	0.878	-17.22978	20.16887
Female#Large	22.24119	9.291424	2.39	0.017	4.03033	40.45204
_cons	394.9727	5.136907	76.89	0.000	384.9045	405.0408

Random-effects Parameters	Estimate	Std. Err.	[95% Conf.	Interval]
schoolid: Unstructured				
var(gender)	38.51799	205.2415	.0011222	1322074
var(_cons)	155.777	115.8815	36.24936	669.432
cov(gender,_cons)	17.51639	128.5056	-234.35	269.3828
var(Residual)	9772.298	271.3318	9254.708	10318.83

```
LR test vs. linear model: chi2(3) = 12.15                 Prob > chi2 = 0.0069
Note: LR test is conservative and provided only for reference.
```

The **contrast** command is used to test the overall **gender#schsize** interaction. This test is significant.

```
. contrast gender#schsize
Contrasts of marginal linear predictions
Margins      : asbalanced
```

	df	chi2	P>chi2
science			
gender#schsize	2	7.17	0.0278

To help interpret this interaction, we can use the **margins** command to display the adjusted mean of science scores as a function of **gender** and **schsize**, as shown below.

```
. margins schsize#gender, nopvalues vsquish
Adjusted predictions                              Number of obs     =      2,764
Expression   : Linear prediction, fixed portion, predict()

------------------------------------------------------------------------
                  |            Delta-method
                  |     Margin   Std. Err.     [95% Conf. Interval]
------------------+-----------------------------------------------------
   schsize#gender |
       Small#Male |   394.9727   5.136907      384.9045    405.0408
     Small#Female |   375.3249   5.462731      364.6182    386.0317
      Medium#Male |   413.5333    5.27121      403.2019    423.8647
    Medium#Female |   395.3551   5.562012      384.4537    406.2564
       Large#Male |   435.1356   5.039913      425.2576    445.0137
     Large#Female |   437.7291   5.357067      427.2294    448.2287
------------------------------------------------------------------------
```

We can create a graph of these adjusted means using the `marginsplot` command. This creates the graph shown in figure 15.5.

```
. marginsplot, noci
  Variables that uniquely identify margins: schsize gender
```

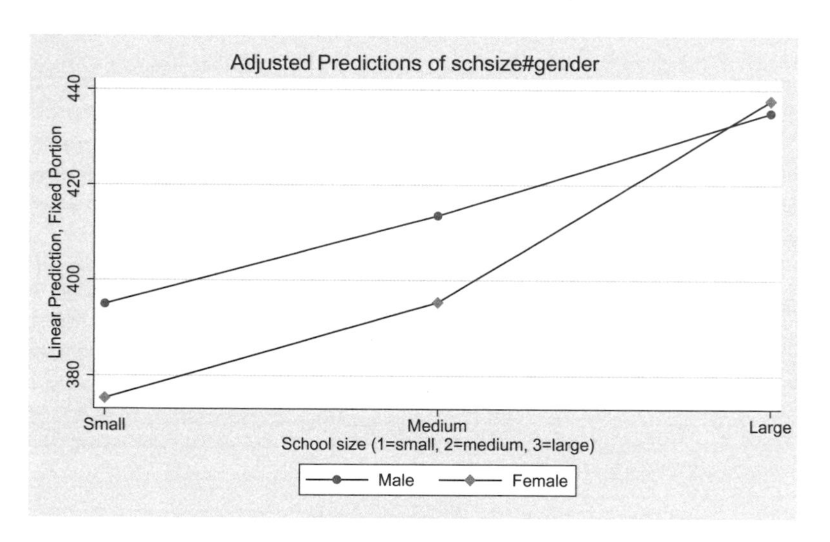

Figure 15.5. Science scores by gender and school size

One way to further understand this interaction is by testing the simple effect of `gender` for each school size. This is performed using the `contrast` command, as shown below. The adjusted mean of science scores for females is significantly lower for females in small schools ($p = 0.003$) and in medium schools ($p = 0.008$). This difference is not significant for females in large schools ($p = 0.690$).

```
. contrast gender@schsize, nowald pveffects
Contrasts of marginal linear predictions
Margins      : asbalanced

                           Contrast   Std. Err.        z    P>|z|

science
         gender@schsize
 (Female vs base) Small   -19.64777     6.64678    -2.96    0.003
(Female vs base) Medium   -18.17822    6.844287    -2.66    0.008
 (Female vs base) Large    2.593417     6.49237     0.40    0.690
```

We can also further probe the interaction by using partial interaction tests. Let's apply the `ar.` contrast operator to school size (comparing each group with the previous group) and interact that with gender. These partial interactions are tested using the `contrast` command, as shown below.

```
. contrast gender#ar.schsize, nowald pveffects
Contrasts of marginal linear predictions
Margins      : asbalanced

                                       Contrast   Std. Err.        z    P>|z|

science
                         gender#schsize
(Female vs base) (Medium vs Small)     1.469545    9.540647     0.15    0.878
(Female vs base) (Large vs Medium)     20.77164    9.433723     2.20    0.028
```

The interaction of the comparison of medium schools with small schools by gender is not significant ($p = 0.878$). This is illustrated in figure 15.5 where we can see that the difference in the adjusted means for medium versus small schools is roughly the same for males and females.

The interaction of the comparison of large schools with medium schools by gender is significant ($p = 0.028$). In figure 15.5, we can see how the difference in adjusted means for large schools versus medium schools for females is substantially greater than the corresponding difference for males. In fact, the output of the `contrast` command tells us that the female science scores increase by 20.77 points more than male science scores when comparing large with medium schools.

This section has illustrated some of the ways that you can interpret a cross-level interaction involving two categorical variables. See chapter 8 for more information about how to dissect and understand interactions involving two categorical variables.

15.6 Summary

This chapter has shown four examples illustrating how to interpret cross-level interactions from a multilevel model. Although cross-level interactions involve a combination of level-1 and level-2 variables, they can be visualized and interpreted using the same techniques illustrated in chapters 5, 8, and 10.

I recommend that you consult your favorite book about multilevel modeling for more background about this modeling technique. If you are new to multilevel modeling, you might prefer an introductory book rather than some of the advanced books. For an introduction to multilevel modeling, I recommend Kreft and De Leeuw (1998), Raudenbush and Bryk (2002), and Snijders and Bosker (1999). Rabe-Hesketh and Skrondal (2012) provide advanced details about multilevel modeling with examples using Stata.

16 Time as a continuous predictor

16.1 Chapter overview

This chapter considers models involving the analysis of longitudinal data. Such designs involve participants that are observed at more than one time point and time is generally treated as one of the important predictors in the model. Like any predictor, we need to ask ourselves how we want to model the relationship between time and the outcome. A key distinction is whether time will be treated as a continuous variable or as a categorical variable. This chapter includes four examples of modeling time where time is treated as a continuous variable, and chapter 17 shows examples where time is treated as a categorical variable.

The four examples in this chapter illustrate modeling of time as a linear predictor (see section 16.2), modeling an interaction of the linear effect of time with a categorical predictor (see section 16.3), modeling time using a piecewise model (see section 16.4), and modeling the piecewise effects of time interacted with a categorical predictor (see section 16.5).

There are several approaches that can be used for modeling longitudinal data, including repeated-measures analysis of variance (ANOVA), generalized estimating equations (GEE), and multilevel modeling. This chapter will focus on using multilevel modeling for analyzing longitudinal models where time is treated as level 1 and the person will be treated as level 2. In such a model, characteristics that change as a function of time are level-1 predictors and characteristics that are a property of the person that do not change over time are level-2 predictors.

This chapter assumes that you are already familiar with multilevel modeling as well as application of such models for the analysis of longitudinal data. (To gain such knowledge, I would highly recommend Singer and Willett [2003].) This chapter aims to build upon that knowledge by illustrating how to fit such models using Stata. Furthermore, the chapter aims to relate the fitting of longitudinal models to techniques illustrated in previous chapters (especially chapters 10 and 12).

16.2 Example 1: Linear effect of time

Let's begin by considering a simple model in which we look at the linear effect of time on the outcome variable. For example, let's consider a study in which we are looking at the number of minutes people sleep at night over a seven-week period. For the sake of this example, assume that the people were selected for the study because they have recently experienced a stressful event in their life, and the purpose of the study is to understand the natural course of sleep change over the seven weeks after a stressful event. In this hypothetical example, 75 people were enrolled and each person's nightly sleep time (in minutes) was recorded on eight occasions; approximately once every seven days.

The dataset for this study is organized in a long format, with one observation per person per night of observation. There are 75 people who were each observed eight times; thus the dataset has 600 observations. Let's look at the first five observations in the dataset.

```
. use sleep_conlin
. list in 1/5

     ┌─────────────────────┐
     │ id   obsday   sleep │
     ├─────────────────────┤
  1. │  1        1     382 │
  2. │  1        6     382 │
  3. │  1       13     390 │
  4. │  1       21     378 │
  5. │  1       27     401 │
     └─────────────────────┘
```

The day of observation is stored in the variable `obsday`. The first person slept 382 minutes on day 1. When observed on day 6, the person again slept 382 minutes. When observed on day 13, the person slept 390 minutes.

Let's now look at the summary statistics for the variables in this dataset.

```
. summarize

    Variable │       Obs        Mean    Std. Dev.       Min        Max
─────────────┼────────────────────────────────────────────────────────
          id │       600          38    21.66677          1         75
      obsday │       600      23.565    14.82854          1         52
       sleep │       600     360.785    48.13086        175        528
```

The variable `id` identifies each person and ranges from 1 to 75, representing the 75 people in this study. The variable `obsday` indicates the day of observation and ranges from 1 to 52. The variable `sleep` represents the number of minutes the person slept on a particular night. The average of this variable is 360.79, with a minimum of 175 and a maximum of 528.

If there were no repeated observations in the dataset (that is, the residuals were independent), we could examine the linear relationship between time (`obsday`) and the number of minutes of sleep (`sleep`) using the linear regression methods described in chapter 2. Instead, we can fit a random-intercept model that accounts for the nonindependence of the residuals within each person. Such a model is fit below first using the `xtset` command to specify that `id` is the panel variable and `obsday` is the time variable. We can then use the `xtreg` command to fit a random-intercept model predicting `sleep` from `obsday`.

```
. xtset id obsday
       panel variable:  id (weakly balanced)
        time variable:  obsday, 1 to 52, but with gaps
                delta:  1 unit

. xtreg sleep obsday
Random-effects GLS regression                   Number of obs      =       600
Group variable: id                              Number of groups   =        75

R-sq:                                           Obs per group:
     within  = 0.1652                                          min =         8
     between = 0.0062                                          avg =       8.0
     overall = 0.0218                                          max =         8

                                                Wald chi2(1)       =    103.56
corr(u_i, X)   = 0 (assumed)                    Prob > chi2        =    0.0000

------------------------------------------------------------------------------
       sleep |      Coef.   Std. Err.      z    P>|z|     [95% Conf. Interval]
-------------+----------------------------------------------------------------
      obsday |   .5086482   .0499837    10.18   0.000     .4106821    .6066144
       _cons |   348.7987   5.311844    65.66   0.000     338.3877    359.2097
-------------+----------------------------------------------------------------
     sigma_u |  44.412437
     sigma_e |  18.051717
         rho |  .85821678   (fraction of variance due to u_i)
------------------------------------------------------------------------------
```

The interpretation of the `obsday` coefficient is straightforward. For each additional day in the study, nightly minutes of sleep increased by, on average, by 0.51 minutes. Multiplying this by seven yields perhaps a simpler interpretation. For each additional week in the study, participants slept, on average, an additional 3.6 minutes per night. This coefficient describes the trajectory of sleep durations across the days of the study.

Terminology! Slopes

In this chapter, I will refer to the `obsday` coefficient in two different ways. For example, I will call it the `obsday` slope, indicating that it is the slope associated with the day of observation. Sometimes, for simplicity, I will refer to this as the *slope*.

In this model, the coefficient for obsday is treated as a fixed effect. The model fits one fixed trajectory of sleep durations across days. The model recognizes that people randomly vary in terms of their average sleep time at the start of the study (represented by the random intercept). But perhaps people also vary individually in their trajectory of sleep times across the weeks of the study.[1] By adding obsday as a random coefficient (that is, a random slope), the model can account for both individual differences in the average length of sleep at the start of the study (that is, a random intercept) and individual differences in the trajectory of sleep times across the weeks of the study (that is, a random slope for obsday). We can fit such a model using the mixed command, shown below. (Note that the nolog option is used to suppress the iteration log to save space.)

```
. mixed sleep obsday || id: obsday, covariance(un) nolog
Mixed-effects ML regression                     Number of obs      =       600
Group variable: id                              Number of groups   =        75
                                                Obs per group:
                                                              min =         8
                                                              avg =       8.0
                                                              max =         8
                                                Wald chi2(1)       =     20.54
Log likelihood = -2488.5378                     Prob > chi2        =    0.0000

------------------------------------------------------------------------------
       sleep |      Coef.   Std. Err.      z    P>|z|     [95% Conf. Interval]
-------------+----------------------------------------------------------------
      obsday |   .5104496   .1126409     4.53   0.000     .2896774    .7312217
       _cons |   348.8205   3.308517   105.43   0.000     342.3359     355.305
------------------------------------------------------------------------------

------------------------------------------------------------------------------
  Random-effects Parameters  |   Estimate   Std. Err.     [95% Conf. Interval]
-----------------------------+------------------------------------------------
id: Unstructured             |
                 var(obsday) |   .8926483   .1552909      .6347458    1.255339
                  var(_cons) |   775.8412    134.126      552.8646    1088.747
           cov(obsday,_cons) |   14.29639   3.556285      7.326203    21.26658
-----------------------------+------------------------------------------------
               var(Residual) |   101.2256   6.745405      88.83181    115.3485
------------------------------------------------------------------------------
LR test vs. linear model: chi2(3) = 1360.11               Prob > chi2 = 0.0000
Note: LR test is conservative and provided only for reference.
```

The slope of the relationship between sleep duration and observation day is significant ($z = 4.53$, $p = 0.000$). For each additional day in the study, the participants slept an additional 0.51 minutes. (Note that this estimate is not much different from the coefficient estimated by the xtreg command.)

This model added a random effect for obsday, which we see reflected in the estimate of var(obsday) in the random-effects portion of the output. This represents the degree to which the sleep trajectories vary between individuals. As described in Singer and

1. Some people might show less sleep over time because of a lingering or increasing impact of a stressor, while others may experience a diminished impact of a stressor and more sleep over time.

Willett (2003), we can form a simple one-sided z test to determine if this value is significantly greater than 0. We can compute a z statistic by dividing the estimate (0.8926) by its standard error (0.1553), which yields a z-value of 5.75. This value well exceeds 1.645 (the 5% cutoff for a one-sided z test), suggesting that there is significant variation in the sleep trajectories among individuals.[2]

We might be interested in estimating the predicted mean number of minutes slept at night across the weeks of the study. Let's estimate this for the first 50 days of observation in 7-day increments. These predicted means are computed using the **margins** command below.

```
. margins, nopvalues at(obsday=(1(7)50)) vsquish
Adjusted predictions                            Number of obs     =        600
Expression   : Linear prediction, fixed portion, predict()
1._at        : obsday          =           1
2._at        : obsday          =           8
3._at        : obsday          =          15
4._at        : obsday          =          22
5._at        : obsday          =          29
6._at        : obsday          =          36
7._at        : obsday          =          43
8._at        : obsday          =          50

             |            Delta-method
             |     Margin   Std. Err.     [95% Conf. Interval]
-------------+------------------------------------------------
         _at |
          1  |   349.3309   3.362049      342.7414    355.9204
          2  |   352.9041   3.809664      345.4373    360.3709
          3  |   356.4772   4.355127      347.9413    365.0131
          4  |   360.0504   4.966301      350.3166    369.7841
          5  |   363.6235   5.621795       352.605     374.642
          6  |   367.1967   6.307808      354.8336    379.5597
          7  |   370.7698   7.015392      357.0199    384.5197
          8  |    374.343   7.738632      359.1755    389.5104
```

The predicted mean of sleep for the first day of the study (**obsday** = 1) is 349.33 minutes. The 95% confidence interval for this estimate is [342.74, 355.92]. On the 50th day of the study, the predicted mean of sleep is 374.34 minutes. We can graph the predicted mean sleep duration by the day of observation using the **marginsplot** command (shown below). The graph, shown in figure 16.1, also includes a 95% confidence interval for the predicted mean for the days specified in the **at()** option of the **margins** command.

2. Singer and Willett (2003) note that although this is a simple test to perform, there is disagreement over its appropriateness and it is probably best treated as a quick, but imprecise, test. They provide a useful discussion of the concerns surrounding this test and provide an alternative test that may be more preferable. The most appropriate method for performing such tests appears to be an unsettled question and the topic of continued research.

```
. marginsplot
Variables that uniquely identify margins: obsday
```

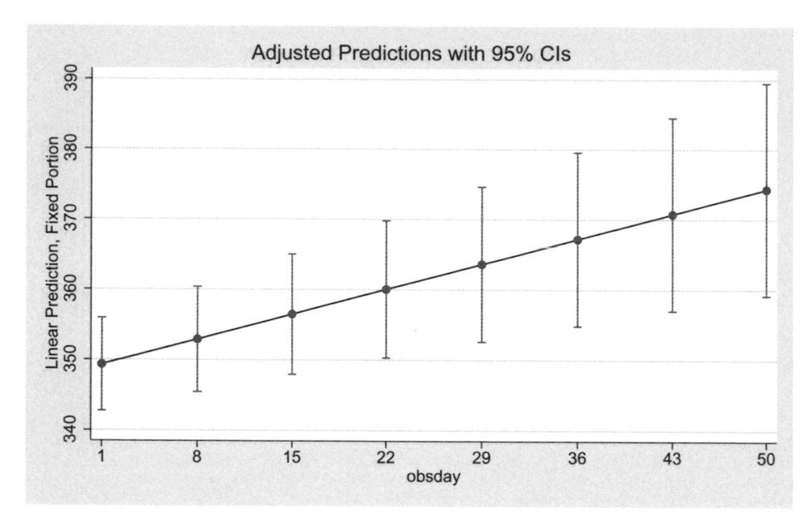

Figure 16.1. Minutes of sleep at night by time

This model could have included covariates at the person level (for example, gender or education) and could have included covariates at the time level (for example, day of week or number of hours worked before going to bed). The `margins` command would have estimated the effect of the day of observation after holding the covariates constant, and you could have used the `at()` option to hold the covariates constant at a specified level.

Let's next consider an example that builds upon this model, adding a categorical predictor.

16.3 Example 2: Linear effect of time by a categorical predictor

Let's consider another sleep study in which participants diagnosed with insomnia were randomly assigned to one of three different treatments to increase the number of minutes of sleep at night. The three different treatments were 1) control group (no treatment), 2) medication group (where a sleep medication is given), or 3) education group (where the participants receive education about how to sleep better and longer).

Like the previous example, we can study the sleep trajectories of the participants across the weeks of the study. This study will explore whether the sleep trajectories vary as a function of treatment group assignment. For example, the average sleep trajectory for the medication group might be steeper than the sleep trajectory for the control group.

This model involves a combination of a continuous predictor (time) and a three-level categorical predictor (treatment group). This is akin to the model we saw in section 10.4 in which age was treated as a continuous predictor and the three-level grouping of education was treated as a categorical variable. Although we will use the `mixed` command instead of the `regress` command, the logic, interpretation, and postestimation commands for understanding the results of the model are much the same as we saw in section 10.4. (The rest of this section may make more sense if you revisit section 10.4.)

Let's begin by using the dataset for this example and listing the first five observations and showing summary statistics for the variables in the dataset.

```
. use sleep_cat3conlin, clear
. list in 1/5

     ┌──────────────────────────────────┐
     │ id     group   obsday   sleep   │
     ├──────────────────────────────────┤
  1. │  1   Control        1     370   │
  2. │  1   Control        7     382   │
  3. │  1   Control       13     377   │
  4. │  1   Control       18     408   │
  5. │  1   Control       24     385   │
     └──────────────────────────────────┘

. summarize
    Variable │        Obs        Mean    Std. Dev.       Min        Max
─────────────┼─────────────────────────────────────────────────────────
          id │        600          38    21.66677          1         75
       group │        600           2    .8171778          1          3
      obsday │        600    20.18833    12.76994          1         46
       sleep │        600    358.5433    52.49169        225        504
```

We can see that the variable `id` identifies the person. The variable `group` reflects the person's treatment group assignment, coded: 1 = control, 2 = medication, and 3 = education. The variable `obsday` is the day in which the person's sleep was observed. This represents time in terms of the number of days from the beginning of the study. The variable `sleep` is the number of minutes they slept for the particular day of observation (that is, the sleep duration). Note that the treatment group assignment is a property of the person, making it a level-2 variable, and observation day is a level-1 variable.

We can fit a model that predicts sleep from the observation day, the group assignment, and the interaction of these two variables using the `mixed` command shown below. Note that the random-effects portion of the model specifies that `obsday` is a random effect. [Thinking in terms of a multilevel model, `group#obsday` is a cross-level interaction of a level-2 variable (`group`) with a level-1 variable (`obsday`).]

```
. mixed sleep i.group##c.obsday || id: obsday, covariance(un) nolog
Mixed-effects ML regression                     Number of obs     =        600
Group variable: id                              Number of groups  =         75

                                                Obs per group:
                                                              min =          8
                                                              avg =        8.0
                                                              max =          8

                                                Wald chi2(5)      =      66.55
Log likelihood = -2482.6017                     Prob > chi2       =     0.0000

------------------------------------------------------------------------------
       sleep |      Coef.   Std. Err.      z    P>|z|     [95% Conf. Interval]
-------------+----------------------------------------------------------------
       group |
     Control |          0  (base)
  Medication |   35.48106   9.498675     3.74   0.000     16.86399    54.09812
   Education |   5.552734   9.499246     0.58   0.559    -13.06545    24.17092
             |
      obsday |  -.0657762   .1798984    -0.37   0.715    -.4183707    .2868183
             |
      group# |
    c.obsday |
  Medication |   .1815398   .2539227     0.71   0.475    -.3161395    .6792191
   Education |   .8307836   .2541045     3.27   0.001      .332748    1.328819
             |
       _cons |   339.4458   6.717216    50.53   0.000     326.2803    352.6113
------------------------------------------------------------------------------

------------------------------------------------------------------------------
  Random-effects Parameters  |   Estimate   Std. Err.     [95% Conf. Interval]
-----------------------------+------------------------------------------------
id: Unstructured             |
                 var(obsday) |   .7192364   .1319566          .502    1.03048
                  var(_cons) |   1079.505   184.1759      772.6809   1508.167
          cov(obsday,_cons)  |   22.51981   4.232774      14.22373   30.81589
-----------------------------+------------------------------------------------
               var(Residual) |   108.3664   7.218497      95.10302   123.4795
------------------------------------------------------------------------------
LR test vs. linear model: chi2(3) = 1425.42               Prob > chi2 = 0.0000
Note: LR test is conservative and provided only for reference.
```

We can interpret the results of this model using the same tools and techniques illustrated in section 10.4. Let's begin by making a graph of the predicted means as a function of time and treatment group. We first use the `margins` command to estimate the predicted means by `group` at selected values of `obsday`. We follow that with the `marginsplot` command to graph the predicted means computed by the `margins` command. Figure 16.2 shows the graph of the predicted means produced by the `marginsplot` command.

```
. margins group, at(obsday=(0 45))
  (output omitted)
. marginsplot, noci
  Variables that uniquely identify margins: obsday group
```

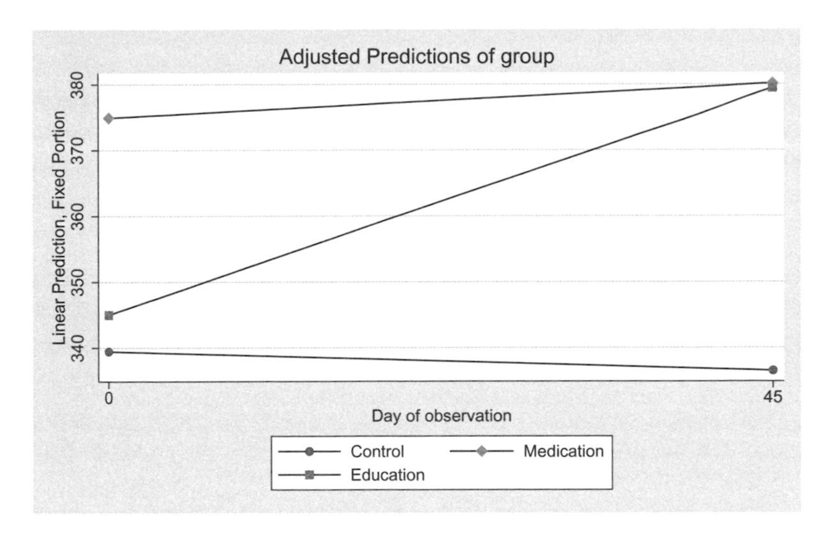

Figure 16.2. Minutes of sleep at night by time and treatment group

Let's focus on the slope of the relationship between sleep durations and day of observation. The slope appears to be slightly negative for the control group. In other words, their sleep durations appear to mildly decrease as a linear function of the observation day. For the medication group, the slope appears to be slightly positive. By contrast, sleep durations increase as a linear function of time for the education group. In other words, the `obsday` slope is positive for the education group.

Let's use the `margins` command to estimate the three slopes depicted in figure 16.2. This is computed using the `margins` command, including the `dydx(obsday)` option to estimate the `obsday` slope.

```
. margins, dydx(obsday) over(group) vsquish
Average marginal effects                        Number of obs     =        600
Expression   : Linear prediction, fixed portion, predict()
dy/dx w.r.t. : obsday
over         : group
```

	dy/dx	Delta-method Std. Err.	z	P>|z|	[95% Conf.	Interval]
obsday						
group						
Control	-.0657762	.1798984	-0.37	0.715	-.4183707	.2868183
Medication	.1157636	.1792018	0.65	0.518	-.2354655	.4669927
Education	.7650074	.1794593	4.26	0.000	.4132737	1.116741

The obsday slope for the control group (group 1) depicted in figure 16.2 is −0.07 but is not significantly different from 0. The slope for the medication group (group 2) is 0.12 and is also not significantly different from 0. For the education group (group 3), the obsday slope is 0.77 and is significantly different from 0. This is consistent with the positive slope for the education group illustrated in figure 16.2. For the education group, sleep durations increased (on average) by 0.77 minutes for every additional day.

Let's now form comparisons of these slopes, comparing the slope for each group against group 1 (the control group). The r. contrast operator is applied to group to form reference group comparisons.

```
. margins r.group, dydx(obsday) contrast(pveffects nowald)
Contrasts of average marginal effects          Number of obs     =        600
Expression   : Linear prediction, fixed portion, predict()
dy/dx w.r.t. : obsday

------------------------------------------------------------------
                          |   Contrast Delta-method
                          |      dy/dx   Std. Err.      z    P>|z|
--------------------------+---------------------------------------
obsday                    |
                    group |
(Medication vs Control)   |   .1815398   .2539227    0.71   0.475
 (Education vs Control)   |   .8307836   .2541045    3.27   0.001
------------------------------------------------------------------
```

The first test compares the obsday slope for the medication group with the control group. The difference in these slopes is 0.18 and is not significant. This is consistent with what we saw in figure 16.2 in which the slopes for the medication group and control group were not very different.

The second test compares the slope for the education group with the slope for the control group. The difference in these slopes is 0.83 and is significant. The slope for the education group is significantly steeper than the slope for the control group (see figure 16.2). For every additional day in the study, the education group sleeps an additional 0.83 more minutes than the control group.

Say that we wanted to compare the predicted mean of sleep duration for each group with the control group across all days of observation. The margins command can be combined with the marginsplot command to create a graph that compares each group with the control group at each of the days of observation (1 to 45 days). The graph created by the marginsplot command, shown in figure 16.3, shows the medication group compared with the control group in the left panel and the comparison of the education group with the control group in the right panel. The confidence interval for each comparison is shown with a reference line at zero. Where the confidence interval excludes zero, the difference is significant at the 5% level.

```
. margins r.group, at(obsday=(1(1)45))
  (output omitted)
. marginsplot, bydimension(group) yline(0) xlabel(1(7)45)
> recast(line) recastci(rarea) ciopts(fcolor(black%20))
  Variables that uniquely identify margins: obsday group
```

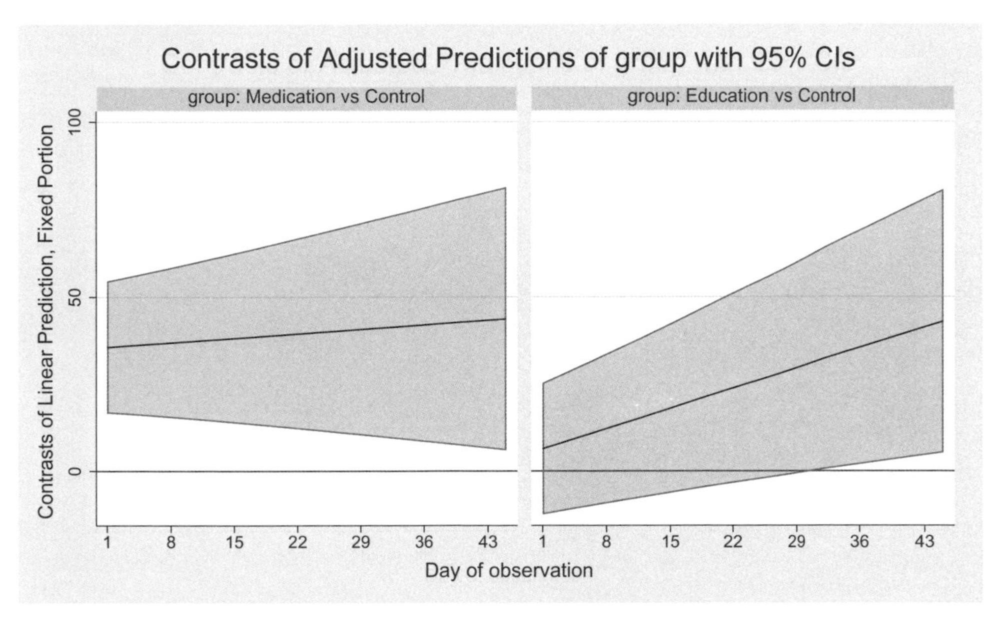

Figure 16.3. Minutes of sleep at night by time and treatment group

Looking at the left panel of figure 16.3, the confidence intervals always exclude zero. This indicates that for every day of observation (that is, from 1 to 45 days), we are 95% confident that the mean number of minutes slept by the medication group is greater than that of the control group. In the right panel, comparing the education group with the control group, the confidence intervals do not exclude zero from day 1 to day 31. Beginning with day 32, the confidence interval excludes zero. We can say with 95% confidence that the education group slept more minutes than the control group beginning on the 32nd day.[3]

16.4 Example 3: Piecewise modeling of time

Let's consider an example similar to the one illustrated in example 1 in section 16.2 earlier in this chapter. Like that example, this example studies sleep over time with 75 participants whose sleep at night is measured approximately every seven days for a total of eight observations. The sample in this study is composed of participants who have

3. If you run the **margins** command for yourself to see the output that was omitted, you would see the *p*-value for each comparison. This shows that the difference between the education group and the control group is significant on the 32nd day and every day thereafter.

been diagnosed with insomnia. In this study, the first 30 days are considered a baseline period during which sleep is simply observed, beginning on the 31st day, participants are given medication aimed to increase their minutes of sleep.

Time can be modeled in a piecewise fashion by breaking up the days of observation into the baseline phase and the treatment phase. This study can be analyzed and interpreted using a model similar to the one illustrated in section 4.5. (This section may make more sense if you revisit section 4.3.) That example showed how income could be modeled as a function of education, where education was divided into two phases, before and after high school graduation. That analysis allowed us to assess the slope of the relationship between education and income at each phase and to also assess the effect of graduating high school on income.

In this example, we can assess the slope of the relationship between sleep duration and time (obsday) during the baseline and treatment phases. We can also test for a sudden jump in sleep duration on the 31st day, corresponding to the start of the treatment phase.

Let's begin by looking at the first five observations from the dataset, as well as summary statistics for the variables in the dataset.

```
. use sleep_conpw, clear
. list in 1/5, sepby(id)

     +----------------------+
     | id   obsday   sleep  |
     |----------------------|
  1. |  1        1     285  |
  2. |  1        6     293  |
  3. |  1       10     267  |
  4. |  1       17     266  |
  5. |  1       23     288  |
     +----------------------+

. summarize

    Variable |        Obs        Mean    Std. Dev.       Min        Max
-------------+---------------------------------------------------------
          id |        600          38    21.66677          1         75
      obsday |        600    23.81667    15.10474          1         55
       sleep |        600     355.095    103.3013        110        654
```

The variable id is the identifier for each person. The variable obsday reflects the day of observation, ranging from 1 to 55. The variable sleep is the duration of sleep (in minutes) for the given day of observation. The mean number of minutes of sleep is 355 and ranges from 110 to 654.

We need to create some variables to prepare for the piecewise analysis. First, we use the mkspline command to create the variables named obsday1m and obsday2m, placing the knot at 31 days (corresponding to the start of the treatment phase). The marginal option is used, so the coefficient for obsday1m will represent the slope during the baseline phase, and the coefficient for obsday2m will represent the slope during the treatment phase minus the slope during the baseline phase.

```
. mkspline obsday1m 31 obsday2m = obsday, marginal
```

Next, we create the variable `trtphase` that is coded 0 for the baseline phase (where `obsday` was 1 to 30) and coded 1 during the treatment phase (where `obsday` is 31 or more).

```
. generate trtphase = 0 if obsday <= 30
(224 missing values generated)
. replace trtphase =  1 if obsday >= 31 & !missing(obsday)
(224 real changes made)
```

We are now ready to run a piecewise model that predicts `sleep` from `obsday1m`, `obsday2m`, and `trtphase`. In this example, `obsday1m`, `obsday2m`, and `trtphase` are also included as random effects.[4] This model is fit using the `mixed` command below.

```
. mixed sleep obsday1m obsday2m trtphase || id: obsday1m obsday2m trtphase,
> covariance(un) nolog
Mixed-effects ML regression                     Number of obs     =        600
Group variable: id                              Number of groups  =         75
                                                Obs per group:
                                                              min =          8
                                                              avg =        8.0
                                                              max =          8
                                                Wald chi2(3)      =      70.48
Log likelihood = -2601.8889                     Prob > chi2       =     0.0000

------------------------------------------------------------------------------
       sleep |      Coef.   Std. Err.      z    P>|z|     [95% Conf. Interval]
-------------+----------------------------------------------------------------
    obsday1m |   -.005795   .1193362    -0.05   0.961    -.2396896    .2280997
    obsday2m |   .5434219   .1663013     3.27   0.001     .2174773    .8693665
    trtphase |   11.13768   1.941483     5.74   0.000     7.332444    14.94292
       _cons |   349.0109    10.6413    32.80   0.000     328.1543    369.8674
------------------------------------------------------------------------------

------------------------------------------------------------------------------
  Random-effects Parameters  |   Estimate   Std. Err.     [95% Conf. Interval]
-----------------------------+------------------------------------------------
id: Unstructured             |
              var(obsday1m)  |   .8271301   .1755854      .5456052    1.253918
              var(obsday2m)  |   .6926937   .3344266      .2688985    1.784408
              var(trtphase)  |   59.53125   48.90214      11.89934    297.8292
                 var(_cons)  |   8425.418    1386.81       6102.18    11633.16
     cov(obsday1m,obsday2m)  |  -.3675087   .1871909     -.7343962   -.0006212
     cov(obsday1m,trtphase)  |  -1.540098   2.138258     -5.731006    2.650811
        cov(obsday1m,_cons)  |   32.85487   11.53306      10.25049    55.45924
     cov(obsday2m,trtphase)  |    2.83779   2.783922     -2.618597    8.294176
        cov(obsday2m,_cons)  |  -5.187991   15.46429     -35.49745    25.12147
        cov(trtphase,_cons)  |  -133.6282   180.1468     -486.7094    219.4529
-----------------------------+------------------------------------------------
              var(Residual)  |   101.8556   8.072371      87.20163    118.9721
------------------------------------------------------------------------------
LR test vs. linear model: chi2(10) = 2057.12              Prob > chi2 = 0.0000
Note: LR test is conservative and provided only for reference.
```

4. See Singer and Willett (2003) for a detailed discussion of how to select the appropriate random effects.

To help with the interpretation of this model, let's begin by graphing the predicted means as a function of observation day. We need to express observation day in terms of obsday1m and obsday2m. First, let's look at the coding for obsday1m and obsday2m at three key days in the study, when observation day is 1, 31, and 49. These correspond to the beginning of the study, the start of the treatment phase, and the approximate end of the study.

```
. showcoding obsday obsday1m obsday2m if inlist(obsday,1,31,49)
```

obsday	obsday1m	obsday2m
1	1	0
31	31	0
49	49	18

We can then use the margins command to compute the predicted means for these key days. When obsday equals 31, we estimate the predicted mean assuming trtphase is 0 and 1, to estimate the jump in the fitted values due to the start of the treatment phase. (The noatlegend option is included to save space.)

```
. margins, at(obsday1m = 1  obsday2m = 0  trtphase=0)
>          at(obsday1m = 31 obsday2m = 0  trtphase=0)
>          at(obsday1m = 31 obsday2m = 0  trtphase=1)
>          at(obsday1m = 49 obsday2m = 18 trtphase=1) nopvalues noatlegend
Adjusted predictions                            Number of obs     =        600
Expression   : Linear prediction, fixed portion, predict()
```

	Margin	Delta-method Std. Err.	[95% Conf.	Interval]
_at				
1	349.0051	10.67886	328.0749	369.9352
2	348.8312	12.30066	324.7224	372.9401
3	359.9689	12.15626	336.1431	383.7947
4	369.6462	13.11127	343.9486	395.3438

Using the techniques from section 4.10, we save the predicted means to the current dataset.

```
. matrix yhat = r(b)'
. matrix day = (1 \ 31 \ 31 \ 49)
. svmat yhat
. svmat day
. list yhat1 day1 in 1/4
```

	yhat1	day1
1.	349.0051	1
2.	348.8312	31
3.	359.9689	31
4.	369.6462	49

We can now graph the fitted values. The graph is shown in figure 16.4.

```
. graph twoway line yhat1 day1, sort xline(31) xlabel(1(7)49 31)
```

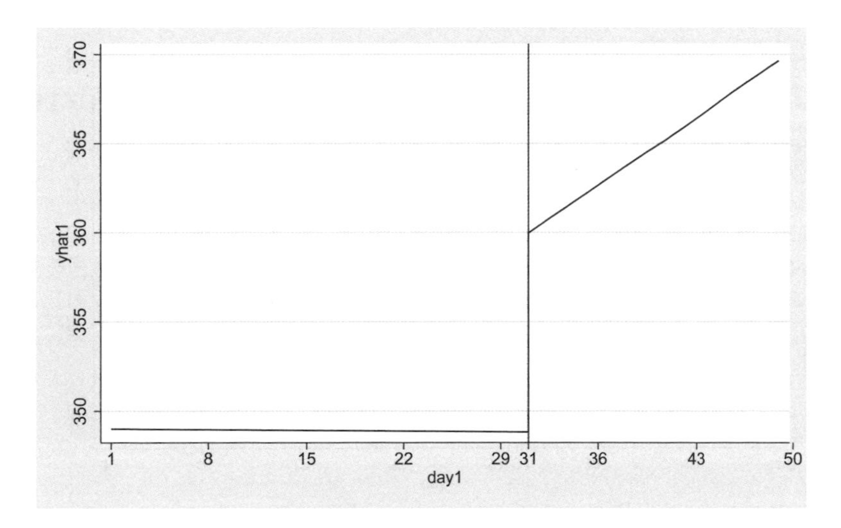

Figure 16.4. Minutes of sleep at night by time

Let's now relate the graph in figure 16.4 to the results from the `mixed` command. The coefficient for `obsday1m` represents the slope during the baseline phase. This value is −0.01 and is not significantly different from 0. We see this in figure 16.4 as a nearly flat line from day 1 to 31. The coefficient for `obsday2m` reflects the change in the slope for the treatment phase minus the baseline phase. The difference in these slopes is 0.543 and is significant. We can estimate the slope during the treatment phase using the `lincom` command.

```
. lincom obsday1m + obsday2m
 ( 1)  [sleep]obsday1m + [sleep]obsday2m = 0
```

sleep	Coef.	Std. Err.	z	P>|z|	[95% Conf.	Interval]
(1)	.5376269	.1607347	3.34	0.001	.2225928	.8526611

During the treatment phase, each additional day of observation is associated with sleeping 0.538 additional minutes of sleep per night.

Finally, consider the coefficient for `trtphase`. This represents the jump in sleep at the start of the treatment phase (when observation day is 31). This coefficient is 11.1 and is statistically significant. We can see this 11.1 minute jump in sleep that occurs when observation day is 31 in figure 16.4.

This example has illustrated the application of a piecewise model with one knot and one jump for a longitudinal model with a treatment phase and a control phase. Chapter 4 illustrates other kinds of piecewise models that could be applied to the analysis of longitudinal data. However, it is important to carefully select the appropriate random effects for such models (see Singer and Willett [2003] for guidance).

16.5 Example 4: Piecewise effects of time by a categorical predictor

Let's consider an extension of the previous example that includes a baseline and treatment phase, but where participants are divided into different groups and receive different kinds of treatments during the treatment phase. Like the previous example, sleep is measured approximately once a week for a total of eight measurements. However, participants are randomly assigned to one of three groups: a control group, a medication group, or a sleep education group.

The first 30 days of the study are a baseline period during which the sleep is observed, but no treatment is administered to any of the groups. Starting on the 31st day, the medication group receives sleep medication, the sleep education group receives education about how to lengthen their sleep, and the control group receives nothing. The first phase of the study (days 1 to 30) is called the baseline phase, and the second phase (day 31 until the end of the study) is called the treatment phase.

In this study, time can be modeled in a piecewise fashion, one piece being the baseline phase and the other piece being the treatment phase. We can study the slope of the relationship between sleep duration and time during each of these phases, as well as the change (jump or drop) in sleep that occurs at the transition from baseline to the treatment phase. Furthermore, we can investigate the impact of the treatment group assignment (control, medication, and education) on the slope in each phase, as well as the jump or drop in sleep due to the start of the treatment phase.

The data for this example are stored in the dataset named sleep_cat3PW.dta. Let's begin by using this dataset, listing the first five observations, and showing summary statistics for the variables.

```
. use sleep_cat3pw, clear
. list in 1/5, sepby(id)
```

	id	group	obsday	sleep
1.	1	Control	1	353
2.	1	Control	9	345
3.	1	Control	15	340
4.	1	Control	21	337
5.	1	Control	26	324

```
. summarize
```

Variable	Obs	Mean	Std. Dev.	Min	Max
id	600	38	21.66677	1	75
group	600	2	.8171778	1	3
obsday	600	23.815	15.02877	1	54
sleep	600	362.4817	39.72333	207	531

The variable id identifies each person and ranges from 1 to 75. The variable group identifies the group assignment, coded: 1 = control, 2 = medication, or 3 = education. The variable obsday is the day of observation, ranging from 1 to 54. The variable sleep is the duration of sleep (in minutes) and ranges from 207 to 531.

The analysis strategy illustrated in section 12.2 can be useful for analyzing this dataset. Imagine the educational periods (before high school graduation and after high school graduation) being replaced with the baseline and treatment phases. Also, imagine gender (male and female) being replaced with treatment group assignment (control, medication, and education).

Before we can run the piecewise analysis, we need to create some variables. First, the mkspline command is used to create the variables obsday1m and obsday2m, specifying 31 as the knot. By including the marginal option, obsday1m will represent the slope for the baseline period and obsday2m will represent the change in the slope for the treatment period compared with the baseline period.

```
. mkspline obsday1m 31 obsday2m = obsday, marginal
```

To account for the jump in sleep at the start of the treatment phase, we use the generate and replace commands to create the variable trtphase that is coded: 0 = baseline and 1 = treatment.

```
. generate trtphase = 0 if obsday < 31
(222 missing values generated)
. replace trtphase  = 1 if obsday >= 31 & !missing(obsday)
(222 real changes made)
. label define trtlab 0 "Baseline" 1 "Treatment"
. label values trtphase trtlab
```

Now, the variables are ready for running the piecewise model. Let's fit a piecewise model like the one described in section 12.2. Specifically, let's use the modeling strategy described as coding scheme #4 from section 12.4.4.[5] In this example, the categorical variable **group** takes the place of **gender**. The variable **trtphase** takes the place of **hsgrad**, and **obsday1m** and **obsday2m** take the place of **ed1m** and **ed2m**. The **mixed** command for fitting this model is shown below. Note the variables **trtphase**, **obsday1m**, and **obsday2m** are specified as random effects. The **nolog**, **noheader**, and **noretable** options are used to suppress the iteration log, header, and random-effects table to save space.[6]

```
. mixed sleep i.group##(i.trtphase c.obsday1m c.obsday2m),
> || id: trtphase obsday1m obsday2m, covariance(un)
> nolog noheader noretable noci
```

sleep	Coef.	Std. Err.	z	P>|z|
group				
Control	0	(base)		
Medication	-2.072872	3.411818	-0.61	0.543
Education	16.31454	3.410371	4.78	0.000
trtphase				
Baseline	0	(base)		
Treatment	-4.203853	2.793214	-1.51	0.132
obsday1m	-.0415024	.2998389	-0.14	0.890
obsday2m	-.0966276	.5077213	-0.19	0.849
group#trtphase				
Medication#Treatment	34.28463	3.997248	8.58	0.000
Education#Treatment	1.647258	3.973666	0.41	0.678
group#c.obsday1m				
Medication	.4016104	.4238518	0.95	0.343
Education	-.1437658	.4240338	-0.34	0.735
group#c.obsday2m				
Medication	-.2913187	.7196931	-0.40	0.686
Education	2.478031	.7176651	3.45	0.001
_cons	351.3643	2.414271	145.54	0.000

5. I chose this coding scheme because I want to 1) focus on the comparison of the treatment phase with the baseline phase and 2) focus on the comparison of each group with the reference group (the control group).
6. The **noci** option is included to make the output more readable for this example. See the callout titled *Using the* **noci** *option for clearer output* in section 2.5.1 for more details.

Note! More on the random effects

Thinking in terms of a multilevel model, the variables **trtphase**, **obsday1m**, and **obsday2m** are level-1 variables. The variable **group** is a level-2 (person level) variable. Cross-level interactions are formed by interacting **group** with **trtphase**, **obsday1m**, and **obsday2m**.

The results from this model can be interpreted in the same manner as the results from coding scheme #4 from section 12.4.4. But before we interpret the results, let's graph the predicted means as a function of observation day and group. However, we need to express observation day in terms of **obsday1m** and **obsday2m**. Below, we see the coding of **obsday1m** and **obsday2m** for three particular days, the first day of the study (when observation day is 1), the first day of the treatment phase (when observation day is 31), and the 49th day of the study (the approximate end of the study).

```
. showcoding obsday obsday1m obsday2m trtphase if inlist(obsday,1,31,49)
```

obsday	obsday1m	obsday2m	trtphase
1	1	0	0
31	31	0	1
49	49	18	1

We can now use the **margins** command to compute the predicted mean of sleep for each group at four key points—for the beginning of the study, the first day of the treatment phase (with and without treatment), and the 49th day of the study. (The **noatlegend** option is specified to save space.)

```
. margins group, at(obsday1m = 1  obsday2m = 0  trtphase=0)
>                at(obsday1m = 31 obsday2m = 0  trtphase=0)
>                at(obsday1m = 31 obsday2m = 0  trtphase=1)
>                at(obsday1m = 49 obsday2m = 18 trtphase=1) nopvalues noatlegend
Adjusted predictions                            Number of obs     =        600
Expression   : Linear prediction, fixed portion, predict()
```

	Margin	Delta-method Std. Err.	[95% Conf.	Interval]
_at#group				
1#Control	351.3228	2.362294	346.6928	355.9528
1#Medication	349.6516	2.359361	345.0273	354.2758
1#Education	367.4936	2.357094	362.8738	372.1134
2#Control	350.0778	9.041145	332.3574	367.7981
2#Medication	360.4548	9.037425	342.7418	378.1678
2#Education	361.9356	9.0434	344.2108	379.6603
3#Control	345.8739	9.052168	328.132	363.6158
3#Medication	390.5356	9.075722	372.7475	408.3237
3#Education	359.379	9.058832	341.624	377.134
4#Control	343.3876	11.2604	321.3176	365.4575
4#Medication	390.0345	11.26474	367.956	412.113
4#Education	398.9094	11.25101	376.8578	420.961

Using the tricks from section 4.10, we can graph the estimates from the `margins` command, creating the graph shown in figure 16.5.

```
. * save predicted means as a matrix named -yhat-
. matrix yhat = r(b)'
. * save the day number in a matrix named -day-
. matrix day = (0 \ 0 \ 0 \ 31 \ 31 \ 31 \ 31 \ 31 \ 31 \ 49 \ 49 \ 49)
. * save the group in a matrix named -gp-
. matrix gp = (1 \ 2 \ 3 \ 1 \ 2 \ 3 \ 1 \ 2 \ 3 \ 1 \ 2 \ 3 )
. * Save the contents of the matrices to the current dataset
. svmat yhat
. svmat day
. svmat gp
. graph twoway (line yhat1 day1 if gp1==1) (line yhat1 day1 if gp1==2)
> (line yhat1 day1 if gp1==3), xline(31) xlabel(1(7)49 30)
> legend(title(Treatment group)
> label(1 "Control") label(2 "Medication") label(3 "Education") rows(1))
```

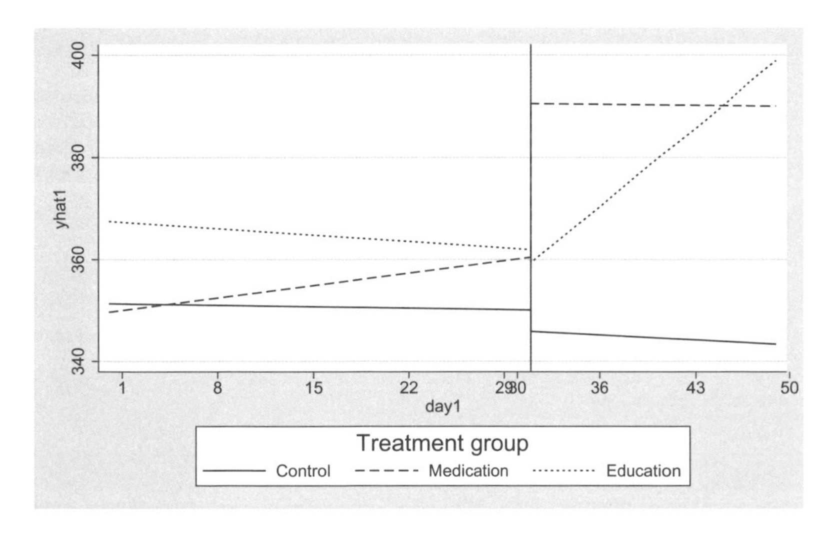

Figure 16.5. Minutes of sleep at night by time and group

Looking at figure 16.5, we see that the sleep durations for the control group remain much the same across the entire study with a tiny (but trivial) drop in sleep at the start of treatment. For the medication group, their sleep durations remain fairly steady during the baseline phase, then jump at the start of treatment, and then remain steady. The sleep durations for the education group are fairly steady during the baseline phase, and then show a small dip at the start of treatment but a substantial increase in slope during the treatment phase. Let's relate the pattern of results we see in figure 16.5 to the output from the `mixed` command.

16.5.1 Baseline slopes

Referring back to the results from the `mixed` command in section 16.5, the coefficient for `obsday1m` (-0.042, $p = 0.890$) represents the slope relating observation day to sleep duration during the baseline phase for the control group (group 1). The coefficient in the section `group#c.obsday1m` labeled `Medication` compares the baseline slope for the medication group with the control group (0.402, $p = 0.343$), while the coefficient labeled `Education` compares the baseline slope for the education group with the control group (-0.143, $p = 0.735$). These results show 1) baseline slope for the control group does not differ from 0; 2) the baseline slopes for the medication and control groups are not significantly different; and 3) the baseline slopes for the education and control groups are not significantly different. We can test the equality of all the baseline slopes using the `contrast` command below. This test is not significant.

```
. contrast group#c.obsday1m
Contrasts of marginal linear predictions
Margins      : asbalanced

                         df        chi2     P>chi2

sleep
group#c.obsday1m          2        1.78     0.4108
```

You can estimate the baseline slope for each group using the `margins` command below. None of these tests are significant. For example, the baseline slope for the medication group is estimated as 0.360 and is not significantly different from 0 (p=0.229).

```
. margins, dydx(obsday1m) over(group)
Average marginal effects                          Number of obs     =        600
Expression   : Linear prediction, fixed portion, predict()
dy/dx w.r.t. : obsday1m
over         : group

                        Delta-method
                  dy/dx   Std. Err.      z    P>|z|     [95% Conf. Interval]

obsday1m
       group
     Control  -.0415024   .2998389    -0.14   0.890    -.6291758    .5461711
  Medication    .360108    .299578     1.20   0.229    -.2270542    .9472702
   Education  -.1852682   .2998354    -0.62   0.537    -.7729347    .4023984
```

16.5.2 Change in slopes: Treatment versus baseline

Because we included the `marginal` option on the `mkspline` command in section 16.5 (using coding scheme #4 from section 12.4.4), the slopes associated with `obsday2m` (from the `mixed` command in section 16.5) reflect the change in the `obsday` slope for the treatment period minus the baseline period. For simplicity, let's call this the change in slope. The `obsday2m` coefficient is the change in slope for the control group. This coefficient is -0.10 and is not significant. The slope coefficient does not significantly change from the baseline to treatment phase. The coefficient in the section `group#c.obsday2m` labeled `Medication` compares the change in slope (treatment versus baseline) for the medication group versus the control group (−0.291, $p = 0.686$), while the coefficient labeled Education compares the change in slope for the education group with the control group (2.48, $p = 0.001$). These results show 1) the change in slope (treatment versus baseline) is not significant; 2) the change in slope for the medication versus control groups is not significantly different; and 3) the change in slope for the education versus control groups is significant ($p = 0.001$).

You can test the equality of the changes in slope coefficients for all three groups using the contrast command below. This test is significant.

```
. contrast group#c.obsday2m
Contrasts of marginal linear predictions
Margins      : asbalanced

                  |         df        chi2     P>chi2
------------------+----------------------------------
sleep             |
group#c.obsday2m  |          2       17.95     0.0001
```

You can estimate the change in slope for each group using the margins command below. The change in slope for the control group and medication groups is not significant. For the education group (group 3), the slope during the treatment phase is significantly greater than the slope during the baseline phase.

```
. margins, dydx(obsday2m) over(group)
Average marginal effects                        Number of obs     =        600
Expression   : Linear prediction, fixed portion, predict()
dy/dx w.r.t. : obsday2m
over         : group

             |            Delta-method
             |      dy/dx   Std. Err.      z    P>|z|     [95% Conf. Interval]
-------------+----------------------------------------------------------------
obsday2m     |
       group |
     Control |  -.0966276   .5077213    -0.19   0.849    -1.091743    .8984879
  Medication |  -.3879464   .5100757    -0.76   0.447    -1.387676    .6117837
   Education |   2.381403   .5072103     4.70   0.000      1.38729    3.375517
```

16.5.3 Jump at treatment

The coefficient for trtphase (from the mixed command in section 16.5) represents the jump (or drop) in sleep coinciding with the start of the treatment phase for the control group. This coefficient is −4.20 but is not significant. As we can see in figure 16.5, there is a slight drop in sleep at the start of the treatment phase for the control group.

The coefficient in the section `group#trtphase` labeled `Medication#Treatment` is the difference in the jump (or drop) for the medication group compared with the control group. For the medication group compared with the control group, sleep durations jump by 34.3 more minutes at the start of the treatment phase ($p < 0.001$).

The `Medication#Treatment` coefficient is the difference in the jump (or drop) for the education group compared with the control group. Sleep durations jumped by 1.65 more minutes for the education group (versus the control group), but this was not significant ($p = 0.735$).

We can test the equality of the jump or drop for all three groups using the `contrast` command below. This test is significant.

```
. contrast group#trtphase
Contrasts of marginal linear predictions
Margins      : asbalanced

                  |         df        chi2     P>chi2
------------------+-----------------------------------
sleep             |
   group#trtphase |          2       92.58     0.0000
------------------------------------------------------
```

We can estimate the size of the jump or drop in sleep durations at the start of the treatment phase for each group using the `contrast` command below. For the control and education groups, there is a drop in the sleep duration at the start of the treatment phase, but neither of these drops are significant. For the medication group, sleep duration jumps by 30.1 minutes at the start of the treatment phase and that jump is significant.

```
. contrast trtphase@group, pveffects nowald
Contrasts of marginal linear predictions
Margins      : asbalanced

                                    |   Contrast   Std. Err.        z    P>|z|
------------------------------------+------------------------------------------
sleep                               |
                     trtphase@group |
   (Treatment vs base) Control      |  -4.203853   2.793214     -1.51   0.132
(Treatment vs base) Medication      |   30.08077   2.859362     10.52   0.000
 (Treatment vs base) Education      |  -2.556595   2.826301     -0.90   0.366
-------------------------------------------------------------------------------
```

16.5.4 Comparisons among groups

We can make comparisons among the groups at particular days within the study. Let's focus on the day before the start of the treatment (when observation day is 30), as well as the days that correspond to the observation day of 35, 40, and 45. The `showcoding` command is used to show the values of `obsday1m` and `obsday2m` for these days.

```
. showcoding obsday obsday1m obsday2m trtphase if inlist(obsday,30,35,40,45)

  obsday   obsday1m   obsday2m   trtphase

      30         30          0          0
      35         35          4          1
      40         40          9          1
      45         45         14          1
```

Let's now use the `margins` command to compare each group with the control group at each of these days. (The `noatlegend` option is specified to save space.)

```
. margins r.group, at(obsday1m=30 obsday2m=0  trtphase=0)
>                  at(obsday1m=35 obsday2m=4  trtphase=1)
>                  at(obsday1m=40 obsday2m=9  trtphase=1)
>                  at(obsday1m=45 obsday2m=14 trtphase=1)
>                  contrast(nowald pveffects) noatlegend
Contrasts of adjusted predictions               Number of obs     =        600
Expression   : Linear prediction, fixed portion, predict()

                                          Delta-method
                              Contrast   Std. Err.        z    P>|z|

                 group@_at
(Medication vs Control) 1      9.97544   12.37464      0.81    0.420
(Medication vs Control) 2     45.10284   12.87497      3.50    0.000
(Medication vs Control) 3      45.6543   13.50462      3.38    0.001
(Medication vs Control) 4     46.20576   14.67722      3.15    0.002
 (Education vs Control) 1     12.00157   12.37867      0.97    0.332
 (Education vs Control) 2     22.84212   12.86972      1.77    0.076
 (Education vs Control) 3     34.51345   13.50243      2.56    0.011
 (Education vs Control) 4     46.18477   14.67256      3.15    0.002
```

Focusing on the comparison of the medication group with the control group, the difference is not significant when observation day is 30 ($p = 0.420$); prior to the treatment phase. However, the difference is significant at each of the time points tested during the treatment phase, when observation day is 35, 40, and 45 ($p = 0.000$, 0.001, and 0.002, respectively).

Shifting our attention to the comparison of the education group with the control group, the comparison is not significant when observation day is 30 ($p = 0.332$); prior to the start of the treatment phase. The difference remains nonsignificant early in the treatment phase when observation day is 35 ($p = 0.076$) but is significant later in the treatment phase when observation day is 40 ($p = 0.011$) and 45 ($p = 0.002$).

This example focused on a model in which time was broken up into two phases (where there was one knot) and was interacted with a categorical variable. For more details about this kind of model, see section 12.2. If you have a model with three phases (for example, baseline, treatment, and return to baseline), see section 12.3 for more details about that kind of model.

16.6 Summary

This chapter has illustrated four different examples of longitudinal models where time was treated as a continuous variable. Instead of using the `regress` command, these models used the `mixed` command. The fixed-effects portion of the `mixed` models (that is, the part before the `||`) was specified the same whether the model was longitudinal or nonlongitudinal. Furthermore, the postestimation commands we used for interpreting and visualizing the results (the `margins`, `marginsplot`, and `contrast` commands) were the same whether the model was longitudinal or nonlongitudinal.

I recommend Singer and Willett (2003) for more information about the use of multilevel models for the analysis of longitudinal data. In addition, Richter (2006) provides a tutorial illustrating the application of multilevel models using reading time data.

17 Time as a categorical predictor

17.1 Chapter overview

The previous chapter (chapter 16) treated time as a continuous variable. However, there are situations when it might be more advantageous to treat time as a categorical variable. For example, when you have few fixed time points, treating time as a categorical variable facilitates comparisons among the different time points. This chapter considers models where time is treated as a categorical variable, using three examples. The first example includes time (treated as a categorical variable) as the only predictor in the model (see section 17.2). The second example illustrates the interaction of time with a two-level categorical variable (see section 17.3), and the third example illustrates time interacted with a three-level categorical variable (see section 17.4).

The models presented in this chapter will use the `mixed` command to fit a model that is a hybrid of a traditional repeated-measures analysis of variance (ANOVA) and a mixed model. Like the repeated-measures ANOVA, there will be one fixed intercept (rather than having random intercepts that we commonly see when using the `mixed` command). By specifying the `noconstant` option in the random-effects portion of the `mixed` command, a fixed intercept will be estimated.

To account for the nonindependence of residuals across time points, we will use the `residuals()` option within the random-effects portion of the `mixed` command. This allows us to model the structure of the residual covariances across time points. The following examples will use an unstructured residual covariance, which estimates a separate residual variance for each time point and a separate residual correlation among each pair of the time points. Section 17.5 provides more details about the nature of residual covariance structures and will describe how you can select among different covariance structures.

17.2 Example 1: Time treated as a categorical variable

Like the examples from the previous chapter, this example studies how many minutes people sleep at night. One hundred people were included in this study, and their sleep was measured once a month for three months (that is, month 1, month 2, and month 3). The measurements at month 1 were a baseline measurement, in that no specific treatment was applied. Prior to the measurement in month 2, participants were given a sleep medication to lengthen their sleep, and the use of this medication continued during month 3. In summary, sleep was measured at three time points, the first a control (baseline) condition and the second and third measurements while taking a sleep medication.

There are two aims of this study. The first aim is to assess the initial impact of sleep medication on duration of sleep. It is predicted the people will sleep longer in month 2 (while on the sleep medication) than in month 1 (before they started the medication). Even if the medication is effective in the second month, there might be a concern of whether it sustains its effectiveness. So the second aim is to the assess the sustained effectiveness of the medication on sleep, comparing the amount of sleep in the second and third months.

Let's begin by looking at the first six observations from this dataset as well as the summary statistics for the variables in the dataset.

```
. use sleep_cat3, clear

. list in 1/6, sepby(id)

     +-------------------+
     | id   month  sleep |
     |-------------------|
  1. |  1       1    303 |
  2. |  1       2    349 |
  3. |  1       3    382 |
     |-------------------|
  4. |  2       1    331 |
  5. |  2       2    380 |
  6. |  2       3    350 |
     +-------------------+

. summarize

    Variable |        Obs        Mean    Std. Dev.       Min        Max
-------------+---------------------------------------------------------
          id |        300        50.5     28.9143          1        100
       month |        300           2    .8178608          1          3
       sleep |        300    361.4067    31.21017        276        466
```

The dataset for this example is in a long format, with one observation per person per month. The variable `id` identifies the person and ranges from 1 to 100, corresponding to the 100 participants. The variable `month` identifies the month in which sleep was observed and ranges from 1 to 3. The variable `sleep` contains the number of minutes the person slept at night and ranges from 276 to 466 minutes.

> **Note! Wide and long datasets**
>
> Data for this kind of study might be stored with one observation per person and three variables representing the different time points. Sometimes, this is called a *multivariate* format, and Stata would call this a *wide* format. If your dataset is in that kind of form, you can use the `reshape` command to convert it to a *long* format.

Let's now use the `mixed` command to predict `sleep` from `month`, treating `month` as a categorical variable. Specifying `|| id:` introduces the random-effects part of the model and indicates that the observations are nested within `id`. Next, the `noconstant` option is specified in the random-effects options so only one fixed intercept is fit. Furthermore, the `residuals()` option is included to specify covariance structure of the residuals between months (within each level of the person's ID). The `residuals()` option specifies an unstructured residual covariance among the different months within each person.[1] This accounts for the nonindependence of the observations among time points for each person.

1. The `unstructured` residual type specified within the `residual()` option specifies an unstructured covariance matrix, and the `t(month)` suboption specifies that observations are repeated across months. Section 17.5 provides more information about how to select among residual covariance structures.

```
. mixed sleep i.month || id:, noconstant residuals(unstructured, t(month)) nolog
Mixed-effects ML regression                     Number of obs     =        300
Group variable: id                              Number of groups  =        100
                                                Obs per group:
                                                              min =          3
                                                              avg =        3.0
                                                              max =          3
                                                Wald chi2(2)      =      35.94
Log likelihood = -1437.3712                     Prob > chi2       =     0.0000

       sleep       Coef.   Std. Err.      z    P>|z|     [95% Conf. Interval]

       month
          1            0  (base)
          2        21.07     3.8331     5.50   0.000     13.55726    28.58274
          3        18.43   3.807288     4.84   0.000     10.96785    25.89215

       _cons      348.24   3.144903   110.73   0.000     342.0761    354.4039


  Random-effects Parameters     Estimate   Std. Err.     [95% Conf. Interval]

id:                  (empty)

Residual: Unstructured
                    var(e1)     989.0415   139.8716      749.6117    1304.946
                    var(e2)     928.5546   131.3175      703.7677    1225.139
                    var(e3)     731.3612   103.4301      554.3114    964.9615
                 cov(e1,e2)     224.1654   98.41878      31.26812    417.0627
                 cov(e1,e3)     135.4291    86.1212     -33.36538    304.2235
                 cov(e2,e3)     107.7223   83.10907     -55.16844    270.6131

LR test vs. linear model: chi2(5) = 11.61                 Prob > chi2 = 0.0405
Note: The reported degrees of freedom assumes the null hypothesis is not on
      the boundary of the parameter space.  If this is not true, then the
      reported test is conservative.
```

Before we interpret the coefficients from this model, let's use the `margins` command to compute the predicted mean of sleep for each level of month, as shown below. The predicted mean of sleep is 348.24 in month 1, 369.31 in month 2, and 366.67 in month 3.

```
. margins month, nopvalues
Adjusted predictions                            Number of obs     =        300
Expression   : Linear prediction, fixed portion, predict()

                          Delta-method
                  Margin   Std. Err.     [95% Conf. Interval]

       month
          1       348.24   3.144903      342.0761    354.4039
          2       369.31   3.047219      363.3376    375.2824
          3       366.67   2.704369      361.3695    371.9705
```

We can use the **marginsplot** command to create a graph showing the predicted mean of sleep across the three months, as shown in figure 17.1.

```
. marginsplot
  Variables that uniquely identify margins: month
```

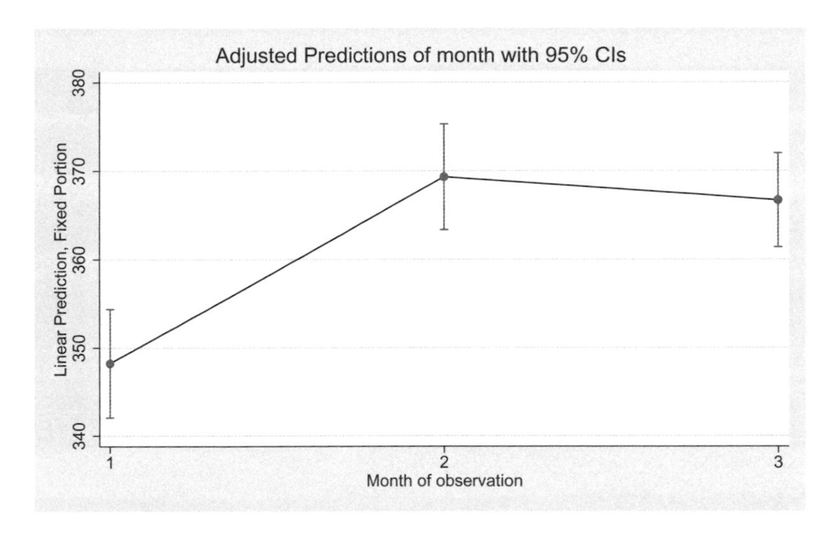

Figure 17.1. Estimated minutes of sleep at night by month

Looking at the predicted means from the **margins** and **marginsplot** commands, it looks like sleep increased from month 1 to month 2 and showed little difference between month 2 and month 3. Before making those specific tests, let's test the overall null hypothesis that the average number of minutes of sleep is equal across all three months. This is tested using the **contrast** command below. This test is significant, indicating that the average number of minutes of sleep is not equal across all months.

```
. contrast month
Contrasts of marginal linear predictions
Margins      : asbalanced
```

	df	chi2	P>chi2
sleep			
month	2	35.94	0.0000

Now, let's use the **contrast** command to perform the specific comparisons of interest. Let's compare the amount of sleep in the second month with the first month to test the initial impact of the sleep medication. Let's also compare the third month with the second month to assess the extent to which any impact was sustained between the second and third month. These comparisons are made using the **contrast** command, applying the **ar.** contrast to **month**.

```
. contrast ar.month, nowald effects
Contrasts of marginal linear predictions
Margins      : asbalanced
```

	Contrast	Std. Err.	z	P>|z|	[95% Conf.	Interval]
sleep						
month						
(2 vs 1)	21.07	3.8331	5.50	0.000	13.55726	28.58274
(3 vs 2)	-2.64	3.80062	-0.69	0.487	-10.08908	4.809078

The first test is significant. In the second month, participants slept (on average) 21.07 minutes longer than in the first month. The second test is not significant. The predicted mean of sleep in the third month was not significantly different from the second month. The effectiveness of the medication did not significantly diminish in the third month compared with the second month.

If we wanted a comparison of every pair of months, we can use the `pwcompare` command.

```
. pwcompare month, effects
Pairwise comparisons of marginal linear predictions
Margins      : asbalanced
```

	Contrast	Std. Err.	Unadjusted z	Unadjusted P>|z|	Unadjusted [95% Conf.	Unadjusted Interval]
sleep						
month						
2 vs 1	21.07	3.8331	5.50	0.000	13.55726	28.58274
3 vs 1	18.43	3.807288	4.84	0.000	10.96785	25.89215
3 vs 2	-2.64	3.80062	-0.69	0.487	-10.08908	4.809078

The `pwcompare` command shows that the comparison of month 2 versus 1 is significant, the comparison of month 3 versus 1 is significant, but the comparison of month 3 versus 2 is not significant.

Before concluding this example, let's look back at the random-effects portion of the output produced by the `mixed` command. The output shows that the residual standard deviation at month 1 is 31.45, at month 2 is 30.47, and at month 3 is 27.04. The output also shows the estimate of the correlation of the residuals between each pair of time points. For example, the correlation of the residuals at month 1 with month 2 is 0.23. This residual covariance structure is called an unstructured covariance and was requested specifying the `unstructured` residual type within the `residuals()` option. Section 17.5 provides more information about ways to select and compare residual covariance structures.

> **Note! Why not use repeated-measures ANOVA?**
>
> Some readers might be asking why this analysis (and the subsequent analyses) are not performed using a traditional repeated-measures analysis of variance. The `mixed` command is more flexible and powerful than a traditional repeated-measures ANOVA. First, a repeated-measures ANOVA omits any observation with missing data (by using listwise deletion). The `mixed` command retains subjects who might have missing data and performs the estimation on the time observations that are present. Second, a traditional repeated-measures ANOVA is restricted to only two different types of covariance structures, exchangeable (compound symmetry) and unstructured (multivariate). There are many situations where an exchangeable structure is overly simple and untenable, and the unstructured covariance is overly complicated and estimates too many superfluous parameters. By contrast, the `mixed` command permits you to choose from a variety of covariance structures such as banded, autoregressive, or Toeplitz, as described in section 17.5. Although the output looks a little bit different, the `mixed` command combined with the `residuals()` option offers all the features from a repeated-measures ANOVA, with the added benefit of retaining subjects who have missing time points and the ability to estimate additional residual covariance structures.

The example in this study had three time points; however, this could be extended to four or more time points. As illustrated in this example, the `contrast` command can be used to test the overall equality of the means at the different time points, as well as form specific comparisons among the time points. This is a generalization of the principles illustrated in chapter 7, and you can borrow and generalize from the tools illustrated in that chapter.

17.3 Example 2: Time (categorical) by two groups

In the previous example, sleep was measured at three time points (month 1, month 2, and month 3). The first month was a baseline measurement, whereas participants received sleep medication during the second and third months. We found that sleep was significantly longer in the second month than in the first month. Furthermore, there was no significant change in sleep comparing the second and third month.

We can augment that design by including a control group that is measured at each of the three months but receives no treatment. Using such a design, let's focus on two main questions. The first question concerns the initial effect of the sleep medication. This would be assessed by comparing the change in sleep from month 1 to month 2 for the treatment group with that for the control group. The second question concerns the sustained effect of the medication. This would be assessed by comparing the change in sleep from month 2 to month 3 for the treatment group with that for the control group.

This study includes 100 participants in the treatment group and 100 participants in the control group. The main predictors for this example are `group` (a two-level categorical variable) and `month` (a three-level categorical variable). We can draw upon the logic illustrated in section 8.3 showing how to interpret the results from a two by three design.

Let's look at the first six observations in the dataset for this study and the summary statistics for the variables in this dataset.

```
. use sleep_catcat23, clear
. list in 1/6, sepby(id)

     +--------------------------------+
     | id   month     group   sleep |
     |--------------------------------|
  1. |  1       1   Control     315 |
  2. |  1       2   Control     379 |
  3. |  1       3   Control     320 |
     |--------------------------------|
  4. |  2       1   Control     392 |
  5. |  2       2   Control     369 |
  6. |  2       3   Control     314 |
     +--------------------------------+

. summarize
    Variable |        Obs        Mean    Std. Dev.       Min        Max
-------------+---------------------------------------------------------
          id |        600       100.5    57.78248          1        200
       month |        600           2    .8171778          1          3
       group |        600         1.5    .5004172          1          2
       sleep |        600      354.49    31.36286        256        448
```

The variable `id` identifies the participant and ranges from 1 to 200. The variable `month` indicates the month in which sleep was measured (month 1, 2, or 3). The variable `group` is coded: 1 = control and 2 = medication. The variable `sleep` is the number of minutes of sleep and ranges from 256 to 448.

Like the previous example, the analysis is conducted using the `mixed` command and uses the `residuals()` option to specify an unstructured residual type. The predictors in the model are `group` (with two levels) and `month` (with three levels). These two terms are entered along with their interaction.

```
. mixed sleep i.group##i.month || id:, noconstant residuals(un, t(month)) nolog
Mixed-effects ML regression                     Number of obs     =        600
Group variable: id                              Number of groups  =        200

                                                Obs per group:
                                                              min =          3
                                                              avg =        3.0
                                                              max =          3

                                                Wald chi2(5)      =      68.47
Log likelihood =   -2879.78                     Prob > chi2       =     0.0000

------------------------------------------------------------------------------
       sleep |      Coef.   Std. Err.      z    P>|z|     [95% Conf. Interval]
-------------+----------------------------------------------------------------
       group |
     Control |          0  (base)
  Medication |      -7.08   4.378767    -1.62   0.106    -15.66223    1.502225
             |
       month |
           1 |          0  (base)
           2 |      -3.73   3.746771    -1.00   0.319    -11.07354    3.613535
           3 |       -.57   4.257299    -0.13   0.893    -8.914152    7.774152
             |
 group#month |
Medication#2 |      28.06   5.298734     5.30   0.000     17.67467    38.44533
Medication#3 |      24.94   6.020729     4.14   0.000     13.13959    36.74041
             |
       _cons |     350.63   3.096256   113.24   0.000     344.5615    356.6985
------------------------------------------------------------------------------

------------------------------------------------------------------------------
  Random-effects Parameters  |   Estimate   Std. Err.     [95% Conf. Interval]
-----------------------------+------------------------------------------------
id:                  (empty) |
-----------------------------+------------------------------------------------
Residual: Unstructured       |
                     var(e1) |     958.68   95.86811      788.0493    1166.256
                     var(e2) |   721.2979   72.12976      592.9179     877.475
                     var(e3) |   980.2552   98.02553      805.7847    1192.503
                 cov(e1,e2)  |   138.0745   59.60521      21.25039    254.8985
                 cov(e1,e3)  |   63.23807   68.69312     -71.39797    197.8741
                 cov(e2,e3)  |   119.7182   60.05779      2.007146    237.4293
------------------------------------------------------------------------------
LR test vs. linear model: chi2(5) = 15.70                 Prob > chi2 = 0.0078

Note: The reported degrees of freedom assumes the null hypothesis is not on
      the boundary of the parameter space.  If this is not true, then the
      reported test is conservative.
```

We can estimate the mean sleep by group and month using the `margins` command below. This is followed by the `marginsplot` command to graph the means computed by the `margins` command (shown in figure 17.2).

```
. margins month#group, nopvalues
Adjusted predictions                          Number of obs     =        600
Expression   : Linear prediction, fixed portion, predict()

------------------------------------------------------------------------------
             |            Delta-method
             |     Margin   Std. Err.      [95% Conf. Interval]
-------------+----------------------------------------------------------------
 month#group |
   1#Control |     350.63    3.096256      344.5615    356.6985
1#Medication |     343.55    3.096256      337.4815    349.6185
   2#Control |      346.9    2.685699      341.6361    352.1639
2#Medication |     367.88    2.685699      362.6161    373.1439
   3#Control |     350.06    3.130903      343.9235    356.1965
3#Medication |     367.92    3.130903      361.7835    374.0565
------------------------------------------------------------------------------

. marginsplot, noci
  Variables that uniquely identify margins: month group
```

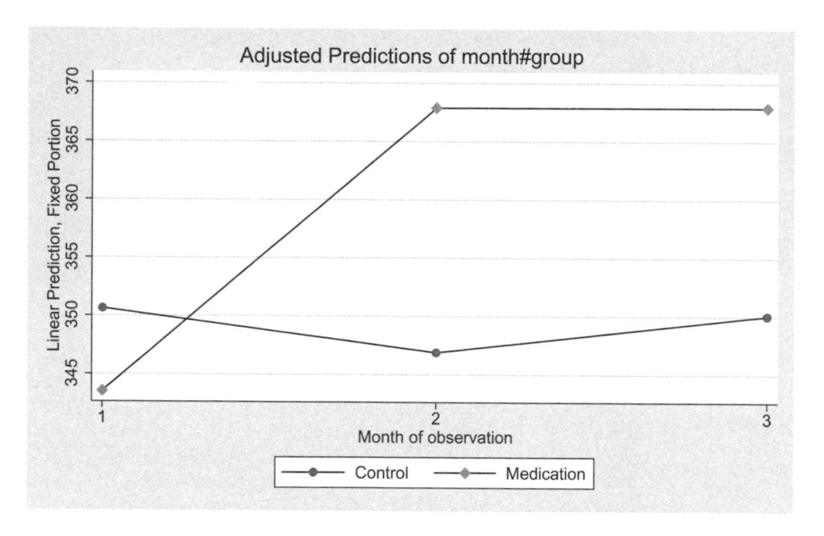

Figure 17.2. Estimated sleep by month and treatment group

The pattern of means shown in figure 17.2 appears to show that the medication increases sleep in month 2 compared with month 1, and that the effect is sustained in month 3. By contrast, the sleep in the control group remains rather steady at each of the three time points.

Before testing our two main questions of interest, let's assess the overall interaction of group by month using the `contrast` command below. The overall interaction is significant.

```
. contrast group#month
Contrasts of marginal linear predictions
Margins      : asbalanced

             |         df        chi2     P>chi2
-------------+----------------------------------
sleep        |
 group#month |          2       30.21     0.0000
------------------------------------------------
```

Now to test our questions of interest regarding the initial effect of medication and the sustained effect of medication, we can apply the `ar.` contrast operator to `month` and interact that with `group`, as shown below.

```
. contrast ar.month#group, nowald pveffects
Contrasts of marginal linear predictions
Margins      : asbalanced

                                     |   Contrast   Std. Err.       z    P>|z|
-------------------------------------+-----------------------------------------
sleep                                |
                         month#group |
  (2 vs 1) (Medication vs base)      |      28.06   5.298734     5.30   0.000
  (3 vs 2) (Medication vs base)      |      -3.12   5.407618    -0.58   0.564
-------------------------------------------------------------------------------
```

The first contrast compares the change in sleep in month 2 versus month 1 for the treatment group versus the control group. This test is significant. The change in sleep for the medication group was 28.06 minutes more than the change for the control group. This is consistent with what we saw in figure 17.2.

The second contrast compares the change for month 3 versus month 2 for the treatment group versus the control group. This comparison is not significant.

For more information about how to dissect a three by two interaction, see section 8.3.

17.4 Example 3: Time (categorical) by three groups

Let's consider one final example that extends the previous example by including a third type of treatment, sleep education. The design for this example now includes three treatment groups (control, medication, and education) and three time points (month 1, month 2, and month 3). Month 1 is a baseline period during which no treatment is administered to any of the groups. During months 2 and 3, the medication group receives sleep medication, and the education group receives sleep education aimed to increase their sleep.

This study includes 300 participants, 100 assigned to each of the three treatment groups. The main variables of interest in this study are treatment group (with three levels) and month (with three levels). We can borrow from the logic and methods of interpretation that were illustrated in section 8.4.

Let's use the dataset for this hypothetical study, and let's list the first six observations and show the summary statistics for the variables in the dataset.

```
. use sleep_catcat33, clear
. list in 1/6, sepby(id)

     +--------------------------------+
     | id   month     group   sleep  |
     |--------------------------------|
  1. |  1       1   Control     305  |
  2. |  1       2   Control     313  |
  3. |  1       3   Control     351  |
     |--------------------------------|
  4. |  2       1   Control     295  |
  5. |  2       2   Control     289  |
  6. |  2       3   Control     374  |
     +--------------------------------+

. summarize
    Variable |        Obs        Mean    Std. Dev.       Min        Max
-------------+---------------------------------------------------------
          id |        900       150.5    86.65021          1        300
       month |        900           2    .8169506          1          3
       group |        900           2    .8169506          1          3
       sleep |        900    362.8378    34.06564        260        479
```

The variable id identifies the person, and month indicates the month in which sleep was observed (1, 2, or 3). The variable group is coded: 1 = control, 2 = medication, and 3 = education. The variable sleep contains the number of minutes of sleep, and ranges from 260 to 479.

The mixed command for analyzing this example is the same as the previous example.

```
. mixed sleep i.group##i.month || id:, noconstant residuals(un, t(month)) nolog
Mixed-effects ML regression                     Number of obs     =        900
Group variable: id                              Number of groups  =        300

                                                Obs per group:
                                                              min =          3
                                                              avg =        3.0
                                                              max =          3

                                                Wald chi2(8)      =     284.84
Log likelihood = -4318.4068                     Prob > chi2       =     0.0000

------------------------------------------------------------------------------
       sleep |      Coef.   Std. Err.      z    P>|z|     [95% Conf. Interval]
-------------+----------------------------------------------------------------
       group |
    Control  |          0  (base)
  Medication |        2.6   4.246231     0.61   0.540     -5.72246    10.92246
   Education |       3.17   4.246231     0.75   0.455     -5.15246    11.49246
             |
       month |
          1  |          0  (base)
          2  |       8.06    3.71971     2.17   0.030     .7695015     15.3505
          3  |      13.35   4.063031     3.29   0.001     5.386605    21.31339
             |
 group#month |
Medication#2 |      21.88   5.260465     4.16   0.000     11.56968    32.19032
Medication#3 |      18.22   5.745994     3.17   0.002     6.958059    29.48194
 Education#2 |       7.11   5.260465     1.35   0.177    -3.200322    17.42032
 Education#3 |      34.49   5.745994     6.00   0.000     23.22806    45.75194
             |
       _cons |      344.7   3.002539   114.80   0.000     338.8151    350.5849
------------------------------------------------------------------------------

------------------------------------------------------------------------------
  Random-effects Parameters  |   Estimate   Std. Err.     [95% Conf. Interval]
-----------------------------+------------------------------------------------
id:                  (empty) |
-----------------------------+------------------------------------------------
Residual: Unstructured       |
                     var(e1) |   901.5239    73.6092      768.2045     1057.98
                     var(e2) |   836.2546   68.27989      712.5875    981.3837
                     var(e3) |   918.4223   74.98886      782.6041    1077.811
                  cov(e1,e2) |    177.077   51.16174      76.80181    277.3521
                  cov(e1,e3) |   84.56206   52.76138     -18.84835    187.9725
                  cov(e2,e3) |   160.4783   51.43886      59.65998    261.2966
------------------------------------------------------------------------------
LR test vs. linear model: chi2(5) = 24.71                 Prob > chi2 = 0.0002

Note: The reported degrees of freedom assumes the null hypothesis is not on
      the boundary of the parameter space.  If this is not true, then the
      reported test is conservative.
```

The `margins` and `marginsplot` commands are used below to estimate the predicted mean of sleep by group and month and to graph the results. The graph is shown in figure 17.3.

```
. margins month#group, nopvalues
Adjusted predictions                              Number of obs     =        900
Expression   : Linear prediction, fixed portion, predict()

------------------------------------------------------------------------------
             |            Delta-method
             |     Margin   Std. Err.     [95% Conf. Interval]
-------------+----------------------------------------------------------------
 month#group |
   1#Control |      344.7   3.002539      338.8151    350.5849
1#Medication |      347.3   3.002539      341.4151    353.1849
 1#Education |     347.87   3.002539      341.9851    353.7549
   2#Control |     352.76   2.891807      347.0922    358.4278
2#Medication |     377.24   2.891807      371.5722    382.9078
 2#Education |     363.04   2.891807      357.3722    368.7078
   3#Control |     358.05   3.030548      352.1102    363.9898
3#Medication |     378.87   3.030548      372.9302    384.8098
 3#Education |     395.71   3.030548      389.7702    401.6498
------------------------------------------------------------------------------

. marginsplot, noci
  Variables that uniquely identify margins: month group
```

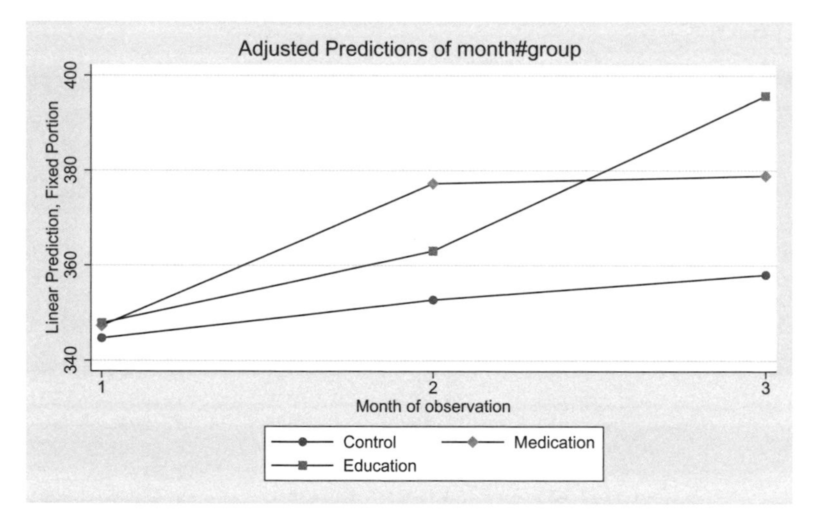

Figure 17.3. Sleep by month and treatment group

Let's now use the contrast command to test the group by month interaction. This overall interaction is significant.

```
. contrast group#month
Contrasts of marginal linear predictions
Margins      : asbalanced

                   df          chi2     P>chi2

sleep
  group#month       4         62.05     0.0000
```

Let's form an interaction contrast in which we apply reference group contrasts to treatment group (r.group) and reverse adjacent group contrasts to month (ar.month). The r.group contrast compares the medication group with the control group and the education group with the control group. The ar.month contrast compares month 2 versus 1 and month 3 versus 2.

```
. contrast ar.month#r.group, nowald pveffects
Contrasts of marginal linear predictions
Margins      : asbalanced

                                        Contrast   Std. Err.        z    P>|z|

sleep
                         month#group
(2 vs 1) (Medication vs Control)           21.88    5.260465     4.16    0.000
 (2 vs 1) (Education vs Control)            7.11    5.260465     1.35    0.177
(3 vs 2) (Medication vs Control)           -3.66    5.354849    -0.68    0.494
 (3 vs 2) (Education vs Control)           27.38    5.354849     5.11    0.000
```

The first two contrasts focus on the comparison of month 2 versus 1. The first contrast shows that the gain in sleep comparing month 2 with month 1 for the medication group is significantly greater (by 21.88 minutes) than the gain for the control group. The second contrast makes the same kind of comparison but focuses on the comparison of the education group with the control group. The gain in sleep in month 2 versus 1 for the education group is 7.11 minutes more than for the control group but is not significant.

The third and fourth contrasts focus on the comparison of month 3 versus 2. The third contrast shows that the change in sleep comparing month 3 with month 2 is not significantly different for the medication group than for the control group. The fourth contrast shows that the change in sleep comparing month 3 with month 2 is significantly greater for the education group than for the control group.

In addition to comparing each treatment (education and medication) with the control group, we may also be interested in comparing education with medication. Let's test the interaction of treatment (education versus medication) by month, but now let's compare each month with the baseline month: month 2 versus 1 and month 3 versus 1.

This is obtained by interacting the comparison of group 3 versus group 2 (`ar3.group`) by each month compared with month 1 (`r.month`). This is tested using the `contrast` command.

```
. contrast ar3.group#r.month, nowald pveffects
Contrasts of marginal linear predictions
Margins      : asbalanced
```

	Contrast	Std. Err.	z	P>|z|
sleep				
group#month				
(Education vs Medication) (2 vs 1)	-14.77	5.260465	-2.81	0.005
(Education vs Medication) (3 vs 1)	16.27	5.745994	2.83	0.005

The first test is significant. The estimate is −14.77, indicating that the gain in sleep comparing month 2 with month 1 for the education group was 14.77 minutes less than the gain for the medication group. In other words, the gain for the medication group was 14.77 minutes greater than the gain for the education group.

The second test is also significant, but in the opposite direction. The estimate is 16.27, indicating that the gain in sleep comparing month 3 with month 1 was 16.27 minutes more for the education group than the gain for the medication group.

These results indicate that at the second month of the study, medication is superior to education in increasing sleep. However, by the third month, education surpasses medication in increasing sleep.

See section 8.4 for more details about how to analyze and dissect interactions from three by three models.

17.5 Comparing models with different residual covariance structures

The examples illustrated in this chapter all modeled the residual covariances using an unstructured covariance matrix. However, Stata allows you to fit a variety of residual covariance structures. The selection of the covariance structure impacts the estimates of the standard errors of the coefficients, but not the point estimates of the coefficients.

When you have only three time points, an unstructured covariance can be a good choice. However, as the number of time points increases, the number of variances and covariances estimated by an unstructured covariance matrix increases dramatically. For example, if you have five time points, an unstructured covariance estimates five variances and 10 covariances (a total of 15 parameters). In such cases, you might consider more parsimonious covariance structures, such as the **exchangeable**, **ar** (autoregressive), or **banded** residual types.[2]

This leads to the question of how to choose among models using different covariance structures. My first recommendation would be to select a residual covariance structure that is grounded in theory or suggested by previous research. However, such information may be scarce or nonexistent. In such cases, you can fit different covariance structures seeking the residual covariance structure that combines the fewest parameters with the best measure of fit—using, for example, Akaike information criterion (AIC) or Bayesian information criterion (BIC). Stata makes this process easy, as illustrated below.

Using the dataset from example 1, models are fit using three different covariance structures: **unstructured**, **exchangeable**, and **ar 1**. After fitting each model, the **estimates store** command is used to store the estimates from the respective model.

```
. use sleep_cat3, clear
. mixed sleep i.month || id:, noconstant residuals(unstructured, t(month))
  (output omitted)
. estimates store m_un
. mixed sleep i.month || id:, noconstant residuals(exchangeable, t(month))
  (output omitted)
. estimates store m_ex
. mixed sleep i.month || id:, noconstant residuals(ar 1, t(month))
  (output omitted)
. estimates store m_ar1
```

Now, the **estimates stats** command can be used to show a table of the fit indices for each of the models.

```
. estimates stats m_un m_ar1 m_ex
Akaike's information criterion and Bayesian information criterion
```

Model	N	ll(null)	ll(model)	df	AIC	BIC
m_un	300	.	-1437.371	9	2892.742	2926.076
m_ar1	300	.	-1439.674	5	2889.348	2907.867
m_ex	300	.	-1438.879	5	2887.758	2906.277

```
Note: BIC uses N = number of observations. See [R] BIC note.
```

Remember that when it comes to AIC and BIC, smaller is better. The **ar 1** and **exchangeable** models have smaller AIC values than the **unstructured** model. Likewise, the **ar 1** and **exchangeable** models also have smaller BIC values than the **unstructured** model. The **ar 1** and **exchangeable** models also have the added benefit of including four fewer residual covariance parameters (5 versus 9). The **ar 1** and **exchangeable**

2. See [ME] **mixed** for a list of all available covariance structures you can choose within the **residuals()** option. Furthermore, chapter 7 of Singer and Willett (2003) provides additional descriptions of these residual covariance structures, including information to help you choose among the different structures.

covariance structure appear to provide a fairly similar quality of fit, and both fit better than the **unstructured** covariance structure.

For the examples in this chapter, the differences in the results are trivial. You can fit the models for yourself and see that the standard errors are similar using the different residual covariance structures. However, if there were more time points and the differences in the covariance structures were more striking, the impact of the choice of the residual covariance structure on the standard errors could be greater.

> **Warning! Likelihood-ratio test**
>
> It is tempting to ask whether the difference in covariance structures is significantly different and to want to use a command like `lrtest` to test whether one covariance structure fits significantly better than another. A key assumption of a likelihood-ratio test is that one model is nested within another model, where one model can be created from the other by omitting one or more parameters. In many (or perhaps most) cases, the models formed by comparing two different residual covariance structures are not nested within each other and the likelihood-ratio test is not valid. However, the AIC and BIC indices can be used even when models are not nested within each other.

17.6 Analyses with small samples

All the examples shown, so far, had large-sample sizes. In particular, the example from section 17.3 had a total of 200 subjects (100 in the treatment condition and 100 in the control condition). Each subject was measured three times. By default, the `mixed` command uses *large-sample* methods. Applying such methods with small-sample sizes may lead to overly liberal statistical tests (that is, greater type I error rates).

The `mixed` command allows you to specify the `dfmethod()` option to select testing methods that are more appropriate for small-sample sizes. Specifying the `dfmethod()` option is easy—knowing the best method to select is not so easy. For an introduction to the different small-sample methods you can choose from, I encourage you to see `help mixed`, especially the section in the PDF documentation titled *Small-sample inference for fixed effects*. That section describes five different small-sample adjustment methods: `residual`, `repeated`, `anova`, `satterthwaite`, and `kroger`. The first three methods are only suitable under limited (and frequently unrealistic) conditions.[3] For the kinds of models discussed in this chapter, I would say that these methods would very rarely be appropriate. This leaves two remaining methods to consider: `dfmethod(kroger)` and `dfmethod(satterthwaite)`. The `dfmethod(kroger)` option (described in Kenward

3. The `residual` and `anova` methods are only appropriate when errors are assumed to be independent, and the `repeated` method is appropriate for balanced repeated-measures designs with spherical correlation structures.

and Roger (1997)) and `dfmethod(satterthwaite)` option (described in Satterthwaite (1946)) offer methods that are more applicable when making inferences with longitudinal models with small-sample sizes. The *Small-sample inference for fixed effects* section in [ME] **mixed** provides a very concise summary that I have found very helpful, quoted below:

> According to Schaalje, McBride, and Fellingham (2002), the Kenward–Roger method should, in general, be preferred to the Satterthwaite method. However, there are situations in which the two methods are expected to perform similarly, such as with compound symmetry covariance structures.

There will likely be continued research in this area, and the potential for new techniques and options to arise. For now, it seems that the Kenward–Roger method is probably the most generally useful method available, although sensitivity analyses considering the Satterthwaite method would seem prudent.

With this advice in mind, consider `sleep_catcat23small.dta` below. It contains a total of 80 observations across three months with 28 observations in month 1, 28 observations in month 2, and 24 observations in month 3.

```
. use sleep_catcat23small, clear
. count
  80
```

The `tabulate` command is used below to show the number of observations by `group` and `month`. The control group has 14, 13, and 11 observations (respectively) across the first three months. The medication group has 14, 15, and 13 observations (respectively) across the first three months.

```
. tab group month
 Treatment |      Month of observation
     group |         1          2          3 |     Total
-----------+---------------------------------+----------
   Control |        14         13         11 |        38
Medication |        14         15         13 |        42
-----------+---------------------------------+----------
     Total |        28         28         24 |        80
```

The `list` command is used below to show the first 20 cases, after sorting the observations on `id` and `month`. Note how `id` number 5 has only two observations (for months 2 and 3) and `id` number 7 has only two observations (for months 1 and 2).

```
. sort id month
. list in 1/20, sepby(id)
```

	id	month	group	sleep
1.	1	1	Control	315
2.	1	2	Control	379
3.	1	3	Control	320
4.	2	1	Control	392
5.	2	2	Control	369
6.	2	3	Control	314
7.	3	1	Control	357
8.	3	2	Control	349
9.	3	3	Control	374
10.	4	1	Control	338
11.	4	2	Control	354
12.	4	3	Control	375
13.	5	2	Control	395
14.	5	3	Control	329
15.	6	1	Control	379
16.	6	2	Control	362
17.	6	3	Control	327
18.	7	1	Control	388
19.	7	2	Control	356
20.	8	1	Control	328

The **summarize** command below shows summary statistics for all the variables in the dataset. Note how sleep time, measured in minutes, ranges from 304 to 413.

```
. summarize
```

Variable	Obs	Mean	Std. Dev.	Min	Max
id	80	15.7625	8.884981	1	30
month	80	1.95	.8097507	1	3
group	80	1.525	.5025253	1	2
sleep	80	358.4625	26.70085	304	413

Although this dataset has a smaller-sample size, the analysis below uses the default, large-sample statistical methods. Note how the significance of each parameter is tested using z tests. This reflects the use of large-sample statistical methods for this analysis.

```
. mixed sleep i.group##i.month || id:, noconstant residuals(un, t(month))
Obtaining starting values by EM:
Performing gradient-based optimization:
Iteration 0:   log likelihood = -364.19065
Iteration 1:   log likelihood = -363.50017
Iteration 2:   log likelihood = -363.12726
Iteration 3:   log likelihood =  -363.1244
Iteration 4:   log likelihood =  -363.1244
Computing standard errors:
Mixed-effects ML regression                     Number of obs      =        80
Group variable: id                              Number of groups   =        30
                                                Obs per group:
                                                              min =         1
                                                              avg =       2.7
                                                              max =         3
                                                Wald chi2(5)       =     27.47
Log likelihood =  -363.1244                     Prob > chi2        =    0.0000

------------------------------------------------------------------------------
       sleep |      Coef.   Std. Err.      z    P>|z|     [95% Conf. Interval]
-------------+----------------------------------------------------------------
       group |
     Control |          0  (base)
  Medication |  -19.76388   9.335211    -2.12   0.034    -38.06055   -1.467197
             |
       month |
           1 |          0  (base)
           2 |  -6.196209   8.577699    -0.72   0.470    -23.00819    10.61577
           3 |  -20.99035   10.16346    -2.07   0.039    -40.91036   -1.070345
             |
 group#month |
Medication#2 |   41.97614   11.93497     3.52   0.000     18.58403    65.36824
Medication#3 |   48.40432   14.14573     3.42   0.001      20.6792    76.12944
             |
       _cons |   361.3839   6.601072    54.75   0.000     348.4461    374.3218
------------------------------------------------------------------------------

------------------------------------------------------------------------------
  Random-effects Parameters  |   Estimate   Std. Err.     [95% Conf. Interval]
-----------------------------+------------------------------------------------
id:                  (empty) |
-----------------------------+------------------------------------------------
Residual: Unstructured       |
                     var(e1) |   612.0411   163.5938      362.4609    1033.475
                     var(e2) |   459.7228   122.7743      272.3785    775.9242
                     var(e3) |   505.4574   145.9643       286.997    890.2084
                  cov(e1,e2) |   40.15257   105.0518     -165.7451    246.0502
                  cov(e1,e3) |  -111.1793   115.7645     -338.0736    115.7149
                  cov(e2,e3) |   60.66916   110.8819     -156.6553    277.9936
------------------------------------------------------------------------------
LR test vs. linear model: chi2(5) = 2.13                  Prob > chi2 = 0.8305
Note: The reported degrees of freedom assumes the null hypothesis is not on
      the boundary of the parameter space.  If this is not true, then the
      reported test is conservative.
```

The **contrast** command is used below to compare the change in sleep across months between the medication group versus the control group. Let's focus our attention on the first result, comparing the change in sleep from month 1 to month 2. From month 1 to month 2, the sleep time increased by 41.98 minutes more for the medication group than for the control group. That increase, tested using a Wald z test, was significant ($z = 3.52$, $p < 0.001$).

```
. contrast ar.month#group, nowald pveffects
Contrasts of marginal linear predictions
Margins      : asbalanced

                                       Contrast   Std. Err.        z    P>|z|

sleep
                        month#group
(2 vs 1) (Medication vs base)          41.97614    11.93497     3.52    0.000
(3 vs 2) (Medication vs base)          6.428186     11.5398     0.56    0.577
```

Given the sample size of this analysis, our results might be more defensible if we chose an analysis technique more appropriate for small-sample sizes. The **mixed** command below adds the **dfmethod(kroger)** option to use the methods presented in Kenward and Roger (1997). This method is only supported with REML[4] estimation, so the **mixed** command also specifies the **reml** option.

4. REML stands for *restricted maximum likelihood*, also known as *residual maximum likelihood*. For more details, see the *Likelihood versus restricted likelihood* section in [ME] **mixed**.

```
. mixed sleep i.group##i.month || id:, noconstant residuals(un, t(month))
> dfmethod(kroger) reml nolog

Mixed-effects REML regression                   Number of obs     =         80
Group variable: id                              Number of groups  =         30

                                                Obs per group:
                                                              min =          1
                                                              avg =        2.7
                                                              max =          3
DF method: Kenward-Roger                        DF:           min =      27.51
                                                              avg =      28.41
                                                              max =      29.07

                                                F(5,      30.73)  =       4.60
Log restricted-likelihood =  -346.5452          Prob > F          =     0.0030

------------------------------------------------------------------------------
       sleep |      Coef.   Std. Err.      t    P>|t|     [95% Conf. Interval]
-------------+----------------------------------------------------------------
       group |
     Control |          0  (base)
  Medication |  -19.75498   9.741694    -2.03   0.052    -39.67701    .1670519
             |
       month |
           1 |          0  (base)
           2 |  -6.185918   9.006856    -0.69   0.498    -24.61691    12.24508
           3 |  -20.97863   10.69184    -1.96   0.060    -42.86182    .9045553
             |
 group#month |
Medication#2 |    41.9638   12.50352     3.36   0.002     16.33777    67.58983
Medication#3 |   48.39221   14.84885     3.26   0.003     17.95127    78.83314
             |
       _cons |   361.3771   6.888292    52.46   0.000     347.2903    375.4638
------------------------------------------------------------------------------

------------------------------------------------------------------------------
  Random-effects Parameters  |   Estimate   Std. Err.     [95% Conf. Interval]
-----------------------------+------------------------------------------------
id:                  (empty) |
-----------------------------+------------------------------------------------
Residual: Unstructured       |
                     var(e1) |   658.9749   182.7475      382.6599    1134.815
                     var(e2) |    495.057   137.1944      287.5829    852.2115
                     var(e3) |   550.8616    165.993      305.1737    994.3471
                  cov(e1,e2) |   42.92493   117.0647     -186.5176    272.3675
                  cov(e1,e3) |  -119.1169   129.7763     -373.4738    135.2401
                  cov(e2,e3) |   65.14252   124.4511     -178.7772    309.0622
------------------------------------------------------------------------------
LR test vs. linear model: chi2(5) = 1.95                  Prob > chi2 = 0.8566

Note: The reported degrees of freedom assumes the null hypothesis is not on
      the boundary of the parameter space.  If this is not true, then the
      reported test is conservative.
```

Like the prior analysis, I again use the `contrast` command to compare the change in sleep across months between the medication and control groups. To obtain inferences using small-sample methods, I included the `small` option on the `contrast` command.[5] As before, let's focus our attention on the first result, comparing the change in sleep from month 1 to month 2. From month 1 to month 2, the sleep time increased by

5. Otherwise, large-sample inference methods would have been used.

41.96 minutes more for the medication group than for the control group. That increase, tested using a t test, was significant ($t(27.7) = 3.36$, $p = 0.002$). Because we specified the `small` option, a t test was used instead of a z test (a large-sample method).

```
. contrast ar.month#group, nowald pveffects small
Contrasts of marginal linear predictions
Margins      : asbalanced
```

	Contrast	Std. Err.	DF	t	P>|t|
sleep					
month#group					
(2 vs 1)					
(Medication vs base)	41.9638	12.50352	27.7	3.36	0.002
(3 vs 2)					
(Medication vs base)	6.428405	12.23524	24.8	0.53	0.604

If you prefer, you could specify the `dfmethod(satterthwaite)` option to specify the Satterthwaite method. This method is only supported with REML estimation, so the `mixed` command also specifies the `reml` option.

```
. mixed sleep i.group##i.month || id:, noconstant residuals(un, t(month))
> reml dfmethod(satterthwaite) nolog

Mixed-effects REML regression                   Number of obs     =         80
Group variable: id                              Number of groups  =         30

                                                Obs per group:
                                                              min =          1
                                                              avg =        2.7
                                                              max =          3
DF method: Satterthwaite                        DF:           min =      27.51
                                                              avg =      28.41
                                                              max =      29.07

                                                F(5,     27.90)   =       5.08
Log restricted-likelihood = -346.5452           Prob > F          =     0.0019

------------------------------------------------------------------------------
       sleep |      Coef.   Std. Err.      t    P>|t|     [95% Conf. Interval]
-------------+----------------------------------------------------------------
       group |
     Control |          0  (base)
  Medication |  -19.75498   9.686877    -2.04   0.051    -39.56491    .0549502
             |
       month |
           1 |          0  (base)
           2 |  -6.185918    8.90339    -0.69   0.493    -24.40519    12.03335
           3 |  -20.97863   10.57139    -1.98   0.057    -42.61528    .6580243
             |
 group#month |
Medication#2 |    41.9638    12.3882     3.39   0.002     16.57411    67.35349
Medication#3 |   48.39221   14.71133     3.29   0.003      18.2332    78.55121
             |
       _cons |   361.3771   6.849739    52.76   0.000     347.3692     375.385
------------------------------------------------------------------------------

------------------------------------------------------------------------------
  Random-effects Parameters  |   Estimate   Std. Err.     [95% Conf. Interval]
-----------------------------+------------------------------------------------
id:                  (empty) |
-----------------------------+------------------------------------------------
Residual: Unstructured       |
                     var(e1) |   658.9749   182.7475      382.6599    1134.815
                     var(e2) |    495.057   137.1944      287.5829    852.2115
                     var(e3) |   550.8616    165.993      305.1737    994.3471
                 cov(e1,e2) |   42.92493   117.0647     -186.5176    272.3675
                 cov(e1,e3) |  -119.1169   129.7763     -373.4738    135.2401
                 cov(e2,e3) |   65.14252   124.4511     -178.7772    309.0622
------------------------------------------------------------------------------
LR test vs. linear model: chi2(5) = 1.95                  Prob > chi2 = 0.8566

Note: The reported degrees of freedom assumes the null hypothesis is not on
      the boundary of the parameter space.  If this is not true, then the
      reported test is conservative.
```

Again, let's use the `contrast` command, as shown below, to compare the change in sleep across months between the medication and control group using small-sample inference methods (by including the `small` option). As before, let's focus our attention on the first result, comparing the change in sleep from month 1 to month 2. From month 1 to month 2, the sleep time increased by 41.96 minutes more for the medication group than for the control group. That increase, tested using a t test, was significant [$t(27.7) = 3.39$, $p = 0.002$].

```
. contrast ar.month#group, nowald pveffects small
Contrasts of marginal linear predictions
Margins      : asbalanced
```

	Contrast	Std. Err.	DF	t	P>|t|
sleep					
month#group					
(2 vs 1) (Medication vs base)	41.9638	12.3882	27.7	3.39	0.002
(3 vs 2) (Medication vs base)	6.428405	12.02356	24.8	0.53	0.598

For overviews of methods of estimation with mixed models with small-sample sizes, see [ME] **mixed**, particularly the *Small-sample inference for fixed effects* section, as well as Yang (2015) and Schaalje, McBride, and Fellingham (2002). For more details about the `dfmethod(kroger)` option, see Kenward and Roger (1997). For more details about the `dfmethod(satterthwaite)` option, see Satterthwaite (1946).

17.7 Summary

This chapter has illustrated three different examples in which time was treated as a categorical variable. These models used the `mixed` command combined with the `residuals()` option to account for the residual covariances across times. The fixed-effects portion of the `mixed` models (that is, the part before the `||`) was specified the same whether the model was longitudinal or nonlongitudinal. Furthermore, the postestimation commands we used for interpreting and visualizing the results (the `margins`, `marginsplot`, and `contrast` commands) were the same whether the model was longitudinal or nonlongitudinal. The chapter also described techniques that you can use when analyzing small-sample sizes and provided examples illustrating the analysis of data with a small-sample size.

For more information about traditional longitudinal models where time is treated categorically, I recommend Keppel and Wickens (2004), Cohen et al. (2003), and Maxwell et al. (2018). Singer and Willett (2003) illustrate longitudinal analysis using the more general multilevel framework, and include more information about residual covariance structures and how to select among them.

18 Nonlinear models

18.1 Chapter overview

The methods that have been illustrated for constructing linear models (that is, with the **regress** and **anova** commands) can also be applied to nonlinear models, such as logistic, multinomial logistic, ordinal logistic, and Poisson models. For example, you can apply the techniques from chapter 7 to fit a logistic regression using a categorical predictor, use what we learned in chapter 4 to fit a piecewise multinomial regression model, or use the techniques from chapter 10 to model continuous by categorical interactions involving an ordinal logistic regression model.

You can also use the **margins**, **marginsplot**, **contrast**, and **pwcompare** commands to interpret and visualize the results from such models. However, for nonlinear models, we need to be attentive to the metric of the outcome. For the **contrast** and **pwcompare** commands, the comparisons are made in the same metric as the estimation model. For example, when using the **contrast** command following a logistic regression, the comparisons are made in the log-odds (logit) metric, the metric in which the model is fit. By contrast, the **margins** command can produce predictive margins in the log-odds metric, in the odds metric, or in the probability metric. The relationship between the predictor and predictive margins is linear in the log-odds metric (but the log-odds metric is more difficult to understand). It can be easier to understand predictive margins expressed in the probability metric, but the model is not linear in that metric.

The issues in fitting and interpreting such nonlinear models are covered in this chapter. The chapter begins with a discussion of logistic regression models (see section 18.2). This section is the most detailed in this chapter, both because logistic models are common and because some issues arise when fitting nonlinear models. Then examples are provided illustrating the analysis and interpretation of multinomial logistic models (see section 18.3), ordinal logistic models (see section 18.4), and Poisson models (see section 18.5). The intent of these sections is to introduce the particular issues involved in using the `margins`, `marginsplot`, `contrast`, and `pwcompare` commands with these kinds of models. Section 18.6 illustrates further applications of nonlinear models, providing examples that apply modeling strategies illustrated in previous chapters to a nonlinear model (specifically, a logistic regression model). Three examples are provided, showing a categorical by categorical interaction (see section 18.6.1), a categorical by continuous interaction (see section 18.6.2), and a piecewise model with two knots (see section 18.6.3).

18.2 Binary logistic regression

18.2.1 A logistic model with one categorical predictor

Let's consider a simple logistic regression model that predicts whether a person smokes (`smoke`) by the person's self-reported social class (`class`). The variable `class` is a categorical variable that is coded: 1 = lower class, 2 = working class, 3 = middle class, and 4 = upper class. Let's use the GSS dataset and fit a logistic regression model predicting smoking by social class.

```
. use gss_ivrm
. logit smoke i.class, nolog
Logistic regression                             Number of obs     =     15,464
                                                LR chi2(3)        =     198.45
                                                Prob > chi2       =     0.0000
Log likelihood = -9904.5707                     Pseudo R2         =     0.0099

------------------------------------------------------------------------------
       smoke |      Coef.   Std. Err.      z    P>|z|     [95% Conf. Interval]
-------------+----------------------------------------------------------------
       class |
 lower class |          0  (base)
working cl.. |   -.318563   .0729422    -4.37   0.000    -.4615271   -.1755989
middle class |  -.7201353   .0734895    -9.80   0.000     -.864172   -.5760985
 upper class |  -.9199569   .1219929    -7.54   0.000    -1.159059   -.6808552
             |
       _cons |  -.1272302   .0687383    -1.85   0.064    -.2619547    .0074944
------------------------------------------------------------------------------
```

The regression coefficients are presented in the log-odds (logit) metric using a dummy coding scheme with lower class as the base (reference) group. So, for example, the coefficient for middle `class` (-0.72) shows the difference in the log odds of smoking for people who identify as middle class versus lower class.

We can interpret and visualize the results of this model using the `contrast`, `pwcompare`, `margins`, and `marginsplot` commands, as described in the following sections.

Using the contrast command

If we want to test the overall equality of the four social class groups in terms of their log odds of smoking, we can use the `contrast` command, as shown below. This test shows that the four social class groups are not all equal in terms of their log odds of smoking.

```
. contrast class
Contrasts of marginal linear predictions
Margins      : asbalanced
```

	df	chi2	P>chi2
class	3	196.71	0.0000

We can also apply contrast operators to form specific comparisons among the levels of the social class. Using the `ar.` contrast operator compares each level of class with the adjacent (previous) level of class.

```
. contrast ar.class, nowald pveffects
Contrasts of marginal linear predictions
Margins      : asbalanced
```

	Contrast	Std. Err.	z	P>|z|
class				
(working class vs lower class)	-.318563	.0729422	-4.37	0.000
(middle class vs working class)	-.4015723	.0356562	-11.26	0.000
(upper class vs middle class)	-.1998216	.104082	-1.92	0.055

The difference in the log odds of smoking for those who identify as working class versus lower class is significant ($z = -4.37$, $p = 0.000$), as is the difference for those who identify as middle class versus working class ($z = -11.26$, $p = 0.000$). However, the log odds of smoking for those who identify as upper class is not significantly different from those who identify as middle class ($z = -1.92$, $p = 0.055$).

We can add the `or` option to the `contrast` command to display the results as odds ratios. The statistical tests are identical; however, the results can now be interpreted using odds ratios. For example, the odds of smoking for a person who identifies as middle class is 0.669 times the odds of smoking for someone who identifies as working class.

```
. contrast ar.class, nowald pveffects or
Contrasts of marginal linear predictions
Margins      : asbalanced
```

	Odds Ratio	Std. Err.	z	P>|z|
class				
(working class vs lower class)	.7271933	.0530431	-4.37	0.000
(middle class vs working class)	.6692669	.0238635	-11.26	0.000
(upper class vs middle class)	.8188768	.0852304	-1.92	0.055

Using the pwcompare command

We can also use the `pwcompare` command to form comparisons among the levels of the social class. Like the `contrast` command, these comparisons are made in the log-odds metric. Five of the six pairwise comparisons are statistically significant. The comparison of class 4 versus 3 is not significant ($z = -1.92$, $p = 0.055$).

```
. pwcompare class, pveffects
Pairwise comparisons of marginal linear predictions
Margins      : asbalanced
```

	Contrast	Std. Err.	Unadjusted z	Unadjusted P>|z|
smoke				
class				
working class vs lower class	-.318563	.0729422	-4.37	0.000
middle class vs lower class	-.7201353	.0734895	-9.80	0.000
upper class vs lower class	-.9199569	.1219929	-7.54	0.000
middle class vs working class	-.4015723	.0356562	-11.26	0.000
upper class vs working class	-.6013939	.1036963	-5.80	0.000
upper class vs middle class	-.1998216	.104082	-1.92	0.055

The `or` option can also be added to the `pwcompare` command to express the results as exponentiated coefficients (that is, odds ratios).

```
. pwcompare class, pveffects or
Pairwise comparisons of marginal linear predictions
Margins      : asbalanced
```

	Odds Ratio	Std. Err.	Unadjusted z	P>|z|
smoke				
class				
working class vs lower class	.7271933	.0530431	-4.37	0.000
middle class vs lower class	.4866864	.0357663	-9.80	0.000
upper class vs lower class	.3985362	.0486186	-7.54	0.000
middle class vs working class	.6692669	.0238635	-11.26	0.000
upper class vs working class	.5480472	.0568305	-5.80	0.000
upper class vs middle class	.8188768	.0852304	-1.92	0.055

Using the margins and marginsplot commands

Unlike the `contrast` and `pwcompare` commands, the `margins` command allows us to choose the metric in which estimates are computed, as illustrated in the following examples.

Let's use the `margins` command to show the predictive margin for the probability of smoking separated by `class`. This shows, for example, that the predictive margin for the probability of smoking for people who describe themselves as lower class is 0.47.

```
. margins class, nopvalues
Adjusted predictions                    Number of obs     =     15,464
Model VCE    : OIM
Expression   : Pr(smoke), predict()
```

	Margin	Delta-method Std. Err.	[95% Conf.	Interval]
class				
lower class	.4682353	.0171152	.4346901	.5017805
working class	.3903614	.0058079	.3789781	.4017448
middle class	.2999858	.0054589	.2892866	.310685
upper class	.2597656	.0193794	.2217827	.2977486

> **Note! Predictive margins**
>
> When discussing linear models, the term *adjusted mean* was used to describe the predicted mean of the outcome after adjusting for the covariates in the model. A predictive margin is a generalization of an adjusted mean applied to a nonlinear model (such as a logistic regression model).

We can use the `marginsplot` command to make a graph of these predictive margins as a function of class. This resulting graph is shown in figure 18.1.

```
. marginsplot, xlabel(, angle(45))
  Variables that uniquely identify margins: class
```

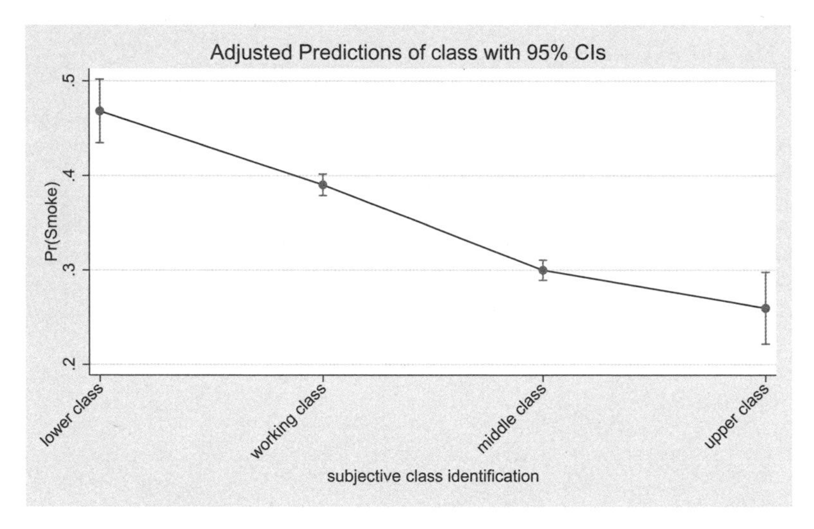

Figure 18.1. Predictive margins for the probability of smoking by social class

Suppose we apply the `ar.` contrast coefficient to `class` using the `margins` command. This forms adjacent group comparisons on `class`, comparing each class group with the previous class.

```
. margins ar.class, contrast(nowald pveffects)
Contrasts of adjusted predictions                 Number of obs     =     15,464
Model VCE    : OIM
Expression   : Pr(smoke), predict()

------------------------------------------------------------------
                                 |            Delta-method
                                 |   Contrast   Std. Err.      z    P>|z|
---------------------------------+--------------------------------
                           class |
   (working class vs lower class) |  -.0778738    .0180738    -4.31   0.000
  (middle class vs working class) |  -.0903756    .0079706   -11.34   0.000
    (upper class vs middle class) |  -.0402202    .0201336    -2.00   0.046
------------------------------------------------------------------
```

Looking at the first contrast, we see that the predictive margin for the probability of smoking is 0.08 lower than for someone who identifies as working class versus lower class. This contrast is significant ($z = -4.31$, $p = 0.000$).

It might be tempting to think that the significance tests produced by the `contrast ar.class` command would be identical to the `margins ar.class` command. Comparing the results of the `margins ar.class` command with the `contrast ar.class` command, we see that the z-values and p-values are similar, but not identical. For example, the p-value is 0.055 for the comparison of upper class versus middle class from the `contrast` command and is 0.046 for the same comparison from the `margins` command. This underscores the fact that the `contrast` command performed its test using the log-odds metric whereas the `margins` command used the probability metric.

You can use the `predict(xb)` option with the `margins` command to request the use of the linear (log-odds) metric, as illustrated below. The results from the following `margins` command are identical to the previous results using the `contrast` command because both are using the log-odds metric.

```
. margins ar.class, contrast(nowald pveffects) predict(xb)
Contrasts of adjusted predictions                 Number of obs     =     15,464
Model VCE    : OIM
Expression   : Linear prediction (log odds), predict(xb)

------------------------------------------------------------------
                                 |            Delta-method
                                 |   Contrast   Std. Err.      z    P>|z|
---------------------------------+--------------------------------
                           class |
   (working class vs lower class) |   -.318563    .0729422    -4.37   0.000
  (middle class vs working class) |  -.4015723    .0356562   -11.26   0.000
    (upper class vs middle class) |  -.1998216     .104082    -1.92   0.055
------------------------------------------------------------------
```

Note! The default metric for the margins command

The `margins` command used predicted probabilities by default following a logistic regression. We can discover the default prediction metric used by `margins` after `logit` by typing `help logit postestimation`, and then clicking on the *Jump to* button and selecting *margins*. This shows us that the default prediction is `pr`, the predicted probability of the outcome. It also shows us that we can specify `xb` to obtain the linear prediction (that is, the logit). You can use this strategy for discovering the default prediction for the `margins` command following any estimation command, as well as discovering other metrics you can use for the `margins` command. Note that some of the options available for the `predict` command (for example, `stdp`) are not appropriate for the `margins` command.

Using the margins command with the pwcompare option

The `margins` command can be combined with the `pwcompare` option to bring you the combination of features provided by the `margins` and `pwcompare` commands. This allows you to form pairwise comparisons in any of the metrics supported by the `margins` command.[1] The `pwcompare` option is added to the `margins` command below to request pairwise comparisons among the four class groups in terms of their probability of smoking.

1. Compare this with the `pwcompare` command that was limited only to the natural metric of the model. In the case of logistic regression, it is limited only to the log-odds metric, or odds-ratios via the `or` option.

```
. margins class, pwcompare

Pairwise comparisons of adjusted predictions     Number of obs     =     15,464
Model VCE    : OIM

Expression   : Pr(smoke), predict()

------------------------------------------------------------------------------
                              |            Delta-method         Unadjusted
                              |   Contrast   Std. Err.     [95% Conf. Interval]
------------------------------+-----------------------------------------------
                        class |
 working class vs lower class |  -.0778738   .0180738     -.1132979   -.0424498
  middle class vs lower class |  -.1682495   .0179647     -.2034596   -.1330394
   upper class vs lower class |  -.2084697   .0258552      -.259145   -.1577944
                 middle class |
                           vs |
                working class |  -.0903756   .0079706     -.1059978   -.0747535
 upper class vs working class |  -.1305958    .020231     -.1702479   -.0909438
  upper class vs middle class |  -.0402202   .0201336     -.0796812   -.0007591
------------------------------------------------------------------------------
```

To display the significance tests of the pairwise comparisons, in lieu of the confidence intervals, we can use the `pwcompare(pveffects)` option. Note that `pveffects` is specified as a suboption to the `pwcompare()` option. All six of the pairwise comparisons among the levels of class are statistically significant in the predicted probability metric.

```
. margins class, pwcompare(pveffects)

Pairwise comparisons of adjusted predictions     Number of obs     =     15,464
Model VCE    : OIM

Expression   : Pr(smoke), predict()

-----------------------------------------------------------------
                               |            Delta-method    Unadjusted
                               |   Contrast   Std. Err.      z    P>|z|
-------------------------------+---------------------------------
                         class |
  working class vs lower class |  -.0778738   .0180738    -4.31   0.000
   middle class vs lower class |  -.1682495   .0179647    -9.37   0.000
    upper class vs lower class |  -.2084697   .0258552    -8.06   0.000
middle class vs working class  |  -.0903756   .0079706   -11.34   0.000
  upper class vs working class |  -.1305958    .020231    -6.46   0.000
   upper class vs middle class |  -.0402202   .0201336    -2.00   0.046
-----------------------------------------------------------------
```

You can request that the table be sorted by the size of the difference in the contrast by adding the `sort` suboption within the `pwcompare()` option, as illustrated below.

```
. margins class, pwcompare(pveffects sort)
Pairwise comparisons of adjusted predictions    Number of obs     =     15,464
Model VCE    : OIM
Expression   : Pr(smoke), predict()
```

	Contrast	Delta-method Std. Err.	Unadjusted z	P>|z|
class				
upper class vs lower class	-.2084697	.0258552	-8.06	0.000
middle class vs lower class	-.1682495	.0179647	-9.37	0.000
upper class vs working class	-.1305958	.020231	-6.46	0.000
middle class vs working class	-.0903756	.0079706	-11.34	0.000
working class vs lower class	-.0778738	.0180738	-4.31	0.000
upper class vs middle class	-.0402202	.0201336	-2.00	0.046

Having considered a logistic model with a categorical predictor, let's turn to a logistic model with a continuous predictor.

18.2.2 A logistic model with one continuous predictor

Let's now briefly consider a model with one continuous predictor, predicting whether a person smokes (`smoke`) from his or her education level. This model is fit below.

```
. use gss_ivrm
. logit smoke educ, nolog
Logistic regression                             Number of obs     =     16,332
                                                LR chi2(1)        =     174.04
                                                Prob > chi2       =     0.0000
Log likelihood = -10483.854                     Pseudo R2         =     0.0082
```

smoke	Coef.	Std. Err.	z	P>|z|	[95% Conf.	Interval]
educ	-.0685603	.0052334	-13.10	0.000	-.0788177	-.058303
_cons	.2216819	.0658944	3.36	0.001	.0925312	.3508326

Let's use the `margins` and `marginsplot` commands to visualize the relationship between education and the log odds of smoking. (Note the inclusion of the `predict(xb)` option on the `margins` command to specify the use of the log-odds metric.) The graph created by the `marginsplot` command is shown in figure 18.2. Note how the relationship between education and the log odds of smoking is linear. For every additional year of education, the log odds of smoking decreases by 0.07.

```
. margins, at(educ=(5(1)20)) predict(xb)
  (output omitted)
. marginsplot
  Variables that uniquely identify margins: educ
```

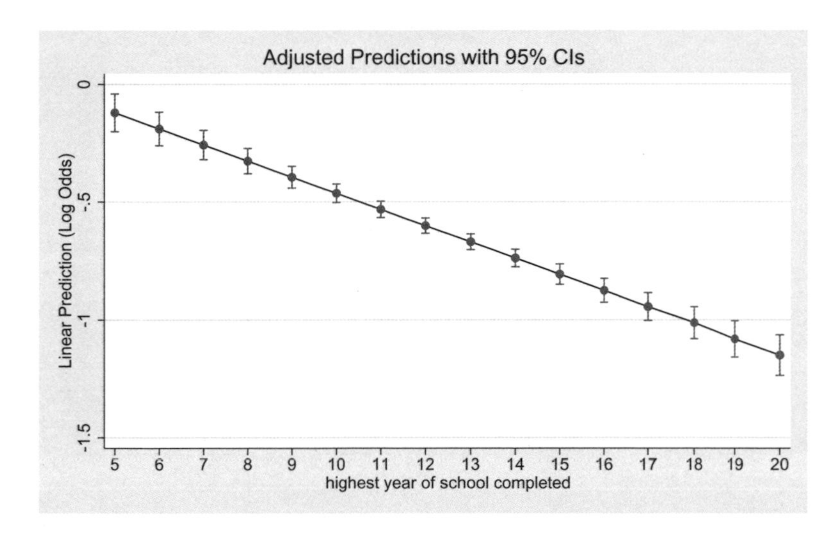

Figure 18.2. Log odds of smoking by education level

Let's now visualize this relationship in terms of the probability of smoking. The `margins` and `marginsplot` commands are used to graph this relationship. (Note the absence of the `predict()` option on the `margins` command, resulting in the use of the predicted probability metric.) The graph created by the `marginsplot` command is shown in figure 18.3. This shows the relationship between the number of years of education and the predictive margin for the probability of smoking. We can see that the probability of smoking decreases as education increases. Although it may be difficult to see, this relationship is not linear.

```
. margins, at(educ=(5(1)20))
  (output omitted)
. marginsplot
  Variables that uniquely identify margins: educ
```

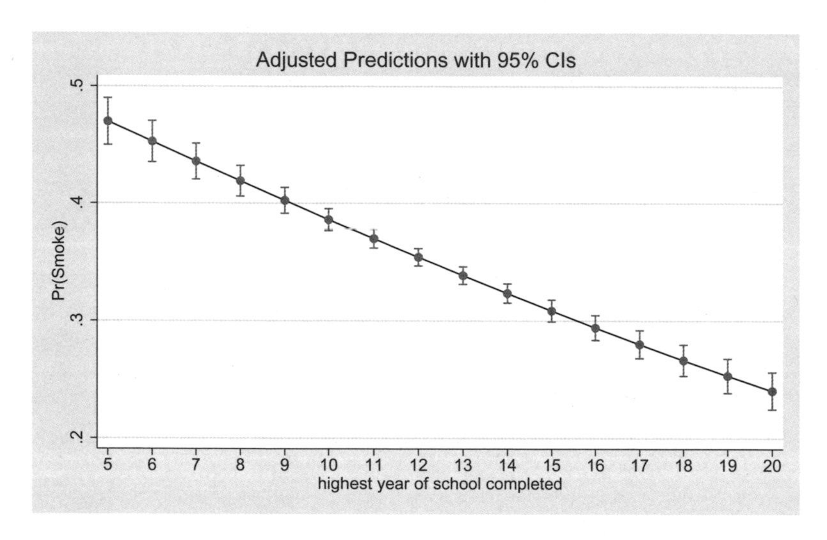

Figure 18.3. Predicted probability of smoking by education level

18.2.3 A logistic model with covariates

Let's add some covariates to this model, predicting smoking from class as well as education, age, and year of interview.

```
. logit smoke i.class educ age yrint, nolog
Logistic regression                             Number of obs     =     15,375
                                                LR chi2(6)        =     742.33
                                                Prob > chi2       =     0.0000
Log likelihood = -9580.0723                     Pseudo R2         =     0.0373

------------------------------------------------------------------------------
       smoke |      Coef.   Std. Err.      z    P>|z|     [95% Conf. Interval]
-------------+----------------------------------------------------------------
       class |
 lower class |          0  (base)
working cl.. |  -.2604145   .0749733    -3.47   0.001    -.4073595   -.1134696
middle class |   -.455007   .0772007    -5.89   0.000    -.6063176   -.3036965
 upper class |  -.5234883   .1274959    -4.11   0.000    -.7733757   -.2736009
             |
        educ |  -.0878902   .0063325   -13.88   0.000    -.1003016   -.0754787
         age |  -.0198399   .0010949   -18.12   0.000    -.0219859   -.0176939
       yrint |  -.0321324   .0033889    -9.48   0.000    -.0387746   -.0254902
       _cons |   65.45743   6.718824     9.74   0.000     52.28877    78.62608
------------------------------------------------------------------------------
```

Let's use the **contrast** command to test the overall effect of class. This test is significant.

```
. contrast class
Contrasts of marginal linear predictions
Margins      : asbalanced

------------------------------------------------
             |         df        chi2     P>chi2
-------------+----------------------------------
       class |          3       49.68     0.0000
------------------------------------------------
```

We can use the **margins** command to help us interpret this effect by computing the predictive margins of the probability of smoking by class, as shown below.

```
. margins class, nopvalues
Predictive margins                              Number of obs     =     15,375
Model VCE    : OIM
Expression   : Pr(smoke), predict()

--------------------------------------------------------------
             |            Delta-method
             |     Margin   Std. Err.     [95% Conf. Interval]
-------------+------------------------------------------------
       class |
 lower class |   .4262118    .0167454      .3933914    .4590323
working class|   .3665092    .0056573      .3554211    .3775974
middle class |   .3242303    .0057422      .3129758    .3354847
 upper class |   .3099447    .0215208      .2677647    .3521246
--------------------------------------------------------------
```

We can use **marginsplot** command to graph these predictive margins, as shown in figure 18.4.

```
. marginsplot, xlabel(, angle(45))
Variables that uniquely identify margins: class
```

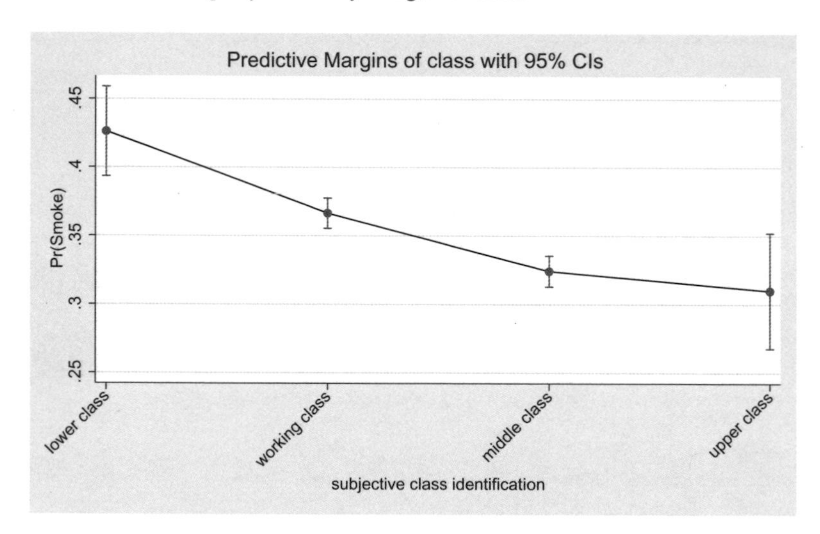

Figure 18.4. The predictive marginal probability of smoking by class

In the predicted probability metric, the size of the effect of a variable can (and will) vary as a function of the value of the covariates. By comparison, in the logit metric (like any linear model), the size of the effect of a variable remains constant regardless of the values of the covariate. Let's explore this point by using the **contrast** command to estimate the effect of class. For example, let's use the **contrast** command to compare each level of class with the previous level of class.

```
. contrast ar.class, nowald pveffects
Contrasts of marginal linear predictions
Margins      : asbalanced
```

	Contrast	Std. Err.	z	P>|z|
class				
(working class vs lower class)	-.2604145	.0749733	-3.47	0.001
(middle class vs working class)	-.1945925	.0378278	-5.14	0.000
(upper class vs middle class)	-.0684813	.1071814	-0.64	0.523

After adjusting for education, age, and year of interview, the first two comparisons were significant, but the third comparison (comparing upper versus middle class) was not significant. These differences are computed and expressed in the log-odds metric, the natural (linear) metric for the model. The magnitude of these group differences and their significance would remain constant at any level of the covariates.

Let's form these same comparisons but instead using the **margins** command, forming the comparisons using the predicted probability metric. This shows the difference in the predictive margins for the probability of smoking between adjacent levels of class, averaging across all values of the covariates in the dataset. (This is often called the average marginal effect.)

```
. margins ar.class, contrast(nowald pveffects)
Contrasts of predictive margins                    Number of obs     =     15,375
Model VCE    : OIM
Expression   : Pr(smoke), predict()

                                               Delta-method
                                   Contrast   Std. Err.        z    P>|z|

                           class
  (working class vs lower class)  -.0597026    .0174984    -3.41    0.001
 (middle class vs working class)  -.0422789    .0082295    -5.14    0.000
   (upper class vs middle class)  -.0142856     .022129    -0.65    0.519
```

Let's repeat this command but specify that we want to estimate the effects at two different values (7 and 18 years) of education.

```
. margins ar.class, contrast(nowald pveffects) at(educ=(7 18))
Contrasts of predictive margins                    Number of obs     =     15,375
Model VCE    : OIM
Expression   : Pr(smoke), predict()
1._at        : educ            =           7
2._at        : educ            =          18

                                                 Delta-method
                                     Contrast   Std. Err.        z    P>|z|

                          class@_at
  (working class vs lower class) 1  -.0626576    .0179638    -3.49    0.000
  (working class vs lower class) 2  -.0519208    .0156562    -3.32    0.001
 (middle class vs working class) 1  -.0464919    .0090132    -5.16    0.000
 (middle class vs working class) 2  -.0351111    .0070012    -5.02    0.000
   (upper class vs middle class) 1  -.0161284    .0251274    -0.64    0.521
   (upper class vs middle class) 2  -.0115633    .0178226    -0.65    0.516
```

Let's focus on the comparison of working class versus lower class. Holding education constant at 7, the effect of comparing working class versus lower class is -0.063 with a z-value of -3.49. The results are slightly different when we hold education constant at 18—the effect of comparing working class versus lower class is -0.052 with a z-value of -3.32. Furthermore, looking at the previous **margins** command (where we averaged across all values of the covariates), this difference is -0.06 with a z-value of -3.41. Although these results are similar, they are not identical.

Although it can be more intuitive to interpret results using probabilities, the effect of any variable can increase or decrease depending on the values of the covariates. Let's assess the contrasts on class, but this time doing so when the covariates education, age, and year of interview are held constant at the 25th percentile. Then, let's perform the same contrasts, holding these covariates each at the 50th percentile and at the 75th percentile. These tests are performed using the `margins` command below.

```
. margins ar.class, contrast(nowald pveffects)
>   at((p25) educ age yrint)
>   at((p50) educ age yrint)
>   at((p75) educ age yrint)
Contrasts of adjusted predictions               Number of obs     =     15,375
Model VCE    : OIM

Expression   : Pr(smoke), predict()

1._at        : educ            =          11 (p25)
               age             =          30 (p25)
               yrint           =        1980 (p25)

2._at        : educ            =          12 (p50)
               age             =          41 (p50)
               yrint           =        1986 (p50)

3._at        : educ            =          14 (p75)
               age             =          59 (p75)
               yrint           =        1989 (p75)

------------------------------------------------------------------------------
                               |            Delta-method
                               |   Contrast   Std. Err.      z     P>|z|
-------------------------------+----------------------------------------------
                      class@_at |
 (working class vs lower class) 1 |  -.0646819   .0184355    -3.51   0.000
 (working class vs lower class) 2 |  -.0630039   .0184179    -3.42   0.001
 (working class vs lower class) 3 |  -.0515222   .0155556    -3.31   0.001
(middle class vs working class) 1 |  -.0485233   .0094069    -5.16   0.000
(middle class vs working class) 2 |  -.0447057   .0086553    -5.17   0.000
(middle class vs working class) 3 |  -.0343771   .0067872    -5.06   0.000
  (upper class vs middle class) 1 |  -.0169205   .0263901    -0.64   0.521
  (upper class vs middle class) 2 |  -.0151053   .0233927    -0.65   0.518
  (upper class vs middle class) 3 |  -.0112338   .0172794    -0.65   0.516
------------------------------------------------------------------------------
```

The first three lines of output show the contrast of those who identify as working class versus lower class at the 25th, 50th, and 75th percentiles. We can see that the size of the contrast is similar across these three percentiles, as is the z test of the contrast. We see similar results for the contrast of those who identify as middle class versus working class, and those who identify as upper class versus middle class.

As we have seen in these examples, the `at()` option allows us to easily explore the extent to which the effect of class (in the probability metric) differs as a function of the value of the covariates.

Note! The margins command and the pwcompare option

You can use the `margins` command and the `at()` option combined with the `pwcompare` option to obtain pairwise comparisons of predicted probabilities while holding predictors constant at the values specified by the `at()` option. For example, the following command computes pairwise comparisons among the class groups for the predicted probability of smoking when education is held constant at 7 years.

```
. margins class, pwcompare(effects) at(educ=7)
```

18.3 Multinomial logistic regression

Let's now consider a multinomial logistic regression model, focusing on how the commands `contrast`, `pwcompare`, `margins`, and `marginsplot` can be used after fitting such a model. For this example, let's use the variable `haprate` as the outcome variable. This variable contains the happiness rating of the respondent on a three point scale: 3 = very happy, 2 = pretty happy, and 1 = not too happy.[2] Let's model this happiness rating as a function of gender, class, education, and year of interview. This is performed using the `mlogit` command shown below. The `mlogit` command chooses the most frequent outcome (which was the second outcome, pretty happy) as the base outcome.

2. For the sake of illustration, this ordered outcome will be analyzed using a multinomial logistic model. In the following section, this outcome will be analyzed using an ordinal logistic model.

```
. use gss_ivrm

. mlogit haprate i.gender i.class educ yrint, nolog

Multinomial logistic regression                 Number of obs     =     48,409
                                                LR chi2(12)       =    2076.07
                                                Prob > chi2       =     0.0000
Log likelihood = -44799.143                     Pseudo R2         =     0.0226

------------------------------------------------------------------------------
     haprate |      Coef.   Std. Err.      z    P>|z|     [95% Conf. Interval]
-------------+----------------------------------------------------------------
not_too_happy|
      gender |
       Male  |          0  (base)
     Female  |   .0196178   .0292278     0.67   0.502    -.0376676    .0769032
             |
       class |
 lower class |          0  (base)
working cl.. |  -.9896612   .0480355   -20.60   0.000    -1.083809   -.8955133
middle class |   -1.19037   .0510237   -23.33   0.000    -1.290374   -1.090365
 upper class |  -.8869056   .1024913    -8.65   0.000    -1.087785   -.6860262
             |
        educ |   -.076872   .0047655   -16.13   0.000    -.0862122   -.0675319
       yrint |   .0034042   .0013375     2.55   0.011     .0007827    .0060257
       _cons |   -6.38401   2.654168    -2.41   0.016    -11.58608   -1.181936
-------------+----------------------------------------------------------------
pretty_happy |  (base outcome)
-------------+----------------------------------------------------------------
very_happy   |
      gender |
       Male  |          0  (base)
     Female  |   .0713498   .0205565     3.47   0.001     .0310597    .1116399
             |
       class |
 lower class |          0  (base)
working cl.. |   .3317557   .0570264     5.82   0.000     .2199859    .4435254
middle class |   .7403293   .0573407    12.91   0.000     .6279435    .8527151
 upper class |    1.17829   .0772902    15.25   0.000     1.026804    1.329776
             |
        educ |  -.0001465   .0034639    -0.04   0.966    -.0069356    .0066427
       yrint |  -.0062441   .0009482    -6.59   0.000    -.0081025   -.0043857
       _cons |   11.27073   1.881358     5.99   0.000     7.583332    14.95812
------------------------------------------------------------------------------
```

I find it hard to interpret the model using the coefficients. I find it far easier to interpret the results using the `contrast`, `pwcompare`, `margins`, and `marginsplot` commands. I will quickly illustrate how these commands can help interpret the results of this model.

We can use the `contrast` command to obtain the overall effect of class. By default, this test is performed for the first equation (that is, for outcome 1, not too happy). This test is significant.

```
. contrast class
Contrasts of marginal linear predictions
Margins      : asbalanced

                  df        chi2     P>chi2

not_too_ha~y
       class       3      558.71     0.0000
```

Adding the `equation(3)` option performs the contrast with respect to the third outcome (that is, very happy). This test is also significant.

```
. contrast class, equation(3)
Contrasts of marginal linear predictions
Margins      : asbalanced

                  df        chi2     P>chi2

very_happy
       class       3      564.26     0.0000
```

The `atequations` option can be used to apply the `contrast` command with respect to all the equations. The output of this command matches what we saw in the previous two `contrast` commands.

```
. contrast class, atequations
Contrasts of marginal linear predictions
Margins      : asbalanced

                  df        chi2     P>chi2

not_too_ha~y
       class       3      558.71     0.0000

pretty_happy
       class   (omitted)

very_happy
       class       3      564.26     0.0000
```

We can use contrast operators with the `contrast` command to make specific comparisons among groups. In the example below, the `ar.` contrast operator is used to compare adjacent levels of class for the third equation. I also included the `rrr` option to interpret the results in terms of relative-risk ratios. Each of these contrasts is significant. Considering the first contrast, the odds of rating oneself as very happy is 1.40 times greater for those who self-identify as working class versus lower class.

```
. contrast ar.class, equation(3) nowald pveffects rrr
Contrasts of marginal linear predictions
Margins      : asbalanced

                                           RRR   Std. Err.       z    P>|z|

very_happy
                                class
  (working class vs lower class)      1.393412   .0794613     5.82   0.000
 (middle class vs working class)       1.50467   .0328338    18.72   0.000
   (upper class vs middle class)      1.549544   .0854613     7.94   0.000
```

The pwcompare command can be used to obtain pairwise comparisons among the four class groups. The equation(3) option specifies that these pairwise comparisons should be made for the third equation, and the rrr option requests the results in terms of relative-risk ratios. Focusing on the third contrast, the odds of rating oneself as very happy is 3.25 times greater for those who self-identify as upper class versus lower class.

```
. pwcompare class, equation(3) pveffects rrr
Pairwise comparisons of marginal linear predictions
Margins      : asbalanced

                                                             Unadjusted
                                         RRR   Std. Err.       z    P>|z|

very_happy
                               class
   working class vs lower class     1.393412   .0794613     5.82   0.000
    middle class vs lower class     2.096626   .1202221    12.91   0.000
     upper class vs lower class     3.248814   .2511013    15.25   0.000
  middle class vs working class      1.50467   .0328338    18.72   0.000
   upper class vs working class     2.331552   .1305295    15.12   0.000
    upper class vs middle class     1.549544   .0854613     7.94   0.000
```

When using the margins command, the default is to compute the predicted probability for every equation. (For more information, see [R] **margins**.) The margins command below computes, by class, the predictive margin of the probability of being not too happy (that is, outcome 1), being pretty happy (that is, outcome 2), and being very happy (that is, outcome 3). For example, the predictive margin of the probability of being very happy (outcome 3) for those who are upper class is 0.461, the last line of the output below.

```
. margins class, nopvalues

Predictive margins                              Number of obs     =     48,409
Model VCE    : OIM

1._predict   : Pr(haprate==not_too_happy), predict(pr outcome(1))
2._predict   : Pr(haprate==pretty_happy), predict(pr outcome(2))
3._predict   : Pr(haprate==very_happy), predict(pr outcome(3))

------------------------------------------------------------------
                  |            Delta-method
                  |     Margin   Std. Err.     [95% Conf. Interval]
------------------+-----------------------------------------------
    _predict#class |
   1#lower class  |   .3006533    .0087292      .2835443    .3177623
 1#working class  |   .1283142     .002215      .1239728    .1326556
  1#middle class  |   .0944107    .0020456      .0904013      .09842
   1#upper class  |   .1034631    .0080951       .087597    .1193292
   2#lower class  |   .5272773     .009724      .5082186     .546336
 2#working class  |   .5992694    .0033154      .5927713    .6057675
  2#middle class  |   .5378745    .0034275      .5311567    .5445924
   2#upper class  |   .4354805    .0125827      .4108189    .4601421
   3#lower class  |   .1720693    .0074623      .1574435    .1866951
 3#working class  |   .2724164     .003024      .2664894    .2783433
  3#middle class  |   .3677148    .0033127       .361222    .3742076
   3#upper class  |   .4610564    .0126679      .4362277     .485885
------------------------------------------------------------------
```

The `marginsplot` command is used below to graph the results of the previous `margins` command. The graph, shown in figure 18.5, shows the predictive margin of each of the three happiness ratings by self-identified class. Using the `s1mono` scheme, different line patterns are used to graph each of the outcomes. The dotted line shows how the rating of being very happy increases as a function of self-identified class.

```
. marginsplot, plotdim( , labels("Not too happy" "Pretty happy" "Very happy"))
> xlabel(, angle(45)) scheme(s1mono)
  Variables that uniquely identify margins: class
```

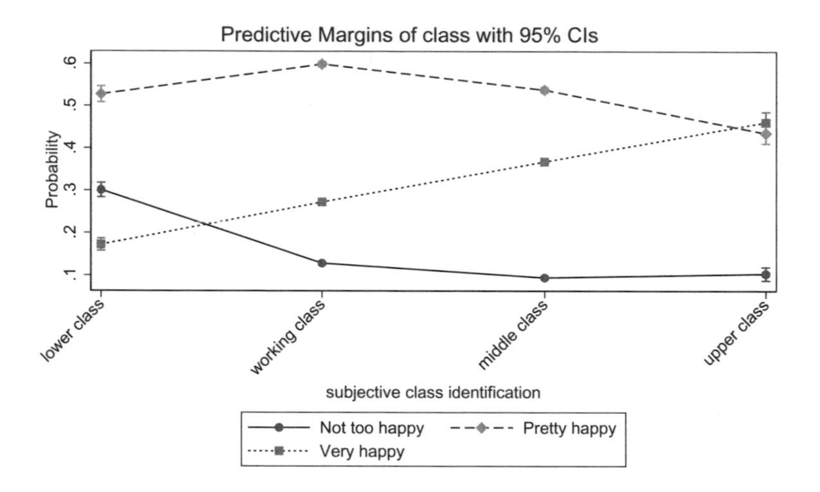

Figure 18.5. The predictive marginal probability of being not too happy, pretty happy, and very happy by self-identified social class

The `margins` command below is used to compute the predictive margin of the probability of being not too happy (the first outcome) for those with 5 to 18 years of education. The `marginsplot` command is then used to graph the results; see figure 18.6. While the relationship between education and ratings of being not too happy is negative, it is not particularly strong. The graph shows that predictive margins of being not too happy diminishes by a small amount with increasing education.

```
. margins, predict(outcome(1)) at(educ=(5(1)18))
  (output omitted)
. marginsplot
  Variables that uniquely identify margins: educ
```

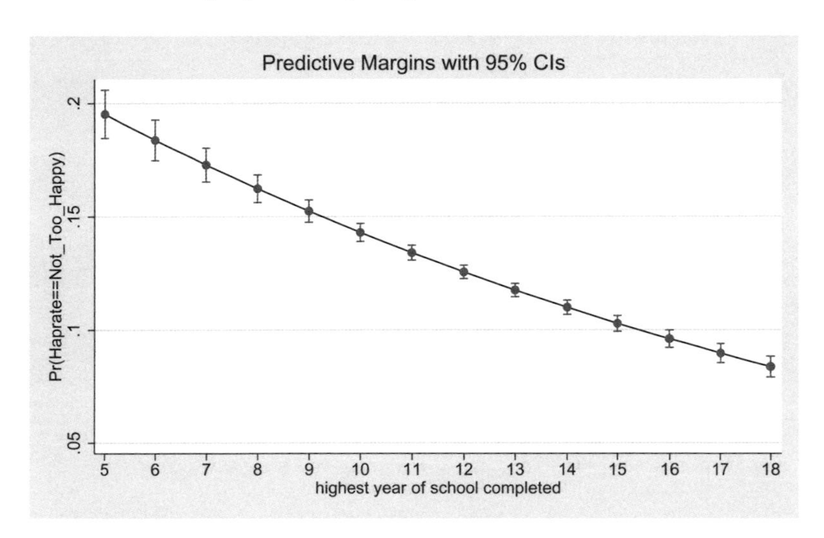

Figure 18.6. The predictive marginal probability of being not too happy by education

18.4 Ordinal logistic regression

Let's use the variable **haprate** from the previous section as the outcome but now model it using an ordinal logistic regression. Let's use the **ologit** command to model the three-level variable **haprate** (1 = not too happy, 2 = pretty happy, and 3 = very happy) as a function of gender, class, education, and year of interview.

```
. use gss_ivrm
. ologit haprate i.gender i.class educ yrint, nolog
Ordered logistic regression                     Number of obs     =     48,409
                                                LR chi2(6)        =    1799.46
                                                Prob > chi2       =     0.0000
Log likelihood = -44937.445                     Pseudo R2         =     0.0196

------------------------------------------------------------------------------
     haprate |      Coef.   Std. Err.      z    P>|z|     [95% Conf. Interval]
-------------+----------------------------------------------------------------
      gender |
       Male  |          0  (base)
     Female  |   .0598592   .0179433     3.34   0.001      .024691    .0950273
             |
       class |
 lower class |          0  (base)
working cl.. |   .9575904   .0413674    23.15   0.000     .8765118    1.038669
middle class |   1.360766   .0423593    32.12   0.000     1.277743    1.443789
 upper class |   1.663033   .0645844    25.75   0.000      1.53645    1.789616
             |
        educ |   .0320956   .0030435    10.55   0.000     .0261304    .0380607
       yrint |  -.0065063   .0008299    -7.84   0.000    -.0081329   -.0048797
-------------+----------------------------------------------------------------
       /cut1 |  -13.41488   1.647082                     -16.64311   -10.18666
       /cut2 |  -10.59686   1.646575                     -13.82409   -7.369634
------------------------------------------------------------------------------
```

As I did in the previous section, I will bypass interpreting the coefficients and briefly illustrate the use of the `contrast`, `pwcompare`, `margins`, and `marginsplot` commands.

First, let's consider the `contrast` command. The `contrast` command below tests the overall effect of class. This shows that the overall test of class is significant ($p < 0.001$).

```
. contrast class
Contrasts of marginal linear predictions
Margins      : asbalanced

------------------------------------------------
             |         df        chi2     P>chi2
-------------+----------------------------------
haprate      |
       class |          3     1286.31     0.0000
------------------------------------------------
```

We can further dissect the overall effect of class through the use of contrast operators. The `contrast` command below uses the `ar.` contrast operator to compare each level of class with the previous level. Focusing on the third contrast, the odds of rating oneself one unit higher in happiness (versus all lower ratings) is 1.35 times greater for those who self-identify as upper class versus middle class.

```
. contrast ar.class, nowald pveffects eform
Contrasts of marginal linear predictions
Margins      : asbalanced
```

	exp(b)	Std. Err.	z	P>|z|
haprate				
class				
(working class vs lower class)	2.605411	.1077791	23.15	0.000
(middle class vs working class)	1.49657	.0288003	20.95	0.000
(upper class vs middle class)	1.352922	.0697364	5.86	0.000

The exponentiated coefficients can be very abstract. Instead, let's compute the predictive marginal probability of being very happy (the third response) as a function of self-identified social class. The predictive marginal probability of rating oneself as very happy was 43.7% for those identifying themselves as upper class.

```
. margins class, predict( pr outcome(3))
Predictive margins                          Number of obs     =     48,409
Model VCE    : OIM
Expression   : Pr(haprate==3), predict(pr outcome(3))
```

	Margin	Delta-method Std. Err.	z	P>|z|	[95% Conf.	Interval]
class						
lower class	.1290581	.0045562	28.33	0.000	.1201282	.137988
working cl..	.2781531	.0028653	97.08	0.000	.2725371	.283769
middle class	.3654943	.0031724	115.21	0.000	.3592765	.3717122
upper class	.4377579	.0122919	35.61	0.000	.4136663	.4618495

By applying the `ar.` contrast operator, we can obtain comparisons among the adjacent levels of social class. Focusing on the third contrast, those who rate themselves as upper class have a predictive marginal probability of rating themselves as very happy that is 7.2% higher than those who describe themselves as middle class.

```
. margins ar.class, predict( pr outcome(3)) contrast(pveffects nowald)
Contrasts of predictive margins             Number of obs     =     48,409
Model VCE    : OIM
Expression   : Pr(haprate==3), predict(pr outcome(3))
```

	Contrast	Delta-method Std. Err.	z	P>|z|
class				
(working class vs lower class)	.1490949	.0050844	29.32	0.000
(middle class vs working class)	.0873413	.0041585	21.00	0.000
(upper class vs middle class)	.0722636	.012603	5.73	0.000

The `margins` command below is used to compute the predictive margin of the probability of each of the three happiness ratings for those with 5 to 18 years of education.

The `marginsplot` command is then used to graph the results. The graph is shown in figure 18.7.

```
. margins, at(educ=(5(1)18))
  (output omitted)
. marginsplot, plotdim( , labels("Not too happy" "Pretty happy" "Very happy"))
  Variables that uniquely identify margins: educ
```

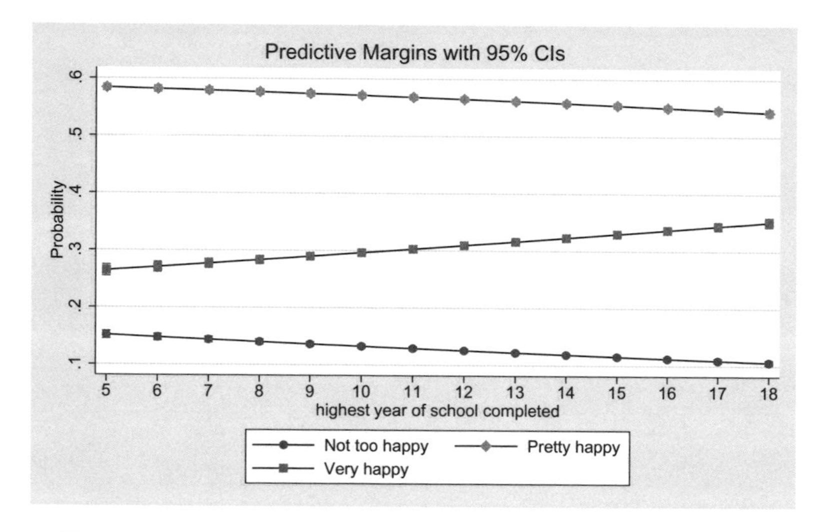

Figure 18.7. Probability of being very unhappy by education

18.5 Poisson regression

Let's now briefly consider a Poisson model, showing the use of the `contrast`, `pwcompare`, `margins`, and `marginsplot` commands following the use of the `poisson` command. Let's fit a model predicting the number of children a person has from gender, class, education, and year of interview.

```
. use gss_ivrm

. poisson children i.gender i.class educ yrint

Iteration 0:   log likelihood =  -95914.71
Iteration 1:   log likelihood = -95914.709

Poisson regression                              Number of obs     =     51,417
                                                LR chi2(6)        =    5773.21
                                                Prob > chi2       =     0.0000
Log likelihood = -95914.709                     Pseudo R2         =     0.0292

------------------------------------------------------------------------------
    children |      Coef.   Std. Err.      z    P>|z|     [95% Conf. Interval]
-------------+----------------------------------------------------------------
      gender |
        Male |          0  (base)
      Female |   .1317294   .0064358    20.47   0.000     .1191154    .1443433
             |
       class |
 lower class |          0  (base)
 working cl.. |  -.0926702   .0129119    -7.18   0.000    -.1179771   -.0673633
 middle class |  -.0472049   .0131897    -3.58   0.000    -.0730562   -.0213536
 upper class |    .061597   .0214595     2.87   0.004     .0195372    .1036568
             |
        educ |  -.0693717   .0010093   -68.73   0.000    -.0713498   -.0673936
       yrint |  -.0010904   .0002909    -3.75   0.000    -.0016606   -.0005203
       _cons |   3.684001   .5773908     6.38   0.000     2.552336    4.815667
------------------------------------------------------------------------------
```

As we have seen before, the **contrast** command can be used to test the overall effect of class. This test is significant ($p = 0.0000$).

```
. contrast class

Contrasts of marginal linear predictions

Margins      : asbalanced

------------------------------------------------
             |         df        chi2     P>chi2
-------------+----------------------------------
       class |          3      129.31     0.0000
------------------------------------------------
```

The **ar.** contrast operator is used to compare adjacent levels of class, comparing each class with the previous class. Each of these tests is significant.

```
. contrast ar.class, nowald pveffects

Contrasts of marginal linear predictions

Margins      : asbalanced

--------------------------------------------------------------------------------
                                |   Contrast   Std. Err.          z      P>|z|
--------------------------------+-----------------------------------------------
                          class |
 (working class vs lower class) |  -.0926702   .0129119      -7.18     0.000
(middle class vs working class) |   .0454653      .0068       6.69     0.000
  (upper class vs middle class) |   .1088019    .018121       6.00     0.000
--------------------------------------------------------------------------------
```

We can use the **pwcompare** command to form pairwise comparisons among the four class groups.

```
. pwcompare class, pveffects
Pairwise comparisons of marginal linear predictions
Margins      : asbalanced
```

	Contrast	Std. Err.	Unadjusted z	P>|z|
children				
class				
working class vs lower class	-.0926702	.0129119	-7.18	0.000
middle class vs lower class	-.0472049	.0131897	-3.58	0.000
upper class vs lower class	.061597	.0214595	2.87	0.004
middle class vs working class	.0454653	.0068	6.69	0.000
upper class vs working class	.1542672	.0181821	8.48	0.000
upper class vs middle class	.1088019	.018121	6.00	0.000

When using the `margins` command, the default is to compute the predicted number of events. The `margins` command below computes the predicted number of children by `class`.

```
. margins class, nopvalues
Predictive margins                              Number of obs     =     51,417
Model VCE    : OIM
Expression   : Predicted number of events, predict()
```

	Margin	Delta-method Std. Err.	[95% Conf.	Interval]
class				
lower class	2.075177	.025154	2.025876	2.124478
working class	1.891511	.0088545	1.874157	1.908866
middle class	1.979494	.0095966	1.960685	1.998303
upper class	2.20702	.0386825	2.131204	2.282837

The `margins` command is used to compute the predicted number of children for those with 5 to 18 years of education. The `marginsplot` command is then used to graph the results, as shown in figure 18.8.

```
. margins, at(educ=(5(1)18))
  (output omitted)
. marginsplot
  Variables that uniquely identify margins: educ
```

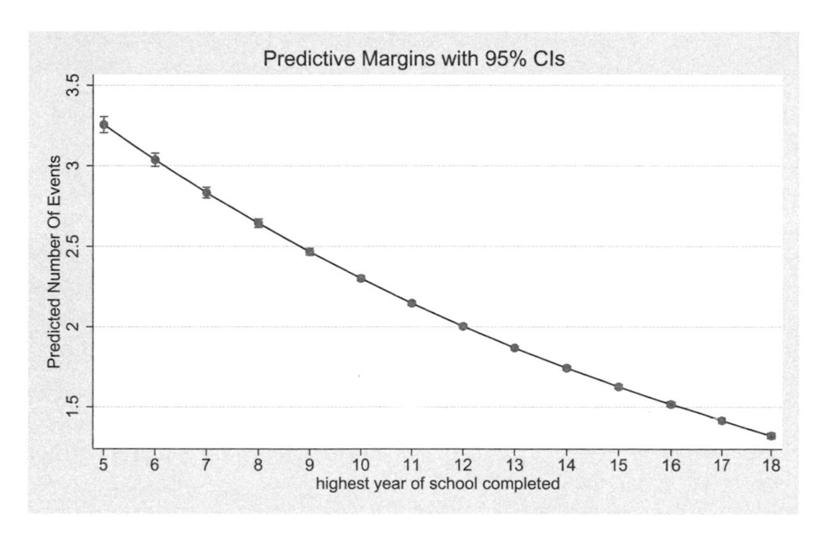

Figure 18.8. Predicted number of children by education

18.6 More applications of nonlinear models

The previous sections have illustrated, using basic models, the ways in which you can use commands like `contrast`, `pwcompare`, `margins`, and `marginsplot` to interpret the results of nonlinear modeling commands like `logit`, `mlogit`, `ologit`, and `poisson`. This section illustrates how you can use the modeling techniques that have been illustrated in parts I, II, and III with these kinds of nonlinear models. Three examples are shown illustrating a categorical by categorical interaction (see section 18.6.1), a categorical by continuous interaction (see section 18.6.2), and a piecewise model with two knots (see section 18.6.3).

18.6.1 Categorical by categorical interaction

This section illustrates a categorical by categorical interaction using a logistic regression. We saw the application of such interactions in the context of a linear model in chapter 8 on categorical by categorical interactions.

For this example, let's use the variable `fepol` as the outcome variable. The respondent was asked if they believed that women are not suited for politics. The variable `fepol` is coded: 1 = yes and 0 = no. Thus using `fepol` as our outcome, we will model endorsement of the statement as a function of two categorical predictors: gender

(gender) and a three-level measure of education (educ3). This analysis is restricted to interviews conducted between 1972 and 1980.

Let's begin by using the dataset and restricting the sample to the interviews that occurred in 1980 and earlier. We then use the logit command to model fepol from educ3, gender, and the interaction of these two categorical variables. The variable age is also included as a control variable.

```
. use gss_ivrm
. keep if yrint <= 1980
(42,967 observations deleted)
. logit fepol i.educ3##gender age, nolog
Logistic regression                             Number of obs     =      5,014
                                                LR chi2(6)        =     310.52
                                                Prob > chi2       =     0.0000
Log likelihood = -3313.7435                     Pseudo R2         =     0.0448
```

fepol	Coef.	Std. Err.	z	P>|z|	[95% Conf.	Interval]
educ3						
not hs	0	(base)				
HS	-.2276976	.1008946	-2.26	0.024	-.4254473	-.0299479
Coll	-.652108	.1308074	-4.99	0.000	-.9084859	-.3957302
gender						
Male	0	(base)				
Female	.1947732	.100224	1.94	0.052	-.0016622	.3912087
educ3#gender						
HS#Female	-.1378429	.12958	-1.06	0.287	-.3918151	.1161293
Coll#Female	-.5742612	.1932399	-2.97	0.003	-.9530045	-.195518
age	.0205419	.0017888	11.48	0.000	.0170359	.024048
_cons	-.8358332	.1196262	-6.99	0.000	-1.070296	-.6013702

The contrast command is used to test the overall interaction of educ3 by gender. This interaction is significant.

```
. contrast educ3#gender
Contrasts of marginal linear predictions
Margins      : asbalanced
```

	df	chi2	P>chi2
educ3#gender	2	8.85	0.0120

To help understand this interaction, let's use the margins command to estimate the log odds of believing that women are not suited for politics by educ3 and gender. Then, let's make a graph of these predicted logits using the marginsplot command. This creates the graph shown in figure 18.9.

```
. margins educ3#gender, nopvalues predict(xb)
Predictive margins                              Number of obs     =      5,014
Model VCE    : OIM

Expression   : Linear prediction (log odds), predict(xb)
```

	Margin	Delta-method Std. Err.	[95% Conf.	Interval]
educ3#gender				
not hs#Male	.0724204	.0763154	-.0771551	.2219959
not hs#Female	.2671936	.0673103	.1352678	.3991195
HS#Male	-.1552772	.0640951	-.2809013	-.0296531
HS#Female	-.0983469	.0523735	-.2009971	.0043034
Coll#Male	-.5796876	.1058136	-.7870783	-.3722968
Coll#Female	-.9591756	.126885	-1.207866	-.7104856

```
. marginsplot, legend(subtitle(Gender))
  Variables that uniquely identify margins: educ3 gender
```

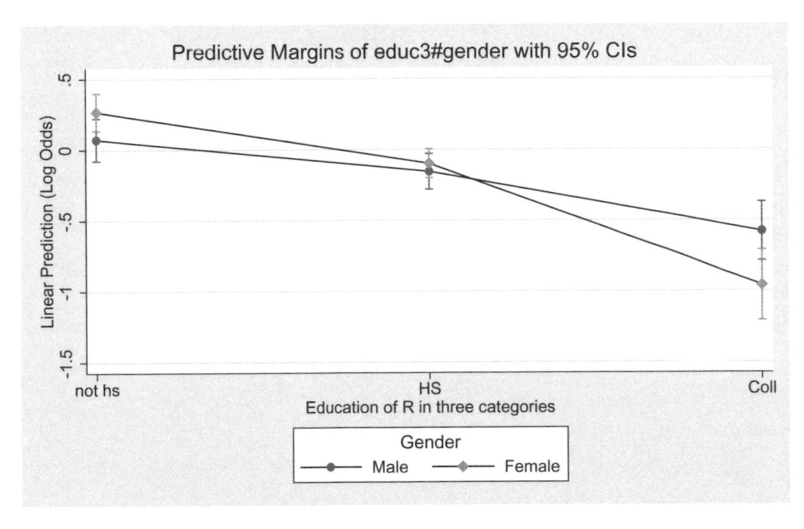

Figure 18.9. Predicted log odds of believing women are not suited for politics by gender and education

The graph in figure 18.9 shows the log odds of believing that women are not suited for politics by gender and education. I think it is important to begin our exploration of this interaction by inspecting the graph that is presented in the metric of the logistic regression analysis. Although the log-odds metric is not easy to interpret, we can still glean the trends implied by the gender by education interaction. The graph suggests that the log odds of agreeing with this statement declines with increasing education, and that this decline appears to be stronger for females than for males. Let's test this by interacting gender with comparisons of adjacent education levels (that is, college graduates with high school graduates, and high school graduates with non–high school graduates). We can test this partial interaction by applying the `ar.` contrast operator to `educ3` and interacting that with `gender`.

```
. contrast ar.educ3#gender
Contrasts of marginal linear predictions
Margins      : asbalanced
```

	df	chi2	P>chi2
educ3#gender			
(HS vs not hs) (joint)	1	1.13	0.2874
(Coll vs HS) (joint)	1	5.60	0.0180
Joint	2	8.85	0.0120

The first contrast estimates the interaction of the comparison of high school graduates with non–high school graduates by gender. This test is not significant ($p = 0.2874$). The second contrast, which interacts the comparison of college graduates to high school graduates by gender, is significant ($p = 0.018$). As we can see in figure 18.9, the log odds of agreeing that women are not suited for politics decreases more for females than for males when comparing college graduates with high school graduates.

Let's now assess the gender difference at each level of education. We can do this by testing the simple effect of gender at each level of education using the `contrast` command below.

```
. contrast gender@educ3
Contrasts of marginal linear predictions
Margins      : asbalanced
```

	df	chi2	P>chi2
gender@educ3			
not hs	1	3.78	0.0520
HS	1	0.48	0.4878
Coll	1	5.28	0.0216
Joint	3	9.54	0.0230

For non–high school graduates, the gender difference is not significant ($p = 0.0520$). For high school graduates, the gender difference is also not significant ($p = 0.4878$). For college graduates, the gender difference is significant ($p = 0.0216$). We can interpret the direction of this significance by looking at figure 18.9, which helps us see that, for college graduates, the log odds of believing that women are not suited for politics is significantly lower for females versus males.

To get a better sense of the magnitude of these effects in a metric that we understand, we can use the `margins` command to estimate the predictive margin of the probability of believing that women are not suited for politics by education and gender. The `marginsplot` command is then used to graph these values (see figure 18.10).

```
. margins educ3#gender, nopvalues
Predictive margins                              Number of obs     =      5,014
Model VCE    : OIM
Expression   : Pr(fepol), predict()

------------------------------------------------------------------
               |            Delta-method
               |     Margin   Std. Err.     [95% Conf. Interval]
---------------+--------------------------------------------------
  educ3#gender |
   not hs#Male |   .5171847   .0185122      .4809015     .553468
 not hs#Female |   .5640704    .016136      .5324444    .5956963
       HS#Male |   .4620812   .0154578      .4317845     .492378
     HS#Female |    .475823   .0126666      .4509969    .5006491
     Coll#Male |   .3627282   .0237663      .3161471    .4093094
   Coll#Female |   .2824397   .0250625       .233318    .3315613
------------------------------------------------------------------

. marginsplot, legend(subtitle(Gender))
  Variables that uniquely identify margins: educ3 gender
```

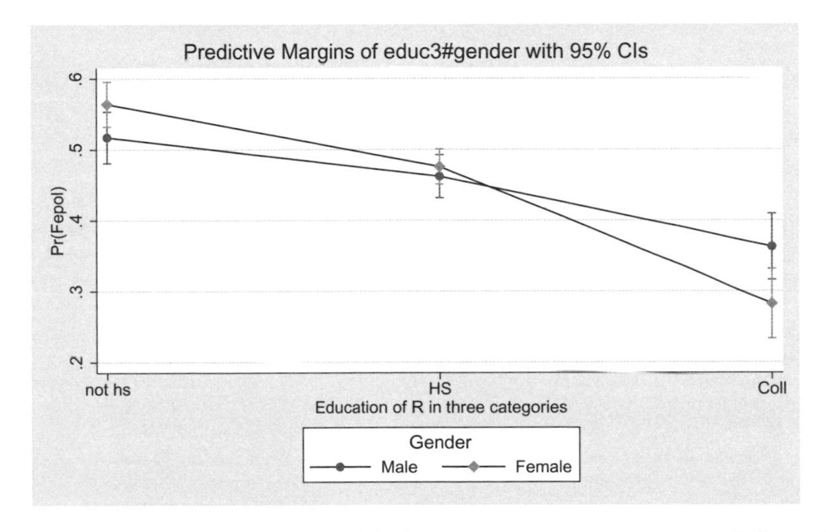

Figure 18.10. Predicted probability of believing women are not suited for politics by gender and education

The pattern of the interaction illustrated in the probability metric in figure 18.10 is similar to the pattern of the interaction shown in the log-odds metric in figure 18.9. Focusing on college graduates, the predictive margins of agreeing with this statement was 36.2% for males and 28.2% for females.

The pattern of the interaction in the probability metric could vary as a function of the other covariates contained in the model. In this case, we have only one covariate, `age`. Let's make two graphs showing the pattern of this interaction in the probability metric, one where age is held constant at 30 and another where age is held constant at 50. The `margins` command below computes the predictive margins by education and gender for 30-year-olds and 50-year-olds. The `marginsplot` command graphs these two sets of results (see figure 18.11). The pattern of the interaction looks similar at each level of age.

```
. margins educ3#gender, nopvalues at(age=(30 50))
Adjusted predictions                            Number of obs     =      5,014
Model VCE    : OIM

Expression   : Pr(fepol), predict()
1._at        : age             =          30
2._at        : age             =          50

------------------------------------------------------------------------
                 |            Delta-method
                 |     Margin   Std. Err.     [95% Conf. Interval]
-----------------+------------------------------------------------------
_at#educ3#gender |
  1#not hs#Male  |   .4453257   .0209418      .4042807    .4863708
1#not hs#Female  |   .4937999   .0189311      .4566955    .5309042
      1#HS#Male  |   .3900094   .0156157      .3594032    .4206156
    1#HS#Female  |   .4036349   .0132895       .377588    .4296818
    1#Coll#Male  |   .2949042   .0224732      .2508576    .3389508
  1#Coll#Female  |   .2224975   .0223342      .1787232    .2662718
  2#not hs#Male  |   .5476707   .0186031      .5112093    .5841321
2#not hs#Female  |   .5953283   .0159992      .5639705    .6266861
      2#HS#Male  |   .4908925   .0165718      .4584124    .5233727
    2#HS#Female  |   .5051239   .0136667      .4783376    .5319103
    2#Coll#Male  |   .3867856   .0252905      .3372172     .436354
  2#Coll#Female  |   .3014648   .0268349      .2488693    .3540603
------------------------------------------------------------------------

. marginsplot, bydimension(age) legend(subtitle(Gender))
  Variables that uniquely identify margins: age educ3 gender
```

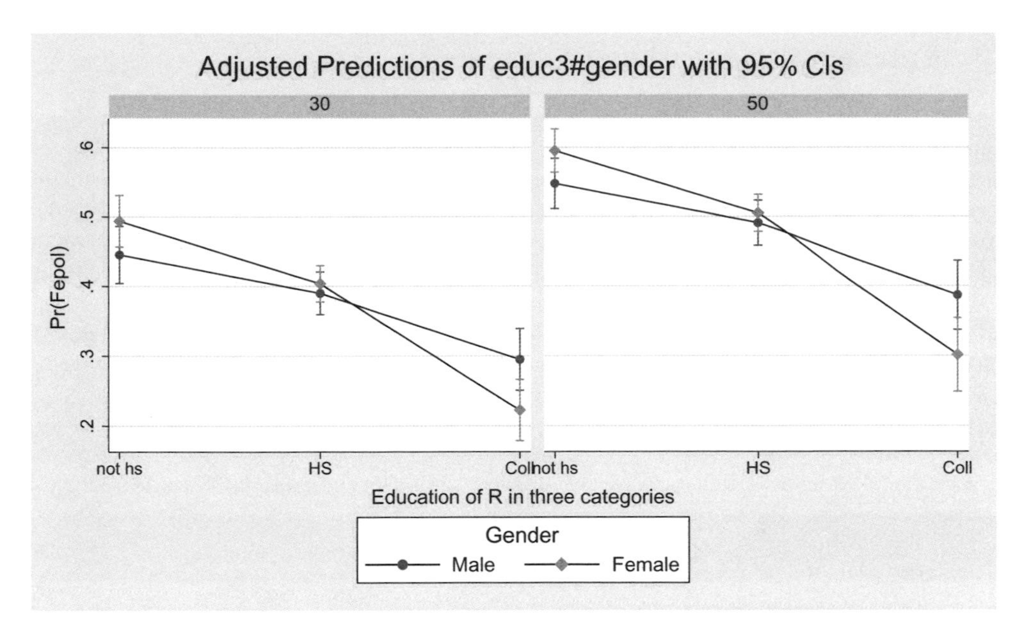

Figure 18.11. Predicted probability of believing women are not suited for politics by gender and education with age held constant at 30 and 50

You could also assess the partial interaction of `ar.educ3#gender`, in the probability metric, for those at different ages. The example below assesses this partial interaction at ages 30 and 50.

```
. margins ar.educ3#gender, at(age=(30 50)) contrast(nowald pveffects)
Contrasts of adjusted predictions                 Number of obs     =      5,014
Model VCE     : OIM
Expression    : Pr(fepol), predict()
1._at         : age            =          30
2._at         : age            =          50

                                       |              Delta-method
                                       |   Contrast   Std. Err.        z    P>|z|
---------------------------------------+------------------------------------------
                     educ3#gender@_at  |
(HS vs not hs) (Female vs base) 1      |  -.0348486   .0316967    -1.10   0.272
(HS vs not hs) (Female vs base) 2      |  -.0334262   .0319873    -1.04   0.296
  (Coll vs HS) (Female vs base) 1      |  -.0860322   .0367553    -2.34   0.019
  (Coll vs HS) (Female vs base) 2      |  -.0995522   .0419946    -2.37   0.018
```

The interaction of gender by education (`HS vs not hs`) is not significant whether age is held constant at 30 or 50. The interaction of gender by education (`Coll vs HS`) is significant whether age is held constant at 30 or 50.

18.6.2 Categorical by continuous interaction

This section illustrates a categorical by continuous interaction using a binary logistic regression. This analysis draws upon the concepts illustrated in chapter 10. The outcome for this example is the variable `fepres`, which is coded: 1 = would vote for a woman president and 0 = would not vote for a woman president. Let's model this as a function of time (that is, year of interview) and education to see if the linear change in this attitude over time differed by education level. With respect to the year of interview, this question was asked in 17 different years ranging from 1972 to 1998, then it was asked again in 2008 and 2010. Because of the large 10-year gap between 1998 and 2008, we will omit the data for 2008 onward. With respect to education, let's use the three-level categorical variable `educ3`, which is coded: 1 = non–high school graduate, 2 = high school graduate, and 3 = college graduate.

The dataset is used below, and the observations for the year 2008 and onward are dropped.

```
. use gss_ivrm
. drop if yrint>=2008
(4,067 observations deleted)
```

Let's begin by assessing the trend in the log odds of the outcome across years. In doing so, let's look at each education group separately so we can assess the trend for each education group. Based on the strategy illustrated in section 2.4.4, we first fit a model using `fepres` as the outcome predicted by `i.yrint72##educ3` as well as `age` and `gender`.

```
. logit fepres i.yrint72##educ3 age gender
  (output omitted)
```

We then use the `margins` command to compute the predicted outcome in the log-odds metric followed by the `marginsplot` command to graph the results. The graph of the log odds of willingness to vote for a woman president by year and education is shown in figure 18.12.

```
. margins yrint72#educ3, predict(xb)
  (output omitted)
. marginsplot, noci
  Variables that uniquely identify margins: yrint72 educ3
```

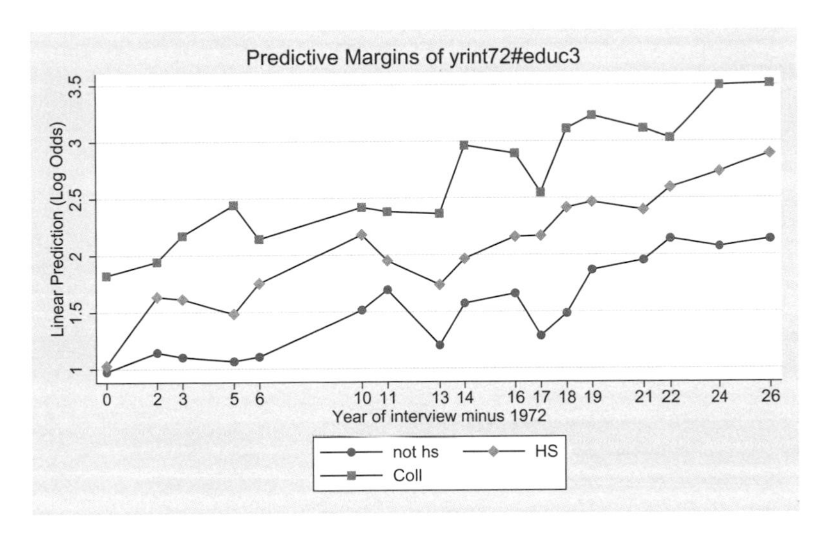

Figure 18.12. Predicted log odds of voting for a woman president by year of interview and education

As we might expect, there are some bumps across time, but the trends seem reasonably linear for each group. We can test for differences in the linear effect of year by education using the `contrast` command below. The `p1` contrast operator indicates to test for the linear effect of `yrint72`. This test is significant ($p < 0.001$).

```
. contrast p1.yrint72
Contrasts of marginal linear predictions
Margins      : asbalanced
```

	df	chi2	P>chi2
yrint72	1	286.09	0.0000

	Contrast	Std. Err.	[95% Conf.	Interval]
yrint72 (linear)	.4248763	.0251197	.3756426	.47411

We can test for a quadratic effect of `yrint72` using the `contrast` command below. This test is not significant ($p = 0.1909$)

```
. contrast p2.yrint72
Contrasts of marginal linear predictions
Margins      : asbalanced
```

	df	chi2	P>chi2
yrint72	1	1.71	0.1909

	Contrast	Std. Err.	[95% Conf.	Interval]
yrint72 (quadratic)	.0331639	.0253555	-.016532	.0828598

Using the `contrast` command below, we can jointly test for quadratic, cubic, quartic, and quintic trends over time. The joint test of these four nonlinear trends is not significant ($p = 0.2364$).

```
. contrast p(2 3 4 5).yrint72
Contrasts of marginal linear predictions
Margins      : asbalanced
```

	df	chi2	P>chi2
yrint72			
(quadratic)	1	1.71	0.1909
(cubic)	1	1.65	0.1994
(quartic)	1	2.09	0.1481
(quintic)	1	0.06	0.8125
Joint	4	5.54	0.2364

	Contrast	Std. Err.	[95% Conf.	Interval]
yrint72				
(quadratic)	.0331639	.0253555	-.016532	.0828598
(cubic)	.0332666	.0259261	-.0175476	.0840807
(quartic)	-.0360718	.0249381	-.0849497	.012806
(quintic)	-.0060709	.0255959	-.0562379	.0440962

We could further analyze the nature of the relationship between time searching for nonlinearities analytically or graphically, but I think these tests give us a sufficient basis to proceed forward with fitting a model where we treat `yrint72` as a continuous variable and interacting it with `educ3`. In this model, let's add `age` and `gender` as covariates.

```
. logit fepres c.yrint72##educ3 age gender, nolog
Logistic regression                             Number of obs     =     23,926
                                                LR chi2(7)        =    1754.77
                                                Prob > chi2       =     0.0000
Log likelihood = -8803.3423                     Pseudo R2         =     0.0906
```

fepres	Coef.	Std. Err.	z	P>|z|	[95% Conf.	Interval]
yrint72	.0435104	.0039697	10.96	0.000	.0357298	.0512909
educ3						
not hs	0	(base)				
HS	.3372813	.0702906	4.80	0.000	.1995143	.4750483
Coll	.9055769	.1205486	7.51	0.000	.669306	1.141848
educ3# c.yrint72						
HS	.0145962	.0052977	2.76	0.006	.0042128	.0249796
Coll	.016982	.008484	2.00	0.045	.0003536	.0336104
age	-.0262412	.0011221	-23.39	0.000	-.0284404	-.024042
gender	-.0204508	.039451	-0.52	0.604	-.0977734	.0568718
_cons	2.163802	.0975701	22.18	0.000	1.972568	2.355036

We can test the overall interaction of `c.yrint72#educ3` using the `contrast` command. This test is significant. The linear trend in the relationship between the log odds of the outcome and the year of interview differs by a three-level measure of education.

```
. contrast c.yrint72#educ3
Contrasts of marginal linear predictions
Margins      : asbalanced
```

	df	chi2	P>chi2
educ3#c.yrint72	2	8.83	0.0121

Let's use the `pwcompare` command to form pairwise comparisons of the effect of `yrint72` among the three levels of education.

```
. pwcompare c.yrint72#educ3, pveffects
Pairwise comparisons of marginal linear predictions
Margins      : asbalanced
```

	Contrast	Std. Err.	Unadjusted z	P>|z|
fepres				
educ3#c.yrint72				
HS vs not hs	.0145962	.0052977	2.76	0.006
Coll vs not hs	.016982	.008484	2.00	0.045
Coll vs HS	.0023858	.0082915	0.29	0.774

The test comparing the effect of the year of interview for high school graduates versus non–high school graduates is significant, and the test comparing the effect of the year of interview for college graduates versus non–high school graduates is also significant. The last test of college graduates versus high school graduates is not significant.

We can visualize the predicted log odds of the willingness to vote for a woman president by year and education using the `margins` and `marginsplot` commands shown below. This creates the graph shown in figure 18.13.

```
. margins educ3, at(yrint72=(0(1)26)) predict(xb)
  (output omitted)
. marginsplot, noci
  Variables that uniquely identify margins: yrint72 educ3
```

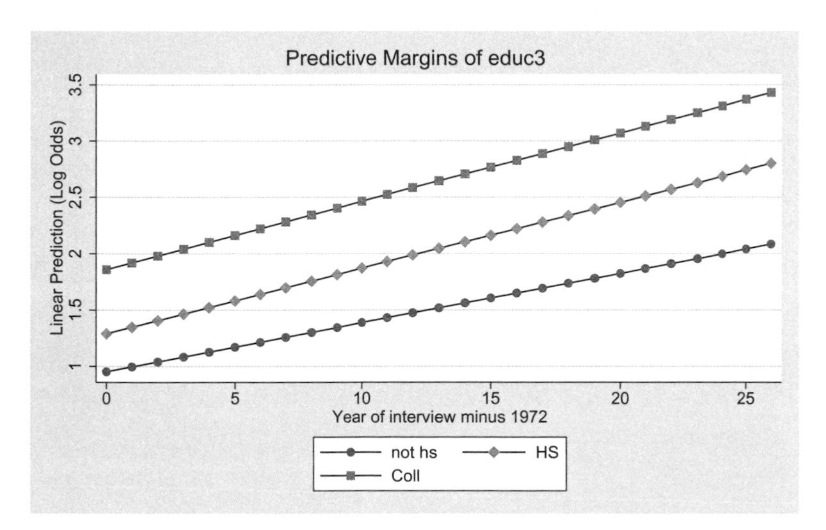

Figure 18.13. Predicted log odds of willingness to vote for a woman president by year of interview and education

If we prefer, we can estimate and graph these results in the predicted probability metric using the `margins` and `marginsplot` commands below. The graph produced by the `margins` command is shown in figure 18.14.

```
. margins educ3, at(yrint72=(0(1)26))
  (output omitted)
. marginsplot, noci
  Variables that uniquely identify margins: yrint72 educ3
```

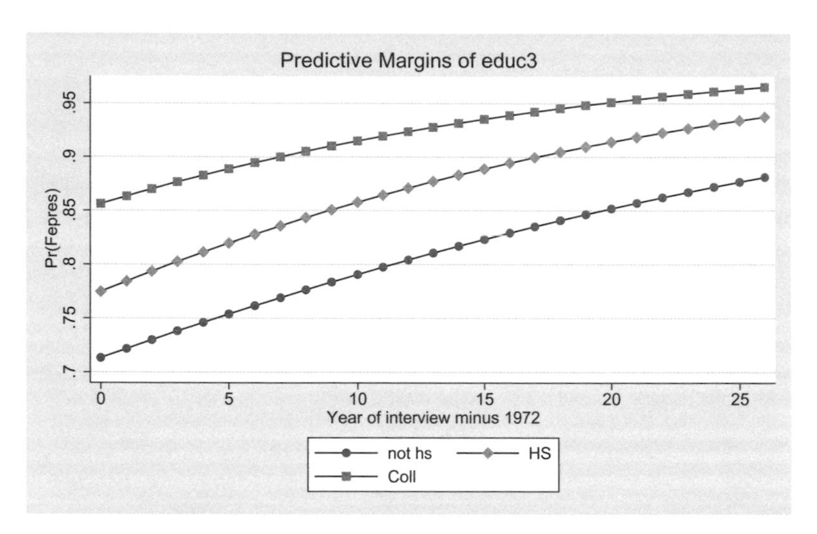

Figure 18.14. Predictive margin of the probability of being willing to vote for a woman president by year of interview and education

This section has illustrated how a categorical by continuous interaction can be fit using a logistic regression model, using the modeling techniques illustrated in chapter 10. You could use other techniques illustrated in that chapter, such as forming comparisons among education groups at different time points.

18.6.3 Piecewise modeling

This section illustrates modeling a continuous predictor fit using piecewise modeling in the context of a logistic regression model. This analysis draws upon the concepts illustrated in chapter 4. The outcome for this example is the variable `smoke`, which measures whether the person smokes, coded: 1 = yes and 0 = no. Let's model smoking as a function of education, modeling education in a piecewise manner as illustrated in section 4.6.

Let's begin by inspecting the nature of the relationship between education and smoking status. Based on the techniques illustrated in section 2.4.4, we first fit a model predicting smoking status from education, treating education as a categorical variable. Age is also included in the model as a covariate. The output of the `logit` command is omitted to save space.

```
. use gss_ivrm
. logit smoke i.educ age
  (output omitted)
```

Let's now make a graph that shows the predicted logit of smoking as a function of education, adjusting for age. This is done using the `margins` command (the output is omitted to save space) followed by the `marginsplot` command. The resulting graph is shown in figure 18.15.

```
. margins educ, predict(xb)
  (output omitted)
. marginsplot, xline(12 16)
  Variables that uniquely identify margins: educ
```

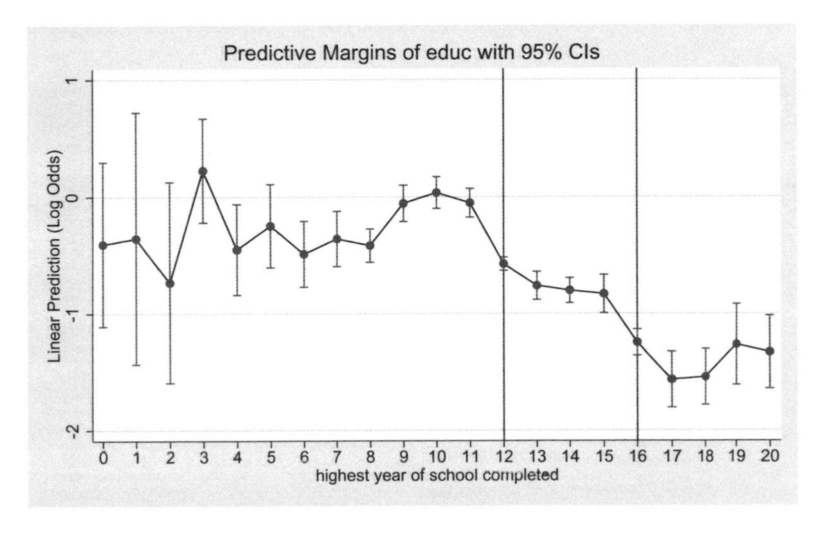

Figure 18.15. Log odds of smoking by education

The graph in figure 18.15 includes vertical lines at 12 and 16 years of education. These junctures seem like excellent candidates for the placement of knots where there is a change in slope and a change in intercept. As we did in section 4.6, let's model education using a piecewise model with two knots at 12 and 16 years of education. Let's fit such a model below.

```
. mkspline edprehsm 12 edhsm 16 edcom = educ, marginal
. logit smoke c.edprehsm c.edhsm c.edcom hsgrad cograd age, nolog
Logistic regression                             Number of obs     =     16,274
                                                LR chi2(6)        =     806.95
                                                Prob > chi2       =     0.0000
Log likelihood = -10136.225                     Pseudo R2         =     0.0383

------------------------------------------------------------------------------
       smoke |      Coef.   Std. Err.      z    P>|z|     [95% Conf. Interval]
-------------+----------------------------------------------------------------
    edprehsm |   .0540074   .0146957     3.68   0.000     .0252044    .0828105
       edhsm |  -.1571181   .0269648    -5.83   0.000    -.2099681   -.1042682
       edcom |   .0589171   .0420088     1.40   0.161    -.0234186    .1412529
      hsgrad |   -.599279    .062409    -9.60   0.000    -.7215984   -.4769597
      cograd |  -.3026171   .0953092    -3.18   0.001    -.4894197   -.1158146
         age |  -.0201919    .001057   -19.10   0.000    -.0222636   -.0181202
       _cons |   .2709491   .1535817     1.76   0.078    -.0300656    .5719638
------------------------------------------------------------------------------
```

Before interpreting the results, let's create a graph of the predicted log odds of smoking as a function of education. This is done using the same techniques illustrated in section 4.10. This produces the graph shown in figure 18.16.

```
. margins,
>   at(edprehsm=0  edhsm=0 edcom=0 hsgrad=0 cograd=0)
>   at(edprehsm=12 edhsm=0 edcom=0 hsgrad=0 cograd=0)
>   at(edprehsm=12 edhsm=0 edcom=0 hsgrad=1 cograd=0)
>   at(edprehsm=16 edhsm=4 edcom=0 hsgrad=1 cograd=0)
>   at(edprehsm=16 edhsm=4 edcom=0 hsgrad=1 cograd=1)
>   at(edprehsm=20 edhsm=8 edcom=4 hsgrad=1 cograd=1) predict(xb) noatlegend
Predictive margins                              Number of obs     =     16,274
Model VCE    : OIM

Expression   : Linear prediction (log odds), predict(xb)

------------------------------------------------------------------------------
             |            Delta-method
             |     Margin   Std. Err.      z    P>|z|     [95% Conf. Interval]
-------------+----------------------------------------------------------------
         _at |
          1  |  -.6360931   .1339523    -4.75   0.000    -.8986348   -.3735515
          2  |   .0119962   .0560904     0.21   0.831     -.097939    .1219315
          3  |  -.5872828   .0278267   -21.11   0.000    -.6418221   -.5327435
          4  |  -.9997256   .0792405   -12.62   0.000    -1.155034    -.844417
          5  |  -1.302343   .0542245   -24.02   0.000    -1.408621   -1.196065
          6  |  -1.479117   .1183285   -12.50   0.000    -1.711037   -1.247197
------------------------------------------------------------------------------

. mat yhat = r(b)'
. mat educ = (0 \ 12 \ 12 \ 16 \ 16 \ 20)
. svmat yhat
. svmat educ
. graph twoway line yhat1 educ1, xline(12 16) title("Piecewise Model")
```

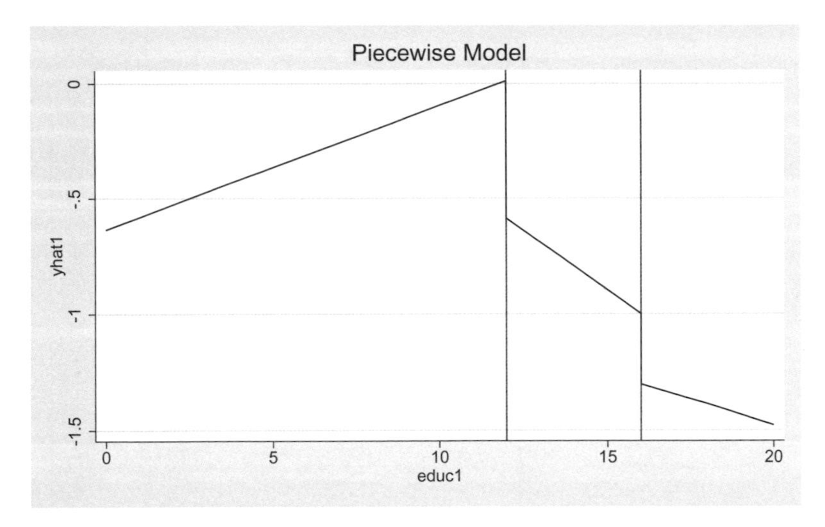

Figure 18.16. Predicted log odds of smoking from education fit using a piecewise model with two knots

We can use the graph of the predicted logits from figure 18.16 to help us interpret the results from the `logit` command. First, let's focus on the slope coefficients (`edprehsm`, `edhsm`, and `edcom`). The coefficient for `edprehsm` is the `educ` slope prior to graduating high school. The coefficient for `edhsm` is the change in the slope due to graduating high school, and `edcom` is the change in the slope due to graduating college. The coefficient for `edprehsm` is 0.054 and is significant ($z = 3.68$, $p = 0.000$). Prior to graduating high school, a one-year increase in education increases the predicted log odds of smoking by 0.054 units. The coefficient for `edhsm` is −0.16 and is significant ($z = -5.83$, $p = 0.000$). Upon graduating high school, the slope relating education to the log odds of smoking decreases by 0.16. The coefficient for `edcom` is 0.06 but is not significant ($z = 1.40$, $p = 0.161$). Upon graduating college, the slope relating education to the log odds of smoking does not significantly change.

The `hsgrad` coefficient shows the change in the log odds of smoking upon graduating high school, showing a significant decrease (portrayed as the sudden drop 12 years in figure 18.16). Upon graduating high school, the log odds of smoking is predicted to drop by 0.60 units ($z = -9.60$, $p = 0.000$). Likewise, the `cograd` coefficient reflects the change in the log odds of smoking upon graduating college. As we see in figure 18.16, the log odds of smoking substantially decreases upon graduating college. The log odds of smoking drop by 0.30 units upon graduating college, a significant decrease ($z = -3.18$, $p = 0.001$).

To assess how well the piecewise model accommodates the relationship between the log odds of smoking and education, let's visualize figures 18.15 and 18.16 side by side. It appears that the piecewise model shown in the right panel of figure 18.17 models the observed data from the left panel of figure 18.17 in a parsimonious manner.

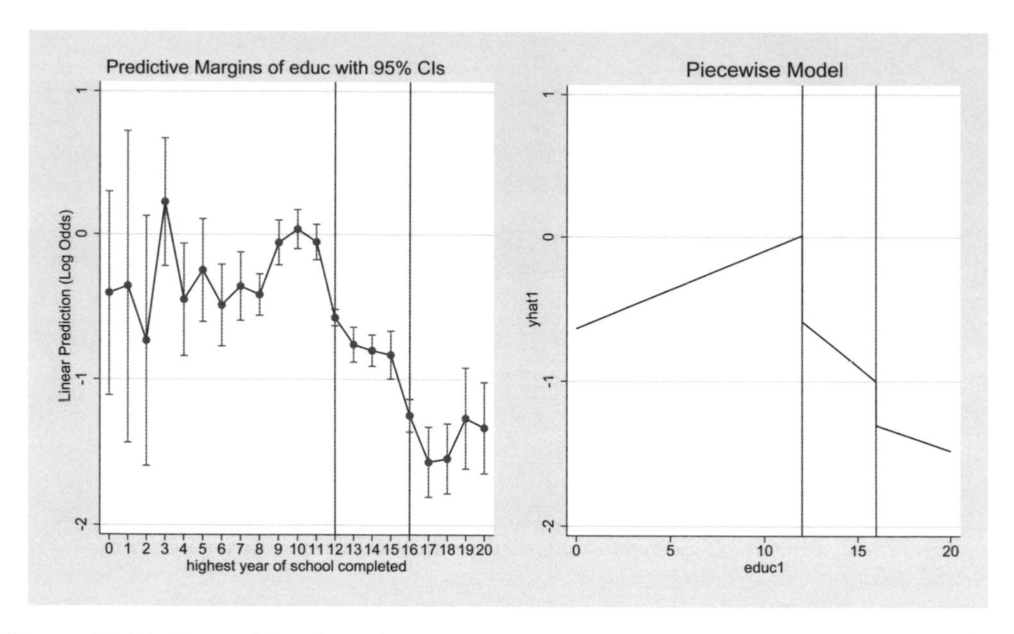

Figure 18.17. Log odds of smoking treating education as a categorical variable (left panel) and fitting education using a piecewise model (right panel)

Let's use the `margins` command to create the predictive margin of the probability of smoking as a function of education, averaging across the covariate `age`. The predictive margins are computed using the `margins` command below, and then they are graphed as a function of education using the `graph twoway` command. The resulting graph is shown in figure 18.18.[3]

3. Note that this graph portrays each piece of the regression model linearly, which is not quite accurate given that the results are graphed in the probability metric. For the most accurate graph, the predictive margin should be computed at each level of education ranging from 5 to 20.

```
. margins,
>   at(edprehsm=5  edhsm=0 edcom=0 hsgrad=0 cograd=0)
>   at(edprehsm=12 edhsm=0 edcom=0 hsgrad=0 cograd=0)
>   at(edprehsm=12 edhsm=0 edcom=0 hsgrad=1 cograd=0)
>   at(edprehsm=16 edhsm=4 edcom=0 hsgrad=1 cograd=0)
>   at(edprehsm=16 edhsm=4 edcom=0 hsgrad=1 cograd=1)
>   at(edprehsm=20 edhsm=8 edcom=4 hsgrad=1 cograd=1) noatlegend
Predictive margins                              Number of obs     =     16,274
Model VCE    : OIM
Expression   : Pr(smoke), predict()
```

	Margin	Delta-method Std. Err.	z	P>|z|	[95% Conf.	Interval]
_at						
1	.4125336	.0153917	26.80	0.000	.3823663	.4427008
2	.5033372	.0135931	37.03	0.000	.4766952	.5299792
3	.3615623	.0062035	58.28	0.000	.3494037	.3737209
4	.2747184	.0153683	17.88	0.000	.2445971	.3048397
5	.2197515	.0090636	24.25	0.000	.2019871	.237516
6	.1914399	.0180026	10.63	0.000	.1561554	.2267244

```
. mat prsmoke = r(b)'
. mat xeduc = (5 \ 12 \ 12 \ 16 \ 16 \ 20)
. svmat prsmoke
. svmat xeduc
. graph twoway line prsmoke1 xeduc1, xline(12 16) name(g3, replace)
```

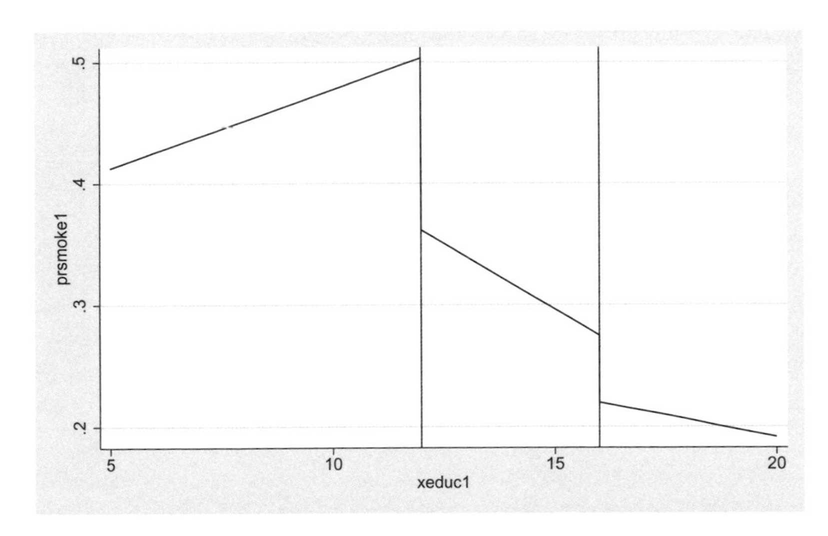

Figure 18.18. Probability of smoking treating by education (fit using a piecewise model)

The predictive margins in figure 18.18 are easier to interpret than the predicted log odds shown in figure 18.16, however we need to remember that the pattern using probabilities could change as a function of the covariates. We could further explore this by creating graphs at different levels of the covariate, `age`.

18.7 Summary

This chapter has illustrated the use of the `contrast`, `pwcompare`, `margins`, and `marginsplot` commands following nonlinear estimation commands like `logit`, `mlogit`, `ologit`, and `poisson`. Furthermore, section 18.6 provided examples showing how the modeling techniques from parts I, II, and III can be applied to nonlinear models. This illustrates the generality of these modeling techniques, showing that the modeling techniques from parts I, II, and III are not restricted to only linear models but can be applied to nonlinear models as well. If we interpret the results using the natural metric of the model (for example, using log odds from a logistic regression model), then the effect of any given predictor remains the same regardless of the value of any of the covariates. However, if we interpret the results in a different metric (for example, using probabilities from a logistic regression model), we need to be mindful that the effect of a predictor can (and will) change depending on the values of the covariates. Such results can be more intuitive, but also may vary as a function of the values of the covariates.

For more information about logistic regression, see Kleinbaum and Klein (2010), Hosmer et al. (2013), Menard (2002), Long (1997), and Hilbe (2009), as well as Long and Freese (2014), who include examples using Stata. Also see Gould (2000) for an excellent tutorial regarding the interpretation of logistic models. For more information about interaction effects in logistic regression models, see Jaccard (2001) and Hilbe (2009).

19 Complex survey data

Stata has extensive capabilities with respect to the analysis of datasets that arise from complex survey designs. In such cases, the svyset command is used to declare the survey sampling design, and the svy prefix is supplied before an estimation command (for example, svy: regress). You will be glad to know that, even when you have complex survey data, you can still use the margins, marginsplot, contrast, and pwcompare commands to interpret your results. The chapter briefly illustrates the mechanics of using these commands in the context of a complex survey.

The example dataset used in this chapter is the nhanes2.dta dataset. This is one of the Stata example datasets and is used via the Internet with the webuse command, shown below.

```
. webuse nhanes2
```

The svyset command has already been used to declare the design for this survey, naming the primary sampling unit, the person weight, and the strata. We can see this, below, by issuing the svyset command. The particular design for this survey is not relevant.

```
. svyset
      pweight: finalwgt
          VCE: linearized
  Single unit: missing
     Strata 1: strata
         SU 1: psu
        FPC 1: <zero>
```

Let's now perform a regression analysis using this dataset. Let's predict systolic blood pressure from the person's age (in six age groups), sex, and weight. We use the svy prefix before the regress command to account for the survey design as specified by the svyset command.

```
. svy: regress bpsystol i.agegrp i.sex c.weight
(running regress on estimation sample)

Survey: Linear regression

Number of strata   =        31              Number of obs     =      10,351
Number of PSUs     =        62              Population size   = 117,157,513
                                            Design df         =          31
                                            F(   7,      25)  =      328.16
                                            Prob > F          =      0.0000
                                            R-squared         =      0.3087

------------------------------------------------------------------------------
             |             Linearized
    bpsystol |      Coef.   Std. Err.      t    P>|t|     [95% Conf. Interval]
-------------+----------------------------------------------------------------
      agegrp |
      20-29  |          0  (base)
      30-39  |   1.204372   .5704928     2.11   0.043     .0408442      2.3679
      40-49  |   6.881622   .7192606     9.57   0.000      5.41468    8.348563
      50-59  |   16.03707   .7148093    22.44   0.000     14.57921    17.49493
      60-69  |   23.38473    .772895    30.26   0.000      21.8084    24.96106
        70+  |   30.28196   .8950843    33.83   0.000     28.45643     32.1075
             |
         sex |
       Male  |          0  (base)
     Female  |   -.648672   .5405893    -1.20   0.239    -1.751211     .453867
             |
      weight |   .4257484   .0171314    24.85   0.000     .3908086    .4606881
       _cons |   88.06503   1.324402    66.49   0.000     85.36389    90.76616
------------------------------------------------------------------------------
```

We can use the **contrast**, **pwcompare**, **margins**, and **marginsplot** commands to interpret these results. The use of these commands is briefly illustrated below.

The **contrast** command can be used to make comparisons among the groups formed by a factor variable. The **contrast** command below tests the equality of the adjusted means for the six age groups. The test shows that the average systolic blood pressure is not equal among the six age groups.

```
. contrast agegrp

Contrasts of marginal linear predictions

                                            Design df         =          31
Margins      : asbalanced

------------------------------------------------
             |         df           F        P>F
-------------+----------------------------------
      agegrp |          5      297.86     0.0000
      Design |         31
------------------------------------------------
Note: F statistics are adjusted for the survey
      design.
```

The output of the **contrast** command indicates the F test is adjusted for the survey design. If you wanted to omit the adjustment for the design degrees of freedom, you could add the **nosvyadjust** option, as shown below. (See [R] **contrast** for more details about this option.)

```
. contrast agegrp, nosvyadjust
Contrasts of marginal linear predictions
                                                 Design df        =        31
Margins      : asbalanced
```

	df	F	P>F
agegrp	5	341.99	0.0000
Design	31		

The `pwcompare` command can also be used to form pairwise comparisons among the different age groups. In the example below, the **`mcompare(sidak)`** option is included to adjust for multiple comparisons. The results show that all the pairwise comparisons are significant except for the first comparison, comparing those `30-39 vs 20-29`.

```
. pwcompare agegrp, pveffects mcompare(sidak)
Pairwise comparisons of marginal linear predictions
                                                 Design df        =        31
Margins      : asbalanced
```

	Number of Comparisons
agegrp	15

	Contrast	Std. Err.	t	Sidak P>|t|
agegrp				
30-39 vs 20-29	1.204372	.5704928	2.11	0.482
40-49 vs 20-29	6.881622	.7192606	9.57	0.000
50-59 vs 20-29	16.03707	.7148093	22.44	0.000
60-69 vs 20-29	23.38473	.772895	30.26	0.000
70+ vs 20-29	30.28196	.8950843	33.83	0.000
40-49 vs 30-39	5.67725	.648312	8.76	0.000
50-59 vs 30-39	14.8327	.7601836	19.51	0.000
60-69 vs 30-39	22.18036	.8176811	27.13	0.000
70+ vs 30-39	29.07759	1.000489	29.06	0.000
50-59 vs 40-49	9.155449	.8885729	10.30	0.000
60-69 vs 40-49	16.50311	.9525501	17.33	0.000
70+ vs 40-49	23.40034	.9574325	24.44	0.000
60-69 vs 50-59	7.347657	.8545896	8.60	0.000
70+ vs 50-59	14.24489	.7472841	19.06	0.000
70+ vs 60-69	6.897234	1.020149	6.76	0.000

We can use the **margins** command to estimate the adjusted means by **agegrp**, as shown below. Note that I include the **vce(unconditional)** option to account for the sampling of covariates not fixed with the **at()** option. The estimates of the adjusted means are not altered by including this option, but the standard errors account for the fact that the values of the covariates were randomly sampled. (See [R] **margins** for more details about this option.)

```
. margins agegrp, nopvalues vce(unconditional)
Predictive margins
Number of strata   =          31            Number of obs     =      10,351
Number of PSUs     =          62            Population size   = 117,157,513
                                            Design df         =          31
Expression   : Linear prediction, predict()

-------------------------------------------------------------
             |            Linearized
             |     Margin   Std. Err.     [95% Conf. Interval]
-------------+-----------------------------------------------
      agegrp |
      20-29  |   118.3389   .6090259      117.0968     119.581
      30-39  |   119.5432   .5907545      118.3384    120.7481
      40-49  |   125.2205    .774779      123.6403    126.8007
      50-59  |   134.3759   .9099662      132.5201    136.2318
      60-69  |   141.7236   .8799832      139.9289    143.5183
        70+  |   148.6208    1.08863      146.4006    150.8411
-------------------------------------------------------------
```

The **marginsplot** command can be used to graph the adjusted means and confidence intervals computed by the **margins** command. Figure 19.1 shows the adjusted means (with confidence intervals) by age group as computed by the most recent **margins** command.

```
. marginsplot
  Variables that uniquely identify margins: agegrp
```

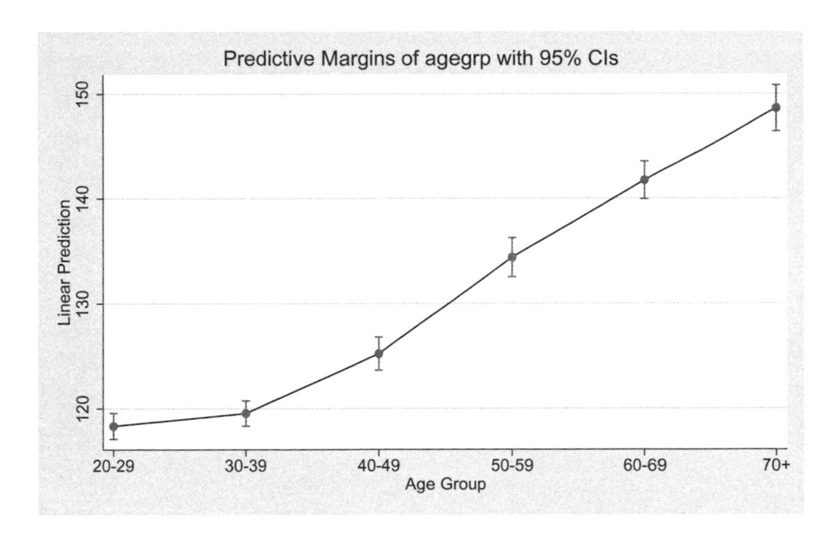

Figure 19.1. Adjusted means of systolic blood pressure by age group

As we have seen in this chapter, the **contrast**, **pwcompare**, **margins**, and **marginsplot** commands can be used to interpret the results from the analysis of complex survey data. When you use the **svy** prefix with an estimation command, the **contrast**, **pwcompare**, and **margins** commands compute appropriate estimates and standard errors based on the survey design. For more information, see [SVY] **svy**.

Part V

Appendices

These appendices take a command-centered focus, illustrating additional options that can be used with estimation commands and the `margins`, `marginsplot`, `contrast`, and `pwcompare` commands.

Appendix A illustrates options that you can use to customize the output from estimation commands, such as the `regress` or `logistic` commands.

Appendix B covers additional features of the `margins` command.

Appendix C covers the `marginsplot`, emphasizing how to customize the look of the graphs created by the `marginsplot` command.

Appendix D illustrates additional options used with the `contrast` command, emphasizing how you can customize the output.

Appendix E covers the `pwcompare` command, including more details regarding the different adjustments that can be used for multiple comparisons as well as how to customize the output of the `pwcompare` command.

A Customizing output from estimation commands

The emphasis of this book is on providing tools and techniques to aid in the interpretation of the results from regression models. Sometimes, interpretation is aided by the format in which the results are displayed. To that end, this appendix describes methods for customizing output from estimation commands, such as the regress or logistic commands. This appendix begins by illustrating how you can selectively omit portions of the statistical output (see section A.1). The following section shows how you can specify the confidence level used in the display of confidence intervals (see section A.2). Next, I illustrate how you can control the formatting of the display of results in the coefficient table (see section A.3). The concluding section of this appendix illustrates how you can customize the display of factor variables (see section A.4).

A.1 Omission of output

Let's consider a linear regression model in which we predict education from the education of the respondent's father, the education of the respondent's mother, and the respondent's age.

```
. use gss_ivrm

. regress educ paeduc maeduc age

      Source |       SS           df       MS      Number of obs   =    36,004
-------------+----------------------------------   F(3, 36000)     =   4681.43
       Model |  92650.2372         3  30883.4124   Prob > F        =    0.0000
    Residual |  237492.286    36,000  6.59700794   R-squared       =    0.2806
-------------+----------------------------------   Adj R-squared   =    0.2806
       Total |  330142.523    36,003  9.16986148   Root MSE        =    2.5685

------------------------------------------------------------------------------
        educ |      Coef.   Std. Err.      t    P>|t|     [95% Conf. Interval]
-------------+----------------------------------------------------------------
      paeduc |   .2135299   .0043313    49.30   0.000     .2050404    .2220195
      maeduc |   .2340485   .0051086    45.81   0.000     .2240355    .2440614
         age |   .0069201   .0008735     7.92   0.000     .0052081    .0086322
       _cons |   8.181127   .0684368   119.54   0.000     8.046989    8.315265
------------------------------------------------------------------------------
```

The output is composed of two parts: the header and the coefficient table. The coefficient table can be suppressed with the notable option. This might be useful if you had a model with dozens (or hundreds) of predictors and you wanted to only display the information regarding the overall model fit, for example, the R-squared.

```
. regress educ paeduc maeduc age, notable

      Source |       SS           df       MS      Number of obs   =    36,004
-------------+----------------------------------   F(3, 36000)     =   4681.43
       Model |  92650.2372         3  30883.4124   Prob > F        =    0.0000
    Residual |  237492.286    36,000  6.59700794   R-squared       =    0.2806
-------------+----------------------------------   Adj R-squared   =    0.2806
       Total |  330142.523    36,003  9.16986148   Root MSE        =    2.5685
```

The header of the output can be suppressed with the `noheader` option. This option allows you to focus solely on the coefficient table. I frequently use this option in this book where I want to focus on the coefficient table and want to save space on the page.

```
. regress educ paeduc maeduc age, noheader

------------------------------------------------------------------------------
        educ |      Coef.   Std. Err.      t    P>|t|     [95% Conf. Interval]
-------------+----------------------------------------------------------------
      paeduc |   .2135299   .0043313    49.30   0.000     .2050404    .2220195
      maeduc |   .2340485   .0051086    45.81   0.000     .2240355    .2440614
         age |   .0069201   .0008735     7.92   0.000     .0052081    .0086322
       _cons |   8.181127   .0684368   119.54   0.000     8.046989    8.315265
------------------------------------------------------------------------------
```

The `noci` option can be used to suppress the display of the columns containing the confidence intervals. (I also included the `noheader` option to save space on the page.) I sometimes use this option in this book to allow more space for the display of the variable names, especially with the inclusion of interaction terms. This can avoid having the description of the interaction span across multiple lines, making the output more readable.

```
. regress educ paeduc maeduc age, noheader noci

-----------------------------------------------------
        educ |      Coef.   Std. Err.      t    P>|t|
-------------+---------------------------------------
      paeduc |   .2135299   .0043313    49.30   0.000
      maeduc |   .2340485   .0051086    45.81   0.000
         age |   .0069201   .0008735     7.92   0.000
       _cons |   8.181127   .0684368   119.54   0.000
-----------------------------------------------------
```

You can suppress the display of the significance tests with the `nopvalues` option. This suppresses the columns displaying the test-statistic and the p-value. In this example, it suppresses the columns showing the t-value and p-value.

```
. regress educ paeduc maeduc age, noheader nopvalues

----------------------------------------------------------------
        educ |      Coef.   Std. Err.     [95% Conf. Interval]
-------------+--------------------------------------------------
      paeduc |   .2135299   .0043313      .2050404    .2220195
      maeduc |   .2340485   .0051086      .2240355    .2440614
         age |   .0069201   .0008735      .0052081    .0086322
       _cons |   8.181127   .0684368      8.046989    8.315265
----------------------------------------------------------------
```

A.2 Specifying the confidence level

Let's continue to use the example we used from the previous section, where we fit a linear regression model in which we predict education from the education of the respondent's father, the education of the respondent's mother, and the respondent's age. To save space, I have also included the **noheader** option.

```
. use gss_ivrm
. regress educ paeduc maeduc age, noheader
```

educ	Coef.	Std. Err.	t	P>|t|	[95% Conf.	Interval]
paeduc	.2135299	.0043313	49.30	0.000	.2050404	.2220195
maeduc	.2340485	.0051086	45.81	0.000	.2240355	.2440614
age	.0069201	.0008735	7.92	0.000	.0052081	.0086322
_cons	8.181127	.0684368	119.54	0.000	8.046989	8.315265

Note how the confidence intervals are displayed using a 95% confidence level. This is the default confidence level used by Stata. Suppose we wanted to display the confidence intervals using a 90% confidence level. We can make this change by adding the `level(90)` option to the **regress** command, shown below.

```
. regress educ paeduc maeduc age, noheader level(90)
```

educ	Coef.	Std. Err.	t	P>|t|	[90% Conf.	Interval]
paeduc	.2135299	.0043313	49.30	0.000	.2064054	.2206545
maeduc	.2340485	.0051086	45.81	0.000	.2256454	.2424515
age	.0069201	.0008735	7.92	0.000	.0054834	.0083569
_cons	8.181127	.0684368	119.54	0.000	8.068556	8.293698

Say that you wanted to change the confidence level to 90% for all subsequent estimation commands. It would be laborious and error prone to try to add the `level(90)` option to all of your estimation commands. Instead, you can use the **set level** command shown below, and then all subsequent estimation commands will be estimated using a confidence level of 90%.

```
. set level 90
```

Now, when I run the regression command from above, it uses the new default confidence level, 90%.

```
. regress educ paeduc maeduc age, noheader
```

educ	Coef.	Std. Err.	t	P>|t|	[90% Conf.	Interval]
paeduc	.2135299	.0043313	49.30	0.000	.2064054	.2206545
maeduc	.2340485	.0051086	45.81	0.000	.2256454	.2424515
age	.0069201	.0008735	7.92	0.000	.0054834	.0083569
_cons	8.181127	.0684368	119.54	0.000	8.068556	8.293698

Suppose I want to use a 90% confidence level as my default confidence level each time I start Stata. I can use the `set level` command with the `permanently` option, shown below.

```
. set level 90, permanently
```

Now, each time I start Stata on my computer, the default confidence level for all estimation commands will be 90%. However, I would like to permanently revert back to the default confidence level of 95%. I do so with the `set level` command below.

```
. set level 95, permanently
(set level preference recorded)
```

A.3 Customizing the formatting of columns in the coefficient table

There are three options that you can use for customizing the display of the columns of results shown in the coefficient table—the `pformat()` option, the `sformat()` option, and the `cformat()` option. To illustrate these options, let's continue to use the regression model from the prior section, repeated below. (I include the `noheader` option to save space.)

```
. use gss_ivrm
. regress educ paeduc maeduc age, noheader
```

educ	Coef.	Std. Err.	t	P>|t|	[95% Conf.	Interval]
paeduc	.2135299	.0043313	49.30	0.000	.2050404	.2220195
maeduc	.2340485	.0051086	45.81	0.000	.2240355	.2440614
age	.0069201	.0008735	7.92	0.000	.0052081	.0086322
_cons	8.181127	.0684368	119.54	0.000	8.046989	8.315265

The `pformat()` option can be used to control the formatting of the p-values. The example below uses the `pformat(%5.2f)` option to display the p-values with a total width of five columns and two digits after the decimal place. You can see `help format` for more details about how to specify formats. Note that the width of the format specified in the `pformat()` option cannot exceed five columns.

```
. regress educ paeduc maeduc age, noheader pformat(%5.2f)
```

educ	Coef.	Std. Err.	t	P>|t|	[95% Conf.	Interval]
paeduc	.2135299	.0043313	49.30	0.00	.2050404	.2220195
maeduc	.2340485	.0051086	45.81	0.00	.2240355	.2440614
age	.0069201	.0008735	7.92	0.00	.0052081	.0086322
_cons	8.181127	.0684368	119.54	0.00	8.046989	8.315265

The `sformat()` option controls the formatting of the test statistics. The example below uses the `sformat(%5.1f)` option to display the *t*-values using a fixed width of five columns with one decimal place. Note that the width of the format specified for the `sformat()` option cannot exceed eight columns.

```
. regress educ paeduc maeduc age, noheader sformat(%5.1f)
```

educ	Coef.	Std. Err.	t	P>|t|	[95% Conf.	Interval]
paeduc	.2135299	.0043313	49.3	0.000	.2050404	.2220195
maeduc	.2340485	.0051086	45.8	0.000	.2240355	.2440614
age	.0069201	.0008735	7.9	0.000	.0052081	.0086322
_cons	8.181127	.0684368	119.5	0.000	8.046989	8.315265

The `cformat()` option controls the format used to display the coefficients, standard errors, and confidence intervals. The format selected via the `cformat()` option is applied to all of these columns. In the example below, I have added the `cformat(%7.4f)` option, which means that the results shown in these columns will be displayed in a fixed format with a total width seven columns and four digits after the decimal point. You can see `help format` for more information about other display formats you could select. Note that the width of the format specified in the `cformat()` option cannot exceed nine columns.

```
. regress educ paeduc maeduc age, noheader cformat(%7.4f)
```

educ	Coef.	Std. Err.	t	P>|t|	[95% Conf.	Interval]
paeduc	0.2135	0.0043	49.30	0.000	0.2050	0.2220
maeduc	0.2340	0.0051	45.81	0.000	0.2240	0.2441
age	0.0069	0.0009	7.92	0.000	0.0052	0.0086
_cons	8.1811	0.0684	119.54	0.000	8.0470	8.3153

Suppose I really liked the display of the *p*-values using the `pformat(%5.2f)` option and wanted to use it for subsequent estimation commands. It would be really laborious to add the `pformat(%5.2f)` to every estimation command just to obtain my desired formatting for the *p*-values. Instead, I could type the following `set pformat` command.

```
. set pformat %5.2f
```

After typing that command, the *p*-values for subsequent estimation commands will be formatted using the specified format. I could return to the default format for the *p*-values with the `set pformat` command shown below.

```
. set pformat
```

Suppose I like `%5.2f` format for the display of *p*-values so much that I would like to adopt that as the permanent default setting. In that case, I can use the `set pformat` command with the `permanently` option, as shown below.

```
. set pformat %5.2f, permanently
```

This immediately changes the default formatting of the p-values for the current Stata session. It also changed the default setting each time I invoke Stata in the future.

I could restore the default setting by typing

```
. set pformat, permanently
```

Stata also includes the `set cformat` command and the `set pformat` commands, which work in the same manner as the `set pformat` command. The `set cformat` command allows you to control the display of the coefficients, standard errors, and confidence intervals, and the `set sformat` command allows you to control the display of test statistics (for example, the t-values).

A.4 Customizing the display of factor variables

This section will illustrate options you can use to customize the manner in which factor variables are used to label the results in the coefficient table of estimation commands. For example, consider this regression model in which income is predicted is a function of father's education, mother's education, the respondent's education (in three categories), race, and gender. Education and race include the `i.` prefix; therefore, they will be treated as factor variables. To save space, these examples will include the `noheader` option to display just the coefficient table. (Throughout this book, I have changed the default settings to explicitly display the base levels of all factor variables. For this section, I have turned that off, resuming the default behavior regarding this setting.)

```
. use gss_ivrm
. regress realrinc paeduc maeduc i.educ3 i.race female, noheader
```

realrinc	Coef.	Std. Err.	t	P>|t|	[95% Conf.	Interval]
paeduc	88.09616	57.41184	1.53	0.125	-24.43498	200.6273
maeduc	-52.00394	69.06015	-0.75	0.451	-187.3665	83.35867
educ3						
HS	6177.731	606.1345	10.19	0.000	4989.666	7365.796
Coll	20604.58	680.0623	30.30	0.000	19271.61	21937.54
race						
black	-3082.098	603.5712	-5.11	0.000	-4265.139	-1899.058
other	-266.8543	836.3088	-0.32	0.750	-1906.077	1372.368
female	-13602.66	350.2554	-38.84	0.000	-14289.19	-12916.14
_cons	19688.6	735.0663	26.78	0.000	18247.82	21129.38

Note how Stata inserts a blank line before and after every factor variable. That can make the output easier to read. However, if you have many factor variables, you might want to omit those blank lines to make the output more compact. To create a more compact table, we can add the `vsquish` option. This vertically squishes the table, getting rid of the blank lines. The output is more compact, but it is also harder to read.

```
. regress realrinc paeduc maeduc i.educ3 i.race female, noheader vsquish

    realrinc        Coef.   Std. Err.      t    P>|t|     [95% Conf. Interval]

      paeduc     88.09616   57.41184     1.53   0.125    -24.43498    200.6273
      maeduc    -52.00394   69.06015    -0.75   0.451    -187.3665    83.35867
       educ3
          HS     6177.731   606.1345    10.19   0.000     4989.666    7365.796
        Coll     20604.58   680.0623    30.30   0.000     19271.61    21937.54
        race
       black    -3082.098   603.5712    -5.11   0.000    -4265.139   -1899.058
       other    -266.8543   836.3088    -0.32   0.750    -1906.077    1372.368
      female    -13602.66   350.2554   -38.84   0.000    -14289.19   -12916.14
       _cons      19688.6   735.0663    26.78   0.000     18247.82    21129.38
```

I happen to like tables that include a row that shows the base level (often called the comparison group or the reference group). By default, Stata omits the display of such a row. But we can add the `baselevels` option to display a row with the base level of each factor variable. By adding the `baselevels` option to the example below, it is clear that the base group for `educ3` is the group without a high school diploma; and the base group for `race` is white.

```
. regress realrinc paeduc maeduc i.educ3 i.race female, noheader baselevels

    realrinc        Coef.   Std. Err.      t    P>|t|     [95% Conf. Interval]

      paeduc     88.09616   57.41184     1.53   0.125    -24.43498    200.6273
      maeduc    -52.00394   69.06015    -0.75   0.451    -187.3665    83.35867

       educ3
     not hs             0  (base)
          HS     6177.731   606.1345    10.19   0.000     4989.666    7365.796
        Coll     20604.58   680.0623    30.30   0.000     19271.61    21937.54

        race
       white            0  (base)
       black    -3082.098   603.5712    -5.11   0.000    -4265.139   -1899.058
       other    -266.8543   836.3088    -0.32   0.750    -1906.077    1372.368

      female    -13602.66   350.2554   -38.84   0.000    -14289.19   -12916.14
       _cons      19688.6   735.0663    26.78   0.000     18247.82    21129.38
```

Note! Using the set command to control reporting of base levels

I prefer output that displays the base (reference) category for factor variables. In fact, I prefer that setting so much, that I have adopted that as the default throughout this book. I did this using the `set baselevels` command, shown below:

```
. set baselevels on
```

After specifying that command, the base (reference) categories are displayed by default. By adding the `permanently` option (shown below), that setting will be the default each time I start Stata.

```
. set baselevels on, permanently
```

You can revert back to the default settings, turning off the display of the base (reference) category with the `set baselevels` command below.

```
. set baselevels off
```

You can add the `permanently` option to make that setting the default each time you invoke Stata.

In the example below, I have combined the `baselevels` and `vsquish` options. These options, when used in combination, include the base (reference) groups for each factor variable and the `vsquish` option helps keep the output compact.

```
. regress realrinc paeduc maeduc i.educ3 i.race female, noheader
> baselevels vsquish
```

realrinc	Coef.	Std. Err.	t	P>|t|	[95% Conf.	Interval]
paeduc	88.09616	57.41184	1.53	0.125	-24.43498	200.6273
maeduc	-52.00394	69.06015	-0.75	0.451	-187.3665	83.35867
educ3						
not hs	0	(base)				
HS	6177.731	606.1345	10.19	0.000	4989.666	7365.796
Coll	20604.58	680.0623	30.30	0.000	19271.61	21937.54
race						
white	0	(base)				
black	-3082.098	603.5712	-5.11	0.000	-4265.139	-1899.058
other	-266.8543	836.3088	-0.32	0.750	-1906.077	1372.368
female	-13602.66	350.2554	-38.84	0.000	-14289.19	-12916.14
_cons	19688.6	735.0663	26.78	0.000	18247.82	21129.38

Sometimes, value labels for factor variables can get very long, and they are hard to read when displayed in the coefficient table. Consider the variable `educ3long`.[1] Using the `tabulate` command, each level is displayed on its own line even though the value labels are rather lengthy.

```
. tabulate educ3long
  Education of R in three
               categories |      Freq.     Percent        Cum.
--------------------------+-----------------------------------
     Less than High School |     13,095       23.84       23.84
      High School Graduate |     29,855       54.36       78.20
College Graduate (or higher) |     11,975       21.80      100.00
--------------------------+-----------------------------------
                     Total |     54,925      100.00
```

Now, let's look at how the labels for this variable are displayed when used as a factor variable in a regression model. For this example, I will use `educ3long` as the only predictor. Note that each level of the factor variable is displayed using only one line, and the value label is abbreviated to fit in that small amount of space.[2]

```
. regress realrinc i.educ3long, noheader baselevel

    realrinc |      Coef.   Std. Err.      t    P>|t|     [95% Conf. Interval]
-------------+----------------------------------------------------------------
   educ3long |
Less than .. |          0  (base)
High Schoo.. |   4847.106   412.6238    11.75   0.000     4038.348    5655.864
College G..) |   20458.81    460.051    44.47   0.000     19557.09    21360.52
             |
       _cons |   13628.71    365.824    37.25   0.000     12911.68    14345.74
```

To help make the output more readable, we can allow the factor variables to wrap across two lines by including the `fvwrap(2)` option. Note how the first two levels of `educ3long` are displayed in their entirety (without abbreviation), but the third level is still so long that it requires abbreviation.

1. This variable is identical to the variable `educ3` except that it is labeled using very long value labels. I created the variable `educ3long` specifically for the following examples, to illustrate ways that you can control the display of long value labels in estimation commands. However, there is another solution that I do not want to belabor but just quickly mention. Say that you want to have long, descriptive value labels for commands like the `tabulate` command but short, pithy value labels for commands like the `regress` command. In such a case, I recommend considering the `label language` command (see `help label language`). You could use that suite of commands to create one set of value labels that are long and elaborate and a second set of value labels that are short and pithy. Before the `tabulate` commands, you could activate the elaborate value labels, and before the `regress` commands, you could activate the pithy labels. For the examples following examples in this chapter, I illustrate options you can use to adjust the display of value labels in estimation commands like the `regress` command.
2. I have included the `baselevel` option to include the display of the base (reference) category.

```
. regress realrinc i.educ3long, noheader baselevel fvwrap(2)

    realrinc       Coef.   Std. Err.      t    P>|t|     [95% Conf. Interval]

   educ3long
   Less than
 High School           0  (base)
 High School
    Graduate    4847.106   412.6238    11.75   0.000     4038.348    5655.864
   College..
  (or hig..)    20458.81    460.051    44.47   0.000     19557.09    21360.52

       _cons    13628.71    365.824    37.25   0.000     12911.68    14345.74
```

The next example uses the `fvwrap(3)` option to indicate that the value labels for factor variables can extend to three lines. By allowing 3 lines, the value label for the third level of education is now displayed in its entirety. Note how the first and second levels of education are displayed using only two lines, because two lines were sufficient for displaying those labels without abbreviation.[3]

```
. regress realrinc i.educ3long, noheader baselevel fvwrap(3)

    realrinc       Coef.   Std. Err.      t    P>|t|     [95% Conf. Interval]

   educ3long
   Less than
 High School           0  (base)
 High School
    Graduate    4847.106   412.6238    11.75   0.000     4038.348    5655.864
     College
Graduate (or
     higher)    20458.81    460.051    44.47   0.000     19557.09    21360.52

       _cons    13628.71    365.824    37.25   0.000     12911.68    14345.74
```

Let's look more carefully at the value labels for `educ3long` in the output above. Notice how each word is always displayed fully on a line. By default, Stata will display the value labels while keeping whole words together. The label skips to the next line at the boundaries between words.

3. I could use the `set fvwrap 3` command, and subsequent estimation commands would allow up to 3 lines for displaying the value labels of factor variable. Specifying `set fvwrap 3, permanently` would make that change permanent, changing the default each time I start Stata.

There may be instances where you would prefer the value label to skip to the next line each time the label fills the width of the current line, even if that means skipping to the next line in the middle of a word. Such behavior is requested by using the `fvwrap(width)` option, shown below. Note how the value labels are wrapped based on the width of the column—for instance, the first label breaks the first and second lines in the middle of the word `High`, showing `Hi` at the end of the first line and `gh` at the start of the second line.[4]

```
. regress realrinc i.educ3long, noheader baselevel fvwrap(3) fvwrapon(width)

    realrinc       Coef.   Std. Err.      t    P>|t|     [95% Conf. Interval]

   educ3long
 Less than Hi
    gh School          0  (base)
 High School
     Graduate   4847.106   412.6238    11.75   0.000     4038.348    5655.864
 College Grad
 uate (or hig
         her)   20458.81    460.051    44.47   0.000     19557.09    21360.52

        _cons   13628.71    365.824    37.25   0.000     12911.68    14345.74
```

Labeling the values of factor variables with value labels creates output that is easier to interpret and helps to avoid mistakes when interpreting the levels of the variables. There can be instances, namely, when running models with interactions, where the output could be much cleaner by displaying the levels of the factor variables instead of their labels. For starters, consider this model. It includes two factor variables, `class` and `race`. I have included the `baselevels` option because I like seeing the base (reference) category.

4. You can type the command `set fvwrapon width` to select this wrapping behavior for subsequent estimation commands. The command `set fvwrapon word` command would revert back to the default wrapping behavior (wrapping on word boundaries). You can add the `permanently` option to the `set fvwrapon` command to make your choice permanent, changing the default whenever you start Stata.

```
. regress realrinc i.class##i.race, noheader baselevels

    realrinc        Coef.   Std. Err.      t    P>|t|     [95% Conf. Interval]

       class
 lower class            0  (base)
working cl..     7171.359   1020.419     7.03   0.000     5171.296    9171.422
middle class      18356.8   1021.293    17.97   0.000     16355.03    20358.58
 upper class     52336.15   1371.341    38.16   0.000     49648.27    55024.04

        race
       white            0  (base)
       black    -502.2865   1857.073    -0.27   0.787    -4142.226    3137.652
       other     134.8222   3186.173     0.04   0.966    -6110.208    6379.853

  class#race
working cl.. #
       black    -1075.158   1936.064    -0.56   0.579    -4869.923    2719.608
working cl.. #
       other    -1930.474   3307.904    -0.58   0.559    -8414.102    4553.155
middle class #
       black    -6903.451   2036.796    -3.39   0.001    -10895.66   -2911.248
middle class #
       other     2609.323   3371.359     0.77   0.439    -3998.679    9217.325
 upper class #
       black     -46158.4   3492.687   -13.22   0.000    -53004.21   -39312.59
 upper class #
       other     16828.87   6516.882     2.58   0.010     4055.515    29602.23

       _cons     9427.408   993.1766     9.49   0.000     7480.741    11374.08
```

Now, let's introduce an interaction of `class` and `gender` in this model. The output with respect to the interactions can be more difficult to follow.

```
. regress realrinc i.class##i.gender, noheader baselevels

    realrinc        Coef.   Std. Err.      t    P>|t|     [95% Conf. Interval]

       class
 lower class            0  (base)
working cl..     8820.579   1190.235     7.41   0.000     6487.669    11153.49
middle class     23876.35   1193.938    20.00   0.000     21536.18    26216.52
 upper class     64264.12   1596.847    40.24   0.000     61134.23       67394

      gender
        Male            0  (base)
      Female    -3213.274   1572.964    -2.04   0.041    -6296.347   -130.2006

class#gender
working cl.. #
      Female    -3982.516   1622.652    -2.45   0.014     -7162.98   -802.0522
middle class #
      Female    -12777.71   1631.547    -7.83   0.000    -15975.61   -9579.814
 upper class #
      Female    -41899.07   2357.007   -17.78   0.000     -46518.9   -37279.24

       _cons     11039.23   1156.036     9.55   0.000     8773.353    13305.11
```

To simplify the output, I added the **nofvlabel** option. Now, the factor variables are displayed using the levels of the factor variable instead of the factor-variable labels. The advantage of this option is that each row of output is compactly displayed usually with one line per result, but it sacrifices the clarity achieved by labeling each value with its corresponding value label.

```
. regress realrinc i.class##i.gender, noheader baselevels nofvlabel

    realrinc       Coef.   Std. Err.      t    P>|t|     [95% Conf. Interval]

       class
          1            0  (base)
          2     8820.579   1190.235     7.41   0.000     6487.669    11153.49
          3     23876.35   1193.938    20.00   0.000     21536.18    26216.52
          4     64264.12   1596.847    40.24   0.000     61134.23       67394

      gender
          1            0  (base)
          2    -3213.274   1572.964    -2.04   0.041    -6296.347   -130.2006

class#gender
         2 2   -3982.516   1622.652    -2.45   0.014     -7162.98   -802.0522
         3 2   -12777.71   1631.547    -7.83   0.000    -15975.61   -9579.814
         4 2   -41899.07   2357.007   -17.78   0.000     -46518.9   -37279.24

       _cons    11039.23   1156.036     9.55   0.000     8773.353    13305.11
```

If you like this style of output—showing the values instead of the factor-variable labels—you might want to use it as the default setting. You can type the following **set fvlabel** command, and the default setting will be to show the levels of factor variables.

```
. set fvlabel off
```

You could make that setting permanent by adding the **permanently** option to the command above.

You can revert back to the default, labeling factor variables with value labels using the **set fvlabel** command below.

```
. set fvlabel on
```

Adding the **permanently** option to the command above would make that setting permanent, changing the default for future Stata sessions.

Note! Showing all base levels

When fitting models with interactions, there are base levels with respect to the main effect terms, and there are base levels with respect to the interaction terms as well. Including the **baselevels** option displays the base levels for the main effects (but not interactions). If you want to display the base levels with respect to the interactions as well, then you can specify the **allbaselevels** option.

B The margins command

This appendix illustrates some of the options that you can use with the **margins** command. In particular, this appendix illustrates the **predict()** and **expression()** options (see section B.1), the **at()** option (see section B.2), computing margins with factor variables (see section B.3), computing margins with factor variables and the **at()** option (see section B.4), and the **dydx()** and related options (see section B.5). This chapter also illustrates how you can specify the confidence level (see section B.6) and customize column formatting (see section B.7). For complete details about the **margins** command, see [R] **margins**.

B.1 The predict() and expression() options

Consider the example below that uses a logistic regression model to predict whether a person is a smoker based on the person's self-reported social class, education, age, and year of interview.

```
. use gss_ivrm
. logit smoke i.class educ age yrint
Iteration 0:   log likelihood = -9951.2382
Iteration 1:   log likelihood = -9582.9808
Iteration 2:   log likelihood = -9580.0732
Iteration 3:   log likelihood = -9580.0723
Logistic regression                             Number of obs     =     15,375
                                                LR chi2(6)        =     742.33
                                                Prob > chi2       =     0.0000
Log likelihood = -9580.0723                     Pseudo R2         =     0.0373

------------------------------------------------------------------------------
       smoke |      Coef.   Std. Err.      z    P>|z|     [95% Conf. Interval]
-------------+----------------------------------------------------------------
       class |
 lower class |          0  (base)
working cl.. |  -.2604145   .0749733    -3.47   0.001    -.4073595   -.1134696
middle class |   -.455007   .0772007    -5.89   0.000    -.6063176   -.3036965
 upper class |  -.5234883   .1274959    -4.11   0.000    -.7733757   -.2736009
             |
        educ |  -.0878902   .0063325   -13.88   0.000    -.1003016   -.0754787
         age |  -.0198399   .0010949   -18.12   0.000    -.0219859   -.0176939
       yrint |  -.0321324   .0033889    -9.48   0.000    -.0387746   -.0254902
       _cons |   65.45743   6.718824     9.74   0.000     52.28877    78.62608
------------------------------------------------------------------------------
```

Suppose we specify the **margins** command shown below. This computes the predictive margin holding age constant at 40, adjusting for all the other covariates. The portion of the output titled `Expression:` tells us that the predictive margin is computed in terms of the probability of smoking. The predictive marginal probability of smoking for a 40-year-old is 0.368.

```
. margins, at(age=40)

Predictive margins                              Number of obs     =     15,375
Model VCE    : OIM

Expression   : Pr(smoke), predict()
at           : age             =          40

------------------------------------------------------------------------------
             |            Delta-method
             |     Margin   Std. Err.      z    P>|z|     [95% Conf. Interval]
-------------+----------------------------------------------------------------
       _cons |   .3684519   .0039911    92.32   0.000     .3606294    .3762743
------------------------------------------------------------------------------
```

Say that we add the **predict(xb)** option to the **margins** command. The output below shows that the predictive margin is computed in terms of the log odds of the outcome. The average adjusted log odds of a 40-year-old smoking is −0.558.

```
. margins, at(age=40) predict(xb)

Predictive margins                              Number of obs     =     15,375
Model VCE    : OIM

Expression   : Linear prediction (log odds), predict(xb)
at           : age             =          40

------------------------------------------------------------------------------
             |            Delta-method
             |     Margin   Std. Err.      z    P>|z|     [95% Conf. Interval]
-------------+----------------------------------------------------------------
       _cons |  -.5576398   .0177666   -31.39   0.000    -.5924618   -.5228179
------------------------------------------------------------------------------
```

Instead of the `predict()` option, you can specify the **expression()** option to specify the metric used by the **margins** command. Let's first consider a few basic examples using the **expression()** option. For example, the **expression()** option below is used to compute the predicted probability of a 40-year-old smoking, adjusting for all other covariates. This is equivalent to specifying the **predict(pr)** option.

```
. margins, at(age=40) expression(predict(pr))

Predictive margins                              Number of obs     =     15,375
Model VCE    : OIM

Expression   : Pr(smoke), predict(pr)
at           : age             =          40

------------------------------------------------------------------------------
             |            Delta-method
             |     Margin   Std. Err.      z    P>|z|     [95% Conf. Interval]
-------------+----------------------------------------------------------------
       _cons |   .3684519   .0039911    92.32   0.000     .3606294    .3762743
------------------------------------------------------------------------------
```

The following example shows the predicted log odds of a 40-year-old smoking, adjusting for all other covariates. This is equivalent to using the `predict(xb)` option.

```
. margins, at(age=40) expression(predict(xb))
Predictive margins                              Number of obs     =     15,375
Model VCE    : OIM
Expression   : Linear prediction (log odds), predict(xb)
at           : age             =          40

------------------------------------------------------------------------------
             |            Delta-method
             |     Margin   Std. Err.      z    P>|z|     [95% Conf. Interval]
-------------+----------------------------------------------------------------
       _cons |  -.5576398   .0177666   -31.39   0.000    -.5924618   -.5228179
------------------------------------------------------------------------------
```

The **expression()** option is more flexible than the **predict()** option because we can specify an expression. Imagine we wanted to compute the odds of a 40-year-old smoking, adjusting for all other covariates. The odds of a person smoking can be computed by taking the exponential of the log odds of smoking. We can compute this by specifying **exp(predict(xb))** (that is, the exponential of the predicted log odds) within the **expression()** option, as shown below. The predictive marginal odds of smoking for a 40-year-old is 0.620.

```
. margins, at(age=40) expression(exp(predict(xb)))
Predictive margins                              Number of obs     =     15,375
Model VCE    : OIM
Expression   : exp(predict(xb))
at           : age             =          40

------------------------------------------------------------------------------
             |            Delta-method
             |     Margin   Std. Err.      z    P>|z|     [95% Conf. Interval]
-------------+----------------------------------------------------------------
       _cons |   .6196325    .011756    52.71   0.000     .5965911    .6426738
------------------------------------------------------------------------------
```

We also could have specified the odds of smoking by taking the probability of smoking divided by the probability of not smoking. The **margins** command below specifies **predict(pr)/(1-predict(pr))** within the **expression()** option. Note that this yields the same predictive margin as the previous **margins** command.

```
. margins, at(age=40) expression(predict(pr)/(1-predict(pr)))
Predictive margins                              Number of obs     =     15,375
Model VCE    : OIM
Expression   : predict(pr)/(1-predict(pr))
at           : age             =          40

------------------------------------------------------------------------------
             |            Delta-method
             |     Margin   Std. Err.      z    P>|z|     [95% Conf. Interval]
-------------+----------------------------------------------------------------
       _cons |   .6196325    .011756    52.71   0.000     .5965911    .6426738
------------------------------------------------------------------------------
```

As you can see, the `expression()` option allows us to specify the metric of the predictive margin in a more general way than the `predict()` option. The examples illustrated in chapter 18 provide more details about the `predict()` option in the context of nonlinear models.

B.2 The at() option

The `at()` option allows us to specify the values of covariates when computing the margins. For example, we can estimate the probability of smoking for a person who describes themselves as upper class (that is, `class=4`), has 15 years of education, is 40 years old, and was interviewed in 1980, using the `at()` option shown below. For someone with these characteristics, the conditional probability of smoking is 0.310.

```
. margins, at(class=4 educ=15 age=40 yrint=1980)
Adjusted predictions                            Number of obs     =     15,375
Model VCE    : OIM

Expression   : Pr(smoke), predict()
at           : class           =           4
               educ            =          15
               age             =          40
               yrint           =        1980

------------------------------------------------------------------------------
             |            Delta-method
             |     Margin   Std. Err.      z    P>|z|     [95% Conf. Interval]
-------------+----------------------------------------------------------------
       _cons |   .3099763   .0225373    13.75   0.000     .2658039    .3541486
------------------------------------------------------------------------------
```

Suppose we want to compute the predictive marginal probability of smoking at 20 to 80 years of age in 10-year increments, while adjusting for all other covariates. We can compute this using the `at()` option shown in the following `margins` command. (The `vsquish` option is included to suppress the display of empty lines in the output.)

```
. margins, at(age=(20 30 40 50 60 70 80)) vsquish
Predictive margins                              Number of obs     =     15,375
Model VCE    : OIM
Expression   : Pr(smoke), predict()
1._at        : age             =          20
2._at        : age             =          30
3._at        : age             =          40
4._at        : age             =          50
5._at        : age             =          60
6._at        : age             =          70
7._at        : age             =          80
```

	Margin	Delta-method Std. Err.	z	P>|z|	[95% Conf.	Interval]
_at						
1	.4610941	.0073885	62.41	0.000	.4466128	.4755753
2	.4140318	.0053163	77.88	0.000	.403612	.4244517
3	.3684519	.0039911	92.32	0.000	.3606294	.3762743
4	.3250497	.0039668	81.94	0.000	.3172748	.3328245
5	.2843881	.0049038	57.99	0.000	.2747768	.2939995
6	.2468764	.0060491	40.81	0.000	.2350203	.2587325
7	.2127647	.0070357	30.24	0.000	.1989749	.2265545

Rather than typing the values 20 30 40 50 60 70 80, we can specify 20(10)80, as shown in the example below.

```
. margins, at(age=(20(10)80))
  (output omitted)
```

Rather than specifying an exact age, we can specify that we want age held constant at the 25th percentile using the at() option shown below.

```
. margins, at((p25) age) vsquish
Predictive margins                              Number of obs     =     15,375
Model VCE    : OIM
Expression   : Pr(smoke), predict()
at           : age             =          30 (p25)
```

	Margin	Delta-method Std. Err.	z	P>|z|	[95% Conf.	Interval]
_cons	.4140318	.0053163	77.88	0.000	.403612	.4244517

In place of p25, we could specify any value ranging from p1 (the first percentile) to p99 (the 99th percentile). We can also specify min, max, or mean.

The at() option below holds the variables age and educ constant at the 25th percentile and the year of interview at 1990.

```
. margins, at((p25) age educ yrint=1990) vsquish
Predictive margins                              Number of obs     =     15,375
Model VCE    : OIM

Expression   : Pr(smoke), predict()
at           : educ            =          11 (p25)
               age             =          30 (p25)
               yrint           =        1990

------------------------------------------------------------------------------
             |            Delta-method
             |     Margin   Std. Err.      z    P>|z|     [95% Conf. Interval]
-------------+----------------------------------------------------------------
       _cons |   .4036987   .0076903    52.49   0.000     .3886259    .4187715
------------------------------------------------------------------------------
```

We can specify the at() option multiple times to compute predictive margins for different combinations of covariate values. For example, the **margins** command below computes the predictive margin once holding age constant at 20 and education constant at 10 and then again holding age constant at 50 and education constant at 15.

```
. margins, at(age=20 educ=10) at(age=50 educ=15) vsquish
Predictive margins                              Number of obs     =     15,375
Model VCE    : OIM

Expression   : Pr(smoke), predict()
1._at        : educ            =          10
               age             =          20
2._at        : educ            =          15
               age             =          50

------------------------------------------------------------------------------
             |            Delta-method
             |     Margin   Std. Err.      z    P>|z|     [95% Conf. Interval]
-------------+----------------------------------------------------------------
         _at |
          1  |   .5123491   .0094082    54.46   0.000     .4939094    .5307888
          2  |   .2739494    .005451    50.26   0.000     .2632656    .2846333
------------------------------------------------------------------------------
```

The legend describing the at() values can take up a lot of space. We can suppress the display of this with the **noatlegend** option, as shown below.

```
. margins, at(age=20 educ=10) at(age=50 educ=15) vsquish noatlegend
Predictive margins                              Number of obs     =     15,375
Model VCE    : OIM

Expression   : Pr(smoke), predict()

------------------------------------------------------------------------------
             |            Delta-method
             |     Margin   Std. Err.      z    P>|z|     [95% Conf. Interval]
-------------+----------------------------------------------------------------
         _at |
          1  |   .5123491   .0094082    54.46   0.000     .4939094    .5307888
          2  |   .2739494    .005451    50.26   0.000     .2632656    .2846333
------------------------------------------------------------------------------
```

The example below computes predictive margins holding the variables **educ**, **age**, and **yrint** constant at the 25th percentile, at the 50th percentile, and at the 75th percentile.

```
. margins, at((p25) educ age yrint)
>          at((p50) educ age yrint)
>          at((p75) educ age yrint) noatlegend vsquish

Predictive margins                              Number of obs     =     15,375
Model VCE    : OIM

Expression   : Pr(smoke), predict()

------------------------------------------------------------------------------
             |            Delta-method
             |     Margin   Std. Err.      z    P>|z|     [95% Conf. Interval]
-------------+----------------------------------------------------------------
         _at |
          1  |   .4824943   .0073814    65.37   0.000      .468027    .4969616
          2  |    .361934   .0041778    86.63   0.000     .3537457    .3701224
          3  |   .2325308   .0056269    41.33   0.000     .2215023    .2435592
------------------------------------------------------------------------------
```

Instead of specifying **educ age yrint**, we can specify the keyword **_continuous**, which refers to all continuous variables in the model.

```
. margins, at((p25) _continuous)
>          at((p50) _continuous)
>          at((p75) _continuous) noatlegend vsquish

Predictive margins                              Number of obs     =     15,375
Model VCE    : OIM

Expression   : Pr(smoke), predict()

------------------------------------------------------------------------------
             |            Delta-method
             |     Margin   Std. Err.      z    P>|z|     [95% Conf. Interval]
-------------+----------------------------------------------------------------
         _at |
          1  |   .4824943   .0073814    65.37   0.000      .468027    .4969616
          2  |    .361934   .0041778    86.63   0.000     .3537457    .3701224
          3  |   .2325308   .0056269    41.33   0.000     .2215023    .2435592
------------------------------------------------------------------------------
```

B.3 Margins with factor variables

Suppose we want to compute the predictive margin of the probability of smoking for the four levels of class, while adjusting for all the other covariates. We could compute this using the following **margins** command:

```
. margins, at(class=(1 2 3 4)) vsquish

Predictive margins                              Number of obs     =     15,375
Model VCE    : OIM

Expression   : Pr(smoke), predict()
1._at        : class           =           1
2._at        : class           =           2
3._at        : class           =           3
4._at        : class           =           4

------------------------------------------------------------------------------
             |            Delta-method
             |     Margin   Std. Err.      z    P>|z|     [95% Conf. Interval]
-------------+----------------------------------------------------------------
         _at |
          1  |   .4262118   .0167454    25.45   0.000     .3933914    .4590323
          2  |   .3665092   .0056573    64.78   0.000     .3554211    .3775974
          3  |   .3242303   .0057422    56.46   0.000     .3129758    .3354847
          4  |   .3099447   .0215208    14.40   0.000     .2677647    .3521246
------------------------------------------------------------------------------
```

Because the original `logit` command specified `i.class`, this variable is treated as a factor variable. Instead of using the `margins` command above, we can use the `margins` command as shown below. This relieves us of the need to manually specify the levels of `class`. Better yet, the output below displays the value label associated with each level of class (for example, showing `lower class` instead of `1`).

```
. margins class

Predictive margins                              Number of obs     =     15,375
Model VCE    : OIM

Expression   : Pr(smoke), predict()

------------------------------------------------------------------------------
             |            Delta-method
             |     Margin   Std. Err.      z    P>|z|     [95% Conf. Interval]
-------------+----------------------------------------------------------------
       class |
 lower class |   .4262118   .0167454    25.45   0.000     .3933914    .4590323
working cl.. |   .3665092   .0056573    64.78   0.000     .3554211    .3775974
middle class |   .3242303   .0057422    56.46   0.000     .3129758    .3354847
 upper class |   .3099447   .0215208    14.40   0.000     .2677647    .3521246
------------------------------------------------------------------------------
```

In the output above, the columns labeled `z` and `P>|z|` concern provide the test statistic and p-value testing the null hypothesis that the predictive margin of the probability of smoking equals 0. For instance, among those who self-identify as `upper class`, predictive margin of the probability of smoking is 0.310 and this is significantly different from 0 ($z = 14.40$, $p < 0.001$). The results of this test are usually not of interest, and often I have found that people can misinterpret these results as indicating support their research hypothesis. In such instances, I like to include the `nopvalues` option, as shown below. For examples like this throughout the book, I will often include the `nopvalues` option.[1]

1. If you wanted to suppress the display of the confidence intervals, you could specify the `noci` option.

```
. margins class, nopvalues

Predictive margins                              Number of obs     =     15,375
Model VCE    : OIM

Expression   : Pr(smoke), predict()

------------------------------------------------------------------
              |            Delta-method
              |     Margin   Std. Err.     [95% Conf. Interval]
--------------+---------------------------------------------------
        class |
  lower class |   .4262118   .0167454      .3933914    .4590323
working class |   .3665092   .0056573      .3554211    .3775974
 middle class |   .3242303   .0057422      .3129758    .3354847
  upper class |   .3099447   .0215208      .2677647    .3521246
------------------------------------------------------------------
```

We can apply contrast operators to factor variables. For example, the `ar.` contrast operator computes reverse adjacent group contrasts. Note that these contrasts are performed in the predicted probability metric, adjusting for all the covariates.

```
. margins ar.class

Contrasts of predictive margins                 Number of obs     =     15,375
Model VCE    : OIM

Expression   : Pr(smoke), predict()

------------------------------------------------------------------------
                                  |         df        chi2     P>chi2
----------------------------------+-------------------------------------
                            class |
  (working class vs lower class)  |          1       11.64     0.0006
 (middle class vs working class)  |          1       26.39     0.0000
   (upper class vs middle class)  |          1        0.42     0.5186
                            Joint |          3       48.49     0.0000
------------------------------------------------------------------------

-------------------------------------------------------------------
                |            Delta-method
                |   Contrast   Std. Err.     [95% Conf. Interval]
----------------+--------------------------------------------------
          class |
 (working class |
             vs |
   lower class) |  -.0597026   .0174984     -.0939988   -.0254064
  (middle class |
             vs |
 working class) |  -.0422789   .0082295     -.0584085   -.0261494
   (upper class |
             vs |
  middle class) |  -.0142856    .022129     -.0576577    .0290865
-------------------------------------------------------------------
```

We can include the option `contrast(effects)` to modify the output to include both confidence intervals and significance tests.

```
. margins ar.class, contrast(effects)
Contrasts of predictive margins                 Number of obs     =     15,375
Model VCE    : OIM

Expression   : Pr(smoke), predict()

------------------------------------------------------------
                                  |         df        chi2     P>chi2
----------------------------------+-------------------------------------
                            class |
  (working class vs lower class)  |          1       11.64     0.0006
 (middle class vs working class)  |          1       26.39     0.0000
   (upper class vs middle class)  |          1        0.42     0.5186
                            Joint |          3       48.49     0.0000
------------------------------------------------------------

--------------------------------------------------------------------------------
              |            Delta-method
              |   Contrast   Std. Err.      z    P>|z|     [95% Conf. Interval]
--------------+-----------------------------------------------------------------
        class |
  (working c..|
           vs |
 lower class) |  -.0597026   .0174984    -3.41   0.001    -.0939988   -.0254064
  (middle cl..|
           vs |
  working c..)|  -.0422789   .0082295    -5.14   0.000    -.0584085   -.0261494
  (upper class|
           vs |
  middle cl..)|  -.0142856    .022129    -0.65   0.519    -.0576577    .0290865
--------------------------------------------------------------------------------
```

The upper portion of the output now is unnecessary because we can see the significance tests in the lower portion of the output. We can suppress the upper portion of the output by adding `contrast()`'s `nowald` argument.

```
. margins ar.class, contrast(nowald effects)
Contrasts of predictive margins                 Number of obs     =     15,375
Model VCE    : OIM

Expression   : Pr(smoke), predict()

--------------------------------------------------------------------------------
              |            Delta-method
              |   Contrast   Std. Err.      z    P>|z|     [95% Conf. Interval]
--------------+-----------------------------------------------------------------
        class |
  (working c..|
           vs |
 lower class) |  -.0597026   .0174984    -3.41   0.001    -.0939988   -.0254064
  (middle cl..|
           vs |
  working c..)|  -.0422789   .0082295    -5.14   0.000    -.0584085   -.0261494
  (upper class|
           vs |
  middle cl..)|  -.0142856    .022129    -0.65   0.519    -.0576577    .0290865
--------------------------------------------------------------------------------
```

To accommodate the constraints of the width of a table that can be displayed on a page, the **contrast(nowald effects)** option forces the output in the first column to be wrapped in a way that is hard to read. But when performing actual data analysis using a typical computer screen that accommodates wider tables, the **contrast(nowald effects)** option provides output that is very compact and informative, including the point estimate of the contrast, the standard error of the contrast, a test of whether the contrast is significant, and a confidence interval for the contrast.

Using the **contrast(nowald pveffects)** option creates an output that is more amenable to the space constraints of the page. This compact output format is very similar to the output above except that the confidence intervals are omitted. This output sacrifices display of the confidence intervals to provide more room to clearly display the first column showing the description of the contrast being tested.

```
. margins ar.class, contrast(nowald pveffects)
Contrasts of predictive margins                 Number of obs     =     15,375
Model VCE    : OIM
Expression   : Pr(smoke), predict()

------------------------------------------------------------------------------
                                 |            Delta-method
                                 |   Contrast   Std. Err.        z    P>|z|
---------------------------------+--------------------------------------------
                           class |
  (working class vs lower class) |  -.0597026   .0174984     -3.41   0.001
 (middle class vs working class) |  -.0422789   .0082295     -5.14   0.000
   (upper class vs middle class) |  -.0142856    .022129     -0.65   0.519
------------------------------------------------------------------------------
```

The **contrast(nowald pveffects)** option is frequently used with many of the **margins** commands in this book. However, when you analyze your data, I recommend using the **contrast(nowald effects)** option to include confidence intervals in your output.

Suppose that you find yourself in a similar situation, where you are trying to create output that would fit on a narrow page, and you are trying to create results from the **margins** command that is most informative. The prior example achieved a narrower display of the **margins** output with the **contrast(nowald pveffects)**. That included the factor-variable labels but sacrificed the display of the confidence intervals. Another alternative to consider is using the **nofvlabel** option. That option specifies that the level of the factor variable will be displayed in lieu of the factor-variable label, as shown below. If you are familiar with the meanings of the factor-variable codes, this could be a good alternative that allows you to display both significance tests and confidence intervals in a compact format that is easy to read.

```
. margins ar.class, contrast(nowald effects) nofvlabel
Contrasts of predictive margins                 Number of obs     =     15,375
Model VCE    : OIM
Expression   : Pr(smoke), predict()

------------------------------------------------------------------------------
             |            Delta-method
             |   Contrast   Std. Err.      z    P>|z|     [95% Conf. Interval]
-------------+----------------------------------------------------------------
       class |
   (2 vs 1)  |  -.0597026   .0174984    -3.41   0.001    -.0939988   -.0254064
   (3 vs 2)  |  -.0422789   .0082295    -5.14   0.000    -.0584085   -.0261494
   (4 vs 3)  |  -.0142856    .022129    -0.65   0.519    -.0576577    .0290865
------------------------------------------------------------------------------
```

For the sake of completeness, I would like to mention that the `margins` command supports the `fvwrap()` option. In the example below, the `fvwrap(3)` option is used to permit value labels to be displayed using up to 3 lines.[2]

```
. margins ar.class, contrast(nowald effects) fvwrap(3)
Contrasts of predictive margins                 Number of obs     =     15,375
Model VCE    : OIM
Expression   : Pr(smoke), predict()

--------------------------------------------------------------------------------
               |            Delta-method
               |   Contrast   Std. Err.      z    P>|z|     [95% Conf. Interval]
---------------+----------------------------------------------------------------
         class |
      (working |
         class |
            vs |
  lower class) |  -.0597026   .0174984    -3.41   0.001    -.0939988   -.0254064
       (middle |
         class |
            vs |
       working |
        class) |  -.0422789   .0082295    -5.14   0.000    -.0584085   -.0261494
  (upper class |
            vs |
 middle cl..)  |  -.0142856    .022129    -0.65   0.519    -.0576577    .0290865
--------------------------------------------------------------------------------
```

The `mcompare()` option can be used to adjust the displayed significance values to account for multiple comparisons. You can specify `bonferroni`, `sidak`, or `scheffe` within `mcompare()` to obtain Bonferroni's, Šidák's, or Scheffé's method of adjustment, respectively. The example below illustrates using the `mcompare(sidak)` option.

2. The `margins` command also supports the `fvwrapon()` options, as illustrated in section A.4 showing customization of the display of factor variables in estimation commands.

```
. margins ar.class, contrast(nowald pveffects) mcompare(sidak)
Contrasts of predictive margins                 Number of obs     =     15,375
Model VCE    : OIM
Expression   : Pr(smoke), predict()

───────────────────────────
             │  Number of
             │ Comparisons
─────────────┼─────────────
       class │           3
───────────────────────────

──────────────────────────────────────────────────────────────────────────
                                   │            Delta-method       Sidak
                                   │   Contrast   Std. Err.       z    P>|z|
───────────────────────────────────┼──────────────────────────────────────
                             class │
   (working class vs lower class)  │  -.0597026    .0174984   -3.41   0.002
  (middle class vs working class)  │  -.0422789    .0082295   -5.14   0.000
    (upper class vs middle class)  │  -.0142856     .022129   -0.65   0.888
──────────────────────────────────────────────────────────────────────────
```

B.4 Margins with factor variables and the at() option

We can specify both the factor variables and the `at()` option on the `margins` command. The example below estimates the predicted probability of smoking for each level of social class holding age constant at 30 and then again at 50, averaging across all other covariates.

```
. margins class, at(age=(30 50)) nopvalues vsquish
Predictive margins                              Number of obs     =     15,375
Model VCE    : OIM
Expression   : Pr(smoke), predict()
1._at        : age             =          30
2._at        : age             =          50

───────────────────────────────────────────────────────────────
                 │            Delta-method
                 │     Margin   Std. Err.     [95% Conf. Interval]
─────────────────┼─────────────────────────────────────────────
        _at#class │
   1#lower class  │   .4959834    .0175386      .4616085    .5303584
 1#working class  │    .433064     .006578      .4201715    .4459566
  1#middle class  │   .3872891     .007514      .3725619    .4020163
   1#upper class  │   .3715911    .0242098      .3241408    .4190414
   2#lower class  │   .4008165    .0168446      .3678016    .4338314
 2#working class  │    .341714    .0059379      .3300759     .353352
  2#middle class  │   .3003918     .005597      .2894219    .3113617
   2#upper class  │    .286533    .0207909      .2457836    .3272823
───────────────────────────────────────────────────────────────
```

The first four lines of output correspond to the predictive margins for the four levels of class when age is held constant at 30. The second four lines (5 to 8) of the output show the predictive margins for the four levels of class when age is held constant at 50.

The example below shows the predicted probability of smoking by social class at three different covariate values, holding the continuous covariates at the 25th, 50th, and 75th percentiles.

```
. margins class, at((p25) _continuous)
>                at((p50) _continuous)
>                at((p75) _continuous) nopvalues noatlegend vsquish

Adjusted predictions                            Number of obs     =     15,375
Model VCE    : OIM

Expression   : Pr(smoke), predict()

------------------------------------------------------------------------
                  |            Delta-method
                  |     Margin   Std. Err.     [95% Conf. Interval]
------------------+-----------------------------------------------------
         _at#class |
   1#lower class  |   .5678827    .0181575      .5322947    .6034707
 1#working class  |   .5032008    .0081471      .4872327    .5191689
  1#middle class  |   .4546775    .0095224      .4360139     .473341
   1#upper class  |   .4377569    .0267364      .3853546    .4901593
   2#lower class  |   .4438118     .017505      .4095026    .4781211
 2#working class  |   .3808079    .0059541      .3691381    .3924777
  2#middle class  |   .3361022    .0063346      .3236866    .3485178
   2#upper class  |   .3209969    .0228609      .2761903    .3658035
   3#lower class  |   .2983846    .0163392      .2663603    .3304089
 3#working class  |   .2468625    .0072841      .2325859     .261139
  3#middle class  |   .2124853    .0058149      .2010883    .2238824
   3#upper class  |   .2012516    .0169463      .1680373    .2344658
------------------------------------------------------------------------
```

Let's repeat the previous **margins** command but add the **ar.** contrast operator to **class**, performing reverse adjacent contrasts on class. The **margins** command computes the marginal effect of reverse adjacent comparisons on class with the continuous covariates held at the 25th, 50th, and 75th percentiles.

```
. margins ar.class, at((p25) _continuous)
>                   at((p50) _continuous)
>                   at((p75) _continuous)
>                   noatlegend vsquish contrast(nowald pveffects)

Contrasts of adjusted predictions              Number of obs     =     15,375
Model VCE    : OIM

Expression   : Pr(smoke), predict()

------------------------------------------------------------------------------
                                  |            Delta-method
                                  |   Contrast   Std. Err.        z    P>|z|
----------------------------------+-------------------------------------------
                      class@_at   |
 (working class vs lower class) 1 |  -.0646819   .0184355     -3.51   0.000
 (working class vs lower class) 2 |  -.0630039   .0184179     -3.42   0.001
 (working class vs lower class) 3 |  -.0515222   .0155556     -3.31   0.001
(middle class vs working class) 1 |  -.0485233   .0094069     -5.16   0.000
(middle class vs working class) 2 |  -.0447057   .0086553     -5.17   0.000
(middle class vs working class) 3 |  -.0343771   .0067872     -5.06   0.000
  (upper class vs middle class) 1 |  -.0169205   .0263901     -0.64   0.521
  (upper class vs middle class) 2 |  -.0151053   .0233927     -0.65   0.518
  (upper class vs middle class) 3 |  -.0112338   .0172794     -0.65   0.516
------------------------------------------------------------------------------
```

This shows that the effect of comparing `working class` versus `lower class` is significant when holding the covariates at the 25th, 50th, and 75th percentiles. Likewise, the effect of `middle class` versus `working class` is significant at the 25th, 50th, and 75th percentiles of the covariates. The comparison of `upper class` versus `middle class` is not significant when holding the covariates at the 25th, 50th, and 75th percentiles.

B.5 The dydx() and related options

Let's consider a linear regression model in which we predict education from the education of the respondent's father, the education of the respondent's mother, the respondent's age, and the respondent's gender, specified via the variable `gender` (coded: 1 = Male and 2 = Female).

```
. regress educ paeduc maeduc age i.gender

      Source |       SS           df       MS      Number of obs   =    36,004
-------------+----------------------------------   F(4, 35999)     =   3527.52
       Model |  92963.8807         4  23240.9702   Prob > F        =    0.0000
    Residual |  237178.642    35,999  6.58847863   R-squared       =    0.2816
-------------+----------------------------------   Adj R-squared   =    0.2815
       Total |  330142.523    36,003  9.16986148   Root MSE        =    2.5668

------------------------------------------------------------------------------
        educ |      Coef.   Std. Err.      t    P>|t|     [95% Conf. Interval]
-------------+----------------------------------------------------------------
      paeduc |   .2136448   .0043286    49.36   0.000     .2051607    .2221289
      maeduc |   .2335704   .0051057    45.75   0.000      .223563    .2435778
         age |    .007092   .0008733     8.12   0.000     .0053804    .0088036
             |
      gender |
        Male |          0  (base)
      Female |  -.1876547   .0271978    -6.90   0.000    -.2409632   -.1343462
             |
       _cons |   8.280247    .069885   118.48   0.000      8.14327    8.417223
------------------------------------------------------------------------------
```

Suppose we use the **margins** command with the **dydx(paeduc)** option. This gives us the same coefficient that we saw from the **regress** output. This shows us that for a one-unit increase in the father's education, the respondent's education is expected to increase, on average, by 0.214, after adjusting for all other predictors in the model.

```
. margins, dydx(paeduc)

Average marginal effects                        Number of obs     =     36,004
Model VCE    : OLS

Expression   : Linear prediction, predict()
dy/dx w.r.t. : paeduc

------------------------------------------------------------------------------
             |            Delta-method
             |      dy/dx   Std. Err.      t    P>|t|     [95% Conf. Interval]
-------------+----------------------------------------------------------------
      paeduc |   .2136448   .0043286    49.36   0.000     .2051607    .2221289
------------------------------------------------------------------------------
```

We can compute the marginal effect of **paeduc** as an elasticity by specifying the **eyex()** option, as shown below. This shows how a 1% change in **paeduc** is related to a percentage change in **educ**.

```
. margins, eyex(paeduc)
Average marginal effects                        Number of obs     =     36,004
Model VCE    : OLS

Expression   : Linear prediction, predict()
ey/ex w.r.t. : paeduc

------------------------------------------------------------------------------
             |            Delta-method
             |      ey/ex   Std. Err.      t    P>|t|     [95% Conf. Interval]
-------------+----------------------------------------------------------------
      paeduc |   .1649225   .0033082    49.85   0.000     .1584383    .1714067
------------------------------------------------------------------------------
```

We can specify the eyex(_continuous) option to see average marginal elasticities for all continuous variables.

```
. margins, eyex(_continuous)
Average marginal effects                        Number of obs     =     36,004
Model VCE    : OLS

Expression   : Linear prediction, predict()
ey/ex w.r.t. : paeduc maeduc age

------------------------------------------------------------------------------
             |            Delta-method
             |      ey/ex   Std. Err.      t    P>|t|     [95% Conf. Interval]
-------------+----------------------------------------------------------------
      paeduc |   .1649225   .0033082    49.85   0.000     .1584383    .1714067
      maeduc |   .1849259   .0040085    46.13   0.000      .177069    .1927827
         age |   .0242471   .0029771     8.14   0.000     .0184119    .0300824
------------------------------------------------------------------------------
```

If you want to express the coefficients in terms of the change in the outcome related to a proportional change in the predictor, you can specify the dyex() option. The example below shows the marginal effect of all continuous predictors showing how a proportion change in the predictor relates to a unit change in the outcome.

```
. margins, dyex(_continuous)
Average marginal effects                        Number of obs     =     36,004
Model VCE    : OLS

Expression   : Linear prediction, predict()
dy/ex w.r.t. : paeduc maeduc age

------------------------------------------------------------------------------
             |            Delta-method
             |      dy/ex   Std. Err.      t    P>|t|     [95% Conf. Interval]
-------------+----------------------------------------------------------------
      paeduc |   2.266644   .0459234    49.36   0.000     2.176633    2.356655
      maeduc |   2.521833   .0551261    45.75   0.000     2.413784    2.629882
         age |   .3122683   .0384502     8.12   0.000     .2369047    .3876318
------------------------------------------------------------------------------
```

You can use the eydx() option to compute the marginal effects in terms of a proportion change in the outcome for a unit change in the predictor. Because the predictor is in its natural metric, we can specify eydx(_all) to display this marginal effect for all predictors. For the factor variable gender, the marginal effect is computed as the discrete change from the base level (that is, male).

```
. margins, eydx(_all)
Average marginal effects                    Number of obs     =     36,004
Model VCE     : OLS

Expression    : Linear prediction, predict()
ey/dx w.r.t.  : paeduc maeduc age 2.gender

------------------------------------------------------------------------------
             |            Delta-method
             |      ey/dx   Std. Err.      t    P>|t|     [95% Conf. Interval]
-------------+----------------------------------------------------------------
      paeduc |   .0163541   .0003336    49.02   0.000     .0157002    .0170081
      maeduc |   .0178794   .0003936    45.43   0.000     .0171079    .0186509
         age |   .0005429   .0000669     8.12   0.000     .0004118    .0006739
             |
      gender |
        Male |          0  (base)
      Female |  -.0143537    .002079    -6.90   0.000    -.0184285   -.0102789
------------------------------------------------------------------------------
Note: ey/dx for factor levels is the discrete change from the base level.
```

Before concluding this section, let's consider an example using a nonlinear model. Let's consider the following logistic regression, which predicts whether a person smokes based on the person's self-reported social class, education, age, and year of interview.

```
. logit smoke i.class educ age yrint
Iteration 0:   log likelihood = -9951.2382
Iteration 1:   log likelihood = -9582.9808
Iteration 2:   log likelihood = -9580.0732
Iteration 3:   log likelihood = -9580.0723
Logistic regression                             Number of obs     =     15,375
                                                LR chi2(6)        =     742.33
                                                Prob > chi2       =     0.0000
Log likelihood = -9580.0723                     Pseudo R2         =     0.0373

------------------------------------------------------------------------------
       smoke |      Coef.   Std. Err.      z    P>|z|     [95% Conf. Interval]
-------------+----------------------------------------------------------------
       class |
 lower class |          0  (base)
 working cl..|  -.2604145   .0749733    -3.47   0.001    -.4073595   -.1134696
middle class |   -.455007   .0772007    -5.89   0.000    -.6063176   -.3036965
 upper class |  -.5234883   .1274959    -4.11   0.000    -.7733757   -.2736009
             |
        educ |  -.0878902   .0063325   -13.88   0.000    -.1003016   -.0754787
         age |  -.0198399   .0010949   -18.12   0.000    -.0219859   -.0176939
       yrint |  -.0321324   .0033889    -9.48   0.000    -.0387746   -.0254902
       _cons |   65.45743   6.718824     9.74   0.000     52.28877    78.62608
------------------------------------------------------------------------------
```

Let's consider the output of the **margins** command using the **dydx(educ)** option.

```
. margins, dydx(educ)
Average marginal effects                        Number of obs     =     15,375
Model VCE    : OIM

Expression   : Pr(smoke), predict()
dy/dx w.r.t. : educ

------------------------------------------------------------------------------
             |            Delta-method
             |      dy/dx   Std. Err.      z    P>|z|     [95% Conf. Interval]
-------------+----------------------------------------------------------------
        educ |  -.0190423   .0013418   -14.19   0.000     -.021672   -.0164125
------------------------------------------------------------------------------
```

This computes the average marginal effect of education (in the predicted probability metric). The average marginal effect of education is −0.019.

We can explore how the size of the average marginal effect varies as a function of the covariates. The example below computes the marginal effect of education when age is held constant at 20 and when age is held constant at 80.

```
. margins, dydx(educ) at(age=(20 80)) vsquish
Average marginal effects                        Number of obs     =     15,375
Model VCE    : OIM

Expression   : Pr(smoke), predict()
dy/dx w.r.t. : educ
1._at        : age             =          20
2._at        : age             =          80

------------------------------------------------------------------------------
             |            Delta-method
             |      dy/dx   Std. Err.      z    P>|z|     [95% Conf. Interval]
-------------+----------------------------------------------------------------
educ         |
         _at |
          1  |  -.0210625   .0014869   -14.17   0.000    -.0239767   -.0181482
          2  |  -.0143383   .0009606   -14.93   0.000     -.016221   -.0124555
------------------------------------------------------------------------------
```

At 20 years of age, the average marginal effect of education is −0.021, but at 80 years of age, the average marginal effect of education is −0.014.

B.6 Specifying the confidence level

Let's continue to use the example from the previous section, where we fit a logistic regression that predicts whether a person smokes based on the person's self-reported social class, education, age, and year of interview.

```
. use gss_ivrm
. logit smoke i.class educ age yrint
  (output omitted)
```

Let's use the **margins** command to compute the predictive margin of the probability of smoking for the four levels of class, while adjusting for all the other covariates. We could compute this using the following **margins** command:

```
. margins class
Predictive margins                              Number of obs     =     15,375
Model VCE    : OIM

Expression   : Pr(smoke), predict()

------------------------------------------------------------------------------
             |            Delta-method
             |     Margin   Std. Err.      z    P>|z|     [95% Conf. Interval]
-------------+----------------------------------------------------------------
       class |
 lower class |   .4262118   .0167454    25.45   0.000     .3933914    .4590323
 working cl..|   .3665092   .0056573    64.78   0.000     .3554211    .3775974
middle class |   .3242303   .0057422    56.46   0.000     .3129758    .3354847
 upper class |   .3099447   .0215208    14.40   0.000     .2677647    .3521246
------------------------------------------------------------------------------
```

Note how the confidence intervals are displayed using a 95% confidence level. This is the default confidence level used by Stata. We could display the confidence intervals using a 90% confidence level by adding the **level(90)** option to the **margins** command, shown below.

```
. margins class, level(90)
Predictive margins                              Number of obs     =     15,375
Model VCE    : OIM

Expression   : Pr(smoke), predict()

------------------------------------------------------------------------------
             |            Delta-method
             |     Margin   Std. Err.      z    P>|z|     [90% Conf. Interval]
-------------+----------------------------------------------------------------
       class |
 lower class |   .4262118   .0167454    25.45   0.000      .398668    .4537556
 working cl..|   .3665092   .0056573    64.78   0.000     .3572037    .3758147
middle class |   .3242303   .0057422    56.46   0.000     .3147852    .3336753
 upper class |   .3099447   .0215208    14.40   0.000     .2745461    .3453432
------------------------------------------------------------------------------
```

The **margins** command respects the settings specified by the **set level** command. You can see section A.2 for more details about the **set level** command.

B.7 Customizing column formatting

The **margins** command supports options for controlling the formatting of the output of the columns displayed by the **margins** command. In fact, these options work in the same manner as the options used for formatting the columns of the coefficient table from estimation commands (as illustrated in section A.3). In that section, we saw that the formatting of results can controlled by three options—the **pformat()** option, the **sformat()** option, and the **cformat()** option. To briefly illustrate these options, let's use the same logistic regression model from the previous section.

```
. use gss_ivrm
. logit smoke i.class educ age yrint
  (output omitted)
```

Let's now repeat the **margins** command we saw on page 555 that performed reverse-adjacent group contrasts on self-identified social class. The output was customized to omit the Wald table, to show both significance tests and confidence intervals, and to omit factor-variable labels.

```
. margins ar.class, contrast(nowald effects) nofvlabel
Contrasts of predictive margins                 Number of obs     =     15,375
Model VCE    : OIM

Expression   : Pr(smoke), predict()

------------------------------------------------------------------------------
             |            Delta-method
             |   Contrast   Std. Err.      z    P>|z|     [95% Conf. Interval]
-------------+----------------------------------------------------------------
       class |
   (2 vs 1)  |  -.0597026   .0174984    -3.41   0.001    -.0939988   -.0254064
   (3 vs 2)  |  -.0422789   .0082295    -5.14   0.000    -.0584085   -.0261494
   (4 vs 3)  |  -.0142856    .022129    -0.65   0.519    -.0576577    .0290865
------------------------------------------------------------------------------
```

The example below adds the **pformat()** option to control the formatting of the *p*-values. The example below uses the **pformat(%5.2f)** option to display the *p*-values with a total width of five columns and two digits after the decimal place. Note that the width of the format specified in the **pformat()** option cannot exceed five columns.

```
. margins ar.class, contrast(nowald effects) nofvlabel pformat(%5.2f)
Contrasts of predictive margins                 Number of obs     =     15,375
Model VCE    : OIM

Expression   : Pr(smoke), predict()

------------------------------------------------------------------------------
             |            Delta-method
             |   Contrast   Std. Err.      z    P>|z|     [95% Conf. Interval]
-------------+----------------------------------------------------------------
       class |
   (2 vs 1)  |  -.0597026   .0174984    -3.41    0.00    -.0939988   -.0254064
   (3 vs 2)  |  -.0422789   .0082295    -5.14    0.00    -.0584085   -.0261494
   (4 vs 3)  |  -.0142856    .022129    -0.65    0.52    -.0576577    .0290865
------------------------------------------------------------------------------
```

The **sformat()** option controls the formatting of the test statistics. The example below uses the **sformat(%5.1f)** option to display the *z*-values using a fixed width of five columns with one decimal place. Note that the width of the format specified for the **sformat()** option cannot exceed eight columns.

```
. margins ar.class, contrast(nowald effects) nofvlabel sformat(%5.1f)
Contrasts of predictive margins                  Number of obs     =     15,375
Model VCE    : OIM
Expression   : Pr(smoke), predict()
```

	Contrast	Delta-method Std. Err.	z	P>|z|	[95% Conf.	Interval]
class						
(2 vs 1)	-.0597026	.0174984	-3.4	0.001	-.0939988	-.0254064
(3 vs 2)	-.0422789	.0082295	-5.1	0.000	-.0584085	-.0261494
(4 vs 3)	-.0142856	.022129	-0.6	0.519	-.0576577	.0290865

The `cformat()` option controls the format used to display the contrast coefficients, standard errors, and confidence intervals. The format selected via the `cformat()` option is applied to all of these columns. In the example below, I have added the `cformat(%7.4f)` option, which means that the results shown in these columns will be displayed in a fixed format with a total width of seven columns and four digits after the decimal point. Note that the width of the format specified in the `cformat()` option cannot exceed nine columns.

```
. margins ar.class, contrast(nowald effects) nofvlabel cformat(%7.4f)
Contrasts of predictive margins                  Number of obs     =     15,375
Model VCE    : OIM
Expression   : Pr(smoke), predict()
```

	Contrast	Delta-method Std. Err.	z	P>|z|	[95% Conf.	Interval]
class						
(2 vs 1)	-0.0597	0.0175	-3.41	0.001	-0.0940	-0.0254
(3 vs 2)	-0.0423	0.0082	-5.14	0.000	-0.0584	-0.0261
(4 vs 3)	-0.0143	0.0221	-0.65	0.519	-0.0577	0.0291

As we saw in section A.3, the `set pformat`, `set sformat`, and `set cformat` commands can be used to customize formatting for subsequent estimation commands. Further, those commands can be used in combination with the `permanently` option to change the default settings each time you start your copy of Stata. The `margins` command respects the settings specified by the `set pformat`, `set sformat`, and `set cformat` commands. For more details on using those commands, see section A.3. Also, you can see `help format` for more information about display formats.

This appendix has certainly not covered all the features of the `margins` command. For more information about the `margins` command, see [R] **margins**.

C The marginsplot command

The body of the book has shown numerous examples using the `marginsplot` command, but those examples have focused on creating utilitarian graphs without much regard for customizing the look of the graphs. This appendix illustrates some of the options you can use to customize the appearance of graphs created by the `marginsplot` command. Some of the options are specific to the `marginsplot` command, and the other options are supported by the `graph twoway` command. You can see [G-2] **graph twoway** for more information about such options, as well as Mitchell (2012).

Consider the multiple regression model that we saw in section 2.3 that predicts the education of the respondent from the father's education, mother's education, and age of the respondent. The `margins` command is then used to compute the adjusted means when the father's education ranges from 0 to 20. The `marginsplot` command can then be used to graph the adjusted means computed by the `margins` command.

```
. use gss_ivrm
. keep if gender==1 & yrint==2008
(54,157 observations deleted)
. regress educ paeduc maeduc age
  (output omitted)
. margins, at(paeduc=(0(1)20))
  (output omitted)
```

```
marginsplot
```

This shows the adjusted means as a function of the father's education. The following examples illustrate how you can use options to customize this graph.

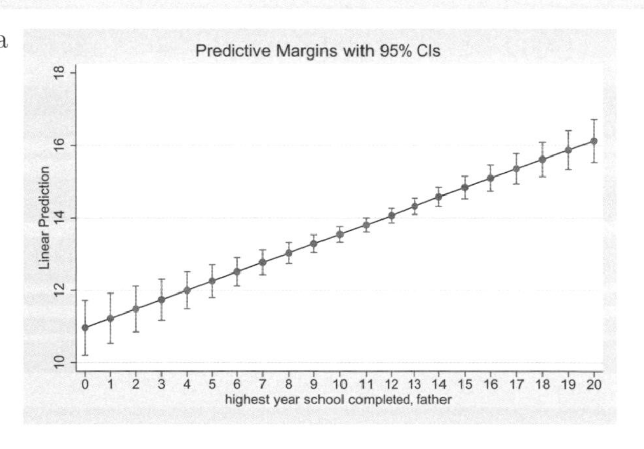

```
marginsplot, title(Title) subtitle(Subtitle) xtitle(X title)
    ytitle(Y title) note(Note) caption(Caption)
```

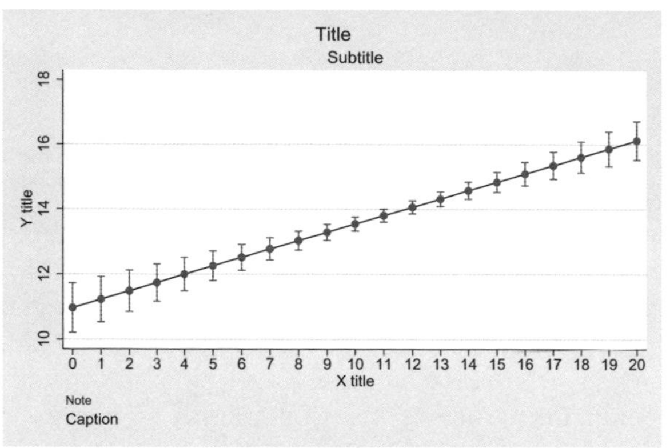

You can add titles to the graph using the **title()**, **subtitle()**, **xtitle()**, and **ytitle()** options. The **note()** and **caption()** options can also be used to annotate the graph.

```
marginsplot, xlabel(0(5)20) ylabel(10(1)18, angle(0))
```

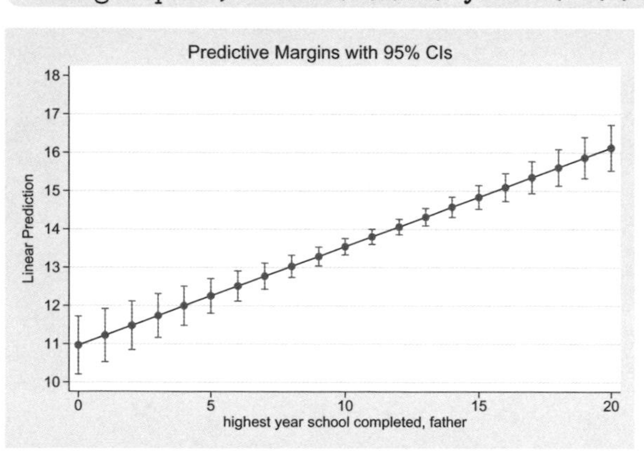

The **xlabel()** and **ylabel()** options can be used to change the labeling of the x and y axes.

```
marginsplot, xscale(range(-1 21)) yscale(range(8 20))
```

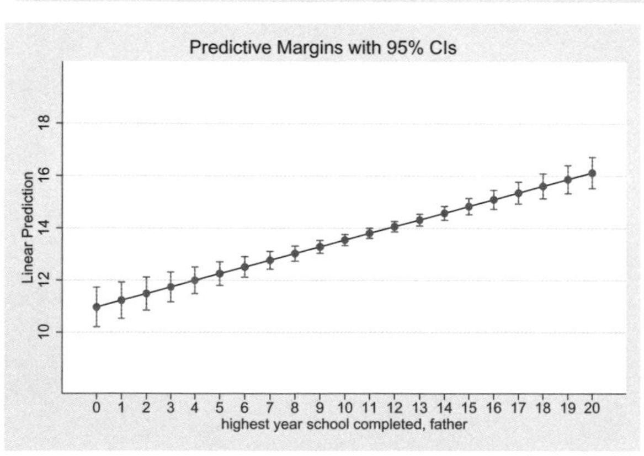

The **xscale()** and **yscale()** options can be used to expand the scale of the x and y axes.

```
marginsplot, plotopts(clwidth(thick) msymbol(Oh) msize(large))
```

The **plotopts()** option allows you to include suboptions that control the look of the line and markers. The **clwidth()** suboption is used to make the fitted line thick, and the **msymbol()** and **msize()** suboptions are used to draw the markers as large hollow circles.

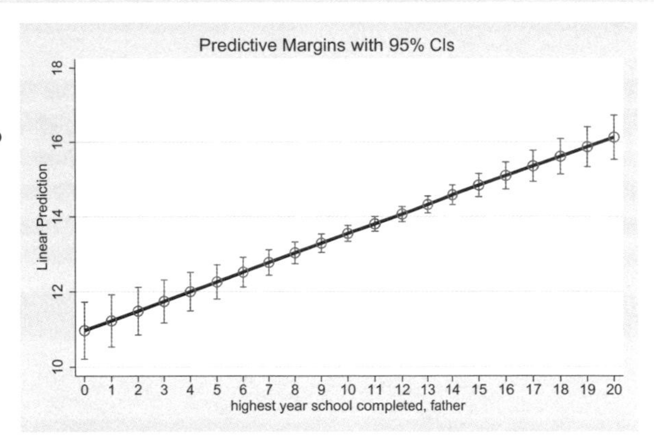


```
marginsplot, ciopts(lwidth(vthick) msize(huge))
```

The **ciopts()** option allows you to include suboptions that control the look of the confidence interval. The **lwidth()** suboption makes the lines for the confidence intervals thick, and the **msize()** suboption makes the caps of each confidence interval huge.

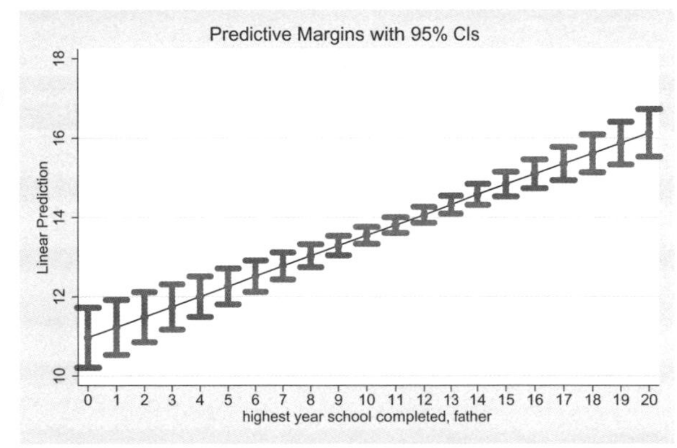


```
marginsplot, recast(line) recastci(rarea)
```

The **recast(line)** option specifies that the fitted line be drawn like a `twoway line` graph. The **recastci(rarea)** option specifies that the confidence interval be drawn like a `twoway rarea` graph.

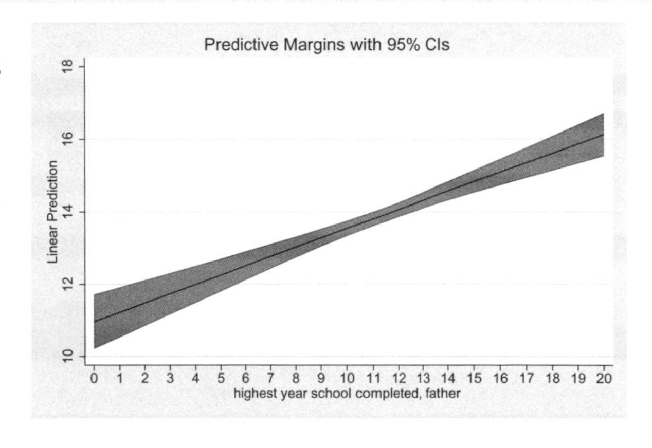

```
marginsplot, recast(line) recastci(rarea), ciopts(fcolor(navy))
```

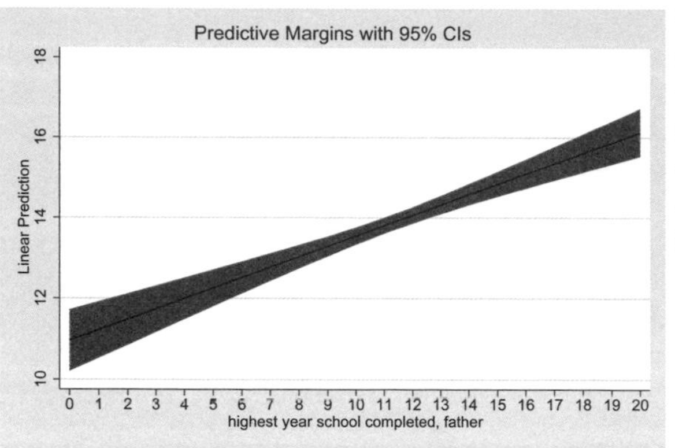

This example adds the **ciopts(fcolor(navy))** option. On the page, the fill color for the confidence interval is displayed in dark gray. If you run this on your computer, you will see the fill color is navy blue.

```
marginsplot, recast(line) recastci(rarea) ciopts(fcolor(navy%20))
```

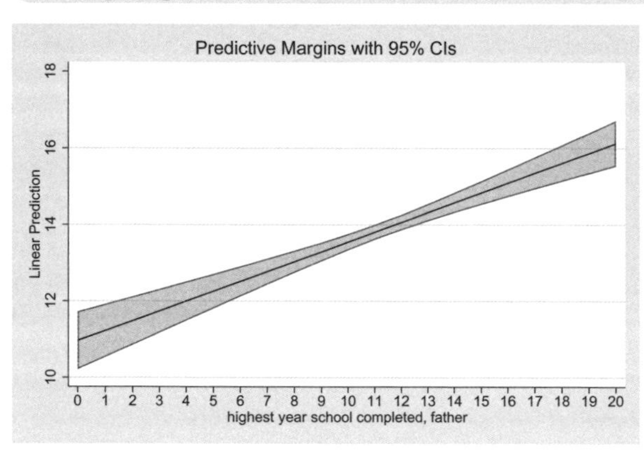

Like the example above, the fill color of the confidence is displayed in dark gray on the page or navy blue on your computer. However, by specifying **fcolor(navy%20)**, the fill color is shown with less opacity (more transparency). If I had specified `fcolor(navy%10)` the fill color would have been even less opaque (more transparent), and if I had specified `fcolor(navy%50)`, the fill color would have been more opaque (less transparent).

```
marginsplot, recast(line) recastci(rarea) ciopts(fcolor(%20))
```

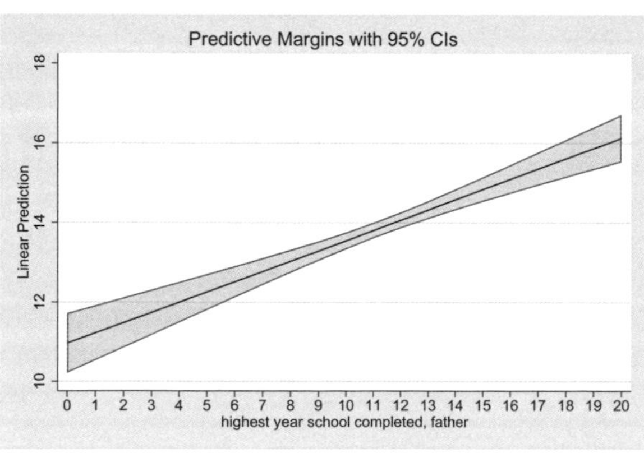

This example customizes the fill color of the confidence interval by specifying **fcolor(%20)**, which uses the default color, displaying it with less opacity (more transparency). If I had specified `fcolor(%10)`, the fill color would have been even less opaque (more transparent), and if I had specified `fcolor(%50)`, the fill color would have been more opaque (less transparent).

```
marginsplot, recast(line) recastci(rarea)
    ciopts(fcolor(navy%20) lcolor(black))
```

In this example, the **lcolor(black)** suboption is given within the ciopts() option to use black when displaying color of the line that surrounds the confidence interval.

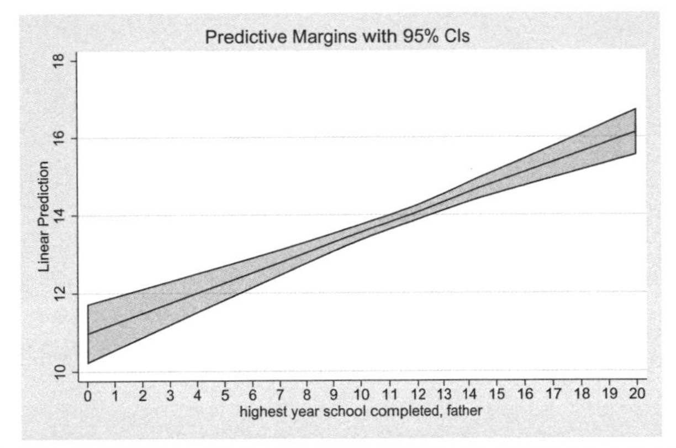


```
marginsplot, recast(line) recastci(rarea) ciopts(fcolor(navy%20)
    lcolor(black%0))
```

The the **lcolor(black%0)** suboption is used within the ciopts() option to specify that the line surrounding the confidence interval will be displayed in black with no opacity (that is, 100% transparency).

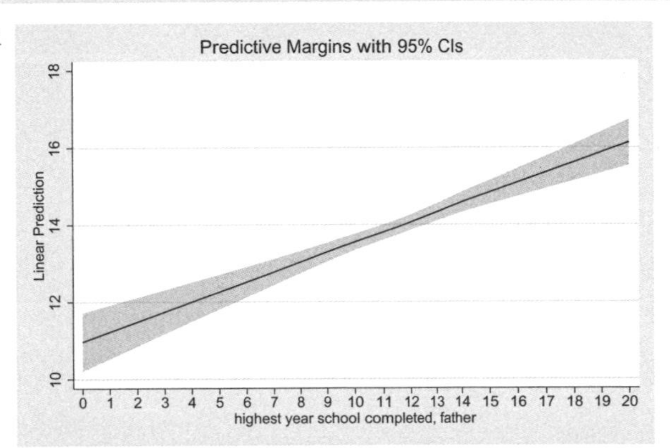


```
marginsplot, recast(line) recastci(rarea) ciopts(fcolor(navy%20)
    lcolor(black%0)) plotopts(lcolor(navy))
```

This example uses the **plotopts(lcolor(navy))** option to specify the color of the fit line, displaying it in dark gray on the page or navy blue on your computer. This illustrates how the lcolor() suboption controls the color of the fit line when specified within the plotopts() option and that same suboption controls the line surrounding the confidence interval when specified within the ciopts() option.

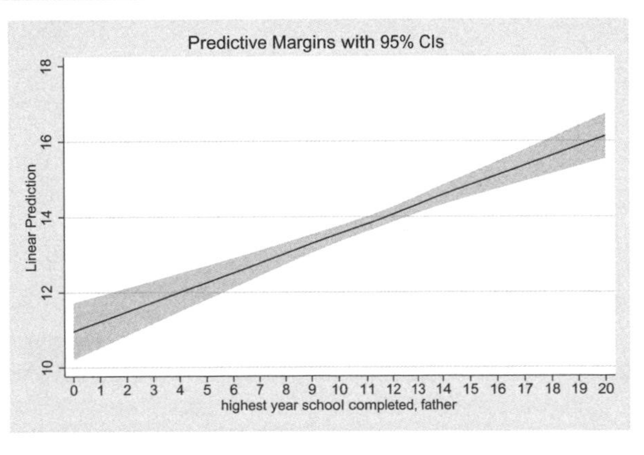

```
marginsplot, noci
```

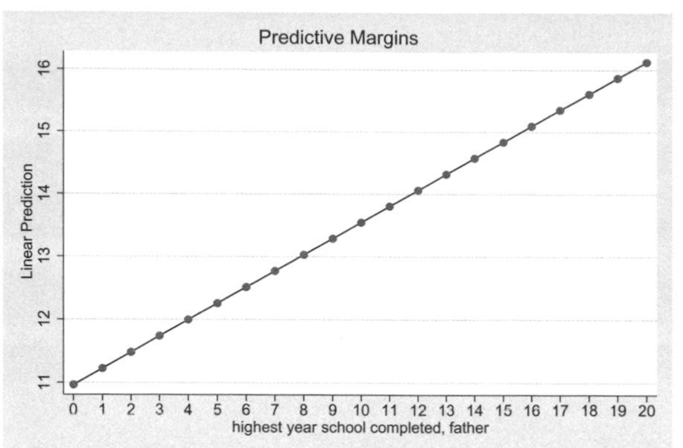

The **noci** option suppresses the display of the confidence interval.

```
marginsplot, noci addplot(scatter educ paeduc, msymbol(o))
```

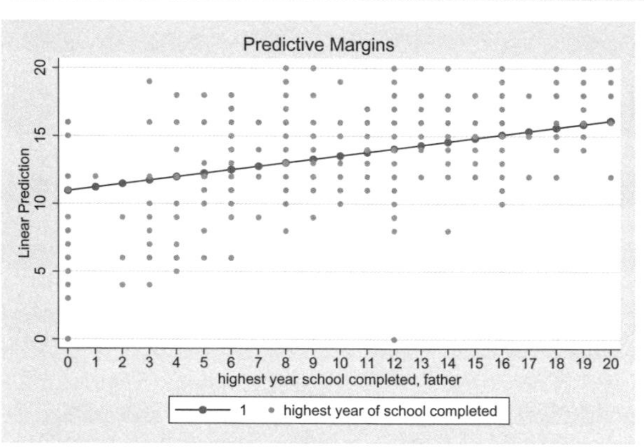

The **addplot()** option can be used to overlay a new graph onto the graph created by the `margins` command. In this case, it overlays a scatterplot of `educ` and `paeduc`.

```
marginsplot, scheme(economist)
```

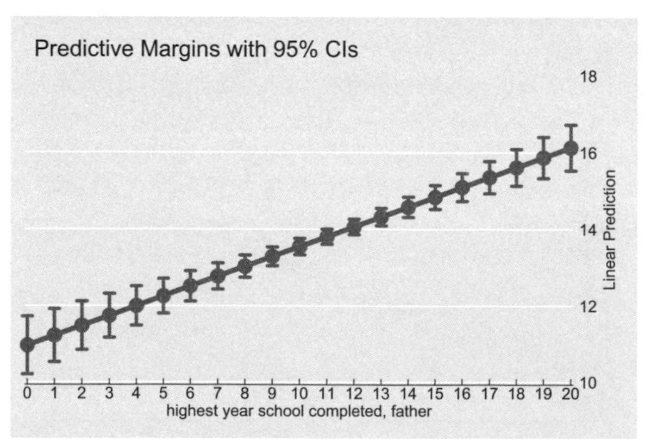

The **scheme()** option can be used to change the overall look of the graph. In this example, the `economist` scheme is used.

Let's now consider another example, this one is based on the three by three design illustrated in section 8.2. The commands below use the dataset, run the `anova` command, and use the `margins` command to obtain the mean of the outcome by `treat` and `depstat`. The output of these commands is omitted to save space. The `marginsplot` command can then be used to graph the results of the interaction as computed by the `margins` command.

```
. use opt-3by3
. anova opt treat##depstat
  (output omitted)
. margins treat##depstat
  (output omitted)
```

`marginsplot`

This shows the graph created by the `marginsplot` command. The following examples illustrate how to customize this graph, focusing on the plot dimension and the legend associated with the plot dimension.

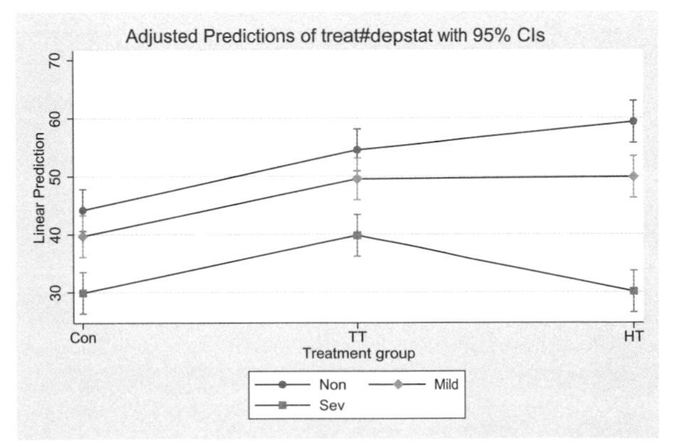

`marginsplot, xdimension(depstat)`

The **`xdimension()`** option controls which variable is placed on the x axis. This option is used to graph `depstat` on the x axis, and hence `treat` is graphed as separate lines.

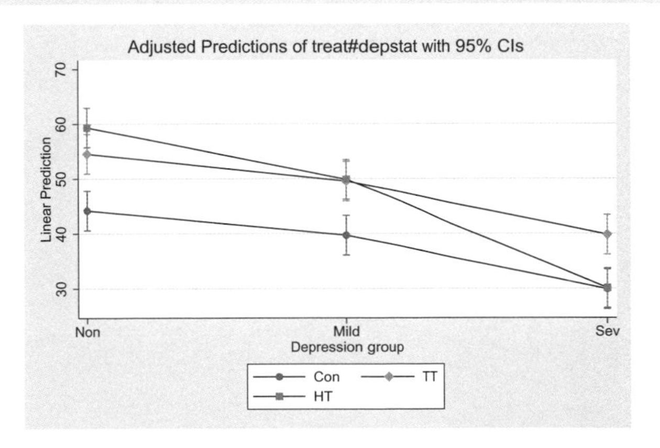

```
marginsplot, plotdimension(depstat)
```

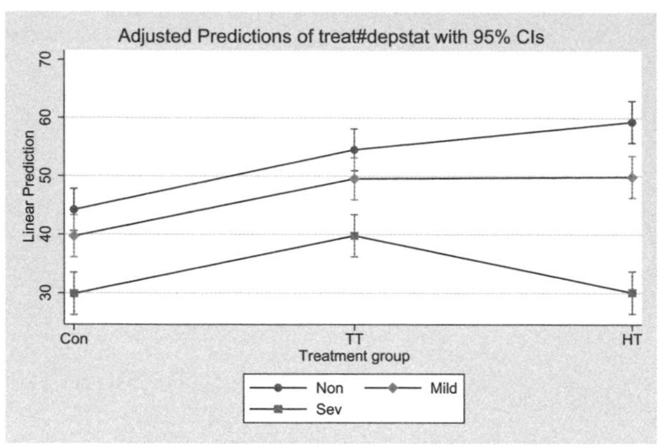

The **plotdimension()** option controls which variable is graphed using the plot dimension—graphed using separate lines. In this example, **depstat** is graphed using separate lines, and thus **treat** is placed on the x axis.

```
marginsplot, plotdim(depstat, labels("Nondep" "Mild dep" "Sev. dep"))
```

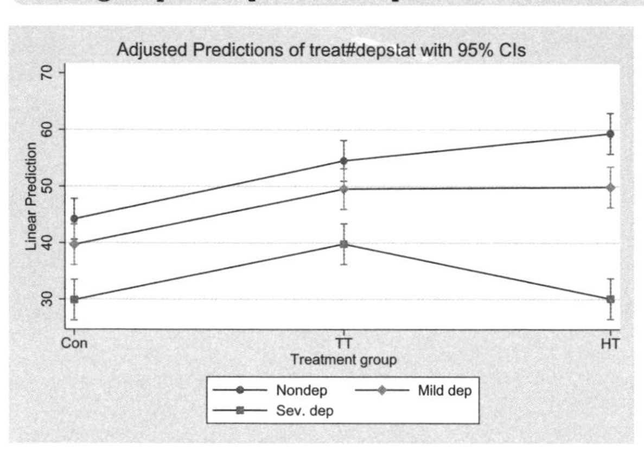

The **labels()** suboption within the **plotdimension()** option changes the labels used for the plot dimension.

```
marginsplot,
    plotdim(depstat, elabels(1 "Nondepressed" 3 "Severely depressed"))
```

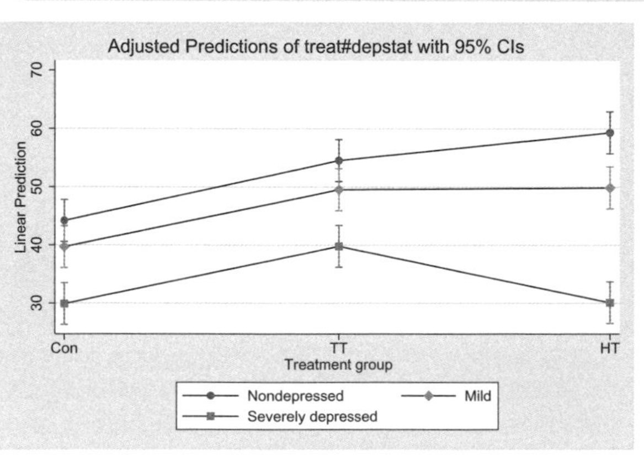

Using the **elabels()** suboption, you can selectively modify the labels of your choice. This example modifies the labels for the first and third group, leaving the label for the second group unchanged.

```
marginsplot, plotdim(depstat, nosimplelabels)
```

Adding the **nosimplelabels** suboption changes the plot label to the variable name, an equal sign, and the value label for the group.

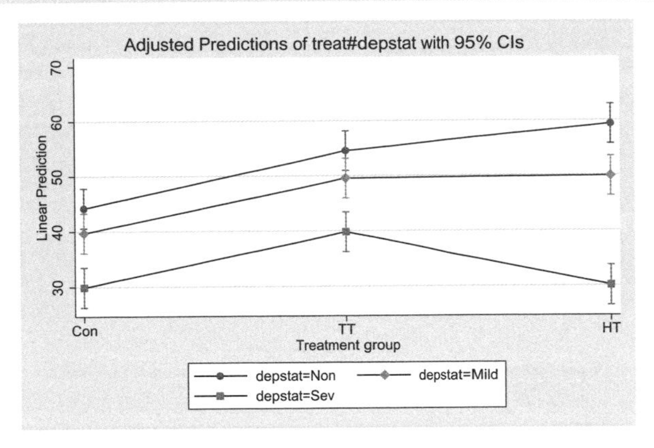


```
marginsplot, plotdim(depstat, nolabels)
```

Adding the **nolabels** suboption changes the plot label to the variable name, an equal sign, and the numeric value for the group.

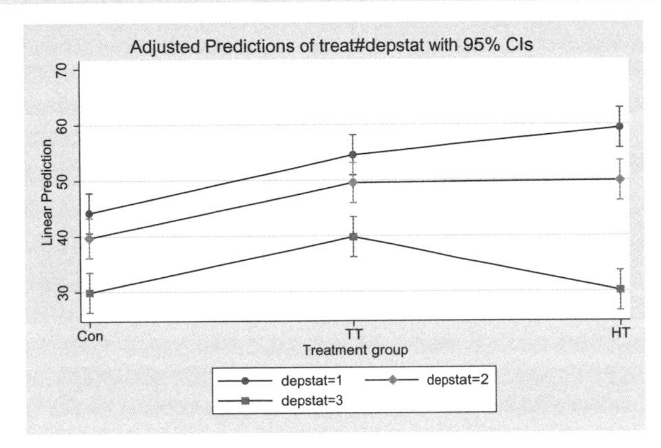


```
marginsplot, plotdim(depstat, allsimplelabels)
```

Using the **allsimplelabels** suboption yields a label that is composed solely of the value label for each group.

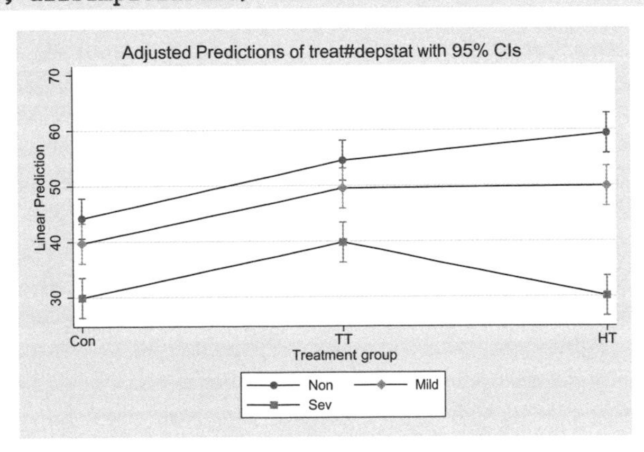

```
marginsplot, plotdim(depstat, allsimplelabels nolabels)
```

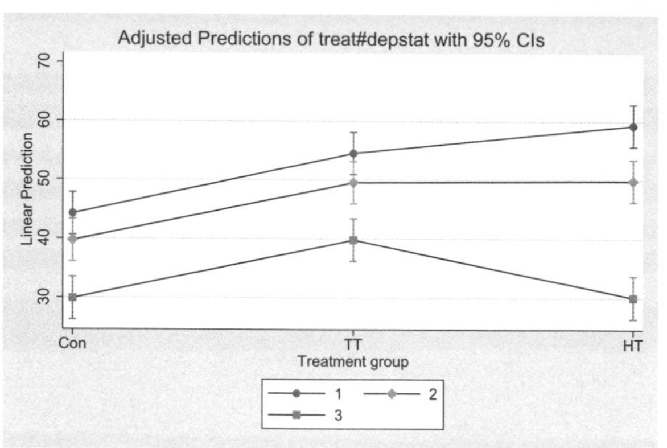

Using the **allsimplelabels** and **nolabels** suboptions displays a label that is composed solely of the numeric value for each group.

```
marginsplot, legend(subtitle("Treatment") rows(1))
```

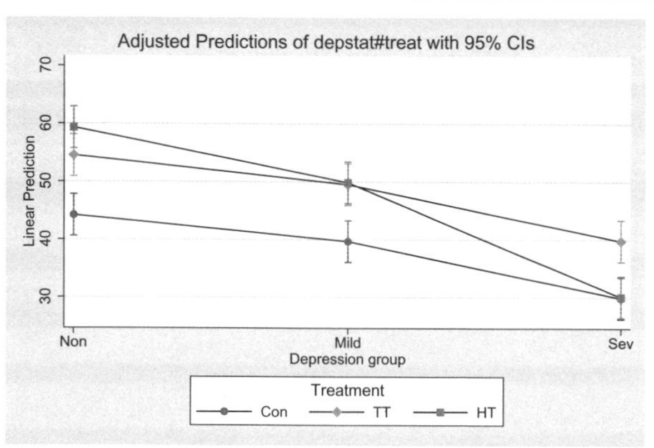

This example includes the **legend()** option to customize the display of the legend, add a subtitle, and display the legend keys in one row.

```
marginsplot, legend(subtitle("Treatment") rows(1) ring(0) position(11))
```

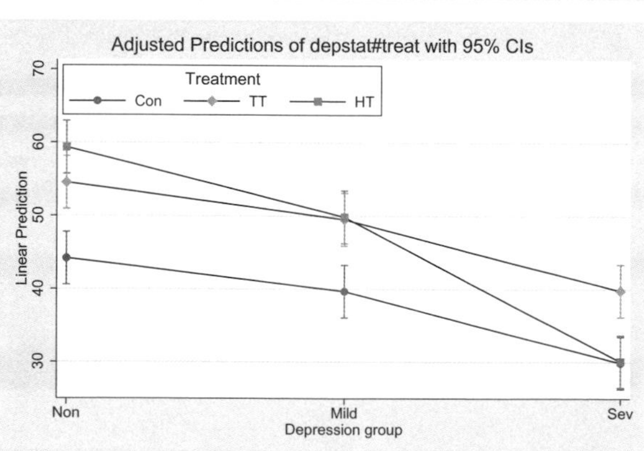

The **ring()** and **position()** suboptions are added to the **legend()** option to display the legend within the graph in the 11 o'clock position.

```
marginsplot, xlabel(1 "Control" 2 "Traditional" 3 "Happiness")
```

The **xlabel()** option is used to control the labeling of the x axis. The next example illustrates how to address the issue of the "Happiness" label being cut off.

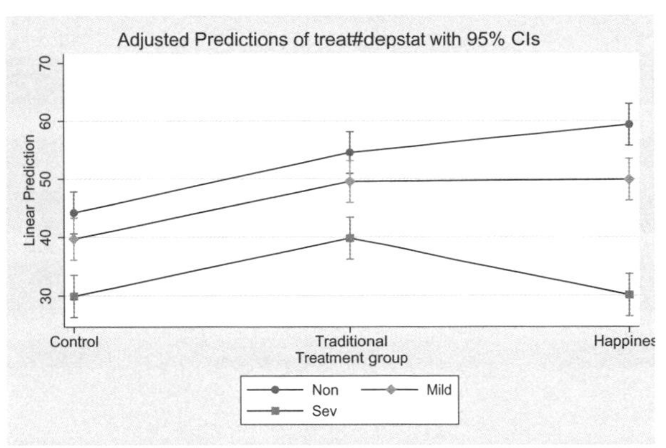


```
marginsplot, xlabel(1 "Control" 2 "Traditional" 3 "Happiness")
    xscale(range(.75 3.25))
```

The **xscale(range())** option is used to expand the range of the x axis to make more room for the longer x-axis labels. When I make graphs with a categorical variables on the x axis, I often use this technique because I think it makes a graph that is more pleasing to the eye.

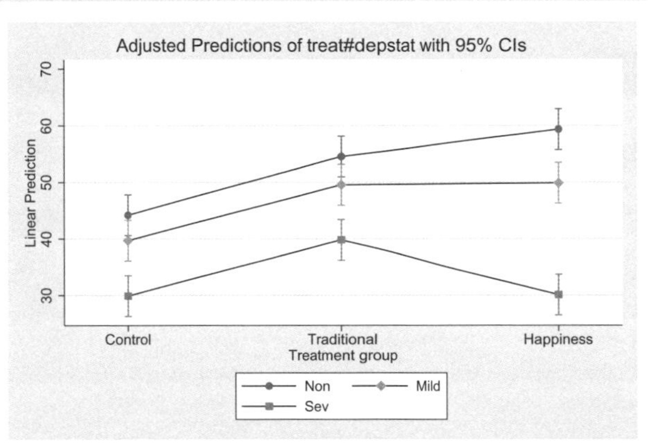

Let's now consider an example where the `marginsplot` command involves by-groups as an additional dimension. This example predicts smoking from education, marital status (married or nonmarried), gender, and all interactions formed among these variables. It also includes age as a covariate. The `c.educ#married#gender` interaction is significant. The **margins** command is used to compute the predictive margin of the probability of smoking as a function of gender, marital status, and education. (The output is omitted to save space.)

```
. use gss_ivrm
. logit smoke c.educ##married##gender age
  (output omitted)
. margins gender#married, at(educ=(0(1)20))
  (output omitted)
```

```
marginsplot, noci
```

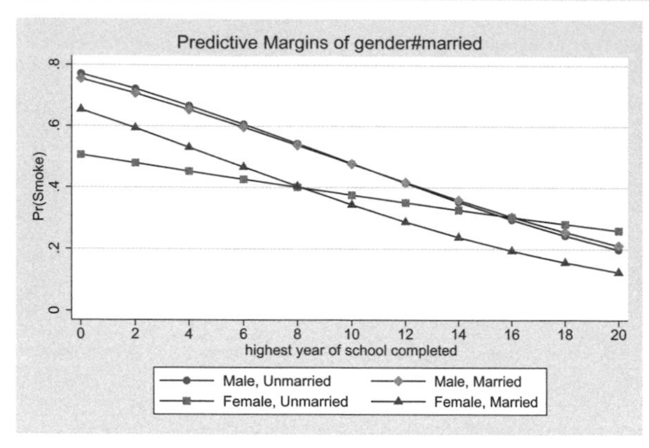

The **marginsplot** command graphs the predicted probabilities computed by the **margins** command. The **noci** option is used to suppress the confidence intervals. The following examples customize the graph, focusing on issues related to the by-dimension.

```
marginsplot, noci bydimension(gender)
```

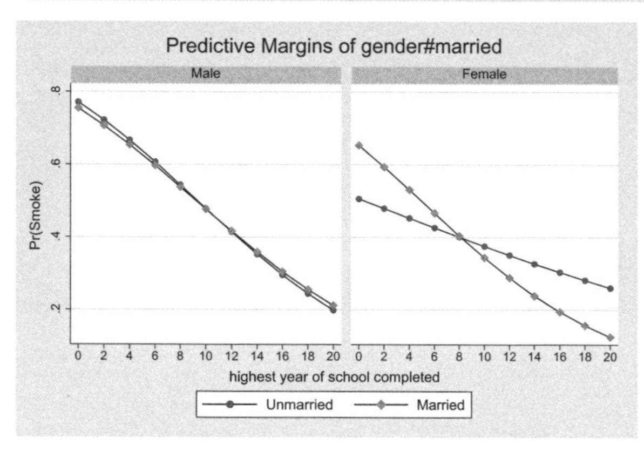

The **bydimension()** option is used to specify that separate graphs should be made by **gender** (that is, gender).

```
marginsplot, noci bydimension(gender, label("Men" "Women"))
```

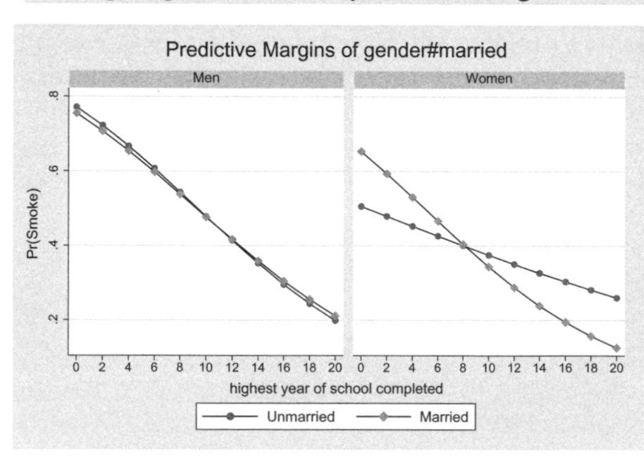

The **label()** suboption can be used to control the labeling of each graph. You can control the labeling of the by-dimension using the **bydimension()** option in the same manner that we controlled the plot dimension using the **plotdimension()** option.

```
marginsplot, noci bydimension(gender) byopts(cols(1) ixaxes)
```

The **byopts()** option allows you to specify suboptions that control the way that the separate graphs are combined. In this example, the cols(1) and ixaxes suboptions are specified to display the graphs in one column, each with its own x axis.

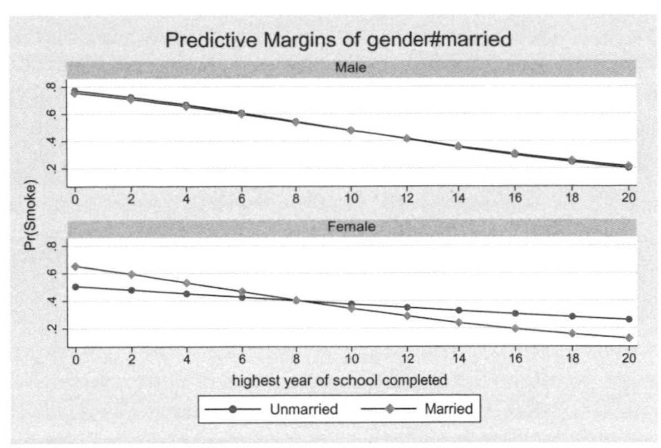


```
marginsplot, noci bydimension(gender)
    byopts(title(Title) subtitle(Subtitle) note(Note) caption(Caption))
```

The **byopts()** option can be used to control the overall title, subtitle, note, and caption for the graph. By placing such options within the **byopts()** option, these options impact the overall title, subtitle, note, and caption for the graph. Contrast this with the next example.

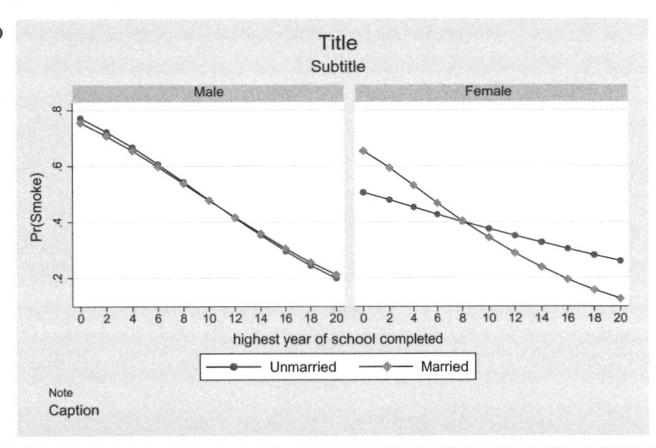


```
marginsplot, noci bydimension(gender)
    title(Title) subtitle(Subtitle) note(Note) caption(Caption)
```

By placing these options outside the byopts() option, these options control the title, subtitle, note, and caption for each graph.

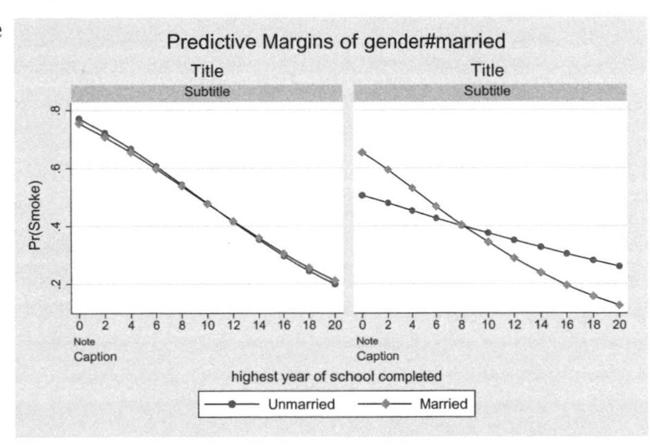

```
marginsplot, bydimension(gender) recast(line)
```

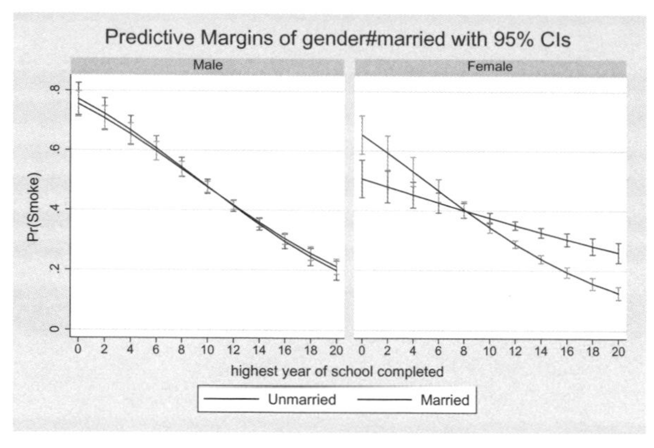

This graph is like the previous graph, but I have added the **recast(line)** option. By specifying `recast(line)`, the graph is displayed like a **graph twoway line** graph—yielding lines without marker symbols. See the next example for an improved version.

```
marginsplot, bydimension(gender) recast(line) scheme(s1mono)
```

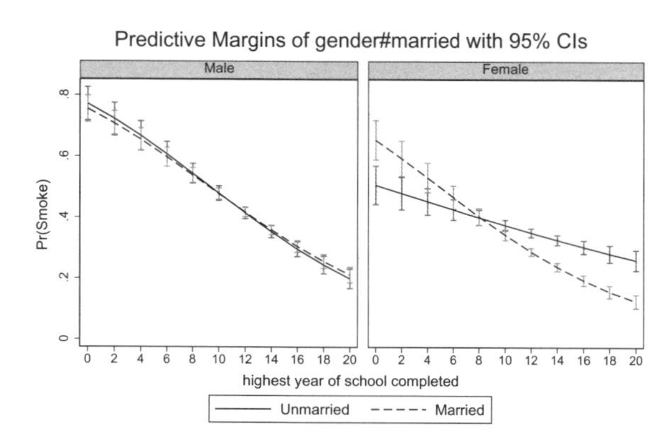

In the prior example, it was impossible to differentiate the lines for the married versus unmarried groups. That graph was displayed using the `s2mono` scheme, the default scheme used in this book, which uses marker symbols to differentiate lines from different groups. By specifying **scheme(s1mono)**, different line patterns differentiate the lines for those married versus those unmarried.

```
marginsplot, bydimension(gender) recast(line) recastci(rarea)
    scheme(s1mono)
```

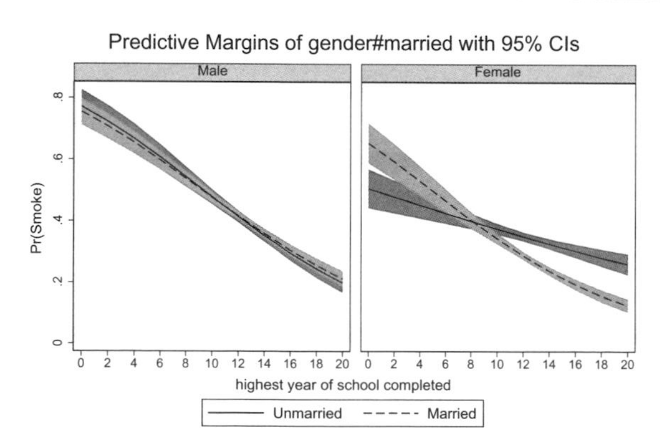

This example adds the **recastci(rarea)** option to the prior example to display the confidence interval using an area plot. The confidence intervals are displayed with full opacity (no transparency). As a consequence, where the confidence intervals overlap, the confidence interval for those who are married obscures the confidence interval for those who are unmarried.

```
marginsplot, bydimension(gender) recast(line) recastci(rarea)
    ciopts(fcolor(%30)) scheme(s1mono)
```

By adding the **`ciopts(fcolor(%30))`** option, the fill color for each confidence interval is displayed with less opacity (more transparency). As a result, the overlap of the confidence intervals shows a blending of the colors from the married and unmarried groups.

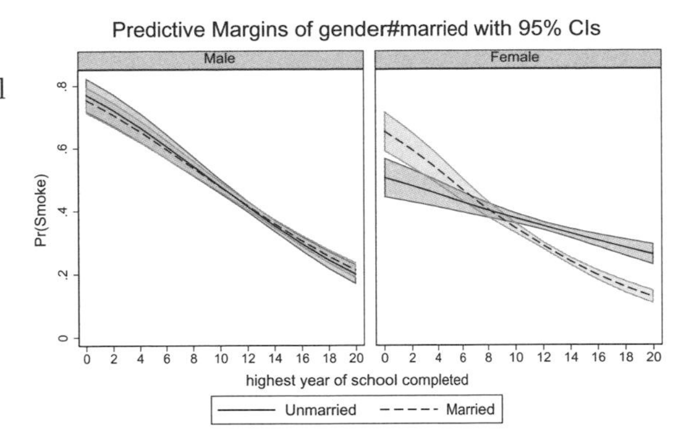

This appendix has illustrated some of the ways that you can customize graphs created by the `marginsplot` command. For more information, see [R] **marginsplot**, [G-2] **graph twoway**, and Mitchell (2012).

> **Tip! Accessing the dialog box for the `marginsplot` command**
>
> I think it is very easy to forget that Stata has excellent point-and-click dialog boxes associated with most commands. The `db` command (short for dialog box) makes it a snap to bring up the dialog box associated with a command. For instance, try typing `db marginsplot` after using the `margins` command, and you can experiment with the dialog box to see what kinds of plots you can produce.

D The contrast command

This appendix discusses additional details about the contrast command, focusing on options that control the display of the output. Let's begin using an example from section 7.5 that predicted happiness from marital status. The dataset for this example is used below. Then, the regress command is used to predict happiness from marital status.

```
. use gss_ivrm, clear
. regress happy7 i.marital
      Source |       SS           df       MS      Number of obs   =     1,160
-------------+----------------------------------   F(4, 1155)      =     11.86
       Model |  43.9299172         4  10.9824793   Prob > F        =    0.0000
    Residual |  1069.44163     1,155  .925923493   R-squared       =    0.0395
-------------+----------------------------------   Adj R-squared   =    0.0361
       Total |  1113.37155     1,159  .960631192   Root MSE        =    .96225

------------------------------------------------------------------------------
      happy7 |      Coef.   Std. Err.      t    P>|t|     [95% Conf. Interval]
-------------+----------------------------------------------------------------
     marital |
     married |          0  (base)
     widowed |  -.2912429    .110942    -2.63   0.009    -.5089133   -.0735724
    divorced |  -.4878643   .0796349    -6.13   0.000    -.6441094   -.3316191
   separated |  -.5579771   .1699952    -3.28   0.001    -.8915112   -.2244431
never marr.. |  -.2453868   .0701012    -3.50   0.000    -.3829269   -.1078468
             |
       _cons |   5.705036   .0408085   139.80   0.000     5.624969    5.785103
------------------------------------------------------------------------------
```

The contrast command can be used to test the overall effect of a factor variable. For example, the contrast command below tests the overall effect of marital.[1]

```
. contrast marital
Contrasts of marginal linear predictions
Margins      : asbalanced
------------------------------------------------
             |         df           F        P>F
-------------+----------------------------------
     marital |          4       11.86     0.0000
             |
 Denominator |       1155
------------------------------------------------
```

1. Had we used the anova command, the main effect of marital from the anova output would have matched the contrast results.

The **contrast** command can also be used to apply contrast operators to factor variables to dissect main effects or interactions. Consider the output we obtain from the **contrast** command when we apply a contrast coefficient, such as **r.marital**.

```
. contrast r.marital
Contrasts of marginal linear predictions
Margins      : asbalanced

------------------------------------------------
                         |         df        F        P>F
-------------------------+----------------------------------
                 marital |
   (widowed vs married)  |          1      6.89     0.0088
  (divorced vs married)  |          1     37.53     0.0000
 (separated vs married)  |          1     10.77     0.0011
(never married vs married)|         1     12.25     0.0005
                   Joint |          4     11.86     0.0000
                         |
             Denominator |       1155
------------------------------------------------

-------------------------------------------------------------------------------
                         |   Contrast   Std. Err.     [95% Conf. Interval]
-------------------------+-----------------------------------------------------
                 marital |
   (widowed vs married)  |  -.2912429    .110942     -.5089133   -.0735724
  (divorced vs married)  |  -.4878643   .0796349     -.6441094   -.3316191
 (separated vs married)  |  -.5579771   .1699952     -.8915112   -.2244431
(never married vs married)| -.2453868   .0701012     -.3829269   -.1078468
-------------------------------------------------------------------------------
```

D.1 Inclusion and omission of output

The output from the prior **contrast** command is divided into two portions. The top portion shows F tests of the specified contrast. The second portion shows the estimates of the contrast, standard error, and confidence interval.

If we wanted to suppress the second portion of the output, we could specify the **noeffects** option, as shown below.

```
. contrast r.marital, noeffects
Contrasts of marginal linear predictions
Margins      : asbalanced

------------------------------------------------
                         |         df        F        P>F
-------------------------+----------------------------------
                 marital |
   (widowed vs married)  |          1      6.89     0.0088
  (divorced vs married)  |          1     37.53     0.0000
 (separated vs married)  |          1     10.77     0.0011
(never married vs married)|         1     12.25     0.0005
                   Joint |          4     11.86     0.0000
                         |
             Denominator |       1155
------------------------------------------------
```

We can suppress the first portion of the output with the `nowald` option, as shown below.

```
. contrast r.marital, nowald
Contrasts of marginal linear predictions
Margins      : asbalanced

                           |  Contrast   Std. Err.     [95% Conf. Interval]
---------------------------+-----------------------------------------------
                   marital |
       (widowed vs married)  | -.2912429    .110942    -.5089133   -.0735724
      (divorced vs married)  | -.4878643   .0796349    -.6441094   -.3316191
     (separated vs married)  | -.5579771   .1699952    -.8915112   -.2244431
 (never married vs married)  | -.2453868   .0701012    -.3829269   -.1078468
```

If we specify the `effects` option, Stata displays both the confidence interval and *p*-value associated with each contrast. This combination of options produces output that is very readable on the computer screen but difficult to read when constrained to fit on a page, as shown below.

```
. contrast r.marital, nowald effects
Contrasts of marginal linear predictions
Margins      : asbalanced

             |  Contrast   Std. Err.      t    P>|t|     [95% Conf. Interval]
-------------+----------------------------------------------------------------
     marital |
    (widowed |
          vs |
    married) | -.2912429    .110942    -2.63   0.009    -.5089133   -.0735724
   (divorced |
          vs |
    married) | -.4878643   .0796349    -6.13   0.000    -.6441094   -.3316191
  (separated |
          vs |
    married) | -.5579771   .1699952    -3.28   0.001    -.8915112   -.2244431
(never mar.. |
          vs |
    married) | -.2453868   .0701012    -3.50   0.000    -.3829269   -.1078468
```

To accommodate the constraints of the width of a table that can be displayed in the book, the `nowald effects` option forces the output in the first column to be wrapped in a way that is hard to read. But when performing actual data analysis using a typical computer screen that accommodates wider tables, the `nowald effects` option provides output that is very compact and informative, including the point estimate of the contrast, the standard error of the contrast, a test of whether the contrast is significant, and a confidence interval for the contrast. When I use the `contrast` command for my analyses, where my computer screen can accommodate wider output, I frequently use the `nowald effects` options because I find the output is both comprehensive and concise.

Using the `nowald` and `pveffects` options, as shown below, creates an output that is more amenable to the space constraints of the page. This compact output format is very similar to the output above, except that the confidence intervals are omitted. This output sacrifices display of the confidence intervals to provide more room to clearly display the first column showing the description of the contrast being tested.

```
. contrast r.marital, nowald pveffects
Contrasts of marginal linear predictions
Margins      : asbalanced
```

	Contrast	Std. Err.	t	P>|t|
marital				
(widowed vs married)	-.2912429	.110942	-2.63	0.009
(divorced vs married)	-.4878643	.0796349	-6.13	0.000
(separated vs married)	-.5579771	.1699952	-3.28	0.001
(never married vs married)	-.2453868	.0701012	-3.50	0.000

The `nowald pveffects` option is frequently used for many of the `contrast` commands in this book. However, when analyzing your data, I recommend using the `nowald effects` option to include confidence intervals in your output.

D.2 Customizing the display of factor variables

Suppose that you find yourself in a similar situation, where you are trying to create output that would fit on a narrow page, and you are trying to create results from the `contrast` command that is most informative. The prior example achieved a narrower display of the `contrast` output with the `nowald` and `pveffects` options. That output included the factor-variable labels but sacrificed the display of the confidence intervals. You could consider narrowing the table via the `nofvlabel` option. That option specifies that the level of the factor variable will be displayed in lieu of the factor-variable label, (as shown below). If you are very familiar with the meanings of the factor-variable codes, this could be a good alternative that allows you to display both significance tests and confidence intervals in a compact format that is easy to read.

```
. contrast r.marital, nowald effects nofvlabel
Contrasts of marginal linear predictions
Margins      : asbalanced
```

	Contrast	Std. Err.	t	P>|t|	[95% Conf.	Interval]
marital						
(2 vs 1)	-.2912429	.110942	-2.63	0.009	-.5089133	-.0735724
(3 vs 1)	-.4878643	.0796349	-6.13	0.000	-.6441094	-.3316191
(4 vs 1)	-.5579771	.1699952	-3.28	0.001	-.8915112	-.2244431
(5 vs 1)	-.2453868	.0701012	-3.50	0.000	-.3829269	-.1078468

Another option you can use for customizing the display of factor-variable labels is the `fvwrap()` option. Consider the example below, which does not include this option. Notice how `never mar..` is used to represent the first label mentioned in the fourth contrast.

```
. contrast r.marital, nowald effects
Contrasts of marginal linear predictions
Margins      : asbalanced

------------------------------------------------------------------------------
             |   Contrast   Std. Err.      t    P>|t|     [95% Conf. Interval]
-------------+----------------------------------------------------------------
     marital |
    (widowed |
          vs |
    married) |  -.2912429    .110942    -2.63   0.009    -.5089133   -.0735724
   (divorced |
          vs |
    married) |  -.4878643   .0796349    -6.13   0.000    -.6441094   -.3316191
  (separated |
          vs |
    married) |  -.5579771   .1699952    -3.28   0.001    -.8915112   -.2244431
 (never mar.. |
          vs |
    married) |  -.2453868   .0701012    -3.50   0.000    -.3829269   -.1078468
------------------------------------------------------------------------------
```

In the example below, the `fvwrap(2)` option is used to permit value labels to be displayed using up to two lines. Notice how the first label in the fourth contrast is now displayed as `never married`. The `fvwrap(2)` option allowed this label to be displayed on two lines.[2]

```
. contrast r.marital, nowald effects fvwrap(2)
Contrasts of marginal linear predictions
Margins      : asbalanced

------------------------------------------------------------------------------
             |   Contrast   Std. Err.      t    P>|t|     [95% Conf. Interval]
-------------+----------------------------------------------------------------
     marital |
    (widowed |
          vs |
    married) |  -.2912429    .110942    -2.63   0.009    -.5089133   -.0735724
   (divorced |
          vs |
    married) |  -.4878643   .0796349    -6.13   0.000    -.6441094   -.3316191
  (separated |
          vs |
    married) |  -.5579771   .1699952    -3.28   0.001    -.8915112   -.2244431
      (never |
     married |
          vs |
    married) |  -.2453868   .0701012    -3.50   0.000    -.3829269   -.1078468
------------------------------------------------------------------------------
```

2. The `contrast` command also supports the `fvwrapon()` options, as illustrated in section A.4 illustrating customizing the display of factor variables in estimation commands.

D.3 Adjustments for multiple comparisons

The `contrast` command can adjust the significance levels to account for multiple comparisons using the `mcompare()` option. You can specify the `bonferroni`, `sidak`, or `scheffe` method within `mcompare()` to obtain Bonferroni's, Šidák's, or Scheffé's method of adjustment, respectively. The example below illustrates using `mcompare(scheffe)`.

```
. contrast r.marital, nowald pveffects mcompare(scheffe)
Contrasts of marginal linear predictions
Margins      : asbalanced
```

	Number of Comparisons
marital	4

	Contrast	Std. Err.	Scheffe t	Scheffe P>|t|
marital				
(widowed vs married)	-.2912429	.110942	-2.63	0.143
(divorced vs married)	-.4878643	.0796349	-6.13	0.000
(separated vs married)	-.5579771	.1699952	-3.28	0.030
(never married vs married)	-.2453868	.0701012	-3.50	0.016

D.4 Specifying the confidence level

By default, 95% confidence intervals are displayed. You can use the `level()` option to specify a different confidence level. The example below illustrates how to show confidence intervals using a 99% confidence interval.

```
. contrast r.marital, nowald effects level(99)
Contrasts of marginal linear predictions
Margins      : asbalanced
```

	Contrast	Std. Err.	t	P>|t|	[99% Conf.	Interval]
marital						
(widowed vs married)	-.2912429	.110942	-2.63	0.009	-.5774835	-.0050022
(divorced vs married)	-.4878643	.0796349	-6.13	0.000	-.6933296	-.2823989
(separated vs married)	-.5579771	.1699952	-3.28	0.001	-.9965806	-.1193737
(never mar.. vs married)	-.2453868	.0701012	-3.50	0.000	-.4262546	-.0645191

The `contrast` command respects the settings specified by the `set level` command. You can see section A.2 for more details about the `set level` command.

D.5 Customizing column formatting

The `contrast` command offers three options for controlling the formatting of the output of the columns displayed by the `contrast` command—the `pformat()` option, the `sformat()` option, and the `cformat()` option. In fact, these options work in the same manner as the options used for formatting the columns of the coefficient table from estimation commands (as illustrated in section A.3). To briefly illustrate these options when using the `contrast` command, let's start with the `contrast` commands illustrated in section D.2 that performed adjacent group contrasts on marital status. The output was customized to omit the Wald table, show both significance tests and confidence intervals, and omit factor-variable labels to produce concise output.

```
. contrast r.marital, nowald effects nofvlabel
Contrasts of marginal linear predictions
Margins      : asbalanced
```

	Contrast	Std. Err.	t	P>|t|	[95% Conf.	Interval]
marital						
(2 vs 1)	-.2912429	.110942	-2.63	0.009	-.5089133	-.0735724
(3 vs 1)	-.4878643	.0796349	-6.13	0.000	-.6441094	-.3316191
(4 vs 1)	-.5579771	.1699952	-3.28	0.001	-.8915112	-.2244431
(5 vs 1)	-.2453868	.0701012	-3.50	0.000	-.3829269	-.1078468

The example below adds the `pformat()` option to control the formatting of the *p*-values. The example below uses the `pformat(%5.2f)` option to display the *p*-values with a total width of five columns and 2 digits after the decimal place. You can see `help format` for more details about specify formats. Note that the width of the format specified in the `pformat()` option cannot exceed 5 columns.

```
. contrast r.marital, nowald effects nofvlabel pformat(%5.2f)
Contrasts of marginal linear predictions
Margins      : asbalanced
```

	Contrast	Std. Err.	t	P>|t|	[95% Conf.	Interval]
marital						
(2 vs 1)	-.2912429	.110942	-2.63	0.01	-.5089133	-.0735724
(3 vs 1)	-.4878643	.0796349	-6.13	0.00	-.6441094	-.3316191
(4 vs 1)	-.5579771	.1699952	-3.28	0.00	-.8915112	-.2244431
(5 vs 1)	-.2453868	.0701012	-3.50	0.00	-.3829269	-.1078468

The `sformat()` option controls the formatting of the test statistics. The example below uses the `sformat(%5.1f)` option to display the z-values using a fixed width of 5 columns with 1 decimal place. Note that the width of the format specified for the `sformat()` option cannot exceed eight columns.

```
. contrast r.marital, nowald effects nofvlabel sformat(%5.1f)
Contrasts of marginal linear predictions
Margins      : asbalanced
```

	Contrast	Std. Err.	t	P>|t|	[95% Conf.	Interval]
marital						
(2 vs 1)	-.2912429	.110942	-2.6	0.009	-.5089133	-.0735724
(3 vs 1)	-.4878643	.0796349	-6.1	0.000	-.6441094	-.3316191
(4 vs 1)	-.5579771	.1699952	-3.3	0.001	-.8915112	-.2244431
(5 vs 1)	-.2453868	.0701012	-3.5	0.000	-.3829269	-.1078468

The `cformat()` option controls the format used to display the contrast coefficients, standard errors, and confidence intervals. The format selected via the `cformat()` option is applied to all of these columns. In the example below, I have added the `cformat(%7.4f)` option which means that these results shown in these columns will be displayed in a fixed format with a total width 7 columns and 4 columns after the decimal point. Note that the width of the format specified in the `cformat()` option cannot exceed nine columns.

```
. contrast r.marital, nowald effects nofvlabel cformat(%7.4f)
Contrasts of marginal linear predictions
Margins      : asbalanced
```

	Contrast	Std. Err.	t	P>|t|	[95% Conf.	Interval]
marital						
(2 vs 1)	-0.2912	0.1109	-2.63	0.009	-0.5089	-0.0736
(3 vs 1)	-0.4879	0.0796	-6.13	0.000	-0.6441	-0.3316
(4 vs 1)	-0.5580	0.1700	-3.28	0.001	-0.8915	-0.2244
(5 vs 1)	-0.2454	0.0701	-3.50	0.000	-0.3829	-0.1078

As we saw in section A.3, the `set pformat`, `set sformat`, and `set cformat` commands can be used to customize formatting for subsequent estimation commands. Further, those commands can be used in combination with the `permanently` option to change the default settings each time you start your copy of Stata. The `contrast` command respects the settings specified by the `set pformat`, `set sformat`, and `set cformat` commands. For more details on using those commands, see section A.3. For more information about specifying formats, see `help format`.

This appendix has certainly not covered all the features of the `contrast` command. For more information about the `contrast` command, see [R] **contrast**.

E The pwcompare command

This appendix covers additional details about the pwcompare command. The pwcompare command allows you to make pairwise comparisons between means from a factor variable.

The mcompare() option permits you to select a method for adjusting for multiple comparisons. The default method is to perform no adjustment for multiple comparisons. Three methods are provided that can be used with balanced or unbalanced data: Bonferroni's method, Šidák's method, and Scheffé's method. These methods can be selected by specifying **bonferroni**, **sidak**, or **scheffe** within the mcompare() option. Four additional methods are provided but require balanced data and can be used only after a linear modeling command (**anova**, **manova**, **regress**, or **mvreg**). These are the Tukey, Student–Newman–Keuls, Duncan, and Dunnett methods, and they can be selected by specifying **tukey**, **snk**, **duncan**, or **dunnett**, respectively, within the mcompare() option.

Let's consider an example using unbalanced data, predicting happiness from marital status.

```
. anova happy7 i.educ4 c.age
                         Number of obs =      1,153    R-squared     =  0.0124
                         Root MSE      =    .975216    Adj R-squared =  0.0089

                Source | Partial SS          df        MS          F    Prob>F
            -----------+-----------------------------------------------------
                 Model |  13.694158           4   3.4235396     3.60   0.0063
                       |
                 educ4 |   10.59956           3   3.5331868     3.72   0.0112
                   age |  4.0091754           1   4.0091754     4.22   0.0403
                       |
              Residual |  1091.8011       1,148   .95104623
            -----------+-----------------------------------------------------
                 Total |  1105.4952       1,152   .95963128
```

The pwcompare command performs pairwise comparisons of the means (after adjusting for age). The default output includes the difference in the means (in the Contrast column) and a confidence interval for the difference. When the confidence interval excludes zero, the difference is significant.

```
. pwcompare educ4
Pairwise comparisons of marginal linear predictions
Margins      : asbalanced
```

	Contrast	Std. Err.	Unadjusted [95% Conf.	Interval]
educ4				
8-11 vs 0-7	.2406146	.2202704	-.1915631	.6727922
12-15 vs 0-7	.4447997	.2073503	.0379717	.8516277
16-20 vs 0-7	.4993586	.2115241	.0843414	.9143758
12-15 vs 8-11	.2041851	.0918713	.0239306	.3844396
16-20 vs 8-11	.258744	.1012196	.060148	.4573401
16-20 vs 12-15	.0545589	.0675454	-.0779674	.1870853

We can use the `pveffects` option to display the significance test of the difference in the means in lieu of the confidence interval. (We could instead specify `effects` to display both the significance tests and confidence interval, not shown.) The results show that all the pairwise comparisons are different except for the comparisons of `8-11 vs 0-7` and `16-20 vs 12-15`.

```
. pwcompare educ4, pveffects
Pairwise comparisons of marginal linear predictions
Margins      : asbalanced
```

	Contrast	Std. Err.	Unadjusted t	P>|t|
educ4				
8-11 vs 0-7	.2406146	.2202704	1.09	0.275
12-15 vs 0-7	.4447997	.2073503	2.15	0.032
16-20 vs 0-7	.4993586	.2115241	2.36	0.018
12-15 vs 8-11	.2041851	.0918713	2.22	0.026
16-20 vs 8-11	.258744	.1012196	2.56	0.011
16-20 vs 12-15	.0545589	.0675454	0.81	0.419

By including the `cimargins` option as well, you get not only the pairwise comparisons of the means but also an estimate of each of the means with a confidence interval.

```
. pwcompare educ4, pveffects cimargins
Pairwise comparisons of marginal linear predictions
Margins      : asbalanced
```

	Margin	Std. Err.	Unadjusted [95% Conf.	Interval]
educ4				
0-7	4.940257	.223449	4.501842	5.378671
8-11	5.180871	.1169722	4.951368	5.410375
12-15	5.385056	.0844101	5.219441	5.550672
16-20	5.439615	.0966675	5.24995	5.62928

	Contrast	Std. Err.	Unadjusted t	P>|t|
educ4				
8-11 vs 0-7	.2406146	.2202704	1.09	0.275
12-15 vs 0-7	.4447997	.2073503	2.15	0.032
16-20 vs 0-7	.4993586	.2115241	2.36	0.018
12-15 vs 8-11	.2041851	.0918713	2.22	0.026
16-20 vs 8-11	.258744	.1012196	2.56	0.011
16-20 vs 12-15	.0545589	.0675454	0.81	0.419

Specifying the **groups** option displays group codes that signify groups that are not significantly different from each other. This output indicates that groups 1 and 2 are not significantly different (because both belong to group A), and groups 3 and 4 are not significantly different (because both belong to group B). This provides a concise summary of which pairwise differences are significant and which ones are nonsignificant.

```
. pwcompare educ4, groups
Pairwise comparisons of marginal linear predictions
Margins      : asbalanced
```

	Margin	Std. Err.	Unadjusted Groups
educ4			
0-7	4.940257	.223449	A
8-11	5.180871	.1169722	A
12-15	5.385056	.0844101	B
16-20	5.439615	.0966675	B

```
Note: Margins sharing a letter in the group label
      are not significantly different at the 5%
      level.
```

So far, these pairwise comparisons have not adjusted for multiple comparisons. Let's adjust for multiple comparisons by adding the `mcompare(sidak)` option. After making this adjustment, none of the pairwise comparisons are significant. (Using the `mcompare(bonferroni)` and `mcompare(scheffe)` options yield similar results, showing all comparisons to be nonsignificant.)

```
. pwcompare educ4, pveffects mcompare(sidak)
Pairwise comparisons of marginal linear predictions
Margins      : asbalanced

             |    Number of
             |  Comparisons
-------------+-------------
       educ4 |            6
---------------------------

---------------------------------------------------------
                 |                             Sidak
                 |   Contrast   Std. Err.      t    P>|t|
-----------------+---------------------------------------
           educ4 |
     8-11 vs 0-7 |   .2406146   .2202704     1.09   0.855
    12-15 vs 0-7 |   .4447997   .2073503     2.15   0.178
    16-20 vs 0-7 |   .4993586   .2115241     2.36   0.105
   12-15 vs 8-11 |   .2041851   .0918713     2.22   0.149
   16-20 vs 8-11 |    .258744   .1012196     2.56   0.063
  16-20 vs 12-15 |   .0545589   .0675454     0.81   0.962
---------------------------------------------------------
```

Finally, let's consider an example using balanced data, in which all groups have equal cell sizes. Let's use the example from section 7.7 that looked at pain ratings as a function of medication dosage group (`dosegrp`), which was coded as 1 = 0 mg, 2 = 50 mg, 3 = 100 mg, 4 = 150 mg, 5 = 200 mg, and 6 = 250 mg. The `anova` command predicting `pain` from `dosegrp` is shown below.

```
. use pain
. anova pain dosegrp

                  Number of obs =        180    R-squared     =  0.4602
                  Root MSE      =    10.4724    Adj R-squared =  0.4447

        Source |  Partial SS         df         MS           F    Prob>F
    -----------+----------------------------------------------------
         Model |   16271.694          5    3254.3389      29.67  0.0000
               |
       dosegrp |   16271.694          5    3254.3389      29.67  0.0000
               |
      Residual |   19082.633        174    109.67031
    -----------+----------------------------------------------------
         Total |   35354.328        179    197.51021
```

Let's use the `pwcompare` command to form pairwise comparisons of all dosage groups, using Tukey's method of adjustment for multiple comparisons. Furthermore, let's sort the results based on the size of difference in means, show the significance levels, and include a table of groups that are not significantly different.

```
. pwcompare dosegrp, mcompare(tukey) sort pveffects groups
Pairwise comparisons of marginal linear predictions
Margins      : asbalanced
```

	Number of Comparisons
dosegrp	15

	Margin	Std. Err.	Tukey Groups
dosegrp			
250mg	48.3	1.911982	B
200mg	54.7	1.911982	B
150mg	70.4	1.911982	A
50mg	70.6	1.911982	A
0mg	71.83333	1.911982	A
100mg	72.13333	1.911982	A

Note: Margins sharing a letter in the group label are not significantly different at the 5% level.

	Number of Comparisons
dosegrp	15

	Contrast	Std. Err.	Tukey t	Tukey P>|t|
dosegrp				
250mg vs 100mg	-23.83333	2.703952	-8.81	0.000
250mg vs 0mg	-23.53333	2.703952	-8.70	0.000
250mg vs 50mg	-22.3	2.703952	-8.25	0.000
250mg vs 150mg	-22.1	2.703952	-8.17	0.000
200mg vs 100mg	-17.43333	2.703952	-6.45	0.000
200mg vs 0mg	-17.13333	2.703952	-6.34	0.000
200mg vs 50mg	-15.9	2.703952	-5.88	0.000
200mg vs 150mg	-15.7	2.703952	-5.81	0.000
250mg vs 200mg	-6.4	2.703952	-2.37	0.174
150mg vs 100mg	-1.733333	2.703952	-0.64	0.988
150mg vs 0mg	-1.433333	2.703952	-0.53	0.995
50mg vs 0mg	-1.233333	2.703952	-0.46	0.997
150mg vs 50mg	-.2	2.703952	-0.07	1.000
100mg vs 0mg	.3	2.703952	0.11	1.000
100mg vs 50mg	1.533333	2.703952	0.57	0.993

The output shows that the groups receiving 250 mg and 200 mg both belong to group B and are not significantly different. Also, the groups receiving 150 mg, 50 mg, 0 mg, and 100 mg all belong to group A and are not significantly different from one another. This is reflected in the more detailed output that shows these groups to be not significantly different. For this example with balanced data, the Student–Newman–Keuls, Duncan, or Dunnett methods could have been selected by specifying the `snk`, `duncan`, or `dunnett` method instead of the `tukey` method.

Although not illustrated here, the `pwcompare` command supports options to control the formatting of *p*-values, test statistics, and margins or contrasts via the `pformat()`, `sformat()`, and `cformat()` options (respectively). These options work in the same manner as illustrated in section A.3, where they are used to customize the formatting of columns of results from estimation commands. Similarly, the `pwcompare` command supports options for controlling the display of factor variables, such as the `vsquish`, `nofvlabel`, `fvwrap()`, and `fvwrapon()` options. Examples of using these options in regards to estimation commands are shown in section A.4. You can find more information about the `pwcompare` command in [R] **pwcompare**.

References

Abelson, R. P., and D. A. Prentice. 1997. Contrast tests of interaction hypotheses. *Psychological Methods* 2: 315–328.

Aiken, L. S., and S. G. West. 1991. *Multiple Regression: Testing and Interpreting Interactions.* Thousand Oaks, CA: SAGE.

Baum, C. F. 2006. *An Introduction to Modern Econometrics Using Stata.* College Station, TX: Stata Press.

Boik, R. J. 1979. Interactions, partial interactions, and interaction contrasts in the analysis of variance. *Psychological Bulletin* 86: 1084–1089.

Cameron, A. C., and P. K. Trivedi. 2010. *Microeconometrics Using Stata.* Rev. ed. College Station, TX: Stata Press.

Chatterjee, S., and A. S. Hadi. 2012. *Regression Analysis by Example.* 5th ed. Hobeken, NJ: Wiley.

Cohen, J., P. Cohen, S. G. West, and L. S. Aiken. 2003. *Applied Multiple Regression/Correlation Analysis for the Behavioral Sciences.* 3rd ed. Hillsdale, NJ: Erlbaum.

Davis, M. J. 2010. Contrast coding in multiple regression analysis: Strengths, weaknesses, and utility of popular coding structures. *Journal of Data Science* 8: 61–73.

Fox, J. 2016. *Applied Regression Analysis and Generalized Linear Models.* 3rd ed. Thousand Oaks, CA: SAGE.

Gould, W. 2000. sg124: Interpreting logistic regression in all its forms. *Stata Technical Bulletin* 53: 19–29. Reprinted in *Stata Technical Bulletin Reprints.* Vol. 9, pp. 257–270. College Station, TX: Stata Press.

Hamilton, L. C. 1992. *Regression with Graphics: A Second Course in Applied Statistics.* Belmont, CA: Duxbury.

Hilbe, J. M. 2009. *Logistic Regression Models.* Boca Raton, FL: Chapman & Hall/CRC.

Hosmer, D. W., Jr., S. Lemeshow, and R. X. Sturdivant. 2013. *Applied Logistic Regression.* 3rd ed. Hoboken, NJ: Wiley.

Jaccard, J. 1998. *Interaction Effects in Factorial Analysis of Variance.* Thousand Oaks, CA: SAGE.

———. 2001. *Interaction Effects in Logistic Regression.* Thousand Oaks, CA: SAGE.

Juul, S., and M. Frydenberg. 2014. *An Introduction to Stata for Health Researchers.* 4th ed. College Station, TX: Stata Press.

Kenward, M. G., and J. H. Roger. 1997. Small sample inference for fixed effects from restricted maximum likelihood. *Biometrics* 53: 983–997.

Keppel, G., and T. D. Wickens. 2004. *Design and Analysis: A Researcher's Handbook.* 4th ed. Englewood Cliffs, NJ: Prentice Hall.

Kleinbaum, D. G., and M. Klein. 2010. *Logistic Regression: A Self-Learning Text.* 3rd ed. New York: Springer.

Kohler, U., and F. Kreuter. 2009. *Data Analysis Using Stata.* 2nd ed. College Station, TX: Stata Press.

Kreft, I., and J. De Leeuw. 1998. *Introducing Multilevel Modeling.* London: SAGE.

Levin, J. R., and L. A. Marascuilo. 1972. Type IV errors and interactions. *Psychological Bulletin* 78: 368–374.

Long, J. S. 1997. *Regression Models for Categorical and Limited Dependent Variables.* Thousand Oaks, CA: SAGE.

Long, J. S., and J. Freese. 2014. *Regression Models for Categorical Dependent Variables Using Stata.* 3rd ed. College Station, TX: Stata Press.

Mander, A. P. 2011. SURFACE: module to draw a 3D wireform surface plot. Statistical Software Components S448501, Department of Economics, Boston College. http://ideas.repec.org/c/boc/bocode/s448501.html.

Marsh, L. C., and D. R. Cormier. 2002. *Spline Regression Models.* Thousand Oaks, CA: SAGE.

Mason, D., A. T. Prevost, and S. Sutton. 2008. Perceptions of absolute versus relative differences between personal and comparison health risk. *Health Psychology* 27: 87–92.

Maxwell, S. E., H. D. Delaney, and K. Kelley. 2018. *Designing Experiments and Analyzing Data: A Model Comparison Perspective.* 3rd ed. New York: Routledge.

Menard, S. 2002. *Applied Logistic Regression Analysis.* 2nd ed. Thousand Oaks, CA: SAGE.

Mitchell, M. 2012. *A Visual Guide to Stata Graphics.* 3rd ed. College Station, TX: Stata Press.

Panis, C. 1994. sg24: The piecewise linear spline transformation. *Stata Technical Bulletin* 18: 27–29. Reprinted in *Stata Technical Bulletin Reprints.* Vol. 3, pp. 146–149. College Station, TX: Stata Press.

Pedhazur, E. J., and L. P. Schmelkin. 1991. *Measurement, Design, and Analysis: An Integrated Approach.* Hillsdale, NJ: Erlbaum.

Rabe-Hesketh, S., and A. Skrondal. 2012. *Multilevel and Longitudinal Modeling Using Stata.* 3rd ed. 2 vols. College Station, TX: Stata Press.

Raudenbush, S. W., and A. S. Bryk. 2002. *Hierarchical Linear Models: Applications and Data Analysis Methods.* 2nd ed. Thousand Oaks, CA: SAGE.

Richter, T. 2006. What is wrong with ANOVA and multiple regression? Analyzing sentence reading times with hierarchical linear models. *Discourse Processes* 41: 221–250.

Royston, P., and W. Sauerbrei. 2008. *Multivariable Model-Building: A Pragmatic Approach to Regression Analysis Based on Fractional Polynomials for Modelling Continuous Variables.* Chichester, UK: Wiley.

Satterthwaite, F. E. 1946. An approximate distribution of estimates of variance components. *Biometrics Bulletin* 2: 110–114.

Schaalje, G. B., J. B. McBride, and G. W. Fellingham. 2002. Adequacy of approximations to distributions of test statistics in complex mixed linear models. *Journal of Agricultural, Biological, and Environmental Statistics* 7: 512–524.

Singer, J. D., and J. B. Willett. 2003. *Applied Longitudinal Data Analysis: Modeling Change and Event Occurrence.* New York: Oxford University Press.

Snijders, T. A. B., and R. J. Bosker. 1999. *Multilevel Analysis: An Introduction to Basic and Advance Multilevel Modeling.* London: SAGE.

Wendorf, C. A. 2004. Primer on multiple regression coding: Common forms and the additional case of repeated contrasts. *Understanding Statistics* 3: 47–57.

West, S. G., L. S. Aiken, and J. L. Krull. 1996. Experimental personality designs: Analyzing categorical by continuous variable interactions. *Journal of Personality* 64: 1–48.

Yang, X. 2015. Small-sample inference for linear mixed-effects models (DDF adjustments). https://www.stata.com/meeting/columbus15/abstracts/materials/columbus15_yang.pdf.

Author index

Subject index

N

O

P

Q

U

W